T0181014

Applied Mathematical Sciences

Volume 82

For further volumes:
http://www.springer.com/series/34

Rainer Kress

Linear Integral Equations

Third Edition

Rainer Kress
Institut für Numerische und Angewandte
Georg-August-Universität Göttingen
Göttingen, Germany

ISSN 0066-5452 ISSN 2196-968X (electronic)
ISBN 978-1-4939-5016-4 ISBN 978-1-4614-9593-2 (eBook)
DOI 10.1007/978-1-4614-9593-2
Springer New York Heidelberg Dordrecht London

Mathematics Subject Classification (2010): 35J05, 35P25, 35R25, 35R30, 45A05, 65M30, 65R20, 65R30, 76Q05, 78A45

Softcover reprint of the hardcover 3rd edition 2014

Printed on acid-free paper

Springer is part of Springer Science+Business Media (www.springer.com)

To the Memory of My Parents

Preface to the Third Edition

In the fourteen years since the second edition of this book appeared, linear integral equations have continued to be an active area of mathematics and they have revealed more of their power and beauty to me. Therefore I am pleased to have the opportunity to make adjustments and additions to the book's contents in this third edition. In the spirit of the two preceding editions, I have kept the balance between theory, applications and numerical methods. To preserve the character of an introduction as opposed to a research monograph, I have made no attempts to include most of the recent developments.

In addition to making corrections and additions throughout the text and updating the references, the following topics have been added. In order to make the introduction to the basic functional analytic tools more complete the Hahn–Banach extension theorem and the Banach open mapping theorem are now included in the text. The treatment of boundary value problems in potential theory has been extended by a more complete discussion of integral equations of the first kind in the classical Hölder space setting and of both integral equations of the first and second kind in the contemporary Sobolev space setting. In the numerical solution part of the book, I included a new collocation method for two-dimensional hypersingular boundary integral equations and the collocation method for the Lippmann–Schwinger equation based on fast Fourier transform techniques due to Vainikko. The final chapter of the book on inverse boundary value problems for the Laplace equation has been largely rewritten with special attention to the trilogy of decomposition, iterative and sampling methods.

Some of the additions to this third edition were written when I was visiting the Institut Mittag-Leffler, Djursholm, Sweden, in spring 2013 during the scientific program on *Inverse Problems and Applications*. I gratefully acknowledge the hospitality and the support.

Over the years most of the thirty-one PhD students that I supervised wrote their thesis on topics related to integral equations. Their work and my discussions with them have had significant influence on my own perspective on integral equations as presented in this book. Therefore, I take this opportunity to thank my PhD students as a group without listing them individually. A special note of thanks is given to my friend David Colton for reading over the new parts of the book and helping me with the English language.

I hope that this new edition of my book continues to attract readers to the field of integral equations and their applications.

Göttingen, Germany Rainer Kress

Preface to the Second Edition

In the ten years since the first edition of this book appeared, integral equations and integral operators have revealed more of their mathematical beauty and power to me. Therefore, I am pleased to have the opportunity to share some of these new insights with the readers of this book. As in the first edition, the main motivation is to present the fundamental theory of integral equations, some of their main applications, and the basic concepts of their numerical solution in a single volume. This is done from my own perspective of integral equations; I have made no attempt to include all of the recent developments.

In addition to making corrections and adjustments throughout the text and updating the references, the following topics have been added: In Section 4.3 the presentation of the Fredholm alternative in dual systems has been slightly simplified and in Section 5.3 the short presentation on the index of operators has been extended. The treatment of boundary value problems in potential theory now includes proofs of the jump relations for single- and double-layer potentials in Section 6.3 and the solution of the Dirichlet problem for the exterior of an arc in two dimensions (Section 7.8). The numerical analysis of the boundary integral equations in Sobolev space settings has been extended for both integral equations of the first kind in Section 13.4 and integral equations of the second kind in Section 12.4. Furthermore, a short outline on fast $O(n \log n)$ solution methods has been added in Section 14.4. Because inverse obstacle scattering problems are now extensively discussed in the monograph [32], in the concluding Chapter 18 the application to inverse obstacle scattering problems has been replaced by an inverse boundary value problem for Laplace's equation.

I would like to thank Peter Hähner and Andreas Vogt for carefully reading the manuscript and for a number of suggestions for improving it. Thanks also go to those readers who helped me by letting me know the errors and misprints they found in the first edition.

I hope that this book continues to attract mathematicians and scientists to the field of integral equations and their applications.

Göttingen, Germany Rainer Kress

Preface to the First Edition

I fell in love with integral equations about twenty years ago when I was working on my thesis, and I am still attracted by their mathematical beauty. This book will try to stimulate the reader to share this love with me.

Having taught integral equations a number of times I felt a lack of a text which adequately combines theory, applications and numerical methods. Therefore, in this book I intend to cover each of these fields with the same weight. The first part provides the basic Riesz–Fredholm theory for equations of the second kind with compact operators in dual systems including all functional analytic concepts necessary for developing this theory. The second part then illustrates the classical applications of integral equation methods to boundary value problems for the Laplace and the heat equation as one of the main historical sources for the development of integral equations, and also introduces Cauchy type singular integral equations. The third part is devoted to describing the fundamental ideas for the numerical solution of integral equations. Finally, in a fourth part, ill-posed integral equations of the first kind and their regularization are studied in a Hilbert space setting.

In order to make the book accessible not only to mathematicians but also to physicists and engineers I have planned it as self-contained as possible by requiring only a solid foundation in differential and integral calculus and, for parts of the book, in complex function theory. Some background in functional analysis will be helpful, but the basic concepts of the theory of normed spaces will be briefly reviewed, and all functional analytic tools which are relevant in the study of integral equations will be developed in the book. Of course, I expect the reader to be willing to accept the functional analytic language for describing the theory and the numerical solution of integral equations. I hope that I succeeded in finding the adequate compromise between presenting integral equations in the proper modern framework and the danger of being too abstract.

An introduction to integral equations cannot present a complete picture of all classical aspects of the theory and of all recent developments. In this sense, this book intends to tell the reader what I think appropriate to teach students in a two-semester course on integral equations. I am willing to admit that the choice of a few of the topics might be biased by my own preferences and that some important subjects are omitted.

I am indebted to Dipl.-Math. Peter Hähner for carefully reading the book, for checking the solutions to the problems and for a number of suggestions for valuable improvements. Thanks also go to Frau Petra Trapp who spent some time assisting me in the preparation of the LaTeX version of the text. And a particular note of thanks is given to my friend David Colton for reading over the book and helping me with the English language. Part of the book was written while I was on sabbatical leave at the Department of Mathematics at the University of Delaware. I gratefully acknowledge the hospitality.

Göttingen, Germany Rainer Kress

Contents

Chapter 1
Introduction and Basic Functional Analysis

The topic of this book is linear integral equations of which

$$\int_a^b K(x,y)\varphi(y)\,dy = f(x), \quad x \in [a,b], \tag{1.1}$$

and

$$\varphi(x) - \int_a^b K(x,y)\varphi(y)\,dy = f(x), \quad x \in [a,b], \tag{1.2}$$

are typical examples. In these equations the function φ is the unknown, and the so-called kernel K and the right-hand side f are given functions. Solving these integral equations amounts to determining a function φ such that (1.1) or (1.2), respectively, are satisfied for all $a \le x \le b$. The term integral equation was first used by du Bois-Reymond [45] in 1888. The equations (1.1) and (1.2) carry the name of Fredholm because of his contributions to the field and are called *Fredholm integral equations* of the *first* and *second kind*, respectively. In the first equation the unknown function only occurs under the integral whereas in the second equation it also appears outside the integral. Later on we will see that this is more than just a formal difference between the two types of equations. A first impression on the difference can be obtained by considering the special case of a constant kernel $K(x,y) = c \neq 0$ for all $x, y \in [a,b]$. On one hand, it is easily seen that the equation of the second kind (1.2) has a unique solution given by

$$\varphi = f + \frac{c}{1 - c(b-a)} \int_a^b f(y)\,dy$$

if $c(b-a) \neq 1$. If $c(b-a) = 1$ then (1.2) is solvable if and only if $\int_a^b f(y)\,dy = 0$ and the general solution is given by $\varphi = f + \gamma$ with an arbitrary constant γ. On the other hand, the equation of the first kind (1.1) is solvable if and only if f is a constant, $f(x) = \alpha$ for all $x \in [a,b]$ with an arbitrary constant α. In this case every function φ with $\int_a^b \varphi(y)\,dy = \alpha/c$ is a solution.

R. Kress, *Linear Integral Equations*, Applied Mathematical Sciences 82,
DOI 10.1007/978-1-4614-9593-2_1,

Of course, the integration domains in (1.1) and (1.2) are not restricted to an interval $[a, b]$. In particular, the integration can be over multi-dimensional domains or surfaces and for the integral equation of the first kind the domain where the equation is required to be satisfied need not coincide with the integration domain.

We will regard the integral equations (1.1) and (1.2) as *operator equations*

$$A\varphi = f$$

and

$$\varphi - A\varphi = f$$

of the *first* and *second kind*, respectively, in appropriate normed function spaces.

The symbol $A : X \to Y$ will mean a single-valued mapping whose domain of definition is a set X and whose range is contained in a set Y, i.e., for every $\varphi \in X$ the mapping A assigns a unique element $A\varphi \in Y$. The *range* $A(X)$ is the set $A(X) := \{A\varphi : \varphi \in X\}$ of all image elements. We will use the terms *mapping, function,* and *operator* synonymously.

Existence and uniqueness of a solution to an operator equation can be equivalently expressed by the existence of the *inverse operator*. If for each $f \in A(X)$ there is only one element $\varphi \in X$ with $A\varphi = f$, then A is said to be *injective* and to have an inverse $A^{-1} : A(X) \to X$ defined by $A^{-1}f := \varphi$. The inverse mapping has domain $A(X)$ and range X. It satisfies $A^{-1}A = I$ on X and $AA^{-1} = I$ on $A(X)$, where I denotes the identity operator mapping each element into itself. If $A(X) = Y$, then the mapping is said to be *surjective*. The mapping is called *bijective* if it is injective and surjective, i.e., if the inverse mapping $A^{-1} : Y \to X$ exists.

In the first part of the book we will present the Riesz–Fredholm theory for compact operators which, in particular, answers the question of existence and uniqueness of solutions to integral equations of the second kind with sufficiently smooth kernels. In order to develop the theory, we will assume that the reader is familiar with the elementary properties of linear spaces, normed spaces, and bounded linear operators. For convenience and to introduce notations, in this chapter, we briefly recall a few basic concepts from the theory of normed spaces, omitting most of the proofs. For a more detailed study, see Aubin [12], Brezis [21], Heuser [94], Kantorovic and Akilov [116], Rudin [209], and Taylor [229] among others.

1.1 Abel's Integral Equation

As an appetizer we consider Abel's integral equation that occurred as one of the first integral equations in mathematical history. A tautochrone is a planar curve for which the time taken by an object sliding without friction in uniform gravity to its lowest point is independent of its starting point. The problem to identify this curve was solved by Huygens in 1659 who, using geometrical tools, established that the tautochrone is a cycloid.

In 1823 Abel [1] attacked the more general problem of determining a planar curve such that the time of descent for a given starting height y coincides with the value $f(y)$ of a given function f. The tautochrone then reduces to the special case when f is a constant. Following Abel we describe the curve by $x = \psi(y)$ (with $\psi(0) = 0$) and, using the principle of conservation of energy, for the velocity v at height $0 \le \eta \le y$ we obtain

$$\frac{1}{2}\,v^2 + g\eta = gy$$

where g denotes the earth's gravity. Therefore, denoting arc length by s, the total time $f(y)$ required for the object to fall from $P = (\psi(y), y)$ to $P_0 = (0, 0)$ is given by

$$f(y) = \int_P^{P_0} \frac{ds}{v} = \int_0^y \sqrt{\frac{1 + [\psi'(\eta)]^2}{2g(y - \eta)}}\, d\eta.$$

Form this, setting

$$\varphi := \sqrt{\frac{1 + [\psi']^2}{2g}}$$

we obtain

$$f(y) = \int_0^y \frac{\varphi(y)}{\sqrt{y - \eta}}\, d\eta, \quad y > 0, \tag{1.3}$$

which is known as *Abel's integral equation*. Given the shape function φ, the falling time f is obtained by simply evaluating the integral on the right-hand side of (1.3). Conversely, given the function f, finding φ requires the solution of the integral equation (1.3) which certainly is a more challenging task.

For any solution φ of (1.3) that is continuous on some interval $(0, a]$ and satisfies $|\varphi(y)| \le C/\sqrt{y}$ for all $y \in (0, a]$ and some constant C, by interchanging the order of integration, we obtain that

$$\begin{aligned}\int_0^z \frac{f(y)}{\sqrt{z - y}}\, dy &= \int_0^z \frac{1}{\sqrt{z - y}} \int_0^y \frac{\varphi(\eta)}{\sqrt{y - \eta}}\, d\eta\, dy \\ &= \int_0^z \varphi(\eta) \int_\eta^z \frac{dy}{\sqrt{(z - y)(y - \eta)}}\, d\eta \\ &= \pi \int_0^z \varphi(\eta)\, d\eta.\end{aligned}$$

Here, we substituted $y = z - (z - \eta)\cos^2 t$ with the result

$$\int_\eta^z \frac{dy}{\sqrt{(z - y)(y - \eta)}} = 2 \int_0^{\pi/2} dt = \pi.$$

Therefore any solution of (1.3) has the form

$$\varphi(z) = \frac{1}{\pi} \frac{d}{dz} \int_0^z \frac{f(y)}{\sqrt{z-y}}\, dy, \quad z \in (0, a]. \tag{1.4}$$

Assuming that f is continuously differentiable on $[0, a]$, by partial integration we obtain

$$\int_0^z \frac{f(y)}{\sqrt{z-y}}\, dy = 2\sqrt{z}\, f(0) + 2 \int_0^z \sqrt{z-y}\, f'(y)\, dy$$

and this transforms (1.4) into

$$\varphi(z) = \frac{1}{\pi} \left\{ \frac{f(0)}{\sqrt{z}} + \int_0^z \frac{f'(y)}{\sqrt{z-y}}\, dy \right\}, \quad z \in (0, a]. \tag{1.5}$$

Inserting (1.5) in (1.3) (after renaming the variables) and interchanging the order of integration as above shows that (1.4) indeed is a solution of (1.3).

For the special case of a constant $f = \pi \sqrt{a/2g}$ with $a > 0$ one obtains from (1.5) that

$$\varphi(y) = \sqrt{\frac{a}{2gy}}, \quad 0 < y \le a. \tag{1.6}$$

Note that the restriction $y \le a$ is a consequence of $2g\,\varphi^2 = 1 + [\psi']^2 \ge 1$. For the arc length s, we have

$$\frac{ds}{dy} = \sqrt{1 + [\psi']^2} = \sqrt{2g}\,\varphi$$

and from (1.6) it follows that

$$s(y) = 2\sqrt{ay}, \quad 0 \le y \le a.$$

For a convenient parameterization we set

$$y(t) = \frac{a}{2}\,(1 - \cos t), \quad 0 \le t \le \pi, \tag{1.7}$$

and obtain first

$$s(t) = 2a \sin \frac{t}{2}, \quad 0 \le t \le \pi,$$

and then with the aid of some trigonometric identities

$$x(t) = \frac{a}{2}\,(t + \sin t), \quad 0 \le t \le \pi. \tag{1.8}$$

Hence, the tautochrone as given by the parameterization (1.7) and (1.8) is the cycloid generated as the trajectory described by a point on the circle of radius $a/2$ when the circle is rolling along the straight line $y = a$. This property of the cycloid, together with the fact that the involute of a cycloid again is a cycloid, was exploited by Huygens to build a cycloidal pendulum for which the frequency of the oscillations does not depend on the amplitude in contrast to the circular pendulum.

1.2 Convergence and Continuity

Definition 1.1. Let X be a complex (or real) linear space (vector space). A function $\|\cdot\| : X \to \mathbb{R}$ with the properties

$$\text{(N1)} \qquad \|\varphi\| \geq 0, \qquad \text{(positivity)}$$

$$\text{(N2)} \qquad \|\varphi\| = 0 \text{ if and only if } \varphi = 0, \qquad \text{(definiteness)}$$

$$\text{(N3)} \qquad \|\alpha\varphi\| = |\alpha|\,\|\varphi\|, \qquad \text{(homogeneity)}$$

$$\text{(N4)} \qquad \|\varphi + \psi\| \leq \|\varphi\| + \|\psi\|, \qquad \text{(triangle inequality)}$$

for all $\varphi, \psi \in X$, and all $\alpha \in \mathbb{C}$ (or $\mathbb{R}$) is called a *norm* on X. A linear space X equipped with a norm is called a *normed space.*

As a consequence of (N3) and (N4) we note the second triangle inequality

$$|\,\|\varphi\| - \|\psi\|\,| \leq \|\varphi - \psi\|. \tag{1.9}$$

For two elements in a normed space $\|\varphi - \psi\|$ is called the *distance* between φ and ψ.

Definition 1.2. A sequence (φ_n) of elements of a normed space X is called *convergent* if there exists an element $\varphi \in X$ such that $\lim_{n\to\infty} \|\varphi_n - \varphi\| = 0$, i.e., if for every $\varepsilon > 0$ there exists an integer $N(\varepsilon)$ such that $\|\varphi_n - \varphi\| < \varepsilon$ for all $n \geq N(\varepsilon)$. The element φ is called the *limit* of the sequence (φ_n), and we write

$$\lim_{n\to\infty} \varphi_n = \varphi \quad \text{or} \quad \varphi_n \to \varphi,\ n \to \infty.$$

Note that by (N4) the limit of a convergent sequence is uniquely determined. A sequence that does not converge is called *divergent.*

Definition 1.3. A function $A : U \subset X \to Y$ mapping a subset U of a normed space X into a normed space Y is called *continuous* at $\varphi \in U$ if $\lim_{n\to\infty} A\varphi_n = A\varphi$ for every sequence (φ_n) from U with $\lim_{n\to\infty} \varphi_n = \varphi$. The function $A : U \subset X \to Y$ is called continuous if it is continuous for all $\varphi \in U$.

An equivalent definition is the following: A function $A : U \subset X \to Y$ is continuous at $\varphi \in U$ if for every $\varepsilon > 0$ there exists $\delta > 0$ such that $\|A\varphi - A\psi\| < \varepsilon$ for all $\psi \in U$ with $\|\varphi - \psi\| < \delta$. Here we have used the same symbol $\|\cdot\|$ for the norms on X and Y. The function A is called *uniformly continuous* if for every $\varepsilon > 0$ there exists $\delta > 0$ such that $\|A\varphi - A\psi\| < \varepsilon$ for all $\varphi, \psi \in U$ with $\|\varphi - \psi\| < \delta$.

Note that by (1.9) the norm is a continuous function.

In our study of integral equations the basic example of a normed space will be the linear space $C[a,b]$ of continuous real- or complex-valued functions φ defined on an interval $[a,b] \subset \mathbb{R}$ furnished either with the *maximum norm*

$$\|\varphi\|_\infty := \max_{x\in[a,b]} |\varphi(x)|$$

or the *mean square norm*

$$\|\varphi\|_2 := \left(\int_a^b |\varphi(x)|^2 dx \right)^{1/2}.$$

Convergence of a sequence of continuous functions in the maximum norm is equivalent to uniform convergence, and convergence in the mean square norm is called *mean square convergence*. Throughout this book, unless stated otherwise, we always assume that $C[a,b]$ (or $C(G)$, i.e., the space of continuous real- or complex-valued functions on compact subsets $G \subset \mathbb{R}^m$) is equipped with the maximum norm.

Definition 1.4. Two norms on a linear space are called *equivalent* if they have the same convergent sequences.

Theorem 1.5. *Two norms $\|\cdot\|_a$ and $\|\cdot\|_b$ on a linear space X are equivalent if and only if there exist positive numbers c and C such that*

$$c\|\varphi\|_a \le \|\varphi\|_b \le C\|\varphi\|_a$$

for all $\varphi \in X$. The limits with respect to the two norms coincide.

Proof. Provided that the conditions are satisfied, from $\|\varphi_n - \varphi\|_a \to 0$, $n \to \infty$, it follows $\|\varphi_n - \varphi\|_b \to 0$, $n \to \infty$, and vice versa.

Conversely, let the two norms be equivalent and assume that there is no $C > 0$ such that $\|\varphi\|_b \le C\|\varphi\|_a$ for all $\varphi \in X$. Then there exists a sequence (φ_n) satisfying $\|\varphi_n\|_a = 1$ and $\|\varphi_n\|_b \ge n^2$. Now, the sequence (ψ_n) with $\psi_n := n^{-1}\varphi_n$ converges to zero with respect to $\|\cdot\|_a$, whereas with respect to $\|\cdot\|_b$ it is divergent because of $\|\psi_n\|_b \ge n$. □

Theorem 1.6. *On a finite-dimensional linear space all norms are equivalent.*

Proof. In a linear space X with finite dimension m and basis $f_1, \dots, f_m$ every element can be expressed in the form

$$\varphi = \sum_{k=1}^m \alpha_k f_k.$$

As is easily verified,

$$\|\varphi\|_\infty := \max_{k=1,\dots,m} |\alpha_k| \tag{1.10}$$

defines a norm on X. Let $\|\cdot\|$ denote any other norm on X. Then, by the triangle inequality we have

$$\|\varphi\| \le C\|\varphi\|_\infty$$

for all $\varphi \in X$, where

$$C := \sum_{k=1}^m \|f_k\|.$$

Assume that there is no $c > 0$ such that $c\|\varphi\|_\infty \le \|\varphi\|$ for all $\varphi \in X$. Then there exists a sequence (φ_n) with $\|\varphi_n\| = 1$ such that $\|\varphi_n\|_\infty \ge n$. Consider the sequence

(ψ_n) with $\psi_n := \|\varphi_n\|_\infty^{-1}\varphi_n$ and write

$$\psi_n = \sum_{k=1}^{m} \alpha_{kn} f_k.$$

Because of $\|\psi_n\|_\infty = 1$ each of the sequences (α_{kn}), $k = 1, \dots, m$, is bounded in $\mathbb{C}$. Hence, by the Bolzano–Weierstrass theorem we can select convergent subsequences $\alpha_{k,n(j)} \to \alpha_k$, $j \to \infty$, for each $k = 1, \dots, m$. This now implies $\|\psi_{n(j)} - \psi\|_\infty \to 0$, $j \to \infty$, where

$$\psi := \sum_{k=1}^{m} \alpha_k f_k,$$

and $\|\psi_{n(j)} - \psi\| \le C\|\psi_{n(j)} - \psi\|_\infty \to 0$, $j \to \infty$. But on the other hand we have $\|\psi_n\| = 1/\|\varphi_n\|_\infty \to 0$, $n \to \infty$. Therefore, $\psi = 0$, and consequently $\|\psi_{n(j)}\|_\infty \to 0$, $j \to \infty$, which contradicts $\|\psi_n\|_\infty = 1$ for all n. □

For an element φ of a normed space X and a positive number r the set $B(\varphi; r) := \{\psi \in X : \|\psi - \varphi\| < r\}$ is called the *open ball* of radius r and center φ, the set $B[\varphi; r] := \{\psi \in X : \|\psi - \varphi\| \le r\}$ is called a *closed ball.*

Definition 1.7. A subset U of a normed space X is called *open* if for each element $\varphi \in U$ there exists $r > 0$ such that $B(\varphi; r) \subset U$.

Obviously, open balls are open.

Definition 1.8. A subset U of a normed space X is called *closed* if it contains all limits of convergent sequences of U.

A subset U of a normed space X is closed if and only if its complement $X \setminus U$ is open. Obviously, closed balls are closed. In particular, using the norm (1.10), it can be seen that finite-dimensional subspaces of a normed space are closed.

Definition 1.9. The *closure* $\overline{U}$ of a subset U of a normed space X is the set of all limits of convergent sequences of U. A set U is called *dense* in another set V if $V \subset \overline{U}$, i.e., if each element in V is the limit of a convergent sequence from U.

A subset U is closed if and only if it coincides with its closure. By the Weierstrass approximation theorem (see [40]) the linear subspace P of polynomials is dense in $C[a, b]$ with respect to the maximum norm and the mean square norm.

Definition 1.10. A subset U of a normed space X is called *bounded* if there exists a positive number C such that $\|\varphi\| \le C$ for all $\varphi \in U$.

Convergent sequences are bounded.

1.3 Completeness

Definition 1.11. A sequence (φ_n) of elements of a normed space is called a *Cauchy sequence* if for each $\varepsilon > 0$ there exists an integer $N(\varepsilon)$ such that $\|\varphi_n - \varphi_m\| < \varepsilon$ for all $n, m \geq N(\varepsilon)$, i.e., if $\lim_{n,m\to\infty} \|\varphi_n - \varphi_m\| = 0$.

Every convergent sequence is a Cauchy sequence, whereas the converse in general is not true. This gives rise to the following definition.

Definition 1.12. A subset U of a normed space X is called *complete* if every Cauchy sequence of elements of U converges to an element in U. A normed space X is called a *Banach space* if it is complete.

Note that in a complete set we can decide on the convergence of a sequence without having to know its limit element. Complete sets are closed, and closed subsets of a complete set are complete. Since the Cauchy criterion is sufficient for convergence of sequences of complex numbers, using the norm (1.10), we observe that finite-dimensional normed spaces are Banach spaces. The Cauchy criterion is also sufficient for uniform convergence of a sequence of continuous functions toward a continuous limit function. Therefore, the space $C[a, b]$ is complete with respect to the maximum norm. As can be seen from elementary counterexamples, $C[a, b]$ is not complete with respect to the mean square norm.

Each normed space can be completed in the following sense.

Theorem 1.13. *For each normed space X there exists a Banach space $\widetilde{X}$ such that X is isomorphic and isometric to a dense subspace of $\widetilde{X}$, i.e., there is a linear bijective mapping I from X into a dense subspace of $\widetilde{X}$ such that $\|I\varphi\|_{\widetilde{X}} = \|\varphi\|_X$ for all $\varphi \in X$. The space $\widetilde{X}$ is uniquely determined up to isometric isomorphisms, i.e., to any two such Banach spaces there exists a linear bijective mapping between the two spaces leaving the norms invariant.*

For a proof of this concept of completion we refer to any introduction to functional analysis. Using Lebesgue integration theory, it can be seen that the completion of $C[a, b]$ with respect to the mean square norm yields the complete space $L^2[a, b]$ of measurable and Lebesgue square-integrable functions, or to be more precise, of equivalence classes of such functions that coincide almost everywhere with respect to the Lebesgue measure (see [12, 209, 229]).

1.4 Compactness

Definition 1.14. A subset U of a normed space X is called *compact* if every open covering of U contains a finite subcovering, i.e., if for every family V_j, $j \in J$, (for some index set J) of open sets with the property

$$U \subset \bigcup_{j \in J} V_j$$

there exists a finite subfamily $V_{j(k)}$, $j(k) \in J$, $k = 1, \ldots, n$, such that

$$U \subset \bigcup_{k=1}^{n} V_{j(k)}.$$

A subset U is called *sequentially compact* if every sequence of elements from U contains a subsequence that converges to an element in U.

A subset U of a normed space is called *totally bounded* if for each $\varepsilon > 0$ there exists a finite number of elements $\varphi_1, \ldots, \varphi_n$ in U such that

$$U \subset \bigcup_{j=1}^{n} B(\varphi_j; \varepsilon),$$

i.e., each element $\varphi \in U$ has a distance less than ε from at least one of the elements $\varphi_1, \ldots, \varphi_n$. Note that each sequentially compact set U is totally bounded. Otherwise there would exist a positive ε and a sequence (φ_n) in U with the property $\|\varphi_n - \varphi_m\| \geq \varepsilon$ for all $n \neq m$. This would imply that the sequence (φ_n) does not contain a convergent subsequence, which contradicts the sequential compactness of U. For each totally bounded set U, letting $\varepsilon = 1/m$, $m = 1, 2, \ldots$, and collecting the corresponding finite systems of elements $\varphi_1, \ldots, \varphi_n$ depending on m, we obtain a sequence that is dense in U.

Theorem 1.15. *A subset of a normed space is compact if and only if it is sequentially compact.*

Proof. Let U be compact and assume that it is not sequentially compact. Then there exists a sequence (φ_n) in U that does not contain a convergent subsequence with limit in U. Consequently, for each $\varphi \in U$ there exists an open ball $B(\varphi; r)$ with center φ and radius $r(\varphi)$ containing at most a finite number of the elements of the sequence (φ_n). The set of these balls clearly is an open covering of U, and since U is compact, it follows that U contains only a finite number of the elements of the sequence (φ_n). This is a contradiction.

Conversely, let U be sequentially compact and let V_j, $j \in J$, be an open covering of U. First, we show that there exists a positive number ε such that for every $\varphi \in U$ the ball $B(\varphi; \varepsilon)$ is contained in at least one of the sets V_j. Otherwise there would exist a sequence (φ_n) in U such that the ball $B(\varphi_n; 1/n)$ is not contained in one of the V_j. Since U is sequentially compact, this sequence (φ_n) contains a convergent subsequence $(\varphi_{n(k)})$ with limit $\varphi \in U$. The element φ is contained in some V_j, and since V_j is open, using the triangle inequality, we see that $B(\varphi_{n(k)}; 1/n(k)) \subset V_j$ for sufficiently large k. This is a contradiction. Now, since the sequentially compact set U is totally bounded, there exists a finite number of elements $\varphi_1, \ldots, \varphi_n$ in U such that the balls $B(\varphi_k; \varepsilon)$, $k = 1, \ldots, n$, cover U. But for each of these balls there exists a set $V_{j(k)}$, $j(k) \in J$, such that $B(\varphi_k; \varepsilon) \subset V_{j(k)}$. Hence, the finite family $V_{j(k)}$, $k = 1, \ldots, n$, covers U. □

In particular, from Theorem 1.15 we observe that compact sets are bounded, closed, and complete.

Definition 1.16. A subset of a normed space is called *relatively compact* if its closure is compact.

As a consequence of Theorem 1.15, a set U is relatively compact if and only if each sequence of elements from U contains a convergent subsequence. Hence, analogous to compact sets, relatively compact sets are totally bounded.

Theorem 1.17. *A bounded and finite-dimensional subset of a normed space is relatively compact.*

Proof. This follows from the Bolzano–Weierstrass theorem using the norm (1.10). □

For the compactness in $C(G)$, the space of continuous real- or complex-valued functions defined on a compact set $G \subset \mathbb{R}^m$, furnished with the maximum norm

$$\|\varphi\|_\infty := \max_{x \in G} |\varphi(x)|,$$

we have the following criterion.

Theorem 1.18 (Arzelà–Ascoli). *A set $U \subset C(G)$ is relatively compact if and only if it is bounded and* equicontinuous, *i.e., if there exists a constant C such that*

$$|\varphi(x)| \leq C$$

for all $x \in G$ and all $\varphi \in U$, and for every $\varepsilon > 0$ there exists $\delta > 0$ such that

$$|\varphi(x) - \varphi(y)| < \varepsilon$$

for all $x, y \in G$ with $|x - y| < \delta$ and all $\varphi \in U$. (Here and henceforth by $|z| := (z_1^2 + \cdots + z_m^2)^{1/2}$ we denote the Euclidean norm of a vector $z = (z_1, \ldots, z_m) \in \mathbb{R}^m$.)

Proof. Let U be bounded and equicontinuous and let (φ_n) be a sequence in U. We choose a sequence (x_i) from the compact and, consequently, totally bounded set G that is dense in G. Since the sequence $(\varphi_n(x_i))$ is bounded in $\mathbb{C}$ for each x_i, by the standard diagonalization procedure we can choose a subsequence $(\varphi_{n(k)})$ such that $(\varphi_{n(k)}(x_i))$ converges in $\mathbb{C}$ as $k \to \infty$ for each x_i. More precisely, since $(\varphi_n(x_1))$ is bounded, we can choose a subsequence $(\varphi_{n_1(k)})$ such that $(\varphi_{n_1(k)}(x_1))$ converges as $k \to \infty$. The sequence $(\varphi_{n_1(k)}(x_2))$ again is bounded and we can choose a subsequence $(\varphi_{n_2(k)})$ of $(\varphi_{n_1(k)})$ such that $(\varphi_{n_2(k)}(x_2))$ converges as $k \to \infty$. Repeating this process of selecting subsequences, we arrive at a double array $(\varphi_{n_i(k)})$ such that each row $(\varphi_{n_i(k)})$ is a subsequence of the previous row $(\varphi_{n_{i-1}(k)})$ and each sequence $(\varphi_{n_i(k)}(x_i))$ converges as $k \to \infty$. For the diagonal sequence $\varphi_{n(k)} := \varphi_{n_k(k)}$ we have that $(\varphi_{n(k)}(x_i))$ converges as $k \to \infty$ for all x_i.

Since the set (x_i), $i = 1, 2, \ldots$, is dense in G, given $\varepsilon > 0$, the balls $B(x_i; \delta)$, $i = 1, 2, \ldots$, cover G. Since G is compact we can choose $i \in \mathbb{N}$ such that each point

$x \in G$ has a distance less than δ from at least one element x_j of the set $x_1, \ldots, x_i$. Next choose $N(\varepsilon) \in \mathbb{N}$ such that

$$|\varphi_{n(k)}(x_j) - \varphi_{n(l)}(x_j)| < \varepsilon$$

for all $k, l \geq N(\varepsilon)$ and all $j = 1, \ldots, i$. From the equicontinuity we obtain

$$|\varphi_{n(k)}(x) - \varphi_{n(l)}(x)| \leq |\varphi_{n(k)}(x) - \varphi_{n(k)}(x_j)| + |\varphi_{n(k)}(x_j) - \varphi_{n(l)}(x_j)|$$

$$+|\varphi_{n(l)}(x_j) - \varphi_{n(l)}(x)| < 3\varepsilon$$

for all $k, l \geq N(\varepsilon)$ and all $x \in G$. This establishes the uniform convergence of the subsequence $(\varphi_{n(k)})$, i.e., convergence in the maximum norm. Hence U is relatively compact.

Conversely, let U be relatively compact. Then U is totally bounded, i.e., given $\varepsilon > 0$ there exist functions $\varphi_1, \ldots, \varphi_i \in U$ such that

$$\min_{j=1,\ldots,i} \|\varphi - \varphi_j\|_\infty < \frac{\varepsilon}{3}$$

for all $\varphi \in U$. Since each of the $\varphi_1, \ldots, \varphi_i$ is uniformly continuous on the compact set G, there exists $\delta > 0$ such that

$$|\varphi_j(x) - \varphi_j(y)| < \frac{\varepsilon}{3}$$

for all $x, y \in G$ with $|x - y| < \delta$ and all $j = 1, \ldots, i$. Then for all $\varphi \in U$, choosing j_0 such that

$$\|\varphi - \varphi_{j_0}\|_\infty = \min_{j=1,\ldots,i} \|\varphi - \varphi_j\|_\infty,$$

we obtain

$$|\varphi(x) - \varphi(y)| \leq |\varphi(x) - \varphi_{j_0}(x)| + |\varphi_{j_0}(x) - \varphi_{j_0}(y)| + |\varphi_{j_0}(y) - \varphi(y)| < \varepsilon$$

for all $x, y \in G$ with $|x - y| < \delta$. Therefore U is equicontinuous. Finally, U is bounded, since relatively compact sets are bounded. □

1.5 Scalar Products

We now consider the important special case where the norm is given in terms of a scalar product.

Definition 1.19. Let X be a complex (or real) linear space. Then a function $(\cdot,\cdot): X \times X \to \mathbb{C}$ (or $\mathbb{R}$) with the properties

$$\text{(H1)} \qquad (\varphi,\varphi) \geq 0, \qquad \text{(positivity)}$$

$$\text{(H2)} \qquad (\varphi,\varphi) = 0 \text{ if and only if } \varphi = 0, \qquad \text{(definiteness)}$$

$$\text{(H3)} \qquad (\varphi,\psi) = \overline{(\psi,\varphi)}, \qquad \text{(symmetry)}$$

$$\text{(H4)} \qquad (\alpha\varphi + \beta\psi, \chi) = \alpha(\varphi,\chi) + \beta(\psi,\chi), \qquad \text{(linearity)}$$

for all $\varphi,\psi,\chi \in X$, and $\alpha,\beta \in \mathbb{C}$ (or $\mathbb{R}$) is called a *scalar product*, or an *inner product*, on X. (By the bar we denote the complex conjugate.)

As a consequence of (H3) and (H4) we note the *antilinearity*

$$(\varphi, \alpha\psi + \beta\chi) = \bar{\alpha}(\varphi,\psi) + \bar{\beta}(\varphi,\chi). \tag{1.11}$$

Theorem 1.20. *A scalar product satisfies the* Cauchy–Schwarz inequality

$$|(\varphi,\psi)|^2 \leq (\varphi,\varphi)(\psi,\psi)$$

for all $\varphi,\psi \in X$, with equality if and only if φ and ψ are linearly dependent.

Proof. The inequality is trivial for $\varphi = 0$. For $\varphi \neq 0$ it follows from

$$\begin{aligned}(\alpha\varphi + \beta\psi, \alpha\varphi + \beta\psi) &= |\alpha|^2(\varphi,\varphi) + 2\,\mathrm{Re}\left\{\alpha\bar{\beta}(\varphi,\psi)\right\} + |\beta|^2(\psi,\psi)\\ &= (\varphi,\varphi)(\psi,\psi) - |(\varphi,\psi)|^2,\end{aligned}$$

where we have set $\alpha = -(\varphi,\varphi)^{-1/2}\,\overline{(\varphi,\psi)}$ and $\beta = (\varphi,\varphi)^{1/2}$. Since $(\cdot,\cdot)$ is positive definite, this expression is nonnegative and is equal to zero if and only if $\alpha\varphi+\beta\psi = 0$. In the latter case φ and ψ are linearly dependent because $\beta \neq 0$. □

Theorem 1.21. *A scalar product $(\cdot,\cdot)$ on a linear space X defines a norm by*

$$\|\varphi\| := (\varphi,\varphi)^{1/2}$$

for all $\varphi \in X$. If X is complete with respect to this norm it is called a Hilbert space*; otherwise it is called a* pre-Hilbert space.

Proof. We leave it as an exercise to verify the norm axioms. The triangle inequality follows from the Cauchy–Schwarz inequality. □

Note that $L^2[a,b]$ is a Hilbert space with the scalar product given by

$$(\varphi,\psi) := \int_a^b \varphi(x)\overline{\psi(x)}\,dx.$$

Definition 1.22. Two elements φ and ψ of a pre-Hilbert space X are called *orthogonal* if

$$(\varphi, \psi) = 0.$$

Two subsets U and V of X are called orthogonal if each pair of elements $\varphi \in U$ and $\psi \in V$ are orthogonal. For two orthogonal elements or subsets we write $\varphi \perp \psi$ and $U \perp V$, respectively. A subset U of X is called an *orthogonal system* if $(\varphi, \psi) = 0$ for all $\varphi, \psi \in U$ with $\varphi \neq \psi$. An orthogonal system U is called an *orthonormal system* if $\|\varphi\| = 1$ for all $\varphi \in U$. The set

$$U^\perp := \{\psi \in X : \psi \perp U\}$$

is called the *orthogonal complement* of the subset U.

1.6 Best Approximation

Definition 1.23. Let $U \subset X$ be a subset of a normed space X and let $\varphi \in X$. An element $v \in U$ is called a *best approximation* to φ with respect to U if

$$\|\varphi - v\| = \inf_{u \in U} \|\varphi - u\|,$$

i.e., $v \in U$ has smallest distance from φ.

Theorem 1.24. *Let U be a finite-dimensional subspace of a normed space X. Then for every element of X there exists a best approximation with respect to U.*

Proof. Let $\varphi \in X$ and choose a minimizing sequence (u_n) for φ, i.e., $u_n \in U$ satisfies

$$\|\varphi - u_n\| \to d := \inf_{u \in U} \|\varphi - u\|, \quad n \to \infty.$$

Because of $\|u_n\| \leq \|\varphi - u_n\| + \|\varphi\|$, the sequence (u_n) is bounded. Since U has finite dimension, it is closed and by Theorem 1.17 the sequence (u_n) contains a convergent subsequence $(u_{n(k)})$ with limit $v \in U$. Then

$$\|\varphi - v\| = \lim_{k \to \infty} \|\varphi - u_{n(k)}\| = d$$

completes the proof. □

Theorem 1.25. *Let U be a linear subspace of a pre-Hilbert space X. An element v is a best approximation to $\varphi \in X$ with respect to U if and only if*

$$(\varphi - v, u) = 0$$

for all $u \in U$, i.e., if and only if $\varphi - v \perp U$. To each $\varphi \in X$ there exists at most one best approximation with respect to U.

Proof. This follows from the equality

$$\|(\varphi - v) + \alpha u\|^2 = \|\varphi - v\|^2 + 2\alpha \operatorname{Re}(\varphi - v, u) + \alpha^2 \|u\|^2,$$

which is valid for all $v, u \in U$ and all $\alpha \in \mathbb{R}$. □

Theorem 1.26. *Let U be a complete linear subspace of a pre-Hilbert space X. Then to every element of X there exists a unique best approximation with respect to U.*

Proof. Let $\varphi \in X$ and choose a sequence (u_n) with

$$\|\varphi - u_n\|^2 \le d^2 + \frac{1}{n}, \quad n \in \mathbb{N}, \tag{1.12}$$

where $d := \inf_{u \in U} \|\varphi - u\|$. Then

$$\|(\varphi - u_n) + (\varphi - u_m)\|^2 + \|u_n - u_m\|^2 = 2\|\varphi - u_n\|^2 + 2\|\varphi - u_m\|^2 \le 4d^2 + \frac{2}{n} + \frac{2}{m}$$

for all $n, m \in \mathbb{N}$, and since $\frac{1}{2}(u_n + u_m) \in U$, it follows that

$$\|u_n - u_m\|^2 \le 4d^2 + \frac{2}{n} + \frac{2}{m} - 4\left\|\varphi - \frac{1}{2}(u_n + u_m)\right\|^2 \le \frac{2}{n} + \frac{2}{m}.$$

Hence, (u_n) is a Cauchy sequence, and because U is complete, there exists an element $v \in U$ such that $u_n \to v$, $n \to \infty$. Passing to the limit $n \to \infty$ in (1.12) shows that v is a best approximation of φ with respect to U. □

We wish to extend Theorem 1.25 to the case of convex subsets of a pre-Hilbert space. A subset U of a linear space X is called *convex* if

$$\lambda\varphi_1 + (1 - \lambda)\varphi_2 \in U$$

for all $\varphi_1, \varphi_2 \in U$ and all $\lambda \in (0, 1)$.

Theorem 1.27. *Let U be a convex subset of a pre-Hilbert space X. An element $v \in U$ is a best approximation to $\varphi \in X$ with respect to U if and only if*

$$\operatorname{Re}(\varphi - v, u - v) \le 0$$

for all $u \in U$. To each $\varphi \in X$ there exists at most one best approximation with respect to U.

Proof. This follows from the equality

$$\|\varphi - [(1 - \lambda)v + \lambda u]\|^2 = \|\varphi - v\|^2 - 2\lambda \operatorname{Re}(\varphi - v, u - v) + \lambda^2 \|u - v\|^2,$$

which is valid for all $u, v \in U$ and all $\lambda \in (0, 1)$. □

It is left to the reader to carry Theorem 1.26 over from the case of a linear subspace to the case of a convex subset.

For a subset U of a linear space X we denote the set *spanned* by all linear combinations of elements of U by span U.

Theorem 1.28. *Let* $\{u_n : n \in \mathbb{N}\}$ *be an orthonormal system in a pre-Hilbert space* X. *Then the following properties are equivalent:*

(a) $\operatorname{span}\{u_n : n \in \mathbb{N}\}$ *is dense in* X.
(b) *Each* $\varphi \in X$ *can be expanded in a* Fourier series

$$\varphi = \sum_{n=1}^{\infty} (\varphi, u_n) u_n.$$

(c) *For each* $\varphi \in X$ *we have* Parseval's equality

$$\|\varphi\|^2 = \sum_{n=1}^{\infty} |(\varphi, u_n)|^2.$$

The properties (a)–(c) *imply that*

(d) $\varphi = 0$ *is the only element in* X *with* $(\varphi, u_n) = 0$ *for all* $n \in \mathbb{N}$,

and (a), (b), (c), *and* (d) *are equivalent if* X *is a Hilbert space. An orthonormal system with the property* (a) *is called* complete.

Proof. What has to be understood by a series is explained in Problem 1.5.
(a) ⇒ (b): By Theorems 1.25 and 1.26, the partial sum

$$\varphi_n = \sum_{k=1}^{n} (\varphi, u_k) u_k$$

is the best approximation to φ with respect to $\operatorname{span}\{u_1, \dots, u_n\}$. Since by assumption $\operatorname{span}\{u_n : n \in \mathbb{N}\}$ is dense in X, the sequence of the best approximations converges, i.e., $\varphi_n \to \varphi$, $n \to \infty$.
(b) ⇒ (c): This follows by taking the scalar product of the Fourier series with φ.
(c) ⇒ (a): This follows from

$$\left\| \varphi - \sum_{k=1}^{n} (\varphi, u_k) u_k \right\|^2 = \|\varphi\|^2 - \sum_{k=1}^{n} |(\varphi, u_k)|^2.$$

(c) ⇒ (d): This is trivial.
(d) ⇒ (a): In the case of a Hilbert space X, we set $U := \overline{\operatorname{span}\{u_n : n \in \mathbb{N}\}}$ and assume that $X \neq U$. Then there exists $\varphi \in X$ with $\varphi \notin U$. Since X is complete, U is also complete. Therefore, by Theorems 1.25 and 1.26 the best approximation v to φ with respect to U exists and satisfies $(v - \varphi, u_n) = 0$ for all $n \in \mathbb{N}$. Hence, we have the contradiction $\varphi = v \in U$. □

Problems

1.1. Show that finite-dimensional subspaces of normed spaces are closed and complete.

1.2. A norm $\|\cdot\|_a$ on a linear space X is called *stronger* than a norm $\|\cdot\|_b$ if every sequence converging with respect to the norm $\|\cdot\|_a$ also converges with respect to the norm $\|\cdot\|_b$. The same fact is also expressed by saying that the norm $\|\cdot\|_b$ is *weaker* than the norm $\|\cdot\|_a$. Show that $\|\cdot\|_a$ is stronger than $\|\cdot\|_b$ if and only if there exists a positive number C such that $\|\varphi_b\| \le C\|\varphi\|_a$ for all $\varphi \in X$. Show that on $C[a,b]$ the maximum norm is stronger than the mean square norm. Construct a counterexample to demonstrate that these two norms are not equivalent.

1.3. Show that a continuous real-valued function on a compact subset of a normed space assumes its supremum and its infimum and that it is uniformly continuous.

1.4. Construct a counterexample to demonstrate that $C[a,b]$ is not complete with respect to the mean square norm.

1.5. Let (φ_n) be a sequence of elements of a normed space X. The *series*

$$\sum_{k=1}^{\infty} \varphi_k$$

is called convergent if the sequence (S_n) of partial sums

$$S_n := \sum_{k=1}^{n} \varphi_k$$

converges. The limit $S = \lim_{n\to\infty} S_n$ is called the *sum* of the series. Show that in a Banach space X the convergence of the series

$$\sum_{k=1}^{\infty} \|\varphi_k\|$$

is a sufficient condition for the convergence of the series $\sum_{k=1}^{\infty} \varphi_k$ and that

$$\left\| \sum_{k=1}^{\infty} \varphi_k \right\| \le \sum_{k=1}^{\infty} \|\varphi_k\|.$$

Hint: Show that (S_n) is a Cauchy sequence.

Chapter 2
Bounded and Compact Operators

In this chapter we briefly review the basic properties of bounded linear operators in normed spaces and then introduce the concept of compact operators that is of fundamental importance in the study of integral equations.

2.1 Bounded Operators

Recall that an operator $A : X \to Y$ mapping a linear space X into a linear space Y is called *linear* if

$$A(\alpha\varphi + \beta\psi) = \alpha A\varphi + \beta A\psi$$

for all $\varphi, \psi \in X$ and all $\alpha, \beta \in \mathbb{C}$ (or $\mathbb{R}$).

Definition 2.1. A linear operator $A : X \to Y$ from a normed space X into a normed space Y is called *bounded* if there exists a positive number C such that

$$\|A\varphi\| \le C\|\varphi\|$$

for all $\varphi \in X$. Each number C for which this inequality holds is called a *bound* for the operator A. (Again we use the same symbol $\|\cdot\|$ for the norms on X and Y.)

Theorem 2.2. *A linear operator $A : X \to Y$ from a normed space X into a normed space Y is bounded if and only if*

$$\|A\| := \sup_{\|\varphi\|=1} \|A\varphi\| < \infty. \tag{2.1}$$

The number $\|A\|$ is the smallest bound for A. It is called the norm *of A and can also be expressed by*

$$\|A\| = \sup_{\|\varphi\|\le 1} \|A\varphi\|.$$

R. Kress, *Linear Integral Equations*, Applied Mathematical Sciences 82,
DOI 10.1007/978-1-4614-9593-2_2, © Springer Science+Business Media New York 2014

Proof. For a bounded linear operator A with bound C we have $\sup_{\|\varphi\|=1} \|A\varphi\| \le C$. Conversely, for all $\psi \in X$ we can estimate

$$\|A\psi\| = \|\psi\| \left\| A\left(\frac{1}{\|\psi\|}\psi\right)\right\| \le \sup_{\|\varphi\|=1} \|A\varphi\| \, \|\psi\|$$

i.e., A has the bound $\sup_{\|\varphi\|=1} \|A\varphi\|$. The equality $\sup_{\|\varphi\|=1} \|A\varphi\| = \sup_{\|\varphi\|\le 1} \|A\varphi\|$ also is a consequence of the last inequality. □

Hence a linear operator is bounded if and only if it maps bounded sets in X into bounded sets in Y. Recall the Definition 1.3 for the continuity of an operator $A : X \to Y$ mapping a normed space X into a normed space Y.

Theorem 2.3. *For a linear operator $A : X \to Y$ mapping a normed space X into a normed space Y the following properties are equivalent:*

1. A is continuous at one element.
2. A is continuous.
3. A is bounded.

Proof. 1. ⇒ 2.: Let A be continuous at $\varphi_0 \in X$. Then for every $\varphi \in X$ and every sequence (φ_n) with $\varphi_n \to \varphi$, $n \to \infty$, we have

$$A\varphi_n = A(\varphi_n - \varphi + \varphi_0) + A(\varphi - \varphi_0) \to A(\varphi_0) + A(\varphi - \varphi_0) = A(\varphi), \quad n \to \infty,$$

since $\varphi_n - \varphi + \varphi_0 \to \varphi_0$, $n \to \infty$. Therefore, A is continuous at all $\varphi \in X$.
2. ⇒ 3.: Let A be continuous and assume there is no $C > 0$ such that $\|A\varphi\| \le C\|\varphi\|$ for all $\varphi \in X$. Then there exists a sequence (φ_n) in X with $\|\varphi_n\| = 1$ and $\|A\varphi_n\| \ge n$. Consider the sequence $\psi_n := \|A\varphi_n\|^{-1}\varphi_n$. Then $\psi_n \to 0$, $n \to \infty$, and since A is continuous $A\psi_n \to A(0) = 0$, $n \to \infty$. This is a contradiction to $\|A\psi_n\| = 1$ for all n.
3. ⇒ 1.: Let A be bounded and let (φ_n) be a sequence in X with $\varphi_n \to 0$, $n \to \infty$. Then from $\|A\varphi_n\| \le C\|\varphi_n\|$ it follows that $A\varphi_n \to 0$, $n \to \infty$. Thus, A is continuous at $\varphi = 0$. □

Theorem 2.4. *Each linear operator $A : X \to Y$ from a finite-dimensional normed space X into a normed space Y is bounded.*

Proof. The mapping $\varphi \mapsto \|A\varphi\| + \|\varphi\|$ is a norm on X. By Theorem 1.6 on the finite-dimensional linear space X all norms are equivalent, i.e., there exists a $C > 0$ such that $\|A\varphi\| \le \|A\varphi\| + \|\varphi\| \le C\|\varphi\|$ for all $\varphi \in X$. □

Remark 2.5. *Let X, Y, and Z be normed spaces and let $A : X \to Y$ and $B : Y \to Z$ be bounded linear operators. Then the product $BA : X \to Z$, defined by $(BA)\varphi := B(A\varphi)$ for all $\varphi \in X$, is a bounded linear operator with $\|BA\| \le \|A\| \, \|B\|$.*

Proof. This follows from $\|(BA)\varphi\| = \|B(A\varphi)\| \le \|B\| \, \|A\| \, \|\varphi\|$. □

Every linear combination of bounded linear operators again is a bounded linear operator, i.e., the set $L(X, Y)$ of bounded linear operators from X into Y forms a linear space.

Theorem 2.6. *The linear space $L(X, Y)$ of bounded linear operators from a normed space X into a normed space Y is a normed space with the norm (2.1). If Y is a Banach space then $L(X, Y)$ also is a Banach space.*

Proof. The proof consists in carrying over the norm axioms and the completeness from Y onto $L(X, Y)$. For the second part, let (A_n) be a Cauchy sequence in $L(X, Y)$, i.e., $\|A_m - A_n\| \to 0$, $m, n \to \infty$. Then for each $\varphi \in X$ the sequence $(A_n\varphi)$ is a Cauchy sequence in Y and converges, since Y is complete. Then $A\varphi := \lim_{n\to\infty} A_n\varphi$ defines a bounded linear operator $A : X \to Y$, which is the limit of the sequence (A_n), i.e., $\|A_n - A\| \to 0$, $n \to \infty$. □

Definition 2.7. Let X and Y be normed spaces. A sequence (A_n) of operators from $L(X, Y)$ is called *norm convergent* to an operator $A \in L(X, Y)$ if $A_n \to A$, $n \to \infty$, in the norm of $L(X, Y)$, i.e., if $\|A_n - A\| \to 0$, $n \to \infty$. It is called *pointwise convergent* to an operator $A \in L(X, Y)$ if $A_n\varphi \to A\varphi$, $n \to \infty$, for every $\varphi \in X$.

Theorem 2.8. *Norm convergence of bounded linear operators implies pointwise convergence, but the converse is not true.*

Proof. The first statement is evident from $\|A_n\varphi - A\varphi\| \le \|A_n - A\| \, \|\varphi\|$ for all $\varphi \in X$. That the converse is not true follows from the following counterexample with $X = C[0, 1]$ and $Y = \mathbb{R}$. For the bounded linear operators given by

$$Q\varphi := \int_0^1 \varphi(x)\, dx \quad \text{and} \quad Q_n\varphi := \frac{1}{n}\sum_{k=1}^{n} \varphi\left(\frac{k}{n}\right)$$

the definition of the Riemann integral implies pointwise convergence $Q_n\varphi \to Q\varphi$, $n \to \infty$, for all $\varphi \in C[0, 1]$. For $0 < \varepsilon < 1/2n$ we choose a function $\psi_\varepsilon \in C[0, 1]$ with $0 \le \psi_\varepsilon(x) \le 1$ for all $x \in [0, 1]$ such that $\psi_\varepsilon(x) = 1$ if $\min_{k=1,\dots,n} |x - x_k| \ge \varepsilon$ and $\psi_\varepsilon(x_k) = 0$, $k = 1, \dots, n$. Then we have

$$\|Q_n - Q\|_\infty = \sup_{\|\varphi\|_\infty = 1} |(Q_n - Q)\varphi| \ge |(Q_n - Q)\psi_\varepsilon| = |Q\psi_\varepsilon| \ge 1 - 2n\varepsilon.$$

Passing to the limit $\varepsilon \to 0$ it follows that $\|Q_n - Q\|_\infty \ge 1$ for all n, i.e., the sequence (Q_n) is not norm convergent. □

Definition 2.9. A bijective bounded linear operator $A : X \to Y$ from a normed space X onto a normed space Y is called an *isomorphism* if its inverse $A^{-1} : Y \to X$ is bounded. Two normed spaces are called *isomorphic* if there exists an isomorphism between them.

Two isomorphic normed spaces do not differ in their linear and topological structure. In particular, they have the same convergent sequences. Equivalent norms create isomorphic spaces.

2.2 The Dual Space

It is an important principle of functional analysis to connect the investigation of a normed space X with that of its *dual space* defined as the space of bounded linear operators $X^* := L(X, \mathbb{C})$ (or $X^* := L(X, \mathbb{R})$). The elements of X^* are called *bounded linear functionals* on X. By Theorem 2.6 the dual space of a normed space X is a Banach space with the norm

$$\|F\| := \sup_{\|\varphi\|=1} |F(\varphi)|$$

for $F \in X^*$.

The following extension theorem ensures the existence of nontrivial bounded linear functionals for each normed space.

Theorem 2.10 (Hahn–Banach). *Let U be a subspace of a normed space X and F a bounded linear functional on U. Then there exists a bounded linear functional G on X with the properties*

$$G(\varphi) = F(\varphi), \quad \varphi \in U,$$

and

$$\|G\| = \|F\|.$$

Proof. We first consider a real normed space X. Assume that X is spanned by U and an element $\varphi_0 \in X$ with $\varphi_0 \notin U$, that is,

$$X = \operatorname{span}\{U, \varphi_0\} = \{\varphi + \alpha\varphi_0 : \varphi \in U,\ \alpha \in \mathbb{R}\}.$$

If for any real number γ we define $G : X \to \mathbb{R}$ by

$$G(\varphi + \alpha\varphi_0) := F(\varphi) + \gamma\alpha, \quad \varphi \in U,\ \alpha \in \mathbb{R},$$

then G is a linear functional on X that is an extension of F, that is, $G(\varphi) = F(\varphi)$ for all $\varphi \in U$. We will now show how to choose γ such that G is bounded with $\|G\| \le \|F\|$. Then also $\|G\| = \|F\|$ is satisfied.

For all pairs $\psi, \chi \in U$ we can estimate

$$F(\chi) - F(\psi) = F(\chi - \psi) \le \|F\|\,\|\chi - \psi\| \le \|F\|\,(\|\chi + \varphi_0\| + \|\psi + \varphi_0\|)$$

whence

$$-F(\psi) - \|F\|\,\|\psi + \varphi_0\| \le -F(\chi) + \|F\|\,\|\chi + \varphi_0\|$$

follows. Therefore there exists $\gamma \in \mathbb{R}$ with the property

$$\sup_{\psi\in U}\,[-F(\psi) - \|F\|\,\|\psi + \varphi_0\|\,] \le \gamma \le \inf_{\chi\in U}\,[-F(\chi) + \|F\|\,\|\chi + \varphi_0\|\,]. \tag{2.2}$$

Now if $\alpha > 0$ we choose $\chi = \varphi/\alpha$ and, multiplying by α, we obtain from (2.2) that

$$G(\varphi + \alpha\varphi_0) = F(\varphi) + \gamma\alpha \le \|F\|\,\|\varphi + \alpha\varphi_0\|.$$

If $\alpha < 0$ we choose $\psi = \varphi/\alpha$ and obtain from (2.2) that again

$$G(\varphi + \alpha\varphi_0) = F(\varphi) + \gamma\alpha \leq \|F\| \, \|\varphi + \alpha\varphi_0\|.$$

Hence $G(\varphi) \leq \|F\| \, \|\varphi\|$ for all $\varphi \in X$ and replacing φ by $-\varphi$ we also have that $-G(\varphi) \leq \|F\| \, \|\varphi\|$ for $\varphi \in X$. Consequently $\|G\| \leq \|F\|$.

Now we consider the family of bounded linear functionals G on subspaces of X that are extensions of F such that the norms of G and F coincide. We make this family into a partially ordered family be defining $G_1 \succ G_2$ to mean that G_1 is an extension of G_2. Then Zorn's lemma (among others, see [40]) ensures the existence of a maximal extension G of F. We have to show that the domain of definition of G coincides with X itself. Assume to the contrary that the domain V of G is a proper subspace of X. Then, taking V as U in the above, we obtain an extension of F in a subspace of X that contains V as a proper subspace which contradicts the maximality of G.

Finally, we can consider a complex normed space X also as a real linear space in which for each $\varphi \in X$ the elements φ and $i\varphi$ are linearly independent. If we define

$$F_r(\varphi) := \operatorname{Re} F(\varphi), \quad \varphi \in U,$$

then $F_r : U \to \mathbb{R}$ is a real bounded linear functional satisfying $\|F_r\| \leq \|F\|$ and

$$F(\varphi) = F_r(\varphi) - iF_r(i\varphi) \tag{2.3}$$

for all $\varphi \in U$. Hence, there exists an extension $G_r : X \to \mathbb{R}$ of F_r with $\|G_r\| = \|F_r\|$. We set

$$G(\varphi) := G_r(\varphi) - iG_r(i\varphi), \quad \varphi \in X.$$

Then $G : X \to \mathbb{C}$ is linear since

$$G(i\varphi) = G_r(i\varphi) - iG_r(-\varphi) = i[G_r(\varphi) - iG_r(i\varphi)] = iG(\varphi)$$

for all $\varphi \in X$ and in view of (2.3) it is an extension of F. Further, abbreviating $\theta = \arg(G(\varphi))$, we have the inequality

$$|G(\varphi)| = e^{-i\theta} G(\varphi) = G(e^{-i\theta}\varphi) = G_r(e^{-i\theta}\varphi) \leq \|G_r\| \, \|e^{-i\theta}\varphi\| = \|G_r\| \, \|\varphi\| \leq \|F\| \, \|\varphi\|$$

for all $\varphi \in X$ whence $\|G\| \leq \|F\|$ follows. This completes the proof. □

We finish the section with two important corollaries of the Hahn–Banach extension theorem.

Corollary 2.11. *For each element $\varphi_0 \neq 0$ in a normed space X there exists a bounded linear functional G with the properties*

$$G(\varphi_0) = \|\varphi_0\| \quad \text{and} \quad \|G\| = 1.$$

Proof. Apply the Hahn–Banach theorem to $U = \operatorname{span}\{\varphi_0\}$ and $F(\alpha\varphi_0) := \alpha \, \|\varphi_0\|$. □

Corollary 2.12. *Each element φ of a normed space X satisfies*

$$\|\varphi\| = \sup_{F \in X^*, \|F\|=1} |F(\varphi)|.$$

Proof. Because of $|F(\varphi)| \le \|F\|\,\|\varphi\|$ clearly $\sup_{\|F\|=1} |F(\varphi)| \le \|\varphi\|$. For $\varphi \neq 0$, by Corollary 2.11, choose a functional F_0 such that $F_0(\varphi) = \|\varphi\|$ and $\|F_0\| = 1$. Then $\sup_{\|F\|=1} |F(\varphi)| \ge |F_0(\varphi)| = \|\varphi\|$. □

2.3 Integral Operators

Now we want to introduce integral operators. We assume that the reader is familiar with the Riemann integral for real- and complex-valued functions in $\mathbb{R}^m$. A set G in $\mathbb{R}^m$ is called *Jordan measurable* if the characteristic function χ_G, given by $\chi_G(x) = 1$ for $x \in G$ and $\chi_G(x) = 0$ for $x \notin G$, is Riemann integrable. The Jordan measure $|G|$ is the integral of χ_G. For each Jordan measurable set G the closure $\bar{G}$ and the boundary ∂G also are Jordan measurable with $|\bar{G}| = |G|$ and $|\partial G| = 0$. If G is compact and Jordan measurable, then each function $f \in C(G)$ is Riemann integrable. In addition, we assume that G is the closure of an open set or, equivalently, that G coincides with the closure of its interior. This ensures that each nonnegative function $f \in C(G)$ satisfying $\int_G f(x)\,dx = 0$ must vanish identically, i.e., $f(x) = 0$ for all $x \in G$.

Theorem 2.13. *Let $G \subset \mathbb{R}^m$ be a nonempty compact and Jordan measurable set that coincides with the closure of its interior. Let $K : G \times G \to \mathbb{C}$ be a continuous function. Then the linear operator $A : C(G) \to C(G)$ defined by*

$$(A\varphi)(x) := \int_G K(x, y)\varphi(y)\,dy, \quad x \in G, \tag{2.4}$$

is called an integral operator *with* continuous kernel K. *It is a bounded linear operator with*

$$\|A\|_\infty = \max_{x \in G} \int_G |K(x, y)|\,dy.$$

Proof. Clearly (2.4) defines a linear operator $A : C(G) \to C(G)$. For each $\varphi \in C(G)$ with $\|\varphi\|_\infty \le 1$ we have

$$|(A\varphi)(x)| \le \int_G |K(x, y)|\,dy, \quad x \in G,$$

and thus

$$\|A\|_\infty = \sup_{\|\varphi\|_\infty \le 1} \|A\varphi\|_\infty \le \max_{x \in G} \int_G |K(x, y)|\,dy.$$

Since K is continuous, there exists $x_0 \in G$ such that

$$\int_G |K(x_0, y)|\,dy = \max_{x \in G} \int_G |K(x, y)|\,dy.$$

For $\varepsilon > 0$ choose $\psi \in C(G)$ by setting

$$\psi(y) := \frac{\overline{K(x_0, y)}}{|K(x_0, y)| + \varepsilon}, \quad y \in G.$$

Then $\|\psi\|_\infty \leq 1$ and

$$\|A\psi\|_\infty \geq |(A\psi)(x_0)| = \int_G \frac{|K(x_0, y)|^2}{|K(x_0, y)| + \varepsilon}\, dy$$

$$\geq \int_G \frac{|K(x_0, y)|^2 - \varepsilon^2}{|K(x_0, y)| + \varepsilon}\, dy = \int_G |K(x_0, y)|\, dy - \varepsilon |G|.$$

Hence

$$\|A\|_\infty = \sup_{\|\varphi\|_\infty \leq 1} \|A\varphi\|_\infty \geq \|A\psi\|_\infty \geq \int_G |K(x_0, y)|\, dy - \varepsilon |G|,$$

and because this holds for all $\varepsilon > 0$, we have

$$\|A\|_\infty \geq \int_G |K(x_0, y)|\, dy = \max_{x \in G} \int_G |K(x, y)|\, dy.$$

This concludes the proof. □

Theorem 2.13 can be extended to the integral operator $A : C(G) \to C(M)$ given by

$$(A\varphi)(x) := \int_G K(x, y)\varphi(y)\, dy, \quad x \in M, \tag{2.5}$$

where $K : M \times G \to \mathbb{C}$ is continuous, $M \subset \mathbb{R}^n$ is a compact set and n can be different from m.

2.4 Neumann Series

For operator equations of the *second kind*

$$\varphi - A\varphi = f$$

existence and uniqueness of a solution can be established by the Neumann series provided that A is a *contraction*, i.e., $\|A\| < 1$.

Theorem 2.14. *Let $A : X \to X$ be a bounded linear operator on a Banach space X with $\|A\| < 1$ and let $I : X \to X$ denote the identity operator. Then $I - A$ has a bounded inverse on X that is given by the* Neumann series

$$(I - A)^{-1} = \sum_{k=0}^{\infty} A^k$$

and satisfies

$$\|(I - A)^{-1}\| \le \frac{1}{1 - \|A\|}\,.$$

(The iterated operators A^k are defined recursively by $A^0 := I$ and $A^k := AA^{k-1}$ for $k \in \mathbb{N}$.)

Proof. Since $\|A\| < 1$, in view of Remark 2.5, we have absolute convergence

$$\sum_{k=0}^{\infty} \|A^k\| \le \sum_{k=0}^{\infty} \|A\|^k = \frac{1}{1 - \|A\|}$$

in the Banach space $L(X, X)$, and therefore, by Problem 1.5, the Neumann series converges in the operator norm and defines a bounded linear operator

$$S := \sum_{k=0}^{\infty} A^k$$

with

$$\|S\| \le \frac{1}{1 - \|A\|}\,.$$

The operator S is the inverse of $I - A$, as can be seen from

$$(I - A)S = (I - A) \lim_{n\to\infty} \sum_{k=0}^{n} A^k = \lim_{n\to\infty} (I - A^{n+1}) = I$$

and

$$S(I - A) = \lim_{n\to\infty} \sum_{k=0}^{n} A^k(I - A) = \lim_{n\to\infty} (I - A^{n+1}) = I,$$

since $\|A^{n+1}\| \le \|A\|^{n+1} \to 0,\ n \to \infty$. □

Obviously, the partial sums

$$\varphi_n := \sum_{k=0}^{n} A^k f$$

of the Neumann series satisfy $\varphi_{n+1} = A\varphi_n + f$ for $n \ge 0$. Hence, the Neumann series is related to successive approximations by the following theorem.

Theorem 2.15. *Under the assumptions of Theorem 2.14 for each $f \in X$ the* successive approximations

$$\varphi_{n+1} := A\varphi_n + f, \quad n = 0, 1, 2, \dots, \tag{2.6}$$

with arbitrary $\varphi_0 \in X$ converge to the unique solution φ of $\varphi - A\varphi = f$.

Proof. By induction it can be seen that

$$\varphi_n = A^n\varphi_0 + \sum_{k=0}^{n-1} A^k f, \quad n = 1, 2, \dots,$$

whence

$$\lim_{n\to\infty} \varphi_n = \sum_{k=0}^{\infty} A^k f = (I - A)^{-1} f$$

follows. □

Corollary 2.16. *Let K be a continuous kernel satisfying*

$$\max_{x\in G} \int_G |K(x, y)|\, dy < 1.$$

Then for each $f \in C(G)$ the integral equation of the second kind

$$\varphi(x) - \int_G K(x, y)\varphi(y)\, dy = f(x), \quad x \in G,$$

has a unique solution $\varphi \in C(G)$. The successive approximations

$$\varphi_{n+1}(x) := \int_G K(x, y)\varphi_n(y)\, dy + f(x), \quad n = 0, 1, 2, \dots,$$

with arbitrary $\varphi_0 \in C(G)$ converge uniformly to this solution.

The method of successive approximations has two drawbacks. First, the Neumann series ensures existence of solutions to integral equations of the second kind only for sufficiently small kernels, and second, in general, it cannot be summed in closed form. Later in the book we will have more to say about using successive approximations to obtain approximate solutions (see Section 10.5).

2.5 Compact Operators

To provide the tools for establishing the existence of solutions to a wider class of integral equations we now turn to the introduction and investigation of compact operators.

Definition 2.17. A linear operator $A : X \to Y$ from a normed space X into a normed space Y is called *compact* if it maps each bounded set in X into a relatively compact set in Y.

Since by Definition 1.16 and Theorem 1.15 a subset U of a normed space Y is relatively compact if each sequence in U contains a subsequence that converges in Y, we have the following equivalent condition for an operator to be compact.

Theorem 2.18. *A linear operator $A : X \to Y$ is compact if and only if for each bounded sequence (φ_n) in X the sequence $(A\varphi_n)$ contains a convergent subsequence in Y.*

We proceed by establishing the basic properties of compact operators.

Theorem 2.19. *Compact linear operators are bounded.*

Proof. This is obvious, since relatively compact sets are bounded (see Theorem 1.15). □

Theorem 2.20. *Linear combinations of compact linear operators are compact.*

Proof. Let $A, B : X \to Y$ be compact linear operators and let $\alpha, \beta \in \mathbb{C}$. Then for each bounded sequence (φ_n) in X, since A and B are compact, we can select a subsequence $(\varphi_{n(k)})$ such that both sequences $(A\varphi_{n(k)})$ and $(B\varphi_{n(k)})$ converge. Hence $(\alpha A + \beta B)\varphi_{n(k)}$ converges, and therefore $\alpha A + \beta B$ is compact. □

Theorem 2.21. *Let X, Y, and Z be normed spaces and let $A : X \to Y$ and $B : Y \to Z$ be bounded linear operators. Then the product $BA : X \to Z$ is compact if one of the two operators A or B is compact.*

Proof. Let (φ_n) be a bounded sequence in X. If A is compact, then there exists a subsequence $(\varphi_{n(k)})$ such that $A\varphi_{n(k)} \to \psi \in Y$, $k \to \infty$. Since B is bounded and therefore continuous, we have $B(A\varphi_{n(k)}) \to B\psi \in Z$, $k \to \infty$. Hence BA is compact. If A is bounded and B is compact, the sequence $(A\varphi_n)$ is bounded in Y, since bounded operators map bounded sets into bounded sets. Therefore, there exists a subsequence $(\varphi_{n(k)})$ such that $(BA)\varphi_{n(k)} = B(A\varphi_{n(k)}) \to \chi \in Z$, $k \to \infty$. Hence, again BA is compact. □

Theorem 2.22. *Let X be a normed space and Y be a Banach space. Let the sequence $A_n : X \to Y$ of compact linear operators be norm convergent to a linear operator $A : X \to Y$, i.e., $\|A_n - A\| \to 0$, $n \to \infty$. Then A is compact.*

Proof. Let (φ_m) be a bounded sequence in X, i.e., $\|\varphi_m\| \le C$ for all $m \in \mathbb{N}$ and some $C > 0$. Because the A_n are compact, by the standard diagonalization procedure (see the proof of Theorem 1.18), we can choose a subsequence $(\varphi_{m(k)})$ such that $(A_n\varphi_{m(k)})$ converges for every fixed n as $k \to \infty$. Given $\varepsilon > 0$, since $\|A_n - A\| \to 0$, $n \to \infty$, there exists $n_0 \in \mathbb{N}$ such that $\|A_{n_0} - A\| < \varepsilon/3C$. Because $(A_{n_0}\varphi_{m(k)})$ converges, there exists $N(\varepsilon) \in \mathbb{N}$ such that

$$\|A_{n_0}\varphi_{m(k)} - A_{n_0}\varphi_{m(l)}\| < \frac{\varepsilon}{3}$$

for all $k, l \ge N(\varepsilon)$. But then we have

$$\|A\varphi_{m(k)} - A\varphi_{m(l)}\| \le \|A\varphi_{m(k)} - A_{n_0}\varphi_{m(k)}\| + \|A_{n_0}\varphi_{m(k)} - A_{n_0}\varphi_{m(l)}\|$$

$$+\|A_{n_0}\varphi_{m(l)} - A\varphi_{m(l)}\| < \varepsilon.$$

Thus $(A\varphi_{m(k)})$ is a Cauchy sequence, and therefore it is convergent in the Banach space Y. □

Theorem 2.23. *Let $A : X \to Y$ be a bounded linear operator with finite-dimensional range $A(X)$. Then A is compact.*

Proof. Let $U \subset X$ be bounded. Then the bounded operator A maps U into the bounded set $A(U)$ contained in the finite-dimensional space $A(X)$. By the Bolzano–Weierstrass Theorem 1.17 the set $A(U)$ is relatively compact. Therefore A is compact. □

Lemma 2.24 (Riesz). *Let X be a normed space, $U \subset X$ a closed subspace with $U \neq X$, and $\alpha \in (0, 1)$. Then there exists an element $\psi \in X$ with $\|\psi\| = 1$ such that $\|\psi - \varphi\| \geq \alpha$ for all $\varphi \in U$.*

Proof. Because $U \neq X$, there exists an element $f \in X$ with $f \notin U$, and because U is closed, we have

$$\beta := \inf_{\varphi \in U} \|f - \varphi\| > 0.$$

We can choose $g \in U$ such that

$$\beta \leq \|f - g\| \leq \frac{\beta}{\alpha} \, .$$

Now we define

$$\psi := \frac{f - g}{\|f - g\|} \, .$$

Then $\|\psi\| = 1$, and for all $\varphi \in U$ we have

$$\|\psi - \varphi\| = \frac{1}{\|f - g\|} \, \|f - \{g + \|f - g\| \varphi\} \| \geq \frac{\beta}{\|f - g\|} \geq \alpha,$$

since $g + \|f - g\| \varphi \in U$. □

Theorem 2.25. *The identity operator $I : X \to X$ is compact if and only if X has finite dimension.*

Proof. Assume that I is compact and X is not finite-dimensional. Choose an arbitrary $\varphi_1 \in X$ with $\|\varphi_1\| = 1$. Then $U_1 := \operatorname{span}\{\varphi_1\}$ is a finite-dimensional and therefore closed subspace of X. By Lemma 2.24 there exists $\varphi_2 \in X$ with $\|\varphi_2\| = 1$ and $\|\varphi_2 - \varphi_1\| \geq 1/2$. Now consider $U_2 := \operatorname{span}\{\varphi_1, \varphi_2\}$. Again by Lemma 2.24 there exists $\varphi_3 \in X$ with $\|\varphi_3\| = 1$ and $\|\varphi_3 - \varphi_1\| \geq 1/2$, $\|\varphi_3 - \varphi_2\| \geq 1/2$. Repeating this procedure, we obtain a sequence (φ_n) with the properties $\|\varphi_n\| = 1$ and $\|\varphi_n - \varphi_m\| \geq 1/2$, $n \neq m$. This implies that the bounded sequence (φ_n) does not contain a convergent subsequence. Hence we have a contradiction to the compactness of I. Therefore, if the identity operator is compact, X has finite dimension. The converse statement is an immediate consequence of Theorem 2.23. □

This theorem, in particular, implies that the converse of Theorem 2.19 is false. It also justifies the distinction between operator equations of the first and second kind, because obviously for a compact operator A the operators A and $I - A$ have different properties. In particular, the following theorem immediately follows from Theorems 2.21 and 2.25.

Theorem 2.26. *A compact linear operator $A : X \to Y$ cannot have a bounded inverse unless X has finite dimension.*

Theorem 2.27. *Integral operators with continuous kernel are compact linear operators on $C(G)$.*

Proof. Let $U \subset C(G)$ be bounded, i.e., $\|\varphi\|_\infty \leq C$ for all $\varphi \in U$ and some $C > 0$. Then for the integral operator A defined by (2.4) we have that

$$|(A\varphi)(x)| \leq C|G| \max_{x,y \in G} |K(x,y)|$$

for all $x \in G$ and all $\varphi \in U$, i.e., $A(U)$ is bounded. Since K is uniformly continuous on the compact set $G \times G$, for every $\varepsilon > 0$ there exists $\delta > 0$ such that

$$|K(x,z) - K(y,z)| < \frac{\varepsilon}{C|G|}$$

for all $x, y, z \in G$ with $|x - y| < \delta$. Then

$$|(A\varphi)(x) - (A\varphi)(y)| < \varepsilon$$

for all $x, y \in G$ with $|x - y| < \delta$ and all $\varphi \in U$, i.e., $A(U)$ is equicontinuous. Hence A is compact by the Arzelà–Ascoli Theorem 1.18. □

We wish to mention that the compactness of the integral operator with continuous kernel also can be established by finite-dimensional approximations using Theorems 2.22 and 2.23 in the Banach space $C(G)$. In this context note that the proofs of Theorems 2.22 and 1.18 are similar in structure. The finite-dimensional operators can be obtained by approximating either the continuous kernel by polynomials through the Weierstrass approximation theorem or the integral through a Riemann sum (see [31]).

For substantial parts of this book we will consider integral operators as operators in classical spaces of continuous functions. However, for certain topics we will find it necessary and convenient to consider integral operators also in Hilbert spaces. Therefore we carry the previous theorem over to the L^2 space of measurable and Lebesgue square integrable functions.

Theorem 2.28. *Integral operators with continuous kernel are compact linear operators on $L^2(G)$.*

Proof. For the integral operator with continuous kernel A defined by (2.4), using the Cauchy–Schwarz inequality and proceeding as in the proof of Theorem 2.27 it can be established that each bounded set $U \subset L^2(G)$ is mapped into a set $A(U) \subset C(G)$ that is bounded with respect to the maximum norm and equicontinuous. Hence, $A : L^2(G) \to C(G)$ is compact and from this the statement of the theorem follows since the maximum norm is stronger than the mean squares norm and $C(G)$ is dense in $L^2(G)$. □

We note that Theorems 2.27 and 2.28 can be extended to the integral operator given by (2.5) where for the L^2 case the set M is required to be Lebesgue measurable.

Now we extend our investigation to integral operators with a *weakly singular kernel*, i.e., the kernel K is defined and continuous for all $x, y \in G \subset \mathbb{R}^m$, $x \neq y$, and there exist positive constants M and $\alpha \in (0, m]$ such that

$$|K(x, y)| \leq M|x - y|^{\alpha - m}, \quad x, y \in G,\ x \neq y. \tag{2.7}$$

Theorem 2.29. *Integral operators with weakly singular kernel are compact linear operators on $C(G)$.*

Proof. The integral in (2.4) defining the operator A exists as an improper integral, since

$$|K(x, y)\varphi(y)| \leq M\|\varphi\|_\infty |x - y|^{\alpha - m}$$

and

$$\int_G |x - y|^{\alpha - m}\, dy \leq \omega_m \int_0^d \rho^{\alpha - m} \rho^{m-1}\, d\rho = \frac{\omega_m}{\alpha}\, d^\alpha,$$

where we have introduced polar coordinates with origin at x, d is the diameter of G, and ω_m denotes the surface area of the unit sphere in $\mathbb{R}^m$.

Now we choose a continuous function $h : [0, \infty) \to \mathbb{R}$ with $0 \leq h(t) \leq 1$ for all $t \geq 0$ such that $h(t) = 0$ for $0 \leq t \leq 1/2$ and $h(t) = 1$ for $t > 1$. For $n \in \mathbb{N}$ we define continuous kernels $K_n : G \times G \to \mathbb{C}$ by

$$K_n(x, y) := \begin{cases} h(n|x - y|)K(x, y), & x \neq y, \\ 0, & x = y. \end{cases}$$

The corresponding integral operators $A_n : C(G) \to C(G)$ are compact by Theorem 2.27. We have the estimate

$$\begin{aligned} |(A\varphi)(x) - (A_n\varphi)(x)| &= \left| \int_{G \cap B[x;1/n]} \{1 - h(n|x - y)\} K(x, y)\varphi(y)\, dy \right| \\ &\leq M\|\varphi\|_\infty \omega_m \int_0^{1/n} \rho^{\alpha - m}\rho^{m-1}\, d\rho = M\|\varphi\|_\infty \frac{\omega_m}{\alpha n^\alpha}, \quad x \in G. \end{aligned}$$

From this we observe that $A_n\varphi \to A\varphi$, $n \to \infty$, uniformly, and therefore $A\varphi \in C(G)$. Furthermore it follows that

$$\|A - A_n\|_\infty \leq M\, \frac{\omega_m}{\alpha n^\alpha} \to 0, \quad n \to \infty,$$

and thus A is compact by Theorem 2.22. □

For the compactness of integral operators with weakly singular kernel on $L^2(G)$ see Problem 4.5.

Finally, we want to expand the analysis to integral operators defined on surfaces in $\mathbb{R}^m$. Having in mind applications to boundary value problems, we will confine our attention to surfaces that are boundaries of smooth domains in $\mathbb{R}^m$. A bounded open domain $D \subset \mathbb{R}^m$ with boundary ∂D is said to be of *class* C^n, $n \in \mathbb{N}$, if the closure $\bar{D}$ admits a finite open covering

$$\bar{D} \subset \bigcup_{q=1}^{p} V_q$$

such that for each of those V_q that intersect with the boundary ∂D we have the properties: The intersection $V_q \cap \bar{D}$ can be mapped bijectively onto the half-ball $H := \{x \in \mathbb{R}^m : |x| < 1, x_m \geq 0\}$ in $\mathbb{R}^m$, this mapping and its inverse are n times continuously differentiable, and the intersection $V_q \cap \partial D$ is mapped onto the disk $H \cap \{x \in \mathbb{R}^m : x_m = 0\}$.

In particular, this implies that the boundary ∂D can be represented locally by a *parametric representation*

$$x(u) = (x_1(u), \ldots, x_m(u))$$

mapping an open parameter domain $U \subset \mathbb{R}^{m-1}$ bijectively onto a *surface patch* S of ∂D with the property that the vectors

$$\frac{\partial x}{\partial u_i}, \quad i = 1, \ldots, m-1,$$

are linearly independent at each point x of S. Such a parameterization we call a *regular parametric representation*. The whole boundary ∂D is obtained by matching a finite number of such surface patches.

On occasion, we will express the property of a domain D to be of class C^n also by saying that its boundary ∂D is of class C^n.

The vectors $\partial x/\partial u_i$, $i = 1, \ldots, m-1$, span the tangent plane to the surface at the point x. The unit *normal* ν is the unit vector orthogonal to the tangent plane. It is uniquely determined up to two opposite directions. The surface element at the point x is given by

$$ds = \sqrt{g}\, du_1 \cdots du_{m-1},$$

where g is the determinant of the positive definite matrix with entries

$$g_{ij} := \frac{\partial x}{\partial u_i} \cdot \frac{\partial x}{\partial u_j}, \quad i, j = 1, \ldots, m-1.$$

In this book, for two vectors $a = (a_1, \ldots, a_m)$ and $b = (b_1, \ldots, b_m)$ in $\mathbb{R}^m$ (or $\mathbb{C}^m$) we denote by $a \cdot b = a_1 b_1 + \cdots + a_m b_m$ the *dot product*.

Assume that ∂D is the boundary of a bounded open domain of class C^1. In the Banach space $C(\partial D)$ of real- or complex-valued continuous functions defined on the surface ∂D and equipped with the maximum norm

$$\|\varphi\|_\infty := \max_{x\in\partial D} |\varphi(x)|,$$

we consider the integral operator $A : C(\partial D) \to C(\partial D)$ defined by

$$(A\varphi)(x) := \int_{\partial D} K(x, y)\varphi(y)\, ds(y), \quad x \in \partial D, \tag{2.8}$$

where K is a continuous or weakly singular kernel. According to the dimension of the surface ∂D, a kernel K is said to be *weakly singular* if it is defined and continuous for all $x, y \in \partial D$, $x \neq y$, and there exist positive constants M and $\alpha \in (0, m-1]$ such that

$$|K(x, y)| \leq M|x - y|^{\alpha-m+1}, \quad x, y \in \partial D,\ x \neq y. \tag{2.9}$$

Analogously to Theorems 2.27 and 2.29 we can prove the following theorem.

Theorem 2.30. *Integral operators with continuous or weakly singular kernel are compact linear operators on $C(\partial D)$ if ∂D is of class C^1.*

Proof. For continuous kernels the proof of Theorem 2.27 essentially remains unaltered. For weakly singular kernels the only major difference in the proof compared with the proof of Theorem 2.29 arises in the verification of the existence of the integral in (2.8). Since the surface ∂D is of class C^1, the normal vector ν is continuous on ∂D. Therefore, we can choose $R \in (0, 1]$ such that

$$\nu(x) \cdot \nu(y) \geq \frac{1}{2} \tag{2.10}$$

for all $x, y \in \partial D$ with $|x-y| \leq R$. Furthermore, we can assume that R is small enough such that the set $S[x; R] := \{y \in \partial D : |y - x| \leq R\}$ is connected for each $x \in \partial D$. Then the condition (2.10) implies that $S[x; R]$ can be projected bijectively onto the tangent plane to ∂D at the point x. By using polar coordinates in the tangent plane with origin in x, we now can estimate

$$\begin{aligned}
\left|\int_{S[x;R]} K(x, y)\varphi(y)\, ds(y)\right| &\leq M\|\varphi\|_\infty \int_{S[x;R]} |x - y|^{\alpha-m+1}\, ds(y) \\
&\leq 2M\|\varphi\|_\infty \omega_{m-1} \int_0^R \rho^{\alpha-m+1}\rho^{m-2}\, d\rho \\
&= 2M\|\varphi\|_\infty \omega_{m-1} \frac{R^\alpha}{\alpha}\,.
\end{aligned}$$

Here we have used the facts that $|x - y| \geq \rho$, that the surface element

$$ds(y) = \frac{\rho^{m-2} d\rho d\omega}{\nu(x) \cdot \nu(y)}\,,$$

expressed in polar coordinates on the tangent plane, can be estimated with the aid of (2.10) by $ds(y) \le 2\rho^{m-2} d\rho d\omega$, and that the projection of $S[x;R]$ onto the tangent plane is contained in the disk of radius R and center x. Furthermore, we have

$$\left| \int_{\partial D \setminus S[x;R]} K(x,y)\varphi(y)\, ds(y) \right| \le M \|\varphi\|_\infty \int_{\partial D \setminus S[x;R]} R^{\alpha-m+1}\, ds(y)$$
$$\le M \|\varphi\|_\infty R^{\alpha-m+1} |\partial D|.$$

Hence, for all $x \in \partial D$ the integral (2.8) exists as an improper integral. For the compactness of A, we now can adopt the proof of Theorem 2.29. □

Problems

2.1. Let $A : X \to Y$ be a bounded linear operator from a normed space X into a normed space Y and let $\widetilde{X}$ and $\widetilde{Y}$ be the completions of X and Y, respectively. Then there exists a uniquely determined bounded linear operator $\widetilde{A} : \widetilde{X} \to \widetilde{Y}$ such that $\widetilde{A}\varphi = A\varphi$ for all $\varphi \in X$. Furthermore, $\|\widetilde{A}\| = \|A\|$. The operator A is called the *continuous extension* of A. (In the sense of Theorem 1.13 the space X is interpreted as a dense subspace of its completion $\widetilde{X}$.)
Hint: For $\varphi \in \widetilde{X}$ define $\widetilde{A}\varphi = \lim_{n\to\infty} A\varphi_n$, where (φ_n) is a sequence from X with $\varphi_n \to \varphi$, $n \to \infty$.

2.2. Show that Theorem 2.15 remains valid for operators satisfying $\|A^k\| < 1$ for some $k \in \mathbb{N}$.

2.3. Write the proofs for the compactness of the integral operator with continuous kernel in $C(G)$ using finite-dimensional approximations as mentioned after the proof of Theorem 2.27.

2.4. Show that the result of Theorem 2.13 for the norm of the integral operator remains valid for weakly singular kernels.
Hint: Use the approximations from the proof of Theorem 2.29.

2.5. Let X be a Hilbert space and $\{u_n : n \in \mathbb{N}\}$ a complete orthonormal system in X. A linear operator $A : X \to X$ is called a Hilbert–Schmidt operator if the series

$$\sum_{n=1}^{\infty} \|Au_n\|^2$$

converges. Show that Hilbert–Schmidt operators are compact and that linear integral operators with continuous kernel $A : L^2[a,b] \to L^2[a,b]$ are Hilbert–Schmidt operators.
Hint: Define operators with finite-dimensional range by

$$A_n\varphi := A \sum_{k=1}^{n} (\varphi, u_k)u_k$$

and use Theorems 2.22 and 2.23.

Chapter 3
Riesz Theory

We now present the basic theory for an operator equation

$$\varphi - A\varphi = f \tag{3.1}$$

of the second kind with a compact linear operator $A : X \to X$ on a normed space X. This theory was developed by Riesz [205] and initiated through Fredholm's [55] work on integral equations of the second kind.

In his work from 1916 Riesz interpreted an integral equation as a special case of (3.1). The notion of a normed space was not yet available in 1916. Riesz set his work up in the function space of continuous real-valued functions on the interval $[0, 1]$. He called the maximum of the absolute value of a function f on $[0, 1]$ the norm of f and confirmed its properties that we now know as the norm axioms of Definition 1.1. Only these axioms, and not the special meaning as the maximum norm, were used by Riesz. The concept of a compact operator also was not yet available in 1916. However, using the notion of compactness (see Definition 1.14) as introduced by Fréchet in 1906, Riesz formulated that the integral operator A defined by (2.4) for the special case $G = [0, 1]$ maps bounded subsets of $C[0, 1]$ into relatively compact sets, i.e., A is a compact operator in the sense of Definition 2.17.

What is fascinating about the work of Riesz is that his proofs are still usable and, as we shall do, can almost literally be transferred from the case of an integral operator on the space of continuous functions to the general case of a compact operator on a normed space.

3.1 Riesz Theory for Compact Operators

Given a compact linear operator $A : X \to X$ on a normed space X, we define

$$L := I - A,$$

where I denotes the identity operator.

R. Kress, *Linear Integral Equations*, Applied Mathematical Sciences 82,
DOI 10.1007/978-1-4614-9593-2_3,

Theorem 3.1 (First Riesz Theorem). *The nullspace of the operator L, i.e.,*

$$N(L) := \{\varphi \in X : L\varphi = 0\},$$

is a finite-dimensional subspace.

Proof. The nullspace of the bounded linear operator L is a closed subspace of X, since for each sequence (φ_n) with $\varphi_n \to \varphi$, $n \to \infty$, and $L\varphi_n = 0$ we have that $L\varphi = 0$. Each $\varphi \in N(L)$ satisfies $A\varphi = \varphi$, and therefore the restriction of A to $N(L)$ coincides with the identity operator on $N(L)$. The operator A is compact on X and therefore also compact from $N(L)$ into $N(L)$, since $N(L)$ is closed. Hence $N(L)$ is finite-dimensional by Theorem 2.25. □

Theorem 3.2 (Second Riesz Theorem). *The range of the operator L, i.e.,*

$$L(X) := \{L\varphi : \varphi \in X\},$$

is a closed linear subspace.

Proof. The range of the linear operator L is a linear subspace. Let f be an element of the closure $\overline{L(X)}$. Then there exists a sequence (φ_n) in X such that $L\varphi_n \to f$, $n \to \infty$. By Theorem 1.24 to each φ_n we choose a best approximation χ_n with respect to $N(L)$, i.e.,

$$\|\varphi_n - \chi_n\| = \inf_{\chi \in N(L)} \|\varphi_n - \chi\|.$$

The sequence defined by

$$\widetilde{\varphi}_n := \varphi_n - \chi_n, \quad n \in \mathbb{N},$$

is bounded. We prove this indirectly, i.e., we assume that the sequence $(\widetilde{\varphi}_n)$ is not bounded. Then there exists a subsequence $(\widetilde{\varphi}_{n(k)})$ such that $\|\widetilde{\varphi}_{n(k)}\| \geq k$ for all $k \in \mathbb{N}$. Now we define

$$\psi_k := \frac{\widetilde{\varphi}_{n(k)}}{\|\widetilde{\varphi}_{n(k)}\|}, \quad k \in \mathbb{N}.$$

Since $\|\psi_k\| = 1$ and A is compact, there exists a subsequence $(\psi_{k(j)})$ such that

$$A\psi_{k(j)} \to \psi \in X, \quad j \to \infty.$$

Furthermore,

$$\|L\psi_k\| = \frac{\|L\widetilde{\varphi}_{n(k)}\|}{\|\widetilde{\varphi}_{n(k)}\|} \leq \frac{\|L\widetilde{\varphi}_{n(k)}\|}{k} \to 0, \quad k \to \infty,$$

since the sequence $(L\varphi_n)$ is convergent and therefore bounded. Hence $L\psi_{k(j)} \to 0$ as $j \to \infty$. Now we obtain

$$\psi_{k(j)} = L\psi_{k(j)} + A\psi_{k(j)} \to \psi, \quad j \to \infty,$$

and since L is bounded, from the two previous equations we conclude that $L\psi = 0$. But then, because $\chi_{n(k)} + \|\widetilde{\varphi}_{n(k)}\|\,\psi \in N(L)$ for all k in $\mathbb{N}$, we find

$$\|\psi_k - \psi\| = \frac{1}{\|\widetilde{\varphi}_{n(k)}\|}\,\|\varphi_{n(k)} - \{\chi_{n(k)} + \|\widetilde{\varphi}_{n(k)}\|\,\psi\}\|$$

$$\geq \frac{1}{\|\widetilde{\varphi}_{n(k)}\|}\inf_{\chi\in N(L)}\|\varphi_{n(k)} - \chi\| = \frac{1}{\|\widetilde{\varphi}_{n(k)}\|}\|\varphi_{n(k)} - \chi_{n(k)}\| = 1.$$

This contradicts the fact that $\psi_{k(j)} \to \psi$, $j \to \infty$.

Therefore the sequence $(\widetilde{\varphi}_n)$ is bounded, and since A is compact, we can select a subsequence $(\widetilde{\varphi}_{n(k)})$ such that $(A\widetilde{\varphi}_{n(k)})$ converges as $k \to \infty$. In view of $L\widetilde{\varphi}_{n(k)} \to f$, $k \to \infty$, from $\widetilde{\varphi}_{n(k)} = L\widetilde{\varphi}_{n(k)} + A\widetilde{\varphi}_{n(k)}$ we observe that $\widetilde{\varphi}_{n(k)} \to \varphi \in X$, $k \to \infty$. But then $L\widetilde{\varphi}_{n(k)} \to L\varphi \in X$, $k \to \infty$, and therefore $f = L\varphi \in L(X)$. Hence $\overline{L(X)} = L(X)$, and the proof is complete. □

For $n \geq 1$ the iterated operators L^n can be written in the form

$$L^n = (I - A)^n = I - A_n,$$

where

$$A_n = \sum_{k=1}^{n}(-1)^{k-1}\binom{n}{k}A^k$$

is compact by Theorems 2.20 and 2.21. Therefore by Theorem 3.1 the nullspaces $N(L^n)$ are finite-dimensional subspaces, and by Theorem 3.2 the ranges $L^n(X)$ are closed subspaces.

Theorem 3.3 (Third Riesz Theorem). *There exists a uniquely determined nonnegative integer r, called the* Riesz number *of the operator A, such that*

$$\{0\} = N(L^0) \subsetneqq N(L^1) \subsetneqq \cdots \subsetneqq N(L^r) = N(L^{r+1}) = \cdots, \tag{3.2}$$

and

$$X = L^0(X) \supsetneqq L^1(X) \supsetneqq \cdots \supsetneqq L^r(X) = L^{r+1}(X) = \cdots. \tag{3.3}$$

Furthermore, we have the direct sum

$$X = N(L^r) \oplus L^r(X),$$

i.e., for each $\varphi \in X$ there exist uniquely determined elements $\psi \in N(L^r)$ and $\chi \in L^r(X)$ such that $\varphi = \psi + \chi$.

Proof. Our proof consists of four steps:

1. Because each φ with $L^n\varphi = 0$ satisfies $L^{n+1}\varphi = 0$, we have

$$\{0\} = N(L^0) \subset N(L^1) \subset N(L^2) \subset \cdots.$$

Now assume that

$$\{0\} = N(L^0) \subsetneqq N(L^1) \subsetneqq N(L^2) \subsetneqq \cdots .$$

Since by Theorem 3.1 the nullspace $N(L^n)$ is finite-dimensional, the Riesz Lemma 2.24 implies that for each $n \in \mathbb{N}$ there exists $\varphi_n \in N(L^{n+1})$ such that $\|\varphi_n\| = 1$ and

$$\|\varphi_n - \varphi\| \geq \frac{1}{2}$$

for all $\varphi \in N(L^n)$. For $n > m$ we consider

$$A\varphi_n - A\varphi_m = \varphi_n - (\varphi_m + L\varphi_n - L\varphi_m).$$

Then $\varphi_m + L\varphi_n - L\varphi_m \in N(L^n)$, because

$$L^n(\varphi_m + L\varphi_n - L\varphi_m) = L^{n-m-1}L^{m+1}\varphi_m + L^{n+1}\varphi_n - L^{n-m}L^{m+1}\varphi_m = 0.$$

Hence

$$\|A\varphi_n - A\varphi_m\| \geq \frac{1}{2}$$

for $n > m$, and thus the sequence $(A\varphi_n)$ does not contain a convergent subsequence. This is a contradiction to the compactness of A.

Therefore in the sequence $N(L^n)$ there exist two consecutive nullspaces that are equal. Define

$$r := \min\left\{k : N(L^k) = N(L^{k+1})\right\}.$$

Now we prove by induction that

$$N(L^r) = N(L^{r+1}) = N(L^{r+2}) = \cdots .$$

Assume that we have proven $N(L^k) = N(L^{k+1})$ for some $k \geq r$. Then for each $\varphi \in N(L^{k+2})$ we have $L^{k+1}L\varphi = L^{k+2}\varphi = 0$. This implies that $L\varphi \in N(L^{k+1}) = N(L^k)$. Hence $L^{k+1}\varphi = L^kL\varphi = 0$, and consequently $\varphi \in N(L^{k+1})$. Therefore, $N(L^{k+2}) \subset N(L^{k+1})$, and we have established that

$$\{0\} = N(L^0) \subsetneqq N(L^1) \subsetneqq \cdots \subsetneqq N(L^r) = N(L^{r+1}) = \cdots .$$

2. Because for each $\psi = L^{n+1}\varphi \in L^{n+1}(X)$ we can write $\psi = L^nL\varphi$, we have

$$X = L^0(X) \supset L^1(X) \supset L^2(X) \supset \cdots .$$

Now assume that

$$X = L^0(X) \supsetneqq L^1(X) \supsetneqq L^2(X) \supsetneqq \cdots .$$

Since by Theorem 3.2 the range $L^n(X)$ is a closed subspace, the Riesz Lemma 2.24 implies that for each $n \in \mathbb{N}$ there exist $\psi_n \in L^n(X)$ such that $\|\psi_n\| = 1$ and

$$\|\psi_n - \psi\| \geq \frac{1}{2}$$

for all $\psi \in L^{n+1}(X)$. We write $\psi_n = L^n\varphi_n$ and for $m > n$ we consider

$$A\psi_n - A\psi_m = \psi_n - (\psi_m + L\psi_n - L\psi_m).$$

Then $\psi_m + L\psi_n - L\psi_m \in L^{n+1}(X)$, because

$$\psi_m + L\psi_n - L\psi_m = L^{n+1}(L^{m-n-1}\varphi_m + \varphi_n - L^{m-n}\varphi_m).$$

Hence

$$\|A\psi_n - A\psi_m\| \geq \frac{1}{2}$$

for $m > n$, and we can derive the same contradiction as before.

Therefore in the sequence $L^n(X)$ there exist two consecutive ranges that are equal. Define

$$q := \min\left\{k : L^k(X) = L^{k+1}(X)\right\}.$$

Now we prove by induction that

$$L^q(X) = L^{q+1}(X) = L^{q+2}(X) = \cdots .$$

Assume that we have proven $L^k(X) = L^{k+1}(X)$ for some $k \geq q$. Then for each $\psi = L^{k+1}\varphi \in L^{k+1}(X)$ we can write $L^k\varphi = L^{k+1}\widetilde{\varphi}$ for some $\widetilde{\varphi} \in X$, because $L^k(X) = L^{k+1}(X)$. Hence $\psi = L^{k+2}\widetilde{\varphi} \in L^{k+2}(X)$, and therefore $L^{k+1}(X) \subset L^{k+2}(X)$, i.e., we have proven that

$$X = L^0(X) \underset{\neq}{\supset} L^1(X) \underset{\neq}{\supset} \cdots \underset{\neq}{\supset} L^q(X) = L^{q+1}(X) = \cdots .$$

3. Now we show that $r = q$. Assume that $r > q$ and let $\varphi \in N(L^r)$. Then, because $L^{r-1}\varphi \in L^{r-1}(X) = L^r(X)$, we can write $L^{r-1}\varphi = L^r\widetilde{\varphi}$ for some $\widetilde{\varphi} \in X$. Since $L^{r+1}\widetilde{\varphi} = L^r\varphi = 0$, we have $\widetilde{\varphi} \in N(L^{r+1}) = N(L^r)$, i.e., $L^{r-1}\varphi = L^r\widetilde{\varphi} = 0$. Thus $\varphi \in N(L^{r-1})$, and hence $N(L^{r-1}) = N(L^r)$. This contradicts the definition of r.

On the other hand, assume that $r < q$ and let $\psi = L^{q-1}\varphi \in L^{q-1}(X)$. Because $L\psi = L^q\varphi \in L^q(X) = L^{q+1}(X)$, we can write $L\psi = L^{q+1}\widetilde{\varphi}$ for some $\widetilde{\varphi} \in X$. Therefore $L^q(\varphi - L\widetilde{\varphi}) = L\psi - L^{q+1}\widetilde{\varphi} = 0$, and from this we conclude that $L^{q-1}(\varphi - L\widetilde{\varphi}) = 0$, because $N(L^{q-1}) = N(L^q)$. Hence $\psi = L^q\widetilde{\varphi} \in L^q(X)$, and consequently $L^{q-1}(X) = L^q(X)$. This contradicts the definition of q.

4. Let $\psi \in N(L^r) \cap L^r(X)$. Then $\psi = L^r\varphi$ for some $\varphi \in X$ and $L^r\psi = 0$. Therefore $L^{2r}\varphi = 0$, whence $\varphi \in N(L^{2r}) = N(L^r)$ follows. This implies $\psi = L^r\varphi = 0$.

Let $\varphi \in X$ be arbitrary. Then $L^r\varphi \in L^r(X) = L^{2r}(X)$ and we can write $L^r\varphi = L^{2r}\widetilde{\varphi}$ for some $\widetilde{\varphi} \in X$. Now define $\psi := L^r\widetilde{\varphi} \in L^r(X)$ and $\chi := \varphi - \psi$. Then $L^r\chi = L^r\varphi - L^{2r}\widetilde{\varphi} = 0$, i.e., $\chi \in N(L^r)$. Therefore the decomposition $\varphi = \chi + \psi$ proves the direct sum $X = N(L^r) \oplus L^r(X)$. □

Operators for which the sequence of the nullspaces of the iterated operators has two subsequent elements that coincide are called operators of *finite ascent*. Analogously, operators for which the sequence of the ranges of the iterated operators has two subsequent elements that coincide are called operators of *finite descent*.

Using this terminology the third Riesz theorem states that for a compact linear operator A the operator $I-A$ has finite ascent and descent. We note that only for these two statements of Theorem 3.3 the proofs make use of the compactness of A whereas all the other statements are just linear algebra.

We are now ready to derive the following fundamental result of the Riesz theory.

Theorem 3.4. *Let $A : X \to X$ be a compact linear operator on a normed space X. Then $I - A$ is injective if and only if it is surjective. If $I - A$ is injective (and therefore also bijective), then the inverse operator $(I - A)^{-1} : X \to X$ is bounded, i.e., $I - A$ is an isomorphism.*

Proof. By (3.2) injectivity of $I - A$ is equivalent to $r = 0$, and by (3.3) surjectivity of $I - A$ is also equivalent to $r = 0$. Therefore injectivity of $I - A$ and surjectivity of $I - A$ are equivalent.

It remains to show that L^{-1} is bounded when $L = I - A$ is injective. Assume that L^{-1} is not bounded. Then there exists a sequence (f_n) in X with $\|f_n\| = 1$ such that $\|L^{-1} f_n\| \geq n$ for all $n \in \mathbb{N}$. Define

$$g_n := \frac{f_n}{\|L^{-1} f_n\|} \quad \text{and} \quad \varphi_n := \frac{L^{-1} f_n}{\|L^{-1} f_n\|}, \quad n \in \mathbb{N}.$$

Then $g_n \to 0$, $n \to \infty$, and $\|\varphi_n\| = 1$ for all n. Since A is compact, we can select a subsequence $(\varphi_{n(k)})$ such that $A\varphi_{n(k)} \to \varphi \in X$, $k \to \infty$. Then, since $\varphi_n - A\varphi_n = g_n$, we observe that $\varphi_{n(k)} \to \varphi$, $k \to \infty$, and $\varphi \in N(L)$. Hence $\varphi = 0$, and this contradicts $\|\varphi_n\| = 1$ for all $n \in \mathbb{N}$. □

If $B : Y \to Z$ is a bounded linear operator mapping a Banach space Y bijectively onto a Banach space Z, then by the Banach open mapping Theorem 10.8 the inverse operator $B^{-1} : Z \to Y$ is bounded. For the Riesz theory we do not need to use this deep result from functional analysis and also do not require completeness of X.

We can rewrite Theorems 3.1 and 3.4 in terms of the solvability of an operator equation of the second kind as follows.

Corollary 3.5. *Let $A : X \to X$ be a compact linear operator on a normed space X. If the homogeneous equation*

$$\varphi - A\varphi = 0 \tag{3.4}$$

only has the trivial solution $\varphi = 0$, then for each $f \in X$ the inhomogeneous equation

$$\varphi - A\varphi = f \tag{3.5}$$

has a unique solution $\varphi \in X$ and this solution depends continuously on f.

If the homogeneous equation (3.4) has a nontrivial solution, then it has only a finite number $m \in \mathbb{N}$ of linearly independent solutions $\varphi_1, \ldots, \varphi_m$ and the inhomogeneous equation (3.5) is either unsolvable or its general solution is of the form

$$\varphi = \widetilde{\varphi} + \sum_{k=1}^{m} \alpha_k \varphi_k,$$

where $\alpha_1, \dots, \alpha_m$ are arbitrary complex numbers and $\widetilde{\varphi}$ denotes a particular solution of the inhomogeneous equation.

The main importance of the Riesz theory for compact operators lies in the fact that it reduces the problem of establishing the existence of a solution to (3.5) to the generally much simpler problem of showing that (3.4) has only the trivial solution $\varphi = 0$.

It is left to the reader to formulate Theorem 3.4 and its Corollary 3.5 for integral equations of the second kind with continuous or weakly singular kernels.

Corollary 3.6. *Theorem 3.4 and its Corollary 3.5 remain valid when $I - A$ is replaced by $S - A$, where $S : X \to Y$ is a bounded linear operator that has a bounded inverse $S^{-1} : Y \to X$, i.e., $S : X \to Y$ is an isomorphism, and $A : X \to Y$ is a compact linear operator from a normed space X into a normed space Y.*

Proof. This follows immediately from the fact that we can transform the equation

$$S\varphi - A\varphi = f$$

into the equivalent form

$$\varphi - S^{-1}A\varphi = S^{-1}f,$$

where $S^{-1}A : X \to X$ is compact by Theorem 2.21. □

The decomposition $X = N(L^r) \oplus L^r(X)$ of Theorem 3.3 generates an operator $P : X \to N(L^r)$ that maps $\varphi \in X$ onto $P\varphi := \psi$ defined by the unique decomposition $\varphi = \psi + \chi$ with $\psi \in N(L^r)$ and $\chi \in L^r(X)$. This operator is called a *projection operator*, because it satisfies $P^2 = P$ (see Chapter 13). We conclude this section with the following lemma on this projection operator that we are going to use in the following chapter. (Note that the proof only uses the finite dimension of $N(L^r)$ and the fact that $L^r(X)$ is closed.)

Lemma 3.7. *The projection operator $P : X \to N(L^r)$ defined by the decomposition $X = N(L^r) \oplus L^r(X)$ is compact.*

Proof. Assume that P is not bounded. Then there exists a sequence (φ_n) in X with $\|\varphi_n\| = 1$ such that $\|P\varphi_n\| \geq n$ for all $n \in \mathbb{N}$. Define

$$\psi_n := \frac{\varphi_n}{\|P\varphi_n\|}, \quad n \in \mathbb{N}.$$

Then $\psi_n \to 0$, $n \to \infty$, and $\|P\psi_n\| = 1$ for all $n \in \mathbb{N}$. Since $N(L^r)$ is finite-dimensional and $(P\psi_n)$ is bounded, by Theorem 1.17 there exists a subsequence $(\psi_{n(k)})$ such that $P\psi_{n(k)} \to \chi \in N(L^r)$, $k \to \infty$. Because $\psi_n \to 0$, $n \to \infty$, we also have $P\psi_{n(k)} - \psi_{n(k)} \to \chi$, $k \to \infty$. This implies that $\chi \in L^r(X)$, since $P\psi_{n(k)} - \psi_{n(k)} \in L^r(X)$ for all k and $L^r(X)$ is closed. Hence $\chi \in N(L^r) \cap L^r(X)$, and therefore $\chi = 0$, i.e., $P\psi_{n(k)} \to 0$, $k \to \infty$. This contradicts $\|P\psi_n\| = 1$ for all $n \in \mathbb{N}$. Hence P is bounded, and because P has finite-dimensional range $P(X) = N(L^r)$, by Theorem 2.23 it is compact. □

3.2 Spectral Theory for Compact Operators

We continue by formulating the results of the Riesz theory in terms of *spectral analysis*.

Definition 3.8. Let $A : X \to X$ be a bounded linear operator on a normed space X. A complex number λ is called an *eigenvalue* of A if there exists an element $\varphi \in X$, $\varphi \neq 0$, such that $A\varphi = \lambda\varphi$. The element φ is called an *eigenelement* of A. A complex number λ is called a *regular value* of A if $(\lambda I - A)^{-1} : X \to X$ exists and is bounded. The set of all regular values of A is called the *resolvent set* $\rho(A)$ and $R(\lambda; A) := (\lambda I - A)^{-1}$ is called the *resolvent* of A. The complement of $\rho(A)$ in $\mathbb{C}$ is called the *spectrum* $\sigma(A)$ and

$$r(A) := \sup_{\lambda \in \sigma(A)} |\lambda|$$

is called the *spectral radius* of A.

For the spectrum of a compact operator we have the following properties.

Theorem 3.9. *Let $A : X \to X$ be a compact linear operator on an infinite-dimensional normed space X. Then $\lambda = 0$ belongs to the spectrum $\sigma(A)$ and $\sigma(A) \setminus \{0\}$ consists of at most a countable set of eigenvalues with no point of accumulation except, possibly, $\lambda = 0$.*

Proof. Theorem 2.26 implies that $\lambda = 0$ belongs to the spectrum $\sigma(A)$. For $\lambda \neq 0$ we can apply the Riesz theory to $\lambda I - A$. Either $N(\lambda I - A) = \{0\}$ and $(\lambda I - A)^{-1}$ exists and is bounded by Corollary 3.6 or $N(\lambda I - A) \neq \{0\}$, i.e., λ is an eigenvalue. Thus each $\lambda \neq 0$ is either a regular value or an eigenvalue of A.

It remains to show that for each $R > 0$ there exist only a finite number of eigenvalues λ with $|\lambda| \geq R$. Assume, on the contrary, that we have a sequence (λ_n) of distinct eigenvalues satisfying $|\lambda_n| \geq R$. Choose eigenelements φ_n such that $A\varphi_n = \lambda_n\varphi_n$ for $n = 0, 1, \ldots$, and define finite-dimensional subspaces

$$U_n := \operatorname{span}\{\varphi_0, \ldots, \varphi_n\}.$$

It is readily verified that eigenelements corresponding to distinct eigenvalues are linearly independent. Hence, we have $U_{n-1} \subset U_n$ and $U_{n-1} \neq U_n$ for $n = 1, 2, \ldots$. Therefore, by the Riesz Lemma 2.24 we can choose a sequence (ψ_n) of elements $\psi_n \in U_n$ such that $\|\psi_n\| = 1$ and

$$\|\psi_n - \psi\| \geq \frac{1}{2}$$

for all $\psi \in U_{n-1}$ and $n = 1, 2, \ldots$. Writing

$$\psi_n = \sum_{k=0}^{n} \alpha_{nk}\varphi_k$$

we obtain

$$\lambda_n\psi_n - A\psi_n = \sum_{k=0}^{n-1}(\lambda_n - \lambda_k)\alpha_{nk}\varphi_k \in U_{n-1}.$$

Therefore, for $m < n$ we have

$$A\psi_n - A\psi_m = \lambda_n\psi_n - (\lambda_n\psi_n - A\psi_n + A\psi_m) = \lambda_n(\psi_n - \psi),$$

where $\psi := \lambda_n^{-1}(\lambda_n\psi_n - A\psi_n + A\psi_m) \in U_{n-1}$. Hence

$$\|A\psi_n - A\psi_m\| \geq \frac{|\lambda_n|}{2} \geq \frac{R}{2}$$

for $m < n$, and the sequence $(A\psi_n)$ does not contain a convergent subsequence. This contradicts the compactness of A. □

3.3 Volterra Integral Equations

Integral equations of the form

$$\int_a^x K(x,y)\varphi(y)\,dy = f(x), \quad x \in [a,b], \tag{3.6}$$

and

$$\varphi(x) - \int_a^x K(x,y)\varphi(y)\,dy = f(x), \quad x \in [a,b], \tag{3.7}$$

with variable limits of integration are called *Volterra integral equations* of the *first* and *second kind*, respectively. Equations of this type were first investigated by Volterra [242]. One can view Volterra equations as special cases of Fredholm equations with $K(x,y) = 0$ for $y > x$, but they have some special properties. In particular, Volterra integral equations of the second kind are always uniquely solvable.

Theorem 3.10. *For each right-hand side $f \in C[a,b]$ the Volterra integral equation of the second kind (3.6) with continuous kernel K has a unique solution $\varphi \in C[a,b]$.*

Proof. We extend the kernel onto $[a,b] \times [a,b]$ by setting $K(x,y) := 0$ for $y > x$. Then K is continuous for $x \neq y$ and

$$|K(x,y)| \leq M := \max_{a \leq y \leq x \leq b} |K(x,y)|$$

for all $x \neq y$. Hence, K is weakly singular with $\alpha = 1$.

Now let $\varphi \in C[a,b]$ be a solution to the homogeneous equation

$$\varphi(x) - \int_a^x K(x,y)\varphi(y)\,dy = 0, \quad x \in [a,b].$$

By induction we show that

$$|\varphi(x)| \le \|\varphi\|_\infty \frac{M^n(x-a)^n}{n!}, \quad x \in [a,b], \tag{3.8}$$

for $n = 0, 1, 2, \dots$. This certainly is true for $n = 0$. Assume that the inequality (3.8) is proven for some $n \ge 0$. Then

$$|\varphi(x)| = \left| \int_a^x K(x,y)\varphi(y)\,dy \right| \le \|\varphi\|_\infty \frac{M^{n+1}(x-a)^{n+1}}{(n+1)!}.$$

Passing to the limit $n \to \infty$ in (3.8) yields $\varphi(x) = 0$ for all $x \in [a,b]$. The statement of the theorem now follows from Theorems 2.29 and 3.4. □

In terms of spectral theory we can formulate the last result as follows: A Volterra integral operator with continuous kernel has no spectral values different from zero.

Analogous to the above estimate (3.8), the integral operator A with the Volterra kernel satisfies

$$|(A^n\varphi)(x)| \le \|\varphi\|_\infty \frac{M^n(x-a)^n}{n!}, \quad x \in [a,b],$$

for $n = 0, 1, 2, \dots$. Consequently

$$\|A^n\|_\infty \le \frac{M^n(b-a)^n}{n!} < 1$$

for sufficiently large n and from Problem 2.2 we can conclude that successive approximations always converge for Volterra integral equations of the second kind with continuous kernels. We will meet Volterra integral equations again in Chapter 9 where we also will consider weakly singular kernels.

To illustrate the close relation between Volterra integral equations and initial value problems for ordinary differential equations we now consider the equation of the second order

$$u'' + a_1 u' + a_2 u = v \tag{3.9}$$

on the interval $[a,b]$ with coefficients $a_1, a_2, v \in C[a,b]$ subject to the initital conditions

$$u(a) = u_0, \quad u'(a) = u'_0 \tag{3.10}$$

for some constants u_0 and u'_0. After setting

$$\varphi := u'' \tag{3.11}$$

two integrations yield

$$u'(x) = u'_0 + \int_a^x \varphi(y)\,dy \tag{3.12}$$

and

$$u(x) = u_0 + (x - a)\, u_0' + \int_a^x (x - y)\varphi(y)\, dy. \tag{3.13}$$

Now let u be a solution to the initial value problem. Then, by combining (3.11)–(3.13) it can be seen that $\varphi = u''$ satisfies the integral equation (3.7) with the kernel given by

$$K(x, y) := -a_1(x) - a_2(x)(x - y)$$

and the right-hand side given by

$$f(x) := v(x) - a_1(x)u_0' - a_2(x)[u_0 + (x - a)\, u_0'].$$

Conversely, for any solution $\varphi \in C[0, 1]$ the function u defined by (3.13) provides a solution to the initial value problem. Thus Theorem 3.10 establishes the Picard–Lindelöf theorem on existence and uniqueness of a solution for the special case of the linear initial value problem (3.9) and (3.10). We note that the proof of the Picard–Lindelöf theorem for nonlinear ordinary differential equations relies on a nonlinear Volterra integral equation.

Despite the fact that, in general, integral equations of the first kind are more delicate than integral equations of the second kind, in some cases Volterra integral equations of the first kind can be treated by reducing them to equations of the second kind. Consider

$$\int_a^x K(x, y)\varphi(y)\, dy = f(x), \quad x \in [a, b], \tag{3.14}$$

and assume that the derivatives $K_x = \partial K/\partial x$ and f' exist and are continuous and that $K(x, x) \neq 0$ for all $x \in [a, b]$. Then differentiating with respect to x reduces (3.14) to

$$\varphi(x) + \int_a^x \frac{K_x(x, y)}{K(x, x)}\, \varphi(y)\, dy = \frac{f'(x)}{K(x, x)}, \quad x \in [a, b]. \tag{3.15}$$

Equations (3.14) and (3.15) are equivalent if $f(a) = 0$. If $K_y = \partial K/\partial y$ exists and is continuous and again $K(x, x) \neq 0$ for all $x \in [a, b]$, then there is a second method to reduce the equation of the first kind to one of the second kind. In this case, setting

$$\psi(x) := \int_a^x \varphi(y)\, dy, \quad x \in [a, b],$$

and performing an integration by parts in (3.14) yields

$$\psi(x) - \int_a^x \frac{K_y(x, y)}{K(x, x)}\, \psi(y)\, dy = \frac{f(x)}{K(x, x)}, \quad x \in [a, b]. \tag{3.16}$$

Abel's integral equation (1.3) is an example of a Volterra integral equation of the first kind for which the above assumptions are not satisfied.

We leave it as an exercise to extend this short discussion of Volterra integral equations to the case of Volterra integral equations for functions of more than one independent variable.

Problems

3.1. Let $A : X \to Y$ be a compact linear operator from a normed space X into a normed space Y. The continuous extension $\widetilde{A} : \widetilde{X} \to \widetilde{Y}$ of A is compact with $\widetilde{A}(\widetilde{X}) \subset Y$ (see Problem 2.1).

3.2. Let X be a linear space, let $A, B : X \to X$ be linear operators satisfying $AB = BA$, and let AB have an inverse $(AB)^{-1} : X \to X$. Then A and B have inverse operators $A^{-1} = B(AB)^{-1}$ and $B^{-1} = A(AB)^{-1}$.

3.3. Prove Theorem 3.4 under the assumption that A^n is compact for some $n \geq 1$.
Hint: Use Problem 3.2 to prove that the set $(\sigma(A))^n := \{\lambda^n : \lambda \in \sigma(A)\}$ is contained in the spectrum $\sigma(A^n)$. Then use Theorem 3.9 to show that there exists an integer $m \geq n$ such that each of the operators

$$L_k := \exp\frac{2\pi ik}{m} I - A, \quad k = 1, \dots, m-1,$$

has a bounded inverse. Then the equations $R(I-A)\varphi = Rf$ and $(I-A)\varphi = f$, where $R := \prod_{k=1}^{m-1} L_k$, are equivalent.

3.4. Let X_i, $i = 1, \dots, n$, be normed spaces. Show that the Cartesian product $X := X_1 \times \cdots \times X_n$ of n-tuples $\varphi = (\varphi_1, \dots, \varphi_n)$ is a normed space with the maximum norm

$$\|\varphi\|_\infty := \max_{i=1,\dots,n} \|\varphi_i\|.$$

Let $A_{ik} : X_k \to X_i$, $i, k = 1, \dots, n$, be linear operators. Show that the matrix operator $A : X \to X$ defined by

$$(A\varphi)_i := \sum_{k=1}^{n} A_{ik}\varphi_k$$

is bounded or compact if and only if each of its components $A_{ik} : X_k \to X_i$ is bounded or compact, respectively. Formulate Theorem 3.4 for systems of operator and integral equations of the second kind.

3.5. Show that the integral operator with continuous kernel

$$K(x, y) := \sum_{k=0}^{\infty} \frac{1}{(k+1)^2} \{\cos(k+1)x \sin ky - \sin(k+1)x \cos ky\}$$

on the interval $[0, 2\pi]$ has no eigenvalues.

Chapter 4
Dual Systems and Fredholm Alternative

In the case when the homogeneous equation has nontrivial solutions, the Riesz theory, i.e., Theorem 3.4 gives no answer to the question of whether the inhomogeneous equation for a given inhomogeneity is solvable. This question is settled by the Fredholm alternative, which we shall develop in this chapter. Rather than presenting it in the context of the Riesz–Schauder theory with the adjoint operator in the dual space we will consider the Fredholm theory for compact adjoint operators in dual systems generated by non-degenerate bilinear or sesquilinear forms. This symmetric version is better suited for applications to integral equations and contains the Riesz–Schauder theory as a special case.

4.1 Dual Systems via Bilinear Forms

Throughout this chapter we tacitly assume that all linear spaces under consideration are complex linear spaces; the case of real linear spaces can be treated analogously.

Definition 4.1. Let X, Y be linear spaces. A mapping

$$\langle \cdot , \cdot \rangle : X \times Y \to \mathbb{C}$$

is called a *bilinear form* if

$$\langle \alpha_1 \varphi_1 + \alpha_2 \varphi_2, \psi \rangle = \alpha_1 \langle \varphi_1, \psi \rangle + \alpha_2 \langle \varphi_2, \psi \rangle,$$

$$\langle \varphi, \beta_1 \psi_1 + \beta_2 \psi_2 \rangle = \beta_1 \langle \varphi, \psi_1 \rangle + \beta_2 \langle \varphi, \psi_2 \rangle$$

for all $\varphi_1, \varphi_2, \varphi \in X$, $\psi_1, \psi_2, \psi \in Y$, and $\alpha_1, \alpha_2, \beta_1, \beta_2 \in \mathbb{C}$. The bilinear form is called *non-degenerate* if for every $\varphi \in X$ with $\varphi \neq 0$ there exists $\psi \in Y$ such that $\langle \varphi, \psi \rangle \neq 0$; and for every $\psi \in Y$ with $\psi \neq 0$ there exists $\varphi \in X$ such that $\langle \varphi, \psi \rangle \neq 0$.

Definition 4.2. Two normed spaces X and Y equipped with a non-degenerate bilinear form $\langle \cdot , \cdot \rangle : X \times Y \to \mathbb{C}$ are called a *dual system* and denoted by $\langle X, Y \rangle$.

R. Kress, *Linear Integral Equations*, Applied Mathematical Sciences 82,
DOI 10.1007/978-1-4614-9593-2_4,

Theorem 4.3. *Each normed space X together with its dual space X^* forms the* canonical dual system $\langle X, X^*\rangle$ *with the bilinear form*

$$\langle \varphi, F\rangle := F(\varphi), \quad \varphi \in X,\ F \in X^*.$$

Proof. The bilinearity is obvious. By the Hahn–Banach Corollary 2.11, for each $\varphi \neq 0$ in X there exists $F \in X^*$ with $F(\varphi) \neq 0$. For each $F \neq 0$ in X^* trivially there exists $\varphi \in X$ with $F(\varphi) \neq 0$. □

Theorem 4.4. *Let $G \subset \mathbb{R}^m$ be as in Theorem 2.13. Then $\langle C(G), C(G)\rangle$ is a dual system with the bilinear form*

$$\langle \varphi, \psi\rangle := \int_G \varphi(x)\psi(x)\,dx, \quad \varphi, \psi \in C(G). \tag{4.1}$$

Proof. The bilinearity is obvious and the non-degenerateness is a consequence of $\int_G |\varphi(x)|^2 dx > 0$ for all $\varphi \in C(G)$ with $\varphi \neq 0$. □

Definition 4.5. Let $\langle X_1, Y_1\rangle$ and $\langle X_2, Y_2\rangle$ be two dual systems. Then two operators $A : X_1 \to X_2$, $B : Y_2 \to Y_1$ are called *adjoint* (with respect to these dual systems) if

$$\langle A\varphi, \psi\rangle = \langle \varphi, B\psi\rangle$$

for all $\varphi \in X_1$, $\psi \in Y_2$. (We use the same symbol $\langle \cdot\,, \cdot\rangle$ for the bilinear forms on $\langle X_1, Y_1\rangle$ and $\langle X_2, Y_2\rangle$.)

Theorem 4.6. *Let $\langle X_1, Y_1\rangle$ and $\langle X_2, Y_2\rangle$ be two dual systems. If an operator $A : X_1 \to X_2$ has an adjoint $B : Y_2 \to Y_1$, then B is uniquely determined, and A and B are linear.*

Proof. Suppose that there exist two adjoint operators to A and denote these by B_1 and B_2. Let $B := B_1 - B_2$. Then

$$\langle \varphi, B\psi\rangle = \langle \varphi, B_1\psi\rangle - \langle \varphi, B_2\psi\rangle = \langle A\varphi, \psi\rangle - \langle A\varphi, \psi\rangle = 0$$

for all $\varphi \in X_1$ and $\psi \in Y_2$. Hence, because $\langle \cdot\,, \cdot\rangle$ is non-degenerate, we have $B\psi = 0$ for all $\psi \in Y_2$, i.e., $B_1 = B_2$.

To show that B is linear we observe that

$$\begin{aligned}
\langle \varphi, \beta_1 B\psi_1 + \beta_2 B\psi_2\rangle &= \beta_1\langle \varphi, B\psi_1\rangle + \beta_2\langle \varphi, B\psi_2\rangle \\
&= \beta_1\langle A\varphi, \psi_1\rangle + \beta_2\langle A\varphi, \psi_2\rangle \\
&= \langle A\varphi, \beta_1\psi_1 + \beta_2\psi_2\rangle \\
&= \langle \varphi, B(\beta_1\psi_1 + \beta_2\psi_2)\rangle
\end{aligned}$$

for all $\varphi \in X_1$, $\psi_1, \psi_2 \in Y_2$, and $\beta_1, \beta_2 \in \mathbb{C}$, i.e., $\beta_1 B\psi_1 + \beta_2 B\psi_2 = B(\beta_1\psi_1 + \beta_2\psi_2)$. In a similar manner, it is seen that A is linear. □

Theorem 4.7. *Let K be a continuous or a weakly singular kernel. Then in the dual system $\langle C(G), C(G)\rangle$ the (compact) integral operators defined by*

$$(A\varphi)(x) := \int_G K(x,y)\varphi(y)\,dy, \quad x \in G,$$

and

$$(B\psi)(x) := \int_G K(y,x)\psi(y)\,dy, \quad x \in G,$$

are adjoint.

Proof. The theorem follows from

$$\begin{aligned}\langle A\varphi, \psi\rangle &= \int_G (A\varphi)(x)\psi(x)\,dx \\ &= \int_G \left(\int_G K(x,y)\varphi(y)\,dy\right)\psi(x)\,dx \\ &= \int_G \varphi(y)\left(\int_G K(x,y)\psi(x)\,dx\right)dy \\ &= \int_G \varphi(y)(B\psi)(y)\,dy = \langle \varphi, B\psi\rangle.\end{aligned}$$

In the case of a weakly singular kernel, interchanging the order of integration is justified by the fact that $A_n\varphi \to A\varphi$, $n \to \infty$, uniformly on G, where A_n is the integral operator with continuous kernel K_n introduced in the proof of Theorem 2.29. □

The operator $A : C[0,1] \to C[0,1]$ defined by $A\varphi := \varphi(1)$ provides an example of a compact operator that does not have an adjoint operator with respect to the dual system $\langle C[0,1], C[0,1]\rangle$ of Theorem 4.4. To the contrary let $B : C[0,1] \to C[0,1]$ be an adjoint of A and choose a function $\psi \in C[0,1]$ with $\int_0^1 \psi(x)\,dx = 1$. With the aid of the Cauchy–Schwarz inequality we estimate

$$1 = |\varphi(1)| = |\langle A\varphi, \psi\rangle| = |\langle \varphi, B\psi\rangle| \le \|\varphi\|_2 \|B\psi\|_2$$

for all $\varphi \in C[0,1]$ with $\varphi(1) = 1$. Considering this inequality for the sequence (φ_n) with $\varphi_n(x) := x^n$ we arrive at a contradiction, since the right-hand side tends to zero as $n \to \infty$.

Theorem 4.8. *Let $A : X \to Y$ be a bounded linear operator mapping a normed space X into a normed space Y. Then the adjoint operator $A^* : Y^* \to X^*$ with respect to the canonical dual systems $\langle X, X^*\rangle$ and $\langle Y, Y^*\rangle$ exists. It is given by*

$$A^*F := FA, \quad F \in Y^*,$$

and is called the dual operator of A. It is bounded with norm $\|A\| = \|A^*\|$.

Proof. Since F and A both are linear and bounded, $A^*F = FA : X \to \mathbb{C}$ also is linear and bounded, i.e., $A^* : Y^* \to X^*$ with

$$\|A^*F\| = \|FA\| \leq \|F\| \, \|A\|. \tag{4.2}$$

The operators A und A^* are adjoint since

$$\langle A\varphi, F\rangle = F(A\varphi) = (A^*F)(\varphi) = \langle \varphi, A^*F\rangle.$$

Furthermore $A^* : Y^* \to X^*$ is linear by Theorem 4.6 and bounded with $\|A^*\| \leq \|A\|$ by (4.2). By Corollary 2.12 we conclude that

$$\|A\varphi\| = \sup_{\|F\|=1} |F(A\varphi)| = \sup_{\|F\|=1} |(A^*F)(\varphi)| \leq \|A^*\| \, \|\varphi\|,$$

i.e., we also have that $\|A\| \leq \|A^*\|$ and from this finally $\|A\| = \|A^*\|$ follows. □

As a result due to Schauder [215] (see also [94, 112, 116, 229]) it can be shown that under the assumptions of Theorem 4.8 the adjoint operator A^* is compact if and only if A is compact, provided that Y is a Banach space.

4.2 Dual Systems via Sesquilinear Forms

Definition 4.9. Let X, Y be linear spaces. A mapping $(\cdot\,,\cdot) : X \times Y \to \mathbb{C}$ is called a *sesquilinear form* if

$$(\alpha_1\varphi_1 + \alpha_2\varphi_2, \psi) = \alpha_1(\varphi_1, \psi) + \alpha_2(\varphi_2, \psi),$$

$$(\varphi, \beta_1\psi_1 + \beta_2\psi_2) = \bar{\beta}_1(\varphi, \psi_1) + \bar{\beta}_2(\varphi, \psi_2)$$

for all $\varphi_1, \varphi_2, \varphi \in X$, $\psi_1, \psi_2, \psi \in Y$, and $\alpha_1, \alpha_2, \beta_1, \beta_2 \in \mathbb{C}$. (Here, the bar indicates the complex conjugate.)

We leave it as an exercise to formulate Definition 4.5 and Theorem 4.6 in dual systems generated by non-degenerate sesquilinear forms. Sesquilinear forms differ from bilinear forms by their anti-linearity with respect to the second space. However, there is a close relation between them. Assume there exists a mapping $* : Y \to Y$ with the properties $(\beta_1\psi_1 + \beta_2\psi_2)^* = \bar{\beta}_1\psi_1^* + \bar{\beta}_2\psi_2^*$ and $(\psi^*)^* = \psi$ for all $\psi_1, \psi_2, \psi \in Y$ and $\beta_1, \beta_2 \in \mathbb{C}$. Such a mapping is called an *involution* and provides a one-to-one correspondence between bilinear and sesquilinear forms by $(\varphi, \psi) = \langle \varphi, \psi^*\rangle$. In the space $C(G)$ the natural involution is given by $\psi^*(x) := \overline{\psi(x)}$ for all $x \in G$ and all $\psi \in C(G)$. Again we leave it as an exercise to formulate Theorems 4.4 and 4.7 for the corresponding sesquilinear form given by the scalar product

$$(\varphi, \psi) := \int_G \varphi(x)\overline{\psi(x)}\,dx, \quad \varphi, \psi \in C(G). \tag{4.3}$$

The operator $A : C[0,1] \to C[0,1]$ considered in the example on p. 47, of course, also serves as an example of a compact operator that does not have an adjoint operator with respect to the dual system $(C[0,1], C[0,1])$ generated by (4.3).

In the sequel, we will demonstrate that in Hilbert spaces for bounded linear operators the adjoint operators always exist. From Definition 1.19 we observe that each scalar product on a linear space X may be considered as a non-degenerate sesquilinear form that is *symmetric*, i.e., $(\varphi, \psi) = \overline{(\psi, \varphi)}$ for all $\varphi, \psi \in X$, and *positive definite*, i.e., $(\varphi, \varphi) > 0$ for all $\varphi \in X$ with $\varphi \neq 0$. Thus each pre-Hilbert space canonically is a dual system.

Theorem 4.10 (Riesz). *Let X be a Hilbert space. Then for each bounded linear functional $F : X \to \mathbb{C}$ there exists a unique element $f \in X$ such that*

$$F(\varphi) = (\varphi, f) \tag{4.4}$$

for all $\varphi \in X$. The norms of the element f and the linear function F coincide, i.e.,

$$\|f\| = \|F\|. \tag{4.5}$$

Proof. Uniqueness follows from the fact that because of the positive definiteness of the scalar product, $f = 0$ is the only representer of the zero function $F = 0$ in the sense of (4.4). For $F \neq 0$ choose $w \in X$ with $F(w) \neq 0$. Since F is continuous, the nullspace $N(F) = \{\varphi \in X : F(\varphi) = 0\}$ can be seen to be a closed, and consequently complete, subspace of the Hilbert space X. By Theorem 1.26 there exists the best approximation v to w with respect to $N(F)$, and by Theorem 1.25 it satisfies

$$w - v \perp N(F).$$

Then for $g := w - v$ we have that

$$(F(g)\varphi - F(\varphi)g, g) = 0, \quad \varphi \in X,$$

because $F(g)\varphi - F(\varphi)g \in N(F)$ for all $\varphi \in X$. Hence,

$$F(\varphi) = \left(\varphi, \frac{\overline{F(g)}g}{\|g\|^2}\right)$$

for all $\varphi \in X$, which completes the proof of (4.4).

From (4.4) and the Cauchy–Schwarz inequality we have that

$$|F(\varphi)| \leq \|f\|\,\|\varphi\|, \quad \varphi \in X,$$

whence $\|F\| \leq \|f\|$ follows. On the other hand, inserting f into (4.4) yields

$$\|f\|^2 = F(f) \leq \|F\|\,\|f\|,$$

and therefore $\|f\| \leq \|F\|$. This concludes the proof of the norm equality (4.5). □

This theorem establishes the existence of a bijective anti-linear mapping between a Hilbert space and its dual space that is isometric in the sense that it preserves the norms. In view of (4.4), this mapping $J : X \to X^*$ is given by

$$(J(f))(\varphi) = (\varphi, f) \tag{4.6}$$

for all $\varphi, f \in X$.

Theorem 4.11. *Let X and Y be Hilbert spaces, and let $A : X \to Y$ be a bounded linear operator. Then there exists a uniquely determined linear operator $A^* : Y \to X$ with the property*

$$(A\varphi, \psi) = (\varphi, A^*\psi)$$

for all $\varphi \in X$ and $\psi \in Y$, i.e., A and A^* are adjoint with respect to the dual systems (X, X) and (Y, Y) generated by the scalar products on X and Y. The operator A^* is bounded and $\|A^*\| = \|A\|$. (Again we use the same symbol $(\cdot\,,\cdot)$ for the scalar products on X and Y.)

Proof. For each $\psi \in Y$ the mapping $\varphi \mapsto (A\varphi, \psi)$ clearly defines a bounded linear function on X, since $|(A\varphi, \psi)| \leq \|A\| \, \|\varphi\| \, \|\psi\|$. By Theorem 4.10 we can write $(A\varphi, \psi) = (\varphi, f)$ for some $f \in X$. Therefore, setting $A^*\psi := f$ we define an operator $A^* : Y \to X$ that is an adjoint of A. By Theorem 4.6, the adjoint is uniquely determined and linear. Using the Cauchy–Schwarz inequality, we derive

$$\|A^*\psi\|^2 = (A^*\psi, A^*\psi) = (AA^*\psi, \psi) \leq \|A\| \, \|A^*\psi\| \, \|\psi\|$$

for all $\psi \in Y$. Hence, A^* is bounded with $\|A^*\| \leq \|A\|$. Conversely, since A is the adjoint of A^*, we also have $\|A\| \leq \|A^*\|$. Hence $\|A^*\| = \|A\|$. □

Using the anti-linear isometric isomorphism J between the Hilbert space and its dual space, an alternative proof of Theorem 4.11 can also be obtained as an application of the corresponding Theorem 4.8 for the dual operator.

Theorem 4.12. *Let X and Y be Hilbert spaces and let $A : X \to Y$ be a compact linear operator. Then the adjoint operator $A^* : Y \to X$ is also compact.*

Proof. Let (ψ_n) be a bounded sequence in Y, i.e, $\|\psi_n\| \leq C$ for all $n \in \mathbb{N}$ and some $C > 0$. By Theorem 4.11 the adjoint operator $A^* : Y \to X$ is bounded and, consequently, the operator $AA^* : Y \to Y$ is compact by Theorem 2.21. Hence there exists a subsequence $(\psi_{n(k)})$ such that $(AA^*\psi_{n(k)})$ converges in Y. But then from

$$\begin{aligned}\|A^*(\psi_{n(k)} - \psi_{n(j)})\|^2 &= (AA^*(\psi_{n(k)} - \psi_{n(j)}), \psi_{n(k)} - \psi_{n(j)}) \\ &\leq 2C\|AA^*(\psi_{n(k)} - \psi_{n(j)})\|\end{aligned}$$

we observe that $(A^*\psi_{n(k)})$ is a Cauchy sequence, and therefore it converges in the Hilbert space X. □

The following theorem is due to Lax [156] and provides a useful tool to extend results on the boundedness and compactness of linear operators from a given norm to a weaker scalar product norm.

Theorem 4.13 (Lax). *Let X and Y be normed spaces, both of which have a scalar product $(\cdot\,,\cdot)$, and assume that there exists a positive constant c such that*

$$|(\varphi, \psi)| \leq c\|\varphi\|\,\|\psi\| \tag{4.7}$$

for all $\varphi, \psi \in X$. Let U be a subspace of X and let $A : X \to Y$ and $B : Y \to X$ be bounded linear operators satisfying

$$(A\varphi, \psi) = (\varphi, B\psi) \tag{4.8}$$

for all $\varphi \in U$ and $\psi \in Y$. Then $A : U \to Y$ is bounded with respect to the norms $\|\cdot\|_s$ induced by the scalar products and

$$\|A\|_s^2 \leq \|A\|\,\|B\|. \tag{4.9}$$

Proof. Consider the bounded operator $M : U \to X$ given by $M := BA$ with $\|M\| \leq \|B\|\,\|A\|$. Then, as a consequence of (4.8), M is self-adjoint, i.e.,

$$(M\varphi, \psi) = (\varphi, M\psi)$$

for all $\varphi, \psi \in U$. Therefore, using the Cauchy–Schwarz inequality, we obtain

$$\|M^n\varphi\|_s^2 = (M^n\varphi, M^n\varphi) = (\varphi, M^{2n}\varphi) \leq \|M^{2n}\varphi\|_s$$

for all $\varphi \in U$ with $\|\varphi\|_s \leq 1$ and all $n \in \mathbb{N}$. From this, by induction, it follows that

$$\|M\varphi\|_s \leq \|M^{2^n}\varphi\|_s^{2^{-n}}.$$

By (4.7) we have $\|\psi\|_s \leq \sqrt{c}\,\|\psi\|$ for all $\psi \in X$. Hence,

$$\|M\varphi\|_s \leq \left\{\sqrt{c}\,\|M^{2^n}\varphi\|\right\}^{2^{-n}} \leq \left\{\sqrt{c}\,\|\varphi\|\,\|M\|^{2^n}\right\}^{2^{-n}} = \left\{\sqrt{c}\,\|\varphi\|\right\}^{2^{-n}} \|M\|.$$

Passing to the limit $n \to \infty$ now yields

$$\|M\varphi\|_s \leq \|M\|$$

for all $\varphi \in U$ with $\|\varphi\|_s \leq 1$. Finally, for all $\varphi \in U$ with $\|\varphi\|_s \leq 1$, we have from the Cauchy–Schwarz inequality that

$$\|A\varphi\|_s^2 = (A\varphi, A\varphi) = (\varphi, M\varphi) \leq \|M\varphi\|_s \leq \|M\|.$$

From this the statement follows. □

For an example of an application of Lax's theorem let $X = Y = C(G)$ be equipped with the maximum norm and the L^2 scalar product (4.3). Using the approximations from the proof of Theorem 2.29, without any further analysis, from Theorems 2.22 and 4.13 it can be seen that integral operators with weakly singular kernels are compact in the completion of $C(G)$ with respect to the scalar product (4.3), i.e., in $L^2(G)$ (see Problem 4.5). Here, completeness is required for the application of Theorem 2.22 on the compactness of the limit of a norm convergent sequence of compact operators.

4.3 The Fredholm Alternative

Now we proceed to develop the Fredholm theory for compact operators, which we will write for a dual system generated by a bilinear form. The analysis for the case of a sesquillinear form is analogous. We begin with the following technical lemma.

Lemma 4.14. *Let $\langle X, Y\rangle$ be a dual system. Then to every set of linearly independent elements $\varphi_1, \dots, \varphi_n \in X$ there exists a set $\psi_1, \dots, \psi_n \in Y$ such that*

$$\langle \varphi_i, \psi_k\rangle = \delta_{ik}, \quad i, k = 1, \dots, n,$$

where $\delta_{ik} = 1$ for $i = k$ and $\delta_{ik} = 0$ for $i \neq k$. The same statement holds with the roles of X and Y interchanged.

Proof. For one linearly independent element the lemma is true, since $\langle \cdot , \cdot\rangle$ is non-degenerate. Assume that the assertion of the lemma has been proven for $n \geq 1$ linearly independent elements. Let $\varphi_1, \dots, \varphi_{n+1}$ be $n+1$ linearly independent elements. Then, by our induction assumption, for every $m = 1, \dots, n + 1$, to the set

$$\varphi_1, \dots, \varphi_{m-1}, \varphi_{m+1}, \dots, \varphi_{n+1}$$

of n elements in X there exists a set of n elements

$$\psi_1^{(m)}, \dots, \psi_{m-1}^{(m)}, \psi_{m+1}^{(m)}, \dots, \psi_{n+1}^{(m)}$$

in Y such that

$$\langle \varphi_i, \psi_k^{(m)}\rangle = \delta_{ik}, \quad i, k = 1, \dots, n + 1, \quad i, k \neq m. \tag{4.10}$$

Since $\langle \cdot , \cdot\rangle$ is non-degenerate, there exists $\chi_m \in Y$ such that

$$\alpha_m := \left\langle \varphi_m, \chi_m - \sum_{\substack{k=1\\k\neq m}}^{n+1} \psi_k^{(m)} \langle \varphi_k, \chi_m\rangle \right\rangle = \left\langle \varphi_m - \sum_{\substack{k=1\\k\neq m}}^{n+1} \langle \varphi_m, \psi_k^{(m)}\rangle \varphi_k, \chi_m \right\rangle \neq 0,$$

because otherwise

$$\varphi_m - \sum_{\substack{k=1 \\ k \neq m}}^{n+1} \langle \varphi_m, \psi_k^{(m)} \rangle \varphi_k = 0,$$

which is a contradiction to the linear independence of the $\varphi_1, \dots, \varphi_{n+1}$. Define

$$\psi_m := \frac{1}{\alpha_m} \left\{ \chi_m - \sum_{\substack{k=1 \\ k \neq m}}^{n+1} \psi_k^{(m)} \langle \varphi_k, \chi_m \rangle \right\}.$$

Then $\langle \varphi_m, \psi_m \rangle = 1$, and for $i \neq m$ we have

$$\langle \varphi_i, \psi_m \rangle = \frac{1}{\alpha_m} \left\{ \langle \varphi_i, \chi_m \rangle - \sum_{\substack{k=1 \\ k \neq m}}^{n+1} \langle \varphi_i, \psi_k^{(m)} \rangle \langle \varphi_k, \chi_m \rangle \right\} = 0$$

because of (4.10). Hence we obtain $\psi_1, \dots, \psi_{n+1}$ such that

$$\langle \varphi_i, \psi_k \rangle = \delta_{ik}, \quad i, k = 1, \dots, n+1,$$

and the lemma is proven. □

Theorem 4.15 (First Fredholm Theorem). *Let $\langle X, Y \rangle$ be a dual system and $A : X \to X$, $B : Y \to Y$ be compact adjoint operators. Then the nullspaces of the operators $I - A$ and $I - B$ have the same finite dimension.*

Proof. By the first Riesz Theorem 3.1 we have

$$m := \dim N(I - A) < \infty$$

and

$$n := \dim N(I - B) < \infty.$$

We assume that $m < n$. Then we choose a basis $\varphi_1, \dots, \varphi_m$ for $N(I-A)$ (if $m > 0$) and a basis $\psi_1, \dots, \psi_n$ for $N(I - B)$. By Lemma 4.14 there exist elements $a_1, \dots, a_m \in Y$ (if $m > 0$) and $b_1, \dots, b_n \in X$ such that

$$\langle \varphi_i, a_k \rangle = \delta_{ik}, \quad i, k = 1, \dots, m,$$

$$\langle b_i, \psi_k \rangle = \delta_{ik}, \quad i, k = 1, \dots, n.$$

Define a linear operator $T : X \to X$ by

$$T\varphi := \sum_{i=1}^{m} \langle \varphi, a_i \rangle b_i, \quad \varphi \in X, \tag{4.11}$$

if $m > 0$ and $T := 0$ if $m = 0$. Recall the compact projection operator

$$P : X \to N[(I - A)^r]$$

from Lemma 3.7. Since $T : N[(I - A)^r] \to X$ is bounded by Theorem 2.4, from Theorem 2.21 we have that $TP : X \to X$ is compact. Therefore, in view of Theorem 2.20, we can apply the Riesz theory to the operator $A - TP$.

For $m > 0$, from

$$\langle \varphi - A\varphi + TP\varphi, \psi_k \rangle = \langle \varphi, \psi_k - B\psi_k \rangle + \langle TP\varphi, \psi_k \rangle = \langle TP\varphi, \psi_k \rangle$$

we find that

$$\langle \varphi - A\varphi + TP\varphi, \psi_k \rangle = \begin{cases} \langle P\varphi, a_k \rangle, & k = 1, \dots, m, \\ 0, & k = m + 1, \dots, n. \end{cases} \tag{4.12}$$

Now let $\varphi \in N(I - A + TP)$. Then from (4.12) we see that

$$\langle P\varphi, a_k \rangle = 0, \quad k = 1, \dots, m, \tag{4.13}$$

and therefore $TP\varphi = 0$. Hence $\varphi \in N(I - A)$ and, consequently, we can write

$$\varphi = \sum_{i=1}^{m} \alpha_i \varphi_i,$$

where $\alpha_i = \langle \varphi, a_i \rangle$ for $i = 1, \dots, m$. Now from $P\varphi = \varphi$ for $\varphi \in N(I - A)$ and (4.13) we conclude that $\varphi = 0$. Thus we have proven that $I - A + TP$ is injective. This, of course, is also true in the case when $m = 0$.

Since $I - A + TP$ is injective, by Theorem 3.4 the inhomogeneous equation

$$\varphi - A\varphi + TP\varphi = b_n$$

has a unique solution φ. Now, with the aid of the second line of (4.12) (which is also true for $m = 0$) we arrive at the contradiction

$$1 = \langle b_n, \psi_n \rangle = \langle \varphi - A\varphi + TP\varphi, \psi_n \rangle = 0.$$

Therefore $m \geq n$. Interchanging the roles of A and B shows that $n \geq m$. Hence $m = n$. □

Theorem 4.16 (Second Fredholm Theorem). *Let $\langle X, Y \rangle$ be a dual system and $A : X \to X$, $B : Y \to Y$ be compact adjoint operators. Then*

$$(I - A)(X) = \{f \in X : \langle f, \psi \rangle = 0, \ \psi \in N(I - B)\}$$

and

$$(I - B)(Y) = \{g \in Y : \langle \varphi, g \rangle = 0, \ \varphi \in N(I - A)\}.$$

Proof. It suffices to carry out the proof for the range of $I - A$, and by Theorems 3.4 and 4.15 we only need to consider the case where $m > 0$. Let $f \in (I - A)(X)$, i.e, $f = \varphi - A\varphi$ for some $\varphi \in X$. Then

$$\langle f, \psi \rangle = \langle \varphi - A\varphi, \psi \rangle = \langle \varphi, \psi - B\psi \rangle = 0$$

for all $\psi \in N(I - B)$.

Conversely, assume that f satisfies $\langle f, \psi \rangle = 0$ for all $\psi \in N(I - B)$. From the proof of the previous theorem we know that the equation

$$\varphi - A\varphi + TP\varphi = f$$

has a unique solution φ. Then, in view of (4.12) and the condition on f, we have

$$\langle P\varphi, a_k \rangle = \langle \varphi - A\varphi + TP\varphi, \psi_k \rangle = \langle f, \psi_k \rangle = 0, \quad k = 1, \dots, m.$$

Hence $TP\varphi = 0$, and thus φ also satisfies $\varphi - A\varphi = f$. □

We now summarize our results in the so-called *Fredholm alternative.*

Theorem 4.17. *Let $A : X \to X$, $B : Y \to Y$ be compact adjoint operators in a dual system $\langle X, Y \rangle$. Then* either *$I - A$ and $I - B$ are bijective* or *$I - A$ and $I - B$ have nontrivial nullspaces with finite dimension*

$$\dim N(I - A) = \dim N(I - B) \in \mathbb{N}$$

and the ranges are given by

$$(I - A)(X) = \{f \in X : \langle f, \psi \rangle = 0, \ \psi \in N(I - B)\}$$

and

$$(I - B)(Y) = \{g \in Y : \langle \varphi, g \rangle = 0, \ \varphi \in N(I - A)\}.$$

We explicitly note that this theorem implies that for the first of the two alternatives each one of the four properties $I - A$ injective, $I - A$ surjective, $I - B$ injective, and $I - B$ surjective implies the three other ones.

Choosing the dual system introduced in Theorem 4.4 and the integral operators with continuous or weakly singular kernels considered in Theorem 4.7, our results include the classical Fredholm alternative for integral equations of the second kind that was first obtained by Fredholm [55] and which we now state as a corollary.

Corollary 4.18. *Let $G \subset \mathbb{R}^m$ be as in Theorem 4.4 and let K be a continuous or weakly singular kernel. Then* either *the homogeneous integral equations*

$$\varphi(x) - \int_G K(x, y)\varphi(y)\, dy = 0, \quad x \in G,$$

and

$$\psi(x) - \int_G K(y, x)\psi(y)\, dy = 0, \quad x \in G,$$

only have the trivial solutions $\varphi = 0$ *and* $\psi = 0$ *and the inhomogeneous integral equations*

$$\varphi(x) - \int_G K(x, y)\varphi(y)\, dy = f(x), \quad x \in G,$$

and

$$\psi(x) - \int_G K(y, x)\psi(y)\, dy = g(x), \quad x \in G,$$

have a unique solution $\varphi \in C(G)$ *and* $\psi \in C(G)$ *for each* $f \in C(G)$ *and* $g \in C(G)$*, respectively,* or *the homogeneous integral equations have the same finite number* $m \in \mathbb{N}$ *of linearly independent solutions and the inhomogeneous integral equations are solvable if and only if the right-hand sides satisfy*

$$\int_G f(x)\psi(x)\, dx = 0$$

for all solutions ψ *of the homogeneous adjoint equation and*

$$\int_G \varphi(x)g(x)\, dx = 0$$

for all solutions φ *of the homogeneous equation, respectively.*

The original proof by Fredholm for this result for continuous kernels is based on considering the integral equations as a limiting case of systems of finite-dimensional linear equations by approximating the integrals by Riemann sums. In Cramer's rule for this linear system Fredholm passes to the limit by using Koch's theory of infinite determinants from 1896 and Hadamard's inequality for determinants from 1893 (see [112]). The idea to view integral equations as the limiting case of linear systems had already been used by Volterra in 1896, but it was Fredholm who completed it successfully. We wish to mention that this original proof by Fredholm is shorter than our more general approach, however, it is restricted to the case of integral equations with continuous kernels.

Our results also include the so-called Riesz–Schauder theory as developed by Schauder [215] (see also [94, 112, 116, 229]) by taking $Y = X^*$, the dual space of X. Our more general form of the Fredholm alternative seems to be more appropriate for the discussion of integral equations because of its symmetric structure. In particular, in our setting the adjoint of an integral equation in a function space again is an integral equation in a function space, whereas in the Schauder theory the adjoint equation is an equation in the dual space of bounded linear functionals. In the case of $C[0, 1]$, for example, the elements of the dual space are functions with bounded variation, i.e., are no longer necessarily continuous functions. Hence, the Schauder theory does not immediately include the classical results of Fredholm on integral equations with continuous kernels. Furthermore, the above presentation of the Fredholm alternative does not make use of the Hahn–Banach theorem.

The Fredholm alternative in dual systems was first proven by Wendland [245, 247] under the assumption that the bilinear or sesquilinear form is bounded. For a pair of normed spaces X and Y, a bilinear form $\langle \cdot , \cdot \rangle : X \times Y \to \mathbb{C}$ is called bounded it there exists a positive real number C such that

$$|\langle \varphi, \psi \rangle| \leq C \|\varphi\| \, \|\psi\|$$

for all $\varphi \in X$ and $\psi \in Y$. Boundedness for a sesquilinear form is defined analogously. An elementary proof, which does not use the Hahn–Banach theorem, i.e., Zorn's lemma, and which also does not require the bilinear or sesquilinear form to be bounded, was first given in [31] (see also [136]). The current version of the proof of Theorem 4.15, which differs slightly from the proof in [31] and in the first edition of this book, is due to Martensen [163]. For a history of the Fredholm alternative in dual systems we refer to [139].

Example 4.19. Consider the integral equation

$$\varphi(x) - \int_a^b e^{x-y} \varphi(y) \, dy = f(x), \quad x \in [a, b]. \tag{4.14}$$

As the simple integral equation on p. 1, this equation belongs to the class of equations with so-called degenerate kernels which can be reduced to solving a finite-dimensional linear system. We shall meet degenerate kernels again later in Chapter 11 on the numerical solution of integral equations. Obviously a solution of (4.14) must be of the form

$$\varphi(x) = f(x) + ce^x, \tag{4.15}$$

where c is a constant. Inserting (4.15) into (4.14), we observe that φ solves the integral equation provided that c satisfies

$$c\{1 - (b - a)\} = \int_a^b e^{-y} f(y) \, dy. \tag{4.16}$$

Now either $b - a \neq 1$ or $b - a = 1$. In the first case (4.16) has a unique solution leading to the unique solution

$$\varphi(x) = f(x) + \frac{1}{1 - (b - a)} \int_a^b e^{-y} f(y) \, dy \; e^x$$

of the integral equation. In the second case, (4.16) has a solution if and only if

$$\int_a^b e^{-y} f(y) \, dy = 0, \tag{4.17}$$

and then any c satisfies (4.16). Note that, for $b - a = 1$, the function

$$\psi(x) = e^{-x}$$

is the solution of the homogeneous adjoint equation

$$\psi(x) - \int_a^b e^{y-x}\psi(y)\,dy = 0, \quad x \in [a,b],$$

and therefore (4.17) coincides with the solvability condition of the Fredholm alternative. □

Theorem 4.20. *Let U be a subspace of a linear space X and assume that U and X are normed spaces carrying two different norms. Let $A : X \to X$ be a linear operator such that $A(U) \subset U$ and $A : U \to U$ and $A : X \to X$ both are compact. Further let Y be a normed space and let $\langle \cdot\,, \cdot \rangle : X \times Y \to \mathbb{C}$ be a non-degenerate bilinear form such that its restriction $\langle \cdot\,, \cdot \rangle : U \times Y \to \mathbb{C}$ is also non-degenerate and let $B : Y \to Y$ be a compact operator that is adjoint to A with respect to this bilinear form. Then the nullspaces of $I - A : U \to U$ and $I - A : X \to X$ coincide.*

Proof. By the first Fredholm Theorem 4.15, applied in the dual system $< U, Y >$, the dimensions of the nullspaces of $I - A : U \to U$ and $I - B : Y \to Y$ are finite and coincide. Again by the first Fredholm Theorem 4.15, applied in the dual system $< X, Y >$, the dimensions of the nullspaces of $I - A : X \to X$ and $I - B : Y \to Y$ are also finite and coincide. Thus the nullspaces of $I - A : U \to U$ and $I - A : X \to X$ must be the same. □

As an example for the application of Theorem 4.20 we consider the integral operator A with weakly singular kernel given by (2.4) both in $C(G)$ and in $L^2(G)$. By Theorem 2.30 and Problem 4.5 we have compactness of A in both spaces. Therefore, Theorem 4.20 implies that all the eigenfunctions, i.e., the eigenelements of an integral operator with a weakly singular kernel are continuous. This idea to use the Fredholm alternative in two different dual systems for showing that the eigenspaces for weakly singular integral operators in the space of continuous functions and in the L^2 space coincide is due to Hähner [83].

4.4 Boundary Value Problems

Following Jörgens [112], we conclude this chapter by giving some flavor of the use of the Riesz–Fredholm theory to solve boundary value problems by considering the ordinary differential equation of the second order

$$u'' + a_1 u' + a_2 u = v \tag{4.18}$$

on the interval $[0,1]$ with coefficients $a_1, a_2, v \in C[0,1]$ subject to the boundary conditions

$$u(0) = u_0, \quad u(1) = u_1. \tag{4.19}$$

The general idea in the application of integral equations in the treatment of boundary value problems is to equivalently transform the boundary value problem into an integral equation and then solve the integral equation.

Let u be twice continuously differentiable and set $\varphi := -u''$. Then, by partial integration we find the relations

$$u(x) = u(0) + u'(0)x - \int_0^x (x-y)\varphi(y)\,dy,$$

$$u(x) = u(1) - u'(1)(1-x) + \int_x^1 (x-y)\varphi(y)\,dy,$$

$$0 = \int_0^1 \varphi(y)\,dy + u'(1) - u'(0).$$

Multiplying the first equation by $(1-x)$, the second by x, the third by $(1-x)x$, and then adding, we obtain

$$u(x) = u(0)(1-x) + u(1)x + \int_0^x (1-x)y\varphi(y)\,dy + \int_x^1 x(1-y)\varphi(y)\,dy. \quad (4.20)$$

Differentiating this equation yields

$$u'(x) = u(1) - u(0) - \int_0^x y\varphi(y)\,dy + \int_x^1 (1-y)\varphi(y)\,dy. \quad (4.21)$$

Now let u be a solution to the boundary value problem. Then, using the differential equation (4.18) and the boundary condition (4.19), from (4.20) and (4.21) we deduce that $\varphi = -u''$ satisfies the integral equation

$$\varphi(x) - \int_0^1 K(x,y)\varphi(y)\,dy = f(x), \quad x \in [0,1], \quad (4.22)$$

with the kernel

$$K(x,y) := \begin{cases} y\{a_2(x)(1-x) - a_1(x)\}, & 0 \le y < x \le 1, \\ (1-y)\{a_2(x)x + a_1(x)\}, & 0 \le x < y \le 1, \end{cases}$$

and the right-hand side

$$f(x) := (u_1 - u_0)a_1(x) + \{u_0(1-x) + u_1 x\}a_2(x) - v(x), \quad x \in [0,1].$$

Conversely, let $\varphi \in C[0,1]$ be a solution to the integral equation (4.22) and define a twice continuously differentiable function u by

$$u(x) := u_0(1-x) + u_1 x + \int_0^x (1-x)y\varphi(y)\,dy + \int_x^1 x(1-y)\varphi(y)\,dy, \quad x \in [0,1].$$

Then $u(0) = u_0$, $u(1) = u_1$, and by construction of the integral equation we have $-u'' = \varphi = a_1 u' + a_2 u - v$. Therefore, the boundary value problem and the integral equation are equivalent. In particular, via

$$w(x) := \int_0^x (1-x)y\varphi(y)\,dy + \int_x^1 (1-y)x\varphi(y)\,dy, \quad x \in [0,1],$$

the homogeneous boundary value problem

$$w'' + a_1 w' + a_2 w = 0, \quad w(0) = w(1) = 0 \tag{4.23}$$

and the homogeneous integral equation

$$\varphi(x) - \int_0^1 K(x,y)\varphi(y)\,dy = 0, \quad x \in [0,1], \tag{4.24}$$

are equivalent. Note that $w \neq 0$ implies that $\varphi \neq 0$ and vice versa.

Because the kernel K is continuous and bounded on $0 \leq y < x \leq 1$ and on $0 \leq x < y \leq 1$, it is weakly singular with $\alpha = 1$. Hence, the Fredholm alternative is valid: Either the inhomogeneous integral equation (4.22) is uniquely solvable for each right-hand side $f \in C[0,1]$ and therefore the boundary value problem itself also is uniquely solvable for all inhomogeneities $v \in C[0,1]$ and all boundary values u_0 and u_1, or the homogeneous integral equation (4.24), and consequently the homogeneous boundary value problem, have nontrivial solutions.

The homogeneous boundary value problem (4.23) has at most one linearly independent solution. To show this, let w_1, w_2 be two solutions of (4.23). Then there exist constants λ_1, λ_2 such that $\lambda_1 w_1'(0) + \lambda_2 w_2'(0) = 0$ and $|\lambda_1| + |\lambda_2| > 0$. From the homogeneous boundary conditions we also have that $\lambda_1 w_1(0) + \lambda_2 w_2(0) = 0$. Now the Picard–Lindelöf uniqueness theorem for initial value problems implies that $\lambda_1 w_1 + \lambda_2 w_2 = 0$ on $[a,b]$, i.e., two solutions to the homogeneous boundary value problem (4.23) are linearly dependent. Therefore, if (4.23) has nontrivial solutions, it has one linearly independent solution and the homogeneous integral equation and its adjoint equation both also have one linearly independent solution.

Let ψ be a solution to the homogeneous adjoint equation

$$\psi(x) - \int_0^1 K(y,x)\psi(y)\,dy = 0, \quad x \in [0,1],$$

i.e.,

$$\psi(x) = \int_0^x (1-x)\{a_2(y)y + a_1(y)\}\psi(y)\,dy + \int_x^1 x\{a_2(y)(1-y) - a_1(y)\}\psi(y)\,dy.$$

Then $\psi(0) = \psi(1) = 0$ and for the derivative we find

$$\psi'(x) - a_1(x)\psi(x) = -\int_0^1 \{a_2(y)y + a_1(y)\}\psi(y)\,dy + \int_x^1 a_2(y)\psi(y)\,dy,$$

whence

$$(\psi' - a_1\psi)' + a_2\psi = 0$$

follows. By the Picard–Lindelöf theorem this homogeneous adjoint boundary value problem again admits at most one linearly independent solution. Therefore, in the case when the homogeneous boundary value problem has a nontrivial solution, the homogeneous adjoint integral equation and the homogeneous adjoint boundary value problem are equivalent. By the Fredholm alternative, the inhomogeneous integral equation (4.22) and therefore the inhomogeneous boundary value problem (4.18) and (4.19) are solvable if and only if

$$\int_0^1 f(x)\psi(x)\,dx = 0 \tag{4.25}$$

is satisfied for the solutions ψ of the homogeneous adjoint boundary value problem. Using the differential equation and the boundary condition for ψ, we find

$$\int_0^1 \{(u_1 - u_0)a_1(x) + [u_0(1-x) + u_1 x]a_2(x)\}\psi(x)\,dx = u_0\psi'(0) - u_1\psi'(1).$$

Hence, the condition (4.25) and

$$\int_0^1 v(x)\psi(x)\,dx = u_0\psi'(0) - u_1\psi'(1)$$

are equivalent, and we can summarize our results in the following form.

Theorem 4.21. Either *the inhomogeneous boundary value problem*

$$u'' + a_1 u' + a_2 u = v, \quad u(0) = u_0, \quad u(1) = u_1,$$

is uniquely solvable for all right-hand sides v *and boundary values* u_0, u_1 or *the homogeneous boundary value problem*

$$w'' + a_1 w' + a_2 w = 0, \quad w(0) = w(1) = 0,$$

and the homogeneous adjoint boundary value problem

$$(\psi' - a_1\psi)' + a_2\psi = 0, \quad \psi(0) = \psi(1) = 0,$$

each have one linearly independent solution w *and* ψ, *respectively. In the latter case, the inhomogeneous boundary value problem is solvable if and only if*

$$\int_0^1 v(x)\psi(x)\,dx = u_0\psi'(0) - u_1\psi'(1).$$

Problems

4.1. Let $\langle\cdot,\cdot\rangle : C[a,b] \times C[a,b] \to \mathbb{R}$ be a degenerate bilinear form. Then there exists a function $f \in C[a,b]$ such that $\langle f,\psi\rangle = 0$ for all $\psi \in C[a,b]$. Since $f \neq 0$, without loss of generality we may assume that $f(a) = 1$. The compact operators $A, B : C[a,b] \to C[a,b]$ defined by $A\varphi := \varphi(a)f$ and $B\psi := 0$ are adjoint with respect to $\langle\cdot,\cdot\rangle$. By showing that $N(I - A) = \text{span}\{f\}$ and $N(I - B) = \{0\}$ demonstrate that for the validity of the Fredholm alternative the bilinear form necessarily must be non-degenerate.

4.2. Let X be the linear space of all functions $\varphi \in C(0,1]$ for which positive numbers M and α (depending on φ) exist such that $|\varphi(x)| \leq Mx^{\alpha-1/2}$ for all $x \in (0,1]$. Then X is a normed space with the norm

$$\|\varphi\| := \sup_{x\in(0,1]} \sqrt{x}\,|\varphi(x)|,$$

and $\langle X, X\rangle$ is a dual system with the bilinear form

$$\langle\varphi,\psi\rangle := \int_0^1 \varphi(x)\psi(x)\,dx.$$

Show that the integral operators $A, B : X \to X$ with continuous kernel K defined as in Theorem 4.7 are compact and adjoint. By using the sequence (φ_n) given by $\varphi_n(x) := x^{1/n-1/2}$ show that the bilinear form is not bounded.

4.3. Formulate and prove the Fredholm alternative for a pair of operators $S - A$ and $T - B$, where S and T each have a bounded inverse and A and B are compact.

4.4. Show that under the assumptions of Theorem 4.17 the operators A and B both have Riesz number one if and only if for each pair of basis $\varphi_1,\dots,\varphi_m$ and $\psi_1,\dots,\psi_m$ of the nullspaces $N(I-A)$ and $N(I - B)$ the matrix $\langle\varphi_i,\psi_k\rangle$, $i,k = 1,\dots,m$, is nonsingular.

4.5. Use Lax's Theorem 4.13 to show that the integral operator with weakly singular kernel of Theorem 2.29 is a compact operator from $L^2(G)$ into $L^2(G)$ (see also Problem 2.3).

Chapter 5
Regularization in Dual Systems

In this chapter we will consider equations that are singular in the sense that they are not of the second kind with a compact operator. We will demonstrate that it is still possible to obtain results on the solvability of singular equations provided that they can be regularized, i.e., they can be transformed into equations of the second kind with a compact operator.

5.1 Regularizers

The following definition will say more precisely what we mean by regularizing a bounded linear operator.

Definition 5.1. Let X_1, X_2 be normed spaces and let $K : X_1 \to X_2$ be a bounded linear operator. A bounded linear operator $R_\ell : X_2 \to X_1$ is called a *left regularizer* of K if

$$R_\ell K = I - A_\ell, \tag{5.1}$$

where $A_\ell : X_1 \to X_1$ is compact; a bounded linear operator $R_r : X_2 \to X_1$ is called a *right regularizer* of K if

$$KR_r = I - A_r, \tag{5.2}$$

where $A_r : X_2 \to X_2$ is compact. A bounded linear operator $R : X_2 \to X_1$ is called a *regularizer* of K if

$$RK = I - A_\ell \quad \text{and} \quad KR = I - A_r, \tag{5.3}$$

where $A_\ell : X_1 \to X_1$ and $A_r : X_2 \to X_2$ are compact.

The difference between a left and a right regularizer is compact, since by multiplying (5.1) from the right by R_r and (5.2) from the left by R_ℓ and then subtracting we obtain $R_r - R_\ell = A_\ell R_r - R_\ell A_r$, which is compact by Theorems 2.20 and 2.21. Again by Theorem 2.21, we observe that adding a compact operator to a regularizer preserves the regularizing property. Therefore, provided that there exist a left and a right regularizer we may always assume that $R_\ell = R_r = R$.

R. Kress, *Linear Integral Equations*, Applied Mathematical Sciences 82,
DOI 10.1007/978-1-4614-9593-2_5, © Springer Science+Business Media New York 2014

Let us first consider regularizing from the left. Any solution to the original equation

$$K\varphi = f \tag{5.4}$$

also solves the regularized equation

$$\varphi - A_\ell \varphi = R_\ell f. \tag{5.5}$$

Therefore, by regularizing from the left we do not lose any solutions. Conversely, let φ be a solution of the regularized equation (5.5). Then

$$R_\ell(K\varphi - f) = 0,$$

and φ solves the original equation (5.4) provided that $N(R_\ell) = \{0\}$. We call a left regularizer an *equivalent left regularizer* if the original and the regularized equation have the same solutions. Then, we have the following theorem.

Theorem 5.2. *A left regularizer is an equivalent left regularizer if and only if it is injective.*

Now let us treat regularizing from the right. Here we have to compare the original equation (5.4) and the regularized equation

$$\psi - A_r \psi = f. \tag{5.6}$$

Provided that ψ solves the regularized equation, $\varphi := R_r\psi$ is a solution to the original equation (5.4). Therefore, by regularizing from the right we do not create additional solutions. Conversely, to each solution φ of the original equation (5.4) there corresponds a solution ψ of the regularized equation (5.6) with $R_r\psi = \varphi$ provided that $R_r(X_2) = X_1$. We call a right regularizer an *equivalent right regularizer* if it maps the solutions of the regularized equation onto the solutions of the original equation. Then, we have the following theorem.

Theorem 5.3. *A right regularizer is an equivalent right regularizer if and only if it is surjective.*

From Theorems 5.2 and 5.3 we conclude that in a situation where we can establish the existence of an injective left regularizer or a surjective right regularizer the results of the Riesz theory partially carry over to the singular equation $K\varphi = f$. In particular, if K is injective and has an equivalent left regularizer, then K is surjective by Theorem 3.4. If K is surjective and has an equivalent right regularizer, then K is injective. The transformation of a Volterra equation of the first kind into Volterra equations of the second kind as described in Section 3.3 may serve as a first example (see Problem 5.1).

5.2 Normal Solvability

In this section we want to demonstrate that it is also possible to obtain solvability results by regularization that is not equivalent. To this end, we first prove the following lemma.

Lemma 5.4. *Under the assumptions of Theorem 4.15 let $I - A$ have a nontrivial nullspace. Then the Riesz number $r(A)$ of A and the Riesz number $r(B)$ of B coincide, i.e., $r(A) = r(B) = r \in \mathbb{N}$. The corresponding projection operators*

$$P : X \to N[(I - A)^r] \quad \textit{and} \quad P' : Y \to N[(I - B)^r]$$

defined by the direct sums

$$X = N[(I - A)^r] \oplus (I - A)^r(X) \quad \textit{and} \quad Y = N[(I - B)^r] \oplus (I - B)^r(Y),$$

respectively, are adjoint. The elements $a_1, \ldots, a_m$ and $b_1, \ldots, b_m$ entering in the definition (4.11) of the operator T can be chosen such that the operator $S := (I - A + TP) : X \to X$ and its inverse $S^{-1} : X \to X$ both have adjoint operators $S' : Y \to Y$ and $[S^{-1}]' : Y \to Y$, respectively. The adjoint S' has an inverse and $[S^{-1}]' = [S']^{-1}$.

Proof. By the Fredholm Theorem 4.15, for $q \in \mathbb{N}$, the two nullspaces $N[(I - A)^q]$ and $N[(I - B)^q]$ have the same finite dimension . Therefore, in view of Theorem 3.3, the Riesz numbers of A and B must coincide.

As a consequence of Theorem 4.16, applied to $(I - A)^r$ and $(I - B)^r$, we have $\langle P\varphi, \psi - P'\psi \rangle = 0$ for $\varphi \in X$ and $\psi \in Y$, since $P\varphi \in N[(I - A)^r]$ and $\psi - P'\psi \in (I - B)^r(Y)$. Therefore

$$\langle P\varphi, \psi \rangle = \langle P\varphi, P'\psi + \psi - P'\psi \rangle = \langle P\varphi, P'\psi \rangle$$

for all $\varphi \in X$ and $\psi \in Y$. Analogously, we have $\langle \varphi, P'\psi \rangle = \langle P\varphi, P'\psi \rangle$, and consequently $\langle P\varphi, \psi \rangle = \langle \varphi, P'\psi \rangle$ for all $\varphi \in X$ and $\psi \in Y$.

Let $\varphi \in N[(I-A)^r]$ such that $\langle \varphi, \psi \rangle = 0$ for all $\psi \in N[(I-B)^r]$. Then, by Theorem 4.16 we also have $\varphi \in (I-A)^r(X)$, and therefore $\varphi = 0$. Analogously, $\psi \in N[(I-B)^r]$ and $\langle \varphi, \psi \rangle = 0$ for all $\varphi \in N[(I-A)^r]$ implies that $\psi = 0$. Therefore, the bilinear form generating the dual system $\langle X, Y \rangle$ is non-degenerate on $N[(I - A)^r] \times N[(I - B)^r]$. Using Lemma 4.14, this implies that we can choose the elements entering in the definition of T such that $a_1, \ldots, a_m \in N[(I - B)^r]$ and $b_1, \ldots, b_m \in N[(I - A)^r]$, and consequently $PT = T$.

The operator $T' : Y \to Y$ defined by

$$T'\psi := \sum_{i=1}^{m} \langle b_i, \psi \rangle a_i, \quad \psi \in Y,$$

clearly is the adjoint of T and $P'T' = T'$. Then we have

$$\langle TP\varphi, \psi\rangle = \langle PTP\varphi, \psi\rangle = \langle \varphi, P'T'P'\psi\rangle = \langle \varphi, T'P'\psi\rangle$$

for all $\varphi \in X$ and $\psi \in Y$, i.e., the operators TP and $T'P'$ are adjoint. Analogous to TP, the operator $T'P'$ is also compact. Therefore, by the Fredholm alternative, the bijectivity of $S := I - A + TP$ implies bijectivity of $S' := I - B + T'P'$. Now the last statement of the lemma follows from

$$\langle S^{-1}f, g\rangle = \langle S^{-1}f, S'[S']^{-1}g\rangle = \langle SS^{-1}f, [S']^{-1}g\rangle = \langle f, [S']^{-1}g\rangle$$

for all $f \in X$ and $g \in Y$, i.e., $[S^{-1}]' = [S']^{-1}$. □

Theorem 5.5. *Let X_1, X_2 be normed spaces, let $K : X_1 \to X_2$ be a bounded linear operator and let $R : X_2 \to X_1$ be a regularizer of K. Then K and R have finite-dimensional nullspaces.*

Proof. By assumption we have that $RK = I - A_\ell$ and $KR = I - A_r$, where A_ℓ and A_r are compact. Let $K\varphi = 0$. Then $RK\varphi = 0$, and therefore $N(K) \subset N(I - A_\ell)$. By Theorem 3.1, $N(I - A_\ell)$ has finite dimension. Therefore $\dim N(K) \le \dim N(I - A_\ell) < \infty$. Analogously $N(R) \subset N(I - A_r)$ implies $\dim N(R) \le \dim N(I - A_r) < \infty$. □

Theorem 5.6. *Assume that $\langle X_1, Y_1\rangle$ and $\langle X_2, Y_2\rangle$ are two dual systems and let $K : X_1 \to X_2$, $K' : Y_2 \to Y_1$ and $R : X_2 \to X_1$, $R' : Y_1 \to Y_2$ be two pairs of bounded and adjoint operators such that R is a regularizer of K and R' is a regularizer of K'. Then the nullspaces of K and K' have finite dimension, and the ranges are given by*

$$K(X_1) = \{f \in X_2 : \langle f, \psi\rangle = 0, \psi \in N(K')\}$$

and

$$K'(Y_2) = \{g \in Y_1 : \langle \varphi, g\rangle = 0, \varphi \in N(K)\}.$$

Proof. The finite dimension of the nullspaces follows from Theorem 5.5. By symmetry it suffices to prove the statement on the ranges for the operator K. By assumption we have

$$RK = I - A, \quad K'R' = I - B,$$

where $A : X_1 \to X_1$ and $B : Y_1 \to Y_1$ are compact and adjoint.

Let $f \in K(X_1)$, i.e., $f = K\varphi$ for some $\varphi \in X_1$. Then

$$\langle f, \psi\rangle = \langle K\varphi, \psi\rangle = \langle \varphi, K'\psi\rangle = 0$$

for all $\psi \in N(K')$, and therefore

$$K(X_1) \subset \{f \in X_2 : \langle f, \psi\rangle = 0, \psi \in N(K')\}.$$

Conversely, assume that $f \in X_2$ satisfies $\langle f, \psi\rangle = 0$ for all $\psi \in N(K')$. Then

$$\langle Rf, \chi\rangle = \langle f, R'\chi\rangle = 0$$

for all $\chi \in N(I - B)$, since $K'R'\chi = \chi - B\chi = 0$. Therefore, by Theorem 4.17, the regularized equation

$$\varphi - A\varphi = Rf \tag{5.7}$$

is solvable. If the regularizer R is injective, then by Theorem 5.2 each solution of (5.7) also satisfies the original equation $K\varphi = f$, and therefore in this case the proof is complete.

If R is not injective, by Theorem 5.5 its nullspace is finite-dimensional. Let $n := \dim N(R)$ and choose a basis $h_1, \ldots, h_n$ of $N(R)$. By Lemma 4.14 there exist elements $c_1, \ldots, c_n \in Y_2$ such that

$$\langle h_i, c_k \rangle = \delta_{ik}, \quad i, k = 1, \ldots, n.$$

In the the case where $I - A$ is injective, for the solution $\varphi = (I - A)^{-1}Rf$ of (5.7) we can write

$$K\varphi - f = \sum_{k=1}^{n} \beta_k h_k,$$

where

$$\beta_k = \langle K(I - A)^{-1}Rf - f, c_k \rangle = \langle f, g_k \rangle$$

and

$$g_k := R'(I - B)^{-1}K'c_k - c_k$$

for $k = 1, \ldots, n$. For all $\psi \in X_1$ we have

$$\langle \psi, K'g_k \rangle = \langle K\psi, g_k \rangle = \langle K(I - A)^{-1}RK\psi - K\psi, c_k \rangle = 0,$$

whence $K'g_k = 0$ follows for $k = 1, \ldots, n$. This implies $\langle f, g_k \rangle = 0$ for $k = 1, \ldots, n$, and therefore $K\varphi = f$, i.e., the proof is also complete in this case.

It remains to consider the case where both R and $I - A$ have a nontrivial nullspace. Let $m := \dim N(I - A)$, choose a basis $\varphi_1, \ldots, \varphi_m$ of $N(I - A)$, and recall the analysis of Theorems 4.15 and 4.16 and Lemma 5.4. The general solution of the regularized equation (5.7) has the form

$$\varphi = \sum_{i=1}^{m} \alpha_i \varphi_i + (I - A + TP)^{-1}Rf \tag{5.8}$$

with complex coefficients $\alpha_1, \ldots, \alpha_m$. Since for each solution φ of (5.7) we have that $K\varphi - f \in N(R)$, as above we can write

$$K\varphi - f = \sum_{k=1}^{n} \beta_k h_k,$$

where

$$\beta_k = \sum_{i=1}^{m} \alpha_i \langle K\varphi_i, c_k \rangle + \langle K(I - A + TP)^{-1}Rf - f, c_k \rangle, \quad k = 1, \ldots, n.$$

Therefore the solution (5.8) of the regularized equation solves the original equation if and only if the coefficients $\alpha_1, \dots, \alpha_m$ solve the linear system

$$\sum_{i=1}^{m} \alpha_i \langle K\varphi_i, c_k\rangle = \langle f - K(I - A + TP)^{-1}Rf, c_k\rangle, \quad k = 1, \dots, n. \tag{5.9}$$

If the matrix $\langle K\varphi_i, c_k\rangle$ of this linear system has rank n, then (5.9) is solvable. Otherwise, i.e., for rank $p < n$, from elementary linear algebra we know that the conditions that are necessary and sufficient for the solvability of (5.9) can be expressed in the form

$$\sum_{k=1}^{n} \rho_{ik} \langle f - K(I - A + TP)^{-1}Rf, c_k\rangle = 0, \quad i = 1, \dots, n - p,$$

for some matrix ρ_{ik}, or in short form

$$\langle f, g_i\rangle = 0, \quad i = 1, \dots, n - p, \tag{5.10}$$

where

$$g_i := \sum_{k=1}^{n} \rho_{ik} \left\{c_k - R'(I - B + T'P')^{-1}K'c_k\right\}, \quad i = 1, \dots, n - p.$$

Now the proof will be completed by showing that $g_i \in N(K')$

For $\psi \in X_1$, observing that $\psi - A\psi = RK\psi$, analogously to (5.8) we can write

$$\psi = \sum_{i=1}^{m} \alpha_i \varphi_i + (I - A + TP)^{-1}RK\psi$$

for some $\alpha_1, \dots, \alpha_m \in \mathbb{C}$, and consequently

$$\sum_{i=1}^{m} \alpha_i K\varphi_i + K(I - A + TP)^{-1}RK\psi - K\psi = 0.$$

This implies that the linear system (5.9) with f replaced by $K\psi$ has a solution. Therefore, since the conditions (5.10) are necessary for the solvability,

$$\langle \psi, K'g_i\rangle = \langle K\psi, g_i\rangle = 0, \quad i = 1, \dots, n - p,$$

for all $\psi \in X_1$. This now implies that $K'g_i = 0$, $i = 1, \dots, n - p$. □

Comparing Theorem 5.6 with the Fredholm alternative for compact operators we note that the only difference lies in the fact that the dimensions of the nullspaces, in general, are no longer the same. In particular, this implies that from injectivity of K, in general, we cannot conclude surjectivity of K as in the Riesz theory. This also will give rise to the introduction of the index of an operator in the following section.

As in Chapter 4 our analysis includes the special case of the canonical dual systems $\langle X_1, X_1^* \rangle$ and $\langle X_2, X_2^* \rangle$ with the dual spaces X_1^* and X_2^* of X_1 and X_2. In this setting the solvability conditions of Theorem 5.6 usually are referred to as *normal solvability* of the operator K, and in this case the results of Theorem 5.6 were first obtained by Atkinson [9]. For the reasons already mentioned in Chapter 4, our more general setting again seems to be more appropriate for the discussion of integral equations. This will become obvious from the various examples discussed in the following chapters, which will include the classical results by Noether [185] on singular integral equations with Hilbert kernels. For convenience, we reformulate Theorem 5.6 in terms of solvability conditions.

Corollary 5.7. *Under the assumptions of Theorem 5.6 each of the homogeneous equations*

$$K\varphi = 0 \quad \text{and} \quad K'\psi = 0$$

has at most a finite number of linearly independent solutions. The inhomogeneous equations

$$K\varphi = f \quad \text{and} \quad K'\psi = g$$

are solvable if and only if the conditions

$$\langle f, \psi \rangle = 0 \quad \text{and} \quad \langle \varphi, g \rangle = 0$$

are satisfied for all solutions ψ and φ of the homogeneous equations

$$K'\psi = 0 \quad \text{and} \quad K\varphi = 0,$$

respectively.

5.3 Index and Fredholm Operators

We conclude this chapter by introducing the concept of the *index* of an operator. Let U and V be subspaces of a linear space X such that

$$X = U \oplus V.$$

Let $n := \dim V < \infty$ and assume that W is another subspace W with the property $X = U \oplus W$. Choose a basis $v_1, \dots, v_n$ of V and let $w_1, \dots, w_{n+1}$ be $n + 1$ elements from W. Then, since $X = U \oplus V$, there exist elements $u_1, \dots, u_{n+1}$ from U and a matrix ρ_{ik} such that

$$w_i = u_i + \sum_{k=1}^{n} \rho_{ik} v_k, \quad i = 1, \dots, n + 1.$$

The homogeneous linear system

$$\sum_{i=1}^{n+1} \rho_{ik}\lambda_i = 0, \quad k = 1, \ldots, n,$$

of n equations for $n+1$ unknowns has a nontrivial solution $\lambda_1, \ldots, \lambda_{n+1}$. Then

$$\sum_{i=1}^{n+1} \lambda_i w_i = \sum_{i=1}^{n+1} \lambda_i u_i,$$

and consequently

$$\sum_{i=1}^{n+1} \lambda_i w_i = 0,$$

since $U \cap W = \{0\}$. Hence the elements $w_1, \ldots, w_{n+1}$ are linearly dependent, and therefore $\dim W \leq \dim V$. Interchanging the roles of V and W we also have $\dim V \leq \dim W$, whence $\dim V = \dim W$ follows.

Therefore, the *codimension* of a subspace U of X is well defined by setting

$$\operatorname{codim} U := \dim V$$

if there exists a finite-dimensional subspace V such that $X = U \oplus V$ and $\operatorname{codim} U = \infty$ otherwise. In particular, $\operatorname{codim} X = 0$. Obviously, the codimension is a measure for the deviation of the subspace U from the whole space X.

Assume that U_1 and U_2 are subspaces of X such that $\operatorname{codim} U_1 < \infty$ and $U_1 \subset U_2$. Then $X = U_1 \oplus V_1$, where V_1 is a finite-dimensional subspace. For the finite-dimensional subspace $V_1 \cap U_2$ of V_1 there exists another finite-dimensional subspace V_2 such that $V_1 = (V_1 \cap U_2) \oplus V_2$. Now let $\varphi \in U_2 \cap V_2$. Then, since $V_2 \subset V_1$, we have $\varphi \in V_1 \cap U_2$. Therefore $\varphi = 0$, since $(V_1 \cap U_2) \cap V_2 = \{0\}$. Now let $\varphi \in X$ be arbitrary. Then, since $X = U_1 \oplus V_1$, we have $\varphi = u_1 + v_1$ for some $u_1 \in U_1$ and $v_1 \in V_1$. Furthermore, since $V_1 = (V_1 \cap U_2) \oplus V_2$, we have $v_1 = w_2 + v_2$ for some $w_2 \in V_1 \cap U_2$ and $v_2 \in V_2$. Then $\varphi = u_2 + v_2$, where $u_2 := u_1 + w_2 \in U_2$. Hence we have proven that $X = U_2 \oplus V_2$, and this implies that $\operatorname{codim} U_2 \leq \operatorname{codim} U_1$ if $U_1 \subset U_2$.

Definition 5.8. A linear operator $K : X \to Y$ from a linear space X into a linear space Y is said to have *finite defect* if its nullspace has finite dimension and its range has finite codimension. The number

$$\operatorname{ind} K := \dim N(K) - \operatorname{codim} K(X)$$

is called the *index* of the operator K.

Note that in the proof of the following theorem we do not use an adjoint operator, i.e., the analysis is based only on the Riesz theory.

Theorem 5.9. *Let A be a compact operator on a normed space X. Then $I - A$ has index zero.*

Proof. Since by Theorem 3.4 the statement is obvious when $I - A$ is injective, we only need to consider the case where $m := \dim N(I - A) \in \mathbb{N}$. From the direct sum $X = N(I - A)^r \oplus (I - A)^r(X)$ and $(I - A)^r(X) \subset (I - A)(X)$ we conclude that there exists a finite-dimensional subspace U of X such that $X = (I - A)(X) \oplus U$. We need to show that $\dim U = m$.

Let $n := \dim U$ and note that $n \neq 0$ by Theorem 3.4. We choose bases $\varphi_1, \dots, \varphi_m$ of $N(I - A)$ and $b_1, \dots, b_n$ of U. In the case where $r > 1$ we choose additional elements $\varphi_{m+1}, \dots, \varphi_{m_r}$ with $m_r > m$ such that $\varphi_1, \dots, \varphi_m, \varphi_{m+1}, \dots, \varphi_{m_r}$ is a basis of $N(I - A)^r$. Then we define a linear operator $T : N(I - A)^r \to U$ by prescribing

$$T : \varphi_k \mapsto \begin{cases} b_k, & k \leq \min(m, n), \\ 0, & k > \min(m, n). \end{cases} \tag{5.11}$$

By Theorem 2.4 the operator T is bounded. Consequently, since the projection operator $P : X \to N(I - A)^r$ is compact, the operator $TP : X \to U$ is compact.

Now assume that $m < n$ and let

$$\varphi - A\varphi + TP\varphi = 0.$$

Then, since $(I - A)(X) \cap U = \{0\}$, we have $\varphi - A\varphi = 0$ and $TP\varphi = 0$. Therefore we can write

$$\varphi = \sum_{k=1}^{m} \alpha_k \varphi_k,$$

and the definition (5.11) of T implies that

$$\sum_{k=1}^{m} \alpha_k b_k = TP\varphi = 0.$$

From this we conclude that $\alpha_k = 0$, $k = 1, \dots, m$, since the b_k are linearly independent. Hence $\varphi = 0$, and therefore the operator $I - A + TP$ is injective, and consequently surjective by Theorem 3.4. Therefore the equation

$$\varphi - A\varphi + TP\varphi - b_n = 0$$

is uniquely solvable. For its solution φ we conclude that $TP\varphi = b_n$, since we have $(I - A)(X) \cap U = \{0\}$. In view of the definition (5.11) of T this is a contradiction to the linear independence of $b_1, \dots, b_n$.

Now assume that $m > n$. Since $X = (I - A)(X) \oplus U$ and $TP(X) \subset U$, we can represent each $f \in X$ in the form

$$f = (I - A + TP)\varphi + \sum_{k=1}^{n} \alpha_k b_k$$

for some $\varphi \in X$ and complex coefficients $\alpha_1, \dots, \alpha_n$. From (5.11) we conclude that $(I - A + TP)\varphi_k = T\varphi_k = b_k$, $k = 1, \dots, n$, and therefore

$$f = (I - A + TP)\left(\varphi + \sum_{k=1}^{n} \alpha_k \varphi_k\right),$$

i.e., the operator $I - A + TP$ is surjective, and consequently injective by Theorem 3.4. Since $m > n$, from (5.11) we have $(I - A + TP)\varphi_m = T\varphi_m = 0$, and the injectivity of $I - A + TP$ leads to the contradiction $\varphi_m = 0$. □

Operators with closed range and finite defect are called *Fredholm operators*. For a detailed analysis of Fredholm operators we refer to [94, 112]. From Theorem 3.3 and Theorem 5.7 we have that for a compact operator $A : X \to X$ the operator $I - A$ is a Fredholm operator.

Theorem 5.10. *Under the assumptions of Theorem 5.6 the operators K and K' have finite defect with index*

$$\operatorname{ind} K = \dim N(K) - \dim N(K') = -\operatorname{ind} K'.$$

Proof. Let $n := \dim N(K')$ and choose a basis $\psi_1, \dots, \psi_n$ of $N(K')$ if $n > 0$. By Lemma 4.14 there exist elements $b_1, \dots, b_n \in X_2$ such that

$$\langle b_i, \psi_k \rangle = \delta_{ik}, \quad i, k = 1, \dots, n. \tag{5.12}$$

Define

$$U := \operatorname{span}\{b_1, \dots, b_n\}.$$

Then $K(X_1) \cap U = \{0\}$ by Theorem 5.6 and (5.12). Furthermore, for $f \in X_2$ we have

$$u := \sum_{k=1}^{n} \langle f, \psi_k \rangle b_k \in U$$

and $f - u \in K(X_1)$ by Theorem 5.6. Hence $X_2 = K(X_1) \oplus U$, and therefore $\operatorname{codim} K(X_1) = \dim U = \dim N(K')$. □

From Theorems 5.6 and 5.10 we observe that under the additional assumption of boundedness for the two bilinear forms that create the dual systems the operators K and K' are Fredholm operators.

Theorem 5.11. *For two operators K_1 and K_2 satisfying the assumptions of Theorem 5.6 we have*

$$\operatorname{ind} K_1 K_2 = \operatorname{ind} K_1 + \operatorname{ind} K_2.$$

Proof. For the sake of notational brevity we confine ourselves to the case of a dual system $\langle X, Y \rangle$ and two operators $K_1, K_2 : X \to X$, which satisfy the assumptions of Theorems 5.6 and 5.10, i.e., together with their adjoint operators $K_1', K_2' : Y \to Y$

they possess regularizers. Then $K_1K_2 : X \to X$ and its adjoint $K_2'K_1' : Y \to Y$ satisfy the assumptions of Theorems 5.6 and 5.10.

Denote $m_j := \dim N(K_j)$, $m_j' := \dim N(K_j')$, and choose bases of the nullspaces

$$N(K_j) = \operatorname{span}\{\varphi_{j,1}, \dots, \varphi_{j,m_j}\}, \quad N(K_j') = \operatorname{span}\{\psi_{j,1}, \dots, \psi_{j,m_j'}\}$$

for $j = 1, 2$. Let $\varphi \in N(K_1K_2)$. Then $K_2\varphi \in N(K_1)$, i.e.,

$$K_2\varphi = \sum_{i=1}^{m_1} \alpha_i \varphi_{1,i}.$$

By Theorem 5.6 this equation is solvable if and only if

$$\sum_{i=1}^{m_1} \alpha_i \langle \varphi_{1,i}, \psi_{2,k} \rangle = 0, \quad k = 1, \dots, m_2'. \tag{5.13}$$

By p we denote the rank of the $m_1 \times m_2'$ matrix $\langle \varphi_{1,i}, \psi_{2,k} \rangle$. Then the solution space of (5.13) has dimension $m_1 - p$. Therefore

$$\dim N(K_1K_2) = \dim N(K_2) + m_1 - p = m_2 + m_1 - p.$$

Similarly, let $\psi \in N(K_2'K_1')$. Then $K_1'\psi \in N(K_2')$, i.e.,

$$K_1'\psi = \sum_{k=1}^{m_2'} \beta_k \psi_{2,k}.$$

This equation is solvable if and only if

$$\sum_{k=1}^{m_2'} \beta_k \langle \varphi_{1,i}, \psi_{2,k} \rangle = 0, \quad i = 1, \dots, m_1. \tag{5.14}$$

The solution space of (5.14) has dimension $m_2' - p$. Therefore

$$\dim N(K_2'K_1') = \dim N(K_1') + m_2' - p = m_1' + m_2' - p.$$

In view of Theorem 5.10, combining the two results yields

$$\operatorname{ind} K_1K_2 = (m_1 - m_1') + (m_2 - m_2') = \operatorname{ind} K_1 + \operatorname{ind} K_2,$$

and the proof is complete. □

Corollary 5.12. *Under the assumptions of Theorem 5.6 the index is stable with respect to compact perturbations, i.e., for compact operators C with a compact adjoint C' we have*

$$\operatorname{ind}(K + C) = \operatorname{ind} K.$$

Proof. Let K and K' be adjoint operators with adjoint regularizers R and R'. Then, since $RK = I - A$, where A is compact, Theorems 5.9 and 5.11 imply that

$$\operatorname{ind} R + \operatorname{ind} K = \operatorname{ind} RK = \operatorname{ind}(I - A) = 0,$$

i.e., $\operatorname{ind} K = -\operatorname{ind} R$. For a compact operator C the operator R also regularizes $K + C$ and the operator R' regularizes $K' + C'$. Therefore

$$\operatorname{ind}(K + C) = -\operatorname{ind} R = \operatorname{ind} K,$$

and the proof is complete. □

For the history of the development of the notion of the index of an operator we refer to [44].

Of course, this chapter can provide only a first glance into the theory of singular operators. For a detailed study, in the canonical dual system $\langle X, X^* \rangle$, we refer to the monograph by Mikhlin and Prössdorf [170].

Problems

5.1. Show that the transformations of the Volterra integral equation of the first kind (3.14) into the Volterra equations of the second kind (3.15) and (3.16) can be interpreted as regularizations from the left and from the right, respectively.
Hint: Use the space $C^1[a, b]$ of continuously differentiable functions furnished with the norm $\|\varphi\|_1 := \|\varphi\|_\infty + \|\varphi'\|_\infty$.

5.2. Convince yourself where in the proof of Theorem 5.6 use is made of the fact that the operators K and K' possess regularizers from the left and from the right.

5.3. Use Theorem 5.9 for an alternative proof of Theorem 4.17.

5.4. Let X_1, X_2 be Banach spaces, let $K : X_1 \to X_2$ be a bounded operator, and let $R : X_2 \to X_1$ be a left (right) regularizer of K. Show that for all operators $C : X_1 \to X_2$ with $\|C\| < \|R\|$ the operator $K + C$ has a left (right) regularizer.

5.5. Use Problem 5.4 to show that in Banach spaces under the assumptions of Theorem 5.6 the index is stable with respect to small perturbations, i.e., there exists a positive number γ (depending on K and K') such that

$$\operatorname{ind}(K + C) = \operatorname{ind} K$$

for all operators C with adjoint C' satisfying $\max(\|C\|, \|C'\|) < \gamma$ (see [9, 43]).

Chapter 6
Potential Theory

The solution of boundary value problems for partial differential equations is one of the most important fields of applications for integral equations. In the second half of the 19th century the systematic development of the theory of integral equations was initiated by the treatment of boundary value problems and there has been an ongoing fruitful interaction between these two areas of applied mathematics. It is the aim of this chapter to introduce the main ideas of this field by studying the basic boundary value problems of potential theory. For the sake of simplicity we shall confine our presentation to the case of two and three space dimensions. The extension to more than three dimensions is straightforward. As we shall see, the treatment of the boundary integral equations for the potential theoretic boundary value problems delivers an instructive example for the application of the Fredholm alternative, since both its cases occur in a natural way. This chapter covers the classical approach to boundary integral equations of the second kind in the space of continuous functions. The treatment of boundary integral equations of the first and of the second kind in Hölder spaces and in Sobolev spaces will be the topic of the two subsequent chapters.

6.1 Harmonic Functions

We begin with a brief outline of the basic properties of harmonic functions going back to the early development of potential theory at the beginning of the 19th century with contributions by Dirichlet, Gauss, Green, Riemann and Weierstrass. For further study of potential theory we refer to Constanda [33], Courant and Hilbert [36], Folland [54], Helms [89], Kanwal [118], Kellogg [119], Martensen [162], and Mikhlin [169].

Definition 6.1. A twice continuously differentiable real-valued function u, defined on a domain $D \subset \mathbb{R}^m$, $m = 2, 3$, is called *harmonic* if it satisfies *Laplace's equation*

$$\Delta u = 0 \quad \text{in } D,$$

R. Kress, *Linear Integral Equations*, Applied Mathematical Sciences 82,
DOI 10.1007/978-1-4614-9593-2_6,

where

$$\Delta u := \sum_{j=1}^{m} \frac{\partial^2 u}{\partial x_j^2} .$$

Harmonic functions describe time-independent temperature distributions, potentials of electrostatic and magnetostatic fields, and velocity potentials of incompressible irrotational fluid flows.

There is a close connection between harmonic functions in $\mathbb{R}^2$ and holomorphic functions in $\mathbb{C}$. From the Cauchy–Riemann equations we readily observe that both the real and imaginary parts of a holomorphic function $f(z) = u(x_1, x_2) + iv(x_1, x_2)$, $z = x_1 + ix_2$, are harmonic functions.

Most of the basic properties of harmonic functions can be deduced from the fundamental solution that is introduced in the following theorem. Recall that by $|x|$ we denote the Euclidean norm of a vector $x \in \mathbb{R}^m$.

Theorem 6.2. *The function*

$$\Phi(x, y) := \begin{cases} \dfrac{1}{2\pi} \ln \dfrac{1}{|x - y|}, & m = 2, \\[2ex] \dfrac{1}{4\pi} \dfrac{1}{|x - y|}, & m = 3, \end{cases}$$

defined for all $x \neq y$ in $\mathbb{R}^m$ is called the fundamental solution *of Laplace's equation. For fixed $y \in \mathbb{R}^m$ it is harmonic in $\mathbb{R}^m \setminus \{y\}$.*

Proof. This follows by straightforward differentiation. □

For $n \in \mathbb{N}$, by $C^n(D)$ we denote the linear space of real- or complex-valued functions defined on the domain D, which are n times continuously differentiable. By $C^n(\bar{D})$ we denote the subspace of all functions in $C^n(D)$, which with all their derivatives up to order n can be extended continuously from D into the closure $\bar{D}$. In this chapter, we mostly deal with real-valued functions but with proper interpretation our results remain valid for complex-valued functions. From p. 30 we recall what is meant by saying a bounded domain D or its boundary ∂D belong to class C^n for $n \in \mathbb{N}$.

One of the basic tools in studying harmonic functions is provided by *Green's integral theorems*. Recall that for two vectors $a = (a_1, \dots, a_m)$ and $b = (b_1, \dots, b_m)$ in $\mathbb{R}^m$ we denote by $a \cdot b = a_1 b_1 + \cdots + a_m b_m$ the dot product.

Theorem 6.3 (Green's Theorem). *Let D be a bounded domain of class C^1 and let ν denote the unit normal vector to the boundary ∂D directed into the exterior of D. Then, for $u \in C^1(\bar{D})$ and $v \in C^2(\bar{D})$ we have* Green's first theorem

$$\int_D (u \Delta v + \operatorname{grad} u \cdot \operatorname{grad} v)\, dx = \int_{\partial D} u \frac{\partial v}{\partial \nu}\, ds \tag{6.1}$$

and for $u, v \in C^2(\bar{D})$ *we have* Green's second theorem

$$\int_D (u\Delta v - v\Delta u)\,dx = \int_{\partial D} \left(u\,\frac{\partial v}{\partial \nu} - v\,\frac{\partial u}{\partial \nu} \right) ds. \tag{6.2}$$

Proof. We apply Gauss' divergence theorem

$$\int_D \operatorname{div} A\,dx = \int_{\partial D} \nu \cdot A\,ds$$

to the vector field $A \in C^1(\bar{D})$ defined by $A := u\operatorname{grad} v$ and use

$$\operatorname{div}(u\operatorname{grad} v) = \operatorname{grad} u \cdot \operatorname{grad} v + u\operatorname{div}\operatorname{grad} v$$

to establish (6.1). To obtain (6.2) we interchange u and v and then subtract. □

Note that our regularity assumptions on D are sufficient conditions for the validity of Gauss' and Green's theorems and can be weakened. In particular, the boundary can be allowed to have edges and corners. For a detailed study see, for example, [169, 177].

Corollary 6.4. *Let* $v \in C^2(\bar{D})$ *be harmonic in D. Then*

$$\int_{\partial D} \frac{\partial v}{\partial \nu}\,ds = 0. \tag{6.3}$$

Proof. This follows by choosing $u = 1$ in (6.1). □

Theorem 6.5 (Green's Formula). *Let D be as in Theorem 6.3 and let* $u \in C^2(\bar{D})$ *be harmonic in D. Then*

$$u(x) = \int_{\partial D} \left\{ \frac{\partial u}{\partial \nu}(y)\,\Phi(x,y) - u(y)\,\frac{\partial \Phi(x,y)}{\partial \nu(y)} \right\} ds(y), \quad x \in D. \tag{6.4}$$

Proof. For $x \in D$ we choose a sphere $\Omega(x;r) := \{y \in \mathbb{R}^m : |y - x| = r\}$ of radius r such that $\Omega(x;r) \subset D$ and direct the unit normal ν to $\Omega(x;r)$ into the interior of $\Omega(x;r)$. Now we apply Green's second theorem (6.2) to the harmonic functions u and $\Phi(x,\cdot)$ in the domain $\{y \in D : |y - x| > r\}$ to obtain

$$\int_{\partial D \cup \Omega(x;r)} \left\{ u(y)\,\frac{\partial \Phi(x,y)}{\partial \nu(y)} - \frac{\partial u}{\partial \nu}(y)\,\Phi(x,y) \right\} ds(y) = 0.$$

Since on $\Omega(x;r)$ we have

$$\operatorname{grad}_y \Phi(x,y) = \frac{\nu(y)}{\omega_m r^{m-1}}, \tag{6.5}$$

where $\omega_2 = 2\pi$, $\omega_3 = 4\pi$, a straightforward calculation, using the mean value theorem and (6.3), shows that

$$\lim_{r\to 0} \int_{\Omega(x;r)} \left\{ u(y) \frac{\partial \Phi(x,y)}{\partial \nu(y)} - \frac{\partial u}{\partial \nu}(y)\, \Phi(x,y) \right\} ds(y) = u(x),$$

whence (6.4) follows. □

From Green's formula we can conclude that harmonic functions are analytic functions of their independent variables.

Theorem 6.6. *Harmonic functions are analytic, i.e., each harmonic function has a local power series expansion.*

Proof. For $f \in C(\partial D)$, we show that the function

$$u(x) := \int_{\partial D} \Phi(x,y) f(y)\, ds(y), \quad x \in D,$$

is analytic in D and confine ourselves to the two-dimensional case. To this end we fix an arbitrary $x_0 \in D$ and choose $R > 0$ such that the disk $B[x_0; 3R] \subset D$ and define $V[x_0; R] := \{(x,y) : |x - x_0| \le R, |y - x_0| \ge 3R\}$. With the aid of

$$|x - y|^2 = |x_0 - y|^2 \left\{ 1 + 2\, \frac{(x_0 - y)\cdot(x - x_0)}{|x_0 - y|^2} + \frac{|x - x_0|^2}{|x_0 - y|^2} \right\}$$

we obtain the series

$$\begin{aligned} \ln \frac{1}{|x-y|} &= \ln \frac{1}{|x_0 - y|} + \frac{1}{2} \sum_{k=1}^{\infty} \frac{(-1)^k}{k}\, t^k \\ &= \ln \frac{1}{|x_0 - y|} + \frac{1}{2} \sum_{k=1}^{\infty} \frac{(-1)^k}{k} \sum_{j=0}^{k} \binom{k}{j} \xi^j \eta^{k-j}, \end{aligned} \tag{6.6}$$

where $t := \xi + \eta$ with

$$\xi := 2\, \frac{(x_0 - y)\cdot(x - x_0)}{|x_0 - y|^2} \quad \text{and} \quad \eta := \frac{|x - x_0|^2}{|x_0 - y|^2}\,.$$

For $(x,y) \in V[x_0; R]$ and $k \in \mathbb{N}$ we can estimate

$$\left| \frac{(-1)^k}{k} \sum_{j=0}^{k} \binom{k}{j} \xi^j \eta^{k-j} \right| \le \sum_{j=0}^{k} \binom{k}{j} |\xi|^j |\eta|^{k-j} = (|\xi| + |\eta|)^k \le \frac{7^k}{9^k}\,.$$

Therefore the series (6.6) has a convergent majorant and consequently it is absolutely and uniformly convergent in $V[x_0; R]$. Hence we can reorder the series and collect powers of ξ^2 and η resulting in

$$\ln \frac{1}{|x-y|} = \ln \frac{1}{|x_0 - y|} + \sum_{k=1}^{\infty} \left\{ \xi \sum_{j=0}^{k} \alpha_{kj} \xi^{2j} \eta^{k-j} + \sum_{j=0}^{k} \beta_{kj} \xi^{2j} \eta^{k-j} \right\} \tag{6.7}$$

with real coefficients α_{kj} and β_{kj}. Since the two inner sums in this expansion are homogeneous polynomials of degree $2k + 1$ and $2k$, respectively, we can rewrite (6.7) in the form

$$\ln \frac{1}{|x-y|} = \ln \frac{1}{|x_0 - y|} + \sum_{k=1}^{\infty} \sum_{j_1+j_2=k} \gamma_{j_1 j_2}(x_0, y)(x_1 - x_{01})^{j_1}(x_2 - x_{02})^{j_2} \tag{6.8}$$

with real coefficients $\gamma_{j_1 j_2}$ depending on $x_0 = (x_{01}, x_{02})$ and y and uniform convergence in $V[x_0; R]$. Thus we may integrate the series (6.8) term by term and obtain

$$u(x) = \sum_{k=0}^{\infty} \sum_{j_1+j_2=k} c_{j_1 j_2}(x_0)(x_1 - x_{01})^{j_1}(x_2 - x_{02})^{j_2}$$

with real coefficients $c_{j_1 j_2}$ depending on x_0 and uniform convergence for all $x = (x_1, x_2) \in B[x_0; R]$.

The proofs for the second term in Green's formula and the three-dimensional case are analogous. □

An alternative proof of Theorem 6.6 makes use of the fact that each holomorphic function of several complex variables, i.e., a function satisfying the Cauchy–Riemann equations with respect to each of the complex variables is also analytic and vice versa (see [64]). Then the theorem follows from the observation that the fundamental solution $\Phi(x, y)$ is an analytic function of the Cartesian coordinates x_j, $j = 1, \ldots, m$, of x and the fact that the integrands in (6.4) and their derivatives with respect to x are continuous with respect to y if x is contained in a compact subset of D. Therefore the Cauchy–Riemann equations for u can be verified by differentiating with respect to x under the integral.

From Theorem 6.6 it follows that a harmonic function that vanishes in an open subset of its domain of definition must vanish identically.

The following theorem is a special case of a more general result for partial differential equations known as *Holmgren's theorem.*

Theorem 6.7. *Let D be as in Theorem 6.5 and let $u \in C^2(D) \cap C^1(\bar{D})$ be harmonic in D such that*

$$u = \frac{\partial u}{\partial \nu} = 0 \quad on\ \Gamma \tag{6.9}$$

for some open subset $\Gamma \subset \partial D$. Then u vanishes identically in D.

Proof. In view of (6.4), we use Green's representation formula to extend the definition of u by setting

$$u(x) := \int_{\partial D \setminus \Gamma} \left\{ \frac{\partial u}{\partial \nu}(y)\, \Phi(x,y) - u(y) \frac{\partial \Phi(x,y)}{\partial \nu(y)} \right\} ds(y)$$

for $x \in (\mathbb{R}^m \setminus \bar{D}) \cup \Gamma$. Then, by Green's second integral theorem (6.2), applied to u and $\Phi(x,\cdot)$, we have $u = 0$ in $\mathbb{R}^m \setminus \bar{D}$. By G we denote a component of $\mathbb{R}^m \setminus \bar{D}$ with $\Gamma \cap \partial G \neq \emptyset$. Clearly u solves the Laplace equation in $(\mathbb{R}^m \setminus \partial D) \cup \Gamma$ and therefore $u = 0$ in D, since D and G are connected through the gap Γ in ∂D. □

Theorem 6.8 (Mean Value Theorem). *Let u be harmonic in an open ball $B(x;r) = \{y \in \mathbb{R}^m : |y - x| < r\}$ with boundary $\Omega(x;r)$ and continuous in the closure $B[x;r]$. Then*

$$u(x) = \frac{m}{\omega_m r^m} \int_{B[x;r]} u(y)\, dy = \frac{1}{\omega_m r^{m-1}} \int_{\Omega(x;r)} u(y)\, ds(y), \tag{6.10}$$

i.e., the value of u at the center of the ball is equal to the integral mean values over both the ball and its boundary surface ($\omega_2 = 2\pi$, $\omega_3 = 4\pi$).

Proof. For each $0 < \rho < r$ we have $u \in C^2(B[x;\rho])$ and can apply (6.3) and (6.4) with the result

$$u(x) = \frac{1}{\omega_m \rho^{m-1}} \int_{|y-x|=\rho} u(y)\, ds(y), \tag{6.11}$$

whence the second mean value formula follows by passing to the limit $\rho \to r$. Multiplying (6.11) by ρ^{m-1} and integrating with respect to ρ from 0 to r we obtain the first mean value formula. □

Theorem 6.9 (Maximum-Minimum Principle). *A harmonic function on a domain cannot attain its maximum or its minimum unless it is constant.*

Proof. It suffices to carry out the proof for the maximum. Let u be a harmonic function in the domain D and assume that it attains its maximum value in D, i.e., the set $D_M := \{x \in D : u(x) = M\}$ where $M := \sup_{x \in D} u(x)$ is not empty. Since u is continuous, D_M is closed relative to D. Let x be any point in D_M and apply the mean value Theorem 6.8 to the harmonic function $M - u$ in a ball $B(x;r)$ with $B[x;r] \subset D$. Then

$$0 = M - u(x) = \frac{m}{\omega_m r^m} \int_{B[x;r]} \{M - u(y)\}\, dy,$$

so that $u = M$ in $B(x;r)$. Therefore D_M is open relative to D. Hence $D = D_M$, i.e., u is constant in D. □

Corollary 6.10. *Let D be a bounded domain and let u be harmonic in D and continuous in $\bar{D}$. Then u attains both its maximum and its minimum on the boundary.*

For the study of exterior boundary value problems we also need to investigate the asymptotic behavior of harmonic functions as $|x| \to \infty$. To this end we extend Green's formula to unbounded domains.

Theorem 6.11. *Assume that D is a bounded domain of class C^1 with a connected boundary ∂D and outward unit normal ν and let $u \in C^2(\mathbb{R}^m \setminus D)$ be a bounded harmonic function. Then*

$$u(x) = u_\infty + \int_{\partial D} \left\{ u(y) \frac{\partial \Phi(x,y)}{\partial \nu(y)} - \frac{\partial u}{\partial \nu}(y)\, \Phi(x,y) \right\} ds(y) \tag{6.12}$$

for $x \in \mathbb{R}^m \setminus \bar{D}$ and some constant u_∞. For $m = 2$, in addition,

$$\int_{\partial D} \frac{\partial u}{\partial \nu}\, ds = 0 \tag{6.13}$$

and the mean value property at infinity

$$u_\infty = \frac{1}{2\pi r} \int_{|y|=r} u(y)\, ds(y) \tag{6.14}$$

for sufficiently large r is satisfied.

Proof. Without loss of generality we may assume that the origin $x = 0$ is contained in D. Since u is bounded, there exists a constant $M > 0$ such that $|u(x)| \leq M$ for all $x \in \mathbb{R}^m \setminus D$. Choose R_0 large enough to ensure that $y \in \mathbb{R}^m \setminus D$ for all $|y| \geq R_0/2$. Then for a fixed x with $|x| \geq R_0$ we can apply the mean value Theorem 6.8 to the components of $\operatorname{grad} u$. From this and Gauss' integral theorem we obtain

$$\operatorname{grad} u(x) = \frac{m}{\omega_m r^m} \int_{B[x;r]} \operatorname{grad} u(y)\, dy = -\frac{m}{\omega_m r^m} \int_{\Omega(x;r)} \nu(y) u(y)\, ds(y),$$

where ν is the unit normal to $\Omega(x;r)$ directed into the interior of $\Omega(x;r)$ and where we choose the radius to be $r = |x|/2$. Then we can estimate

$$|\operatorname{grad} u(x)| \leq \frac{mM}{r} = \frac{2mM}{|x|} \tag{6.15}$$

for all $|x| \geq R_0$.

For $m = 2$, we choose r large enough such that $\Omega_r := \Omega(0;r)$ is contained in $\mathbb{R}^2 \setminus \bar{D}$ and apply Green's second theorem (6.2) to u and $\Phi(0,\cdot)$ in the annulus $r < |y| < R$ and use (6.5) to obtain

$$\frac{1}{r} \int_{\Omega_r} u\, ds - \ln \frac{1}{r} \int_{\Omega_r} \frac{\partial u}{\partial \nu}\, ds = \frac{1}{R} \int_{\Omega_R} u\, ds - \ln \frac{1}{R} \int_{\Omega_R} \frac{\partial u}{\partial \nu}\, ds.$$

(Note that ν is the interior normal to Ω_r and Ω_R.) From this, with the aid of Corollary 6.4 applied in the annulus between ∂D and Ω_r, we find

$$\frac{1}{r} \int_{\Omega_r} u\, ds + \ln \frac{1}{r} \int_{\partial D} \frac{\partial u}{\partial \nu}\, ds = \frac{1}{R} \int_{\Omega_R} u\, ds + \ln \frac{1}{R} \int_{\partial D} \frac{\partial u}{\partial \nu}\, ds. \tag{6.16}$$

Since the first term on the right-hand side is bounded by $2\pi M$, letting $R \to \infty$ in (6.16) implies that the integral in (6.13) must be zero. Note that (6.13) only holds in the two-dimensional case and is a consequence of the fact that in $\mathbb{R}^2$ the fundamental solution is not bounded at infinity.

For $x \in \mathbb{R}^m \setminus \bar{D}$, $m = 2, 3$, we now choose r large enough such that $\bar{D} \subset B(x; r)$. Then by Green's formula (6.4), applied in the domain between ∂D and $\Omega(x; r)$, we have that

$$u(x) = \int_{\partial D \cup \Omega(x;r)} \left\{ u(y) \frac{\partial \Phi(x, y)}{\partial \nu(y)} - \frac{\partial u}{\partial \nu}(y)\, \Phi(x, y) \right\} ds(y). \tag{6.17}$$

With the aid of Corollary 6.4 we find

$$\int_{\Omega(x;r)} \frac{\partial u}{\partial \nu}(y)\, \Phi(x, y)\, ds(y) = \frac{1}{4\pi r} \int_{\partial D} \frac{\partial u}{\partial \nu}(y)\, ds(y) \to 0, \quad r \to \infty,$$

if $m = 3$, and

$$\int_{\Omega(x;r)} \frac{\partial u}{\partial \nu}(y)\, \Phi(x, y)\, ds(y) = \frac{1}{2\pi} \ln \frac{1}{r} \int_{\partial D} \frac{\partial u}{\partial \nu}(y)\, ds(y) = 0$$

if $m = 2$, where we have made use of (6.13). With the aid of (6.5) we can write

$$\int_{\Omega(x;r)} u(y) \frac{\partial \Phi(x, y)}{\partial \nu(y)}\, ds(y) = \frac{1}{\omega_m r^{m-1}} \int_{|y-x|=r} u(y)\, ds(y).$$

From the mean value theorem we have

$$u(x + y) - u(y) = \operatorname{grad} u(y + \theta x) \cdot x$$

for some $\theta \in [0, 1]$, and using (6.15) we can estimate

$$|u(x + y) - u(y)| \leq \frac{2mM\,|x|}{|y| - |x|}$$

provided that $|y|$ is sufficiently large. Therefore

$$\frac{1}{\omega_m r^{m-1}} \left| \int_{|y-x|=r} u(y)\, ds(y) - \int_{|y|=r} u(y)\, ds(y) \right| \leq \frac{C}{r}$$

for some constant $C > 0$ depending on x and all sufficiently large r. Now choose a sequence (r_n) of radii with $r_n \to \infty$. Since the integral mean values

$$\mu_n := \frac{1}{\omega_m r_n^{m-1}} \int_{|y|=r_n} u(y)\, ds(y)$$

are bounded through $|\mu_n| \leq M$, $n \in \mathbb{N}$, by the Bolzano–Weierstrass theorem we may assume that the sequence (μ_n) converges, i.e., $\mu_n \to u_\infty$, $n \to \infty$, for some $u_\infty \in \mathbb{R}$.

From this, in view of the above estimates, we now have that

$$\int_{\Omega(x;r_n)} \left\{ u(y) \frac{\partial \Phi(x,y)}{\partial \nu(y)} - \frac{\partial u}{\partial \nu}(y)\, \Phi(x,y) \right\} ds(y) \to u_\infty, \quad n \to \infty.$$

Hence (6.12) follows by setting $r = r_n$ in (6.17) and passing to the limit $n \to \infty$. Finally, (6.14) follows by setting $R = r_n$ in (6.16), passing to the limit $n \to \infty$, and using (6.13). □

From (6.12), using the asymptotic behavior of the fundamental solution

$$\Phi(x,y) := \begin{cases} \dfrac{1}{2\pi} \ln \dfrac{1}{|x|} + O\left(\dfrac{1}{|x|}\right), & m = 2, \\[2ex] O\left(\dfrac{1}{|x|}\right), & m = 3, \end{cases} \tag{6.18}$$

and

$$\frac{\partial \Phi(x,y)}{\partial x_j} = O\left(\frac{1}{|x|^{m-1}}\right), \quad \frac{\partial^2 \Phi(x,y)}{\partial x_j \partial x_k} = O\left(\frac{1}{|x|^m}\right) \tag{6.19}$$

for $|x| \to \infty$, which holds uniformly for all directions $x/|x|$ and all $y \in \partial D$, and the property (6.13) if $m = 2$, we can deduce that bounded harmonic functions in an exterior domain satisfy

$$u(x) = u_\infty + O\left(\frac{1}{|x|}\right), \quad \operatorname{grad} u(x) = O\left(\frac{1}{|x|^{m-1}}\right), \quad |x| \to \infty, \tag{6.20}$$

uniformly for all directions.

6.2 Boundary Value Problems: Uniqueness

Green's formula (6.4) represents any harmonic function in terms of its boundary values and its normal derivative on the boundary, the so-called *Cauchy data*. In the subsequent analysis we shall see that a harmonic function is already completely determined by either its boundary values or, up to a constant, its normal derivative alone. In the sequel, let $D \subset \mathbb{R}^m$ be a bounded domain of class C^2. For the sake of simplicity for the rest of this chapter we assume that the boundary ∂D is connected. Again by ν we denote the unit normal of ∂D directed into the exterior domain $\mathbb{R}^m \setminus \bar{D}$.

Interior Dirichlet Problem. *Find a function u that is harmonic in D, is continuous in $\bar{D}$, and satisfies the boundary condition*

$$u = f \quad \text{on } \partial D,$$

where f is a given continuous function.

Interior Neumann Problem. *Find a function u that is harmonic in D, is continuous in $\bar{D}$, and satisfies the boundary condition*

$$\frac{\partial u}{\partial \nu} = g \quad \text{on } \partial D$$

in the sense

$$\lim_{h \to +0} \nu(x) \cdot \operatorname{grad} u(x - h\nu(x)) = g(x), \quad x \in \partial D,$$

of uniform convergence on ∂D, where g is a given continuous function.

Exterior Dirichlet Problem. *Find a function u that is harmonic in $\mathbb{R}^m \setminus \bar{D}$, is continuous in $\mathbb{R}^m \setminus D$, and satisfies the boundary condition*

$$u = f \quad \text{on } \partial D,$$

where f is a given continuous function. For $|x| \to \infty$ it is required that

$$u(x) = O(1), \quad m = 2, \quad \text{and} \quad u(x) = o(1), \quad m = 3,$$

uniformly for all directions.

Exterior Neumann Problem. *Find a function u that is harmonic in $\mathbb{R}^m \setminus \bar{D}$, is continuous in $\mathbb{R}^m \setminus D$, and satisfies the boundary condition*

$$\frac{\partial u}{\partial \nu} = g \quad \text{on } \partial D$$

in the sense of uniform convergence on ∂D, where g is a given continuous function. For $|x| \to \infty$ it is required that $u(x) = o(1)$ uniformly for all directions.

Note that for the exterior problems we impose that $u_\infty = 0$, with the exception of the Dirichlet problem in $\mathbb{R}^2$, where u is only required to be bounded.

These boundary value problems carry the names of Dirichlet, who made important contributions to potential theory, and Neumann, who gave the first rigorous existence proof (see Problem 6.5). From the numerous applications we mention:

(1) Determine the stationary temperature distribution in a heat-conducting body from the temperature on the boundary or from the heat flux through the boundary.
(2) Find the potential of the electrostatic field in the exterior of a perfect conductor.
(3) Find the velocity potential of an incompressible irrotational flow around an obstacle.

Our aim is to establish that each of the above potential theoretic boundary value problems has a unique solution depending continuously on the given boundary data, i.e., they are well-posed in the sense of Hadamard (see Section 15.1).

In our uniqueness proofs we need to apply Green's Theorem 6.3. Since for solutions to the boundary value problems we do not assume differentiability up to the boundary, we introduce the concept of *parallel surfaces*. These are described by

$$\partial D_h := \{z = x + h\nu(x) : x \in \partial D\},$$

with a real parameter h. Because ∂D is assumed to be of class C^2, we observe that ∂D_h is of class C^1. For $m = 3$, let $x(u) = (x_1(u), x_2(u), x_3(u))$, $u = (u_1, u_2)$, be a regular parametric representation of a surface patch of ∂D . Then straightforward differential geometric calculations show that the determinants

$$g(u) := \det\left[\frac{\partial x}{\partial u_i} \cdot \frac{\partial x}{\partial u_j}\right] \quad \text{and} \quad g(u;h) := \det\left[\frac{\partial z}{\partial u_i} \cdot \frac{\partial z}{\partial u_j}\right]$$

are related by

$$g(u;h) = g(u)\left\{1 - 2hH(u) + h^2K(u)\right\}^2,$$

where H and K denote the mean and Gaussian curvature of ∂D, respectively (see [162, 177]). This verifies that the parallel surfaces are well defined provided the parameter h is sufficiently small to ensure that $1 - 2hH + h^2K$ remains positive. This also ensures that in a sufficiently small neighborhood of ∂D each point z can be uniquely represented in the form $z = x + h\nu(x)$, where $x \in \partial D$ and $h \in \mathbb{R}$.

In particular, the surface elements ds on ∂D and ds_h on ∂D_h are related by

$$ds_h(z) = \left\{1 - 2hH + h^2K\right\} ds(x). \tag{6.21}$$

Since $\nu(x) \cdot \nu(x) = 1$, we have

$$\frac{\partial \nu(x)}{\partial u_i} \cdot \nu(x) = 0, \quad i = 1, 2,$$

for all $x \in \partial D$, and therefore the tangential vectors

$$\frac{\partial z}{\partial u_i} = \frac{\partial x}{\partial u_i} + h\,\frac{\partial \nu(x)}{\partial u_i}, \quad i = 1, 2,$$

for all (sufficiently small) h lie in the tangent plane to ∂D at the point x, i.e., the normal vector $\nu_h(z)$ of the parallel surface ∂D_h coincides with the normal vector $\nu(x)$ of ∂D for all $x \in \partial D$. Hence, in view of (6.21), Theorems 6.5 and 6.11 remain valid for harmonic functions $u \in C(\bar{D})$ and $u \in C(\mathbb{R}^m \setminus D)$, respectively, provided they have a normal derivative in the sense of uniform convergence.

Note that in two dimensions the equation (6.21) has to be replaced by $ds_h(z) = (1 - \kappa h)ds(x)$, where κ denotes the curvature of ∂D, i.e., for the representation $\partial D = \{x(s) : s_0 \le s \le s_1\}$ in terms of the arc length we have $\kappa = \nu \cdot x''$.

Theorem 6.12. *Both the interior and the exterior Dirichlet problems have at most one solution.*

Proof. The difference $u := u_1 - u_2$ of two solutions to the Dirichlet problem is a harmonic function that is continuous up to the boundary and satisfies the homogeneous boundary condition $u = 0$ on ∂D. Then, from the maximum-minimum principle of Corollary 6.10 we obtain $u = 0$ in D for the interior problem, and observing that $u(x) = o(1)$, $|x| \to \infty$, we also obtain $u = 0$ in $\mathbb{R}^3 \setminus D$ for the exterior problem in three dimensions.

For the exterior problem in two dimensions, by the maximum-minimum principle Theorem 6.9 the supremum and the infimum of the bounded harmonic function u are either attained on the boundary or equal to u_∞. When the maximum and minimum are both attained on the boundary then from the homogeneous boundary condition we immediately have $u = 0$ in $\mathbb{R}^m \setminus \bar{D}$. If the supremum is equal to u_∞, then from $u(x) \le u_\infty$ for all $x \in \mathbb{R}^2 \setminus D$ and the mean value property (6.14) we observe that $u = u_\infty$ in the exterior of some circle. Now we can apply the maximum principle to see that $u = u_\infty$ in all of $\mathbb{R}^2 \setminus D$ and the homogeneous boundary condition finally implies $u = 0$ in $\mathbb{R}^2 \setminus D$. The case where the infimum is equal to u_∞ is settled by the same argument. □

Theorem 6.13. *Two solutions of the interior Neumann problem can differ only by a constant. The exterior Neumann problem has at most one solution.*

Proof. The difference $u := u_1 - u_2$ of two solutions for the Neumann problem is a harmonic function continuous up to the boundary satisfying the homogeneous boundary condition $\partial u/\partial\nu = 0$ on ∂D in the sense of uniform convergence. For the interior problem, suppose that u is not constant in D. Then there exists some closed ball B contained in D such that $\int_B |\operatorname{grad} u|^2 dx > 0$. From Green's first theorem (6.1), applied to the interior D_h of some parallel surface $\partial D_h := \{x - h\nu(x) : x \in \partial D\}$ with sufficiently small $h > 0$, we derive

$$\int_B |\operatorname{grad} u|^2 dx \le \int_{D_h} |\operatorname{grad} u|^2 dx = \int_{\partial D_h} u \frac{\partial u}{\partial \nu}\, ds.$$

Passing to the limit $h \to 0$, we obtain the contradiction $\int_B |\operatorname{grad} u|^2 dx \le 0$. Hence, u must be constant.

For the exterior problem, assume that $\operatorname{grad} u \ne 0$ in $\mathbb{R}^m \setminus \bar{D}$. Then again, there exists some closed ball B contained in $\mathbb{R}^m \setminus \bar{D}$ such that $\int_B |\operatorname{grad} u|^2 dx > 0$. From Green's first theorem, applied to the domain $D_{h,r}$ between some parallel surface $\partial D_h := \{x + h\nu(x) : x \in \partial D\}$ with sufficiently small $h > 0$ and some sufficiently large sphere Ω_r of radius r centered at the origin (with interior normal ν), we obtain

$$\int_B |\operatorname{grad} u|^2 dx \le \int_{D_{h,r}} |\operatorname{grad} u|^2 dx = -\int_{\Omega_r} u \frac{\partial u}{\partial \nu}\, ds - \int_{\partial D_h} u \frac{\partial u}{\partial \nu}\, ds.$$

Letting $r \to \infty$ and $h \to 0$, with the aid of the asymptotics (6.20), we arrive at the contradiction $\int_B |\operatorname{grad} u|^2 dx \le 0$. Therefore, u is constant in $\mathbb{R}^m \setminus \bar{D}$ and the constant must be zero, since $u_\infty = 0$. □

From the proofs it is obvious that our uniqueness results remain valid under weaker regularity conditions on the boundary. Uniqueness for the Dirichlet problem via the maximum-minimum principle needs no regularity of the boundary, and uniqueness for the Neumann problem holds for those boundaries for which Green's integral theorem is valid. We have formulated the boundary value problems for C^2 boundaries, since we shall establish the existence of solutions under these conditions.

6.3 Surface Potentials

Definition 6.14. Given a function $\varphi \in C(\partial D)$, the functions

$$u(x) := \int_{\partial D} \varphi(y)\Phi(x, y)\, ds(y), \quad x \in \mathbb{R}^m \setminus \partial D, \tag{6.22}$$

and

$$v(x) := \int_{\partial D} \varphi(y) \frac{\partial \Phi(x, y)}{\partial \nu(y)}\, ds(y), \quad x \in \mathbb{R}^m \setminus \partial D, \tag{6.23}$$

are called, respectively, *single-layer* and *double-layer potential* with density φ. In two dimensions, occasionally, for obvious reasons we will call them *logarithmic single-layer* and *logarithmic double-layer potential.*

For fixed $y \in \mathbb{R}^m$ the fundamental solution $u = \Phi(\cdot\,, y)$ represents the potential of a unit *point source* located at the point y, i.e., $\operatorname{grad}_x \Phi(x, y)$ gives the force-field of this point source acting at the point x. The single-layer potential is obtained by distributing point sources on the boundary ∂D. For $h > 0$, by the mean value theorem we have

$$\Phi(x, y + h\nu(y)) - \Phi(x, y - h\nu(y)) = 2h\, \nu(y) \cdot \operatorname{grad} \Phi(x, y + \theta h\nu(y))$$

for some $\theta = \theta(y) \in [-1, 1]$. Therefore, the double-layer potential can be interpreted as the limit $h \to 0$ of the superposition of the single-layer potentials u_h and u_{-h} with densities $\varphi/2h$ on ∂D_h and $-\varphi/2h$ on ∂D_{-h}, respectively, i.e., the double-layer potential is obtained by distributing *dipoles* on the boundary ∂D.

Since for points $x \notin \partial D$ we can interchange differentiation and integration, the single- and double-layer potentials represent harmonic functions in D and $\mathbb{R}^m \setminus \bar{D}$. For the solution of the boundary value problems we need to investigate the behavior of the potentials at the boundary ∂D where the integrals become singular. The boundary behavior is expressed by the following so-called *jump relations.*

Theorem 6.15. *Let ∂D be of class C^2 and $\varphi \in C(\partial D)$. Then the single-layer potential u with density φ is continuous throughout $\mathbb{R}^m$. On the boundary we have*

$$u(x) = \int_{\partial D} \varphi(y)\Phi(x, y) ds(y), \quad x \in \partial D, \tag{6.24}$$

where the integral exists as an improper integral.

Proof. Analogous to the proofs of Theorems 2.29 and 2.30, by using the cut-off function h, it can be shown that the single-layer potential u is the uniform limit of a sequence of functions u_n that are continuous in $\mathbb{R}^m$. □

For the further analysis of the jump relations we need the following lemma. The inequality (6.25) expresses the fact that the vector $x - y$ for x close to y is almost orthogonal to the normal vector $\nu(x)$.

Lemma 6.16. *Let ∂D be of class C^2. Then there exists a positive constant L such that*

$$|\nu(x)\cdot\{x-y\}| \le L|x-y|^2 \tag{6.25}$$

and

$$|\nu(x)-\nu(y)| \le L|x-y| \tag{6.26}$$

for all $x, y \in \partial D$.

Proof. We confine ourselves to the two-dimensional case. For the three-dimensional case we refer to [31]. Let $\Gamma = \{x(s) : s \in [0, s_0]\}$ be a regular parameterization of a patch $\Gamma \subset \partial D$, i.e., $x : [0,1] \to \Gamma \subset \partial D$ is injective and twice continuously differentiable with $x'(s) \neq 0$ for all $s \in [0, s_0]$. Then, by Taylor's formula we have

$$|\nu(x(t))\cdot\{x(t)-x(\tau)\}| \le \frac{1}{2}\max_{0\le s\le s_0}|x''(s)|\,|t-\tau|^2,$$

$$|\nu(x(t))-\nu(x(\tau))| \le \max_{0\le s\le s_0}\left|\frac{d}{ds}\,\nu(x(s))\right|\,|t-\tau|,$$

$$|x(t))-x(\tau)| \ge \min_{0\le s\le s_0}|x'(s)|\,|t-\tau|.$$

The statement of the lemma is evident from this. □

Example 6.17. For the double-layer potential with constant density we have

$$2\int_{\partial D}\frac{\partial\Phi(x,y)}{\partial\nu(y)}\,ds(y) = \begin{cases} -2, & x \in D,\\ -1, & x \in \partial D,\\ 0, & x \in \mathbb{R}^m\setminus\bar{D}. \end{cases} \tag{6.27}$$

This follows for $x \in \mathbb{R}^m \setminus \bar{D}$ from (6.3) applied to $\Phi(x,\cdot)$ and for $x \in D$ from (6.4) applied to $u = 1$ in D. The result for $x \in \partial D$ is derived by excluding x from the integration by circumscribing it with a sphere $\Omega(x;r)$ of radius r and center x with the unit normal directed toward the center. Let $H(x;r) := \Omega(x;r)\cap D$. Then, by (6.3) applied to $\Phi(x,\cdot)$, we have

$$\int_{\{y\in\partial D:|y-x|\ge r\}}\frac{\partial\Phi(x,y)}{\partial\nu(y)}\,ds(y) + \int_{H(x;r)}\frac{\partial\Phi(x,y)}{\partial\nu(y)}\,ds(y) = 0,$$

and from

$$\lim_{r\to 0} 2\int_{H(x;r)}\frac{\partial\Phi(x,y)}{\partial\nu(y)}\,ds(y) = \lim_{r\to 0}\frac{2}{\omega_m r^{m-1}}\int_{H(x;r)}ds(y) = 1$$

the result follows. □

Theorem 6.18. *For ∂D of class C^2, the double-layer potential v with continuous density φ can be continuously extended from D to $\bar{D}$ and from $\mathbb{R}^m \setminus \bar{D}$ to $\mathbb{R}^m \setminus D$ with limiting values*

$$v_\pm(x) = \int_{\partial D} \varphi(y) \frac{\partial \Phi(x, y)}{\partial \nu(y)} \, ds(y) \pm \frac{1}{2} \varphi(x), \quad x \in \partial D, \tag{6.28}$$

where

$$v_\pm(x) := \lim_{h \to +0} v(x \pm h\nu(x))$$

and where the integral exists as an improper integral.

Proof. Because of Lemma 6.16 we have the estimate

$$\left| \frac{\partial \Phi(x, y)}{\partial \nu(y)} \right| = \frac{|\nu(y) \cdot \{x - y\}|}{\omega_m |x - y|^m} \leq \frac{L}{\omega_m |x - y|^{m-2}}, \quad x \neq y, \tag{6.29}$$

i.e., the integral in (6.28) has a weakly singular kernel. Therefore, by Theorem 2.30 the integral exists for $x \in \partial D$ as an improper integral and represents a continuous function on ∂D.

As pointed out on p. 85, in a sufficiently small neighborhood U of ∂D we can represent every $x \in U$ uniquely in the form $x = z + h\nu(z)$, where $z \in \partial D$ and $h \in [-h_0, h_0]$ for some $h_0 > 0$. Then we write the double-layer potential v with density φ in the form

$$v(x) = \varphi(z) w(x) + u(x), \quad x = z + h\nu(z) \in U \setminus \partial D,$$

where

$$w(x) := \int_{\partial D} \frac{\partial \Phi(x, y)}{\partial \nu(y)} \, ds(y)$$

and

$$u(x) := \int_{\partial D} \{\varphi(y) - \varphi(z)\} \frac{\partial \Phi(x, y)}{\partial \nu(y)} \, ds(y). \tag{6.30}$$

For $x \in \partial D$, i.e., for $h = 0$, the integral in (6.30) exists as an improper integral and represents a continuous function on ∂D. Therefore, in view of Example 6.17, to establish the theorem it suffices to show that

$$\lim_{h \to 0} u(z + h\nu(z)) = u(z), \quad z \in \partial D,$$

uniformly on ∂D.

From (6.25) we can conclude that

$$|x - y|^2 \geq \frac{1}{2} \left\{ |z - y|^2 + |x - z|^2 \right\}$$

for $x = z + h\nu(z)$ and $h \in [-h_0, h_0]$ provided that h_0 is sufficiently small. Therefore, writing

$$\frac{\partial \Phi(x,y)}{\partial \nu(y)} = \frac{\nu(y) \cdot \{z - y\}}{\omega_m |x-y|^m} + \frac{\nu(y) \cdot \{x - z\}}{\omega_m |x-y|^m},$$

and again using (6.25), we can estimate

$$\left| \frac{\partial \Phi(x,y)}{\partial \nu(y)} \right| \le C_1 \left\{ \frac{1}{|x-y|^{m-2}} + \frac{|x-z|}{[|z-y|^2 + |x-z|^2]^{m/2}} \right\}$$

for some constant $C_1 > 0$. Recalling the proof of Theorem 2.30 and denoting $\partial D(z;r) := \partial D \cap B[z;r]$, for sufficiently small r we project onto the tangent plane and deduce that

$$\begin{aligned} \int_{\partial D(z;r)} \left| \frac{\partial \Phi(x,y)}{\partial \nu(y)} \right| ds(y) &\le C_1 \left\{ \int_0^r d\rho + \int_0^r \frac{|x-z|\rho^{m-2}\, d\rho}{(\rho^2 + |x-z|^2)^{m/2}} \right\} \\ &\le C_1 \left\{ r + \int_0^\infty \frac{\lambda^{m-2}\, d\lambda}{(\lambda^2+1)^{m/2}} \right\}. \end{aligned} \tag{6.31}$$

From the mean value theorem we obtain that

$$\left| \frac{\partial \Phi(x,y)}{\partial \nu(y)} - \frac{\partial \Phi(z,y)}{\partial \nu(y)} \right| \le C_2 \frac{|x-z|}{|z-y|^m}$$

for some constant $C_2 > 0$ and $2|x-z| \le |z-y|$. Hence we can estimate

$$\int_{\partial D \setminus \partial D(z;r)} \left| \frac{\partial \Phi(x,y)}{\partial \nu(y)} - \frac{\partial \Phi(z,y)}{\partial \nu(y)} \right| ds(y) \le C_3 \frac{|x-z|}{r^m} \tag{6.32}$$

for some constant $C_3 > 0$ and $|x-z| \le r/2$. Now we can combine (6.31) and (6.32) to find that

$$|u(x) - u(z)| \le C \left\{ \max_{|y-z| \le r} |\varphi(y) - \varphi(z)| + \frac{|x-z|}{r^m} \right\}$$

for some constant $C > 0$, all sufficiently small r, and $|x-z| \le r/2$. Given $\varepsilon > 0$ we can choose $r > 0$ such that

$$\max_{|y-z| \le r} |\varphi(y) - \varphi(z)| \le \frac{\varepsilon}{2C}$$

for all $z \in \partial D$, since φ is uniformly continuous on ∂D. Then, taking $\delta < \varepsilon r^m / 2C$, we see that $|u(x) - u(z)| < \varepsilon$ for all $|x-z| < \delta$, and the proof is complete. □

Theorem 6.19. *Let ∂D be of class C^2. Then for the single-layer potential u with continuous density φ we have*

$$\frac{\partial u_\pm}{\partial \nu}(x) = \int_{\partial D} \varphi(y) \frac{\partial \Phi(x,y)}{\partial \nu(x)} ds(y) \mp \frac{1}{2} \varphi(x), \quad x \in \partial D, \tag{6.33}$$

where

$$\frac{\partial u_\pm}{\partial \nu}(x) := \lim_{h \to +0} \nu(x) \cdot \operatorname{grad} u(x \pm h\nu(x))$$

is to be understood in the sense of uniform convergence on ∂D and where the integral exists as an improper integral.

Proof. Let v denote the double-layer potential with density φ and let U be as in the proof of Theorem 6.18. Then for $x = z + h\nu(z) \in U \setminus \partial D$ we can write

$$\nu(z) \cdot \operatorname{grad} u(x) + v(x) = \int_{\partial D} \{\nu(y) - \nu(z)\} \cdot \operatorname{grad}_y \Phi(x, y)\, \varphi(y)\, ds(y),$$

where we have made use of $\operatorname{grad}_x \Phi(x, y) = -\operatorname{grad}_y \Phi(x, y)$. Using (6.26), analogous to the single-layer potential in Theorem 6.15, the right-hand side can be seen to be continuous in U. The proof is now completed by applying Theorem 6.18. □

Theorem 6.20. *Let ∂D be of class C^2. Then the double-layer potential v with continuous density φ satisfies*

$$\lim_{h \to +0} \nu(x) \cdot \{\operatorname{grad} v(x + h\nu(x)) - \operatorname{grad} v(x - h\nu(x))\} = 0 \tag{6.34}$$

uniformly for all $x \in \partial D$.

Proof. We omit the rather lengthy proof, which is similar in structure to the proof of Theorem 6.18. For a detailed proof we refer to [31]. □

6.4 Boundary Value Problems: Existence

Green's formula shows that each harmonic function can be represented as a combination of single- and double-layer potentials. For boundary value problems we try to find a solution in the form of one of these two potentials. To this end we introduce two integral operators $K, K' : C(\partial D) \to C(\partial D)$ by

$$(K\varphi)(x) := 2 \int_{\partial D} \varphi(y) \frac{\partial \Phi(x, y)}{\partial \nu(y)}\, ds(y), \quad x \in \partial D, \tag{6.35}$$

and

$$(K'\psi)(x) := 2 \int_{\partial D} \psi(y) \frac{\partial \Phi(x, y)}{\partial \nu(x)}\, ds(y), \quad x \in \partial D. \tag{6.36}$$

Because of (6.29) the integral operators K and K' have weakly singular kernels and therefore are compact by Theorem 2.30. Note that in two dimensions for C^2 boundaries the kernels of K and K' actually turn out to be continuous (see Problem 6.1). As seen by interchanging the order of integration, K and K' are adjoint with respect to the dual system $\langle C(\partial D), C(\partial D)\rangle$ defined by

$$\langle \varphi, \psi \rangle := \int_{\partial D} \varphi \psi\, ds, \quad \varphi, \psi \in C(\partial D).$$

Theorem 6.21. *The operators $I - K$ and $I - K'$ have trivial nullspaces*

$$N(I - K) = N(I - K') = \{0\}.$$

The nullspaces of the operators $I + K$ and $I + K'$ have dimension one and

$$N(I + K) = \text{span}\{1\}, \quad N(I + K') = \text{span}\{\psi_0\}$$

with

$$\int_{\partial D} \psi_0 \, ds = 1, \tag{6.37}$$

i.e., the Riesz number is one. The function ψ_0 is called the natural charge *on ∂D and the single-layer potential with density ψ_0 is constant in D.*

Proof. Let φ be a solution to the homogeneous equation $\varphi - K\varphi = 0$ and define a double-layer potential v by (6.23). Then by (6.28) we have $2v_- = K\varphi - \varphi = 0$ and from the uniqueness for the interior Dirichlet problem (Theorem 6.12) it follows that $v = 0$ in D. From (6.34) we see that $\partial v_+ / \partial \nu = 0$ on ∂D, and since $v(x) = o(1)$, $|x| \to \infty$, from the uniqueness for the exterior Neumann problem (Theorem 6.13) we find that $v = 0$ in $\mathbb{R}^m \setminus \bar{D}$. Hence, from (6.28) we deduce $\varphi = v_+ - v_- = 0$ on ∂D. Thus $N(I - K) = \{0\}$ and, by the Fredholm alternative, $N(I - K') = \{0\}$.

Now let φ be a solution to $\varphi + K\varphi = 0$ and again define v by (6.23). Then by (6.28) we have $2v_+ = K\varphi + \varphi = 0$ on ∂D. Since $v(x) = o(1)$, $|x| \to \infty$, from the uniqueness for the exterior Dirichlet problem it follows that $v = 0$ in $\mathbb{R}^m \setminus \bar{D}$. From (6.34) we see that $\partial v_- / \partial \nu = 0$ on ∂D and from the uniqueness for the interior Neumann problem we find that v is constant in D. Hence, from (6.28) we deduce that φ is constant on ∂D. Therefore, $N(I + K) \subset \text{span}\{1\}$, and since by (6.27) we have $1 + K1 = 0$, it follows that $N(I + K) = \text{span}\{1\}$.

By the Fredholm alternative, $N(I+K')$ also has dimension one. Hence $N(I+K') = \text{span}\{\psi_0\}$ with some function $\psi_0 \in C(\partial D)$ that does not vanish identically. Assume that $\langle 1, \psi_0 \rangle = 0$ and define a single-layer potential u with density ψ_0. Then by (6.24) and (6.33) we have

$$u_+ = u_-, \quad \frac{\partial u_-}{\partial \nu} = 0, \quad \text{and} \quad \frac{\partial u_+}{\partial \nu} = -\psi_0 \quad \text{on } \partial D \tag{6.38}$$

in the sense of uniform convergence. From $\partial u_- / \partial \nu = 0$ on ∂D, by the uniqueness for the interior Neumann problem (Theorem 6.13), we conclude that u is constant in D. Assume that u is not constant in $\mathbb{R}^m \setminus \bar{D}$. Then there exists a closed ball B contained in $\mathbb{R}^m \setminus \bar{D}$ such that $\int_B |\operatorname{grad} u|^2 dx > 0$. By Green's theorem (6.1), using the jump relations (6.38), the assumption $\langle 1, \psi_0 \rangle = 0$ and the fact that u_+ is constant on ∂D, we find

$$\begin{aligned} \int_B |\operatorname{grad} u|^2 dx &\leq - \int_{\Omega_r} u \frac{\partial u}{\partial \nu} \, ds - \int_{\partial D} u_+ \frac{\partial u_+}{\partial \nu} \, ds \\ &= - \int_{\Omega_r} u \frac{\partial u}{\partial \nu} \, ds + \int_{\partial D} u_+ \psi_0 \, ds = - \int_{\Omega_r} u \frac{\partial u}{\partial \nu} \, ds \end{aligned}$$

where Ω_r denotes a sphere with sufficiently large radius r centered at the origin (and interior normal ν). With the help of $\int_{\partial D} \psi_0 ds = 0$, using (6.18) and (6.19), it can be seen that u has the asymptotic behavior (6.20) with $u_\infty = 0$. Therefore, passing to the limit $r \to \infty$, we arrive at the contradiction $\int_B |\operatorname{grad} u|^2 dx \leq 0$. Hence, u is constant in $\mathbb{R}^m \setminus \bar{D}$ and from the jump relation (6.38) we derive the contradiction $\psi_0 = 0$. Therefore, we can normalize such that $\langle 1, \psi_0 \rangle = 1$. The statement on the Riesz number is a consequence of Problem 4.4. □

Theorem 6.22. *The double-layer potential*

$$u(x) = \int_{\partial D} \varphi(y) \frac{\partial \Phi(x, y)}{\partial \nu(y)} \, ds(y), \quad x \in D, \tag{6.39}$$

with continuous density φ is a solution of the interior Dirichlet problem provided that φ is a solution of the integral equation

$$\varphi(x) - 2 \int_{\partial D} \varphi(y) \frac{\partial \Phi(x, y)}{\partial \nu(y)} \, ds(y) = -2f(x), \quad x \in \partial D. \tag{6.40}$$

Proof. This follows from Theorem 6.18. □

Theorem 6.23. *The interior Dirichlet problem has a unique solution.*

Proof. The integral equation $\varphi - K\varphi = -2f$ of the interior Dirichlet problem is uniquely solvable by Theorem 3.4, since $N(I - K) = \{0\}$. □

From Theorem 6.15 we see that in order to obtain an integral equation of the second kind for the Dirichlet problem it is crucial to seek the solution in the form of a double-layer potential rather than a single-layer potential, which would lead to an integral equation of the first kind. Historically, this important observation goes back to Beer [16].

The double-layer potential approach (6.39) for the exterior Dirichlet problem leads to the integral equation $\varphi + K\varphi = 2f$ for the density φ. Since $N(I + K') = \operatorname{span}\{\psi_0\}$, by the Fredholm alternative, this equation is solvable if and only if $\langle f, \psi_0 \rangle = 0$. Of course, for arbitrary boundary data f we cannot expect this condition to be satisfied. Therefore we modify our approach as follows.

Theorem 6.24. *The modified double-layer potential*

$$u(x) = \int_{\partial D} \varphi(y) \left\{ \frac{\partial \Phi(x, y)}{\partial \nu(y)} + \frac{1}{|x|^{m-2}} \right\} ds(y), \quad x \in \mathbb{R}^m \setminus \bar{D}, \tag{6.41}$$

with continuous density φ is a solution to the exterior Dirichlet problem provided that φ is a solution of the integral equation

$$\varphi(x) + 2 \int_{\partial D} \varphi(y) \left\{ \frac{\partial \Phi(x, y)}{\partial \nu(y)} + \frac{1}{|x|^{m-2}} \right\} ds(y) = 2f(x), \quad x \in \partial D. \tag{6.42}$$

Here, we assume that the origin is contained in D.

Proof. This again follows from Theorem 6.18. Observe that u has the required behavior for $|x| \to \infty$, i.e., $u(x) = O(1)$ if $m = 2$ and $u(x) = o(1)$ if $m = 3$. □

Theorem 6.25. *The exterior Dirichlet problem has a unique solution.*

Proof. The integral operator $\widetilde{K} : C(\partial D) \to C(\partial D)$ defined by

$$(\widetilde{K}\varphi)(x) := 2\int_{\partial D} \varphi(y)\left\{\frac{\partial \Phi(x,y)}{\partial \nu(y)} + \frac{1}{|x|^{m-2}}\right\} ds(y), \quad x \in \partial D,$$

is compact, since the difference $\widetilde{K} - K$ has a continuous kernel. Let φ be a solution to the homogeneous equation $\varphi + \widetilde{K}\varphi = 0$ and define u by (6.41). Then $2u = \widetilde{K}\varphi + \varphi = 0$ on ∂D, and by the uniqueness for the exterior Dirichlet problem it follows that $u = 0$ in $\mathbb{R}^m \setminus \bar{D}$. Using (6.19), we deduce the asymptotic behavior

$$|x|^{m-2}u(x) = \int_{\partial D} \varphi\, ds + O\left(\frac{1}{|x|}\right), \quad |x| \to \infty,$$

uniformly for all directions. From this, since $u = 0$ in $\mathbb{R}^m \setminus \bar{D}$, we obtain $\int_{\partial D} \varphi\, ds = 0$. Therefore $\varphi + K\varphi = 0$, and from Theorem 6.21 we conclude that φ is constant on ∂D. Now $\int_{\partial D} \varphi\, ds = 0$ implies that $\varphi = 0$, and the existence of a unique solution to the integral equation (6.42) follows from Theorem 3.4. □

Theorem 6.26. *The single-layer potential*

$$u(x) = \int_{\partial D} \psi(y)\Phi(x,y)\, ds(y), \quad x \in D, \tag{6.43}$$

with continuous density ψ is a solution of the interior Neumann problem provided that ψ is a solution of the integral equation

$$\psi(x) + 2\int_{\partial D} \psi(y)\,\frac{\partial \Phi(x,y)}{\partial \nu(x)}\, ds(y) = 2g(x), \quad x \in \partial D. \tag{6.44}$$

Proof. This follows from Theorem 6.19. □

Theorem 6.27. *The interior Neumann problem is solvable if and only if*

$$\int_{\partial D} g\, ds = 0 \tag{6.45}$$

is satisfied.

Proof. The necessity of condition (6.45) is a consequence of Green's theorem (6.3) applied to a solution u. Its sufficiency follows from the fact that by Theorem 6.21 it coincides with the solvability condition of the Fredholm alternative for the inhomogeneous integral equation (6.44), i.e., for $\psi + K'\psi = 2g$. □

Theorem 6.28. *The single-layer potential*

$$u(x) = \int_{\partial D} \psi(y)\Phi(x, y)\, ds(y), \quad x \in \mathbb{R}^m \setminus \bar{D}, \tag{6.46}$$

with continuous density ψ is a solution of the exterior Neumann problem provided that ψ is a solution of the integral equation

$$\psi(x) - 2\int_{\partial D} \psi(y) \frac{\partial \Phi(x, y)}{\partial \nu(x)}\, ds(y) = -2g(x), \quad x \in \partial D, \tag{6.47}$$

and, if $m = 2$, also satisfies

$$\int_{\partial D} \psi\, ds = 0. \tag{6.48}$$

Proof. Again this follows from Theorem 6.19. Observe that for $m = 2$ the additional condition (6.48) ensures that u has the required behavior $u(x) = o(1)$, $|x| \to \infty$, as can be seen from (6.18). □

Theorem 6.29. *In $\mathbb{R}^3$ the exterior Neumann problem has a unique solution. In $\mathbb{R}^2$ the exterior Neumann problem is uniquely solvable if and only if*

$$\int_{\partial D} g\, ds = 0 \tag{6.49}$$

is satisfied.

Proof. By Theorems 3.4 and 6.21 the equation $\psi - K'\psi = -2g$ is uniquely solvable for each right-hand side g. If (6.49) is satisfied, using the fact that $1 + K1 = 0$, we find

$$2\langle 1, \psi\rangle = \langle 1 - K1, \psi\rangle = \langle 1, \psi - K'\psi\rangle = -2\langle 1, g\rangle = 0.$$

Hence, the additional property (6.48) is satisfied in $\mathbb{R}^2$. That condition (6.49) is necessary for the solvability in $\mathbb{R}^2$ follows from (6.13). □

We finally show that the solutions depend continuously on the given boundary data.

Theorem 6.30. *The solutions to the Dirichlet and Neumann problems depend continuously in the maximum norm on the given data.*

Proof. For the Dirichlet problem the assertion follows from the maximum-minimum principle (Theorem 6.9). In two dimensions, for the exterior problem, from the form (6.41) of the solution u we observe that we have to incorporate the value u_∞ at infinity through $\int_{\partial D} \varphi\, ds$. But this integral depends continuously on the given boundary data, since the inverse $(I + \widetilde{K})^{-1}$ of $I + \widetilde{K}$ is bounded by Theorem 3.4.

For the Neumann problem we first observe that for single-layer potentials u with continuous density ψ for any closed ball B in $\mathbb{R}^m$ we have an estimate of the form

$$\|u\|_{\infty,B} \le \|w\|_{\infty,B}\|\psi\|_{\infty,\partial D},$$

where the function

$$w(x) := \int_{\partial D} |\Phi(x, y)| \, ds(y), \quad x \in \mathbb{R}^m,$$

is continuous in $\mathbb{R}^m$ by Theorem 6.15. Then for the exterior problem choose a sufficiently large ball B and the continuous dependence of the solution on the boundary data in B follows from the boundedness of the inverse $(I - K')^{-1}$ of $I - K'$. In the remaining exterior of B, continuity then follows from the maximum-minimum principle.

For the interior problem we can expect continuity only after making the solution u unique by an additional condition, for example, by requiring that $\int_{\partial D} u \, ds = 0$. From $\langle 1, K'\psi \rangle = \langle K1, \psi \rangle = -\langle 1, \psi \rangle$ we observe that K' maps the closed subspace $C_0(\partial D) := \left\{ \psi \in C(\partial D) : \int_{\partial D} \psi \, ds = 0 \right\}$ into itself. By Theorem 6.21 the operator $I + K'$ has a trivial nullspace in $C_0(\partial D)$. Hence, the inverse $(I + K')^{-1}$ is bounded from $C_0(\partial D)$ onto $C_0(\partial D)$, i.e., the unique solution ψ_0 of $\psi_0 + K'\psi_0 = g$ satisfying $\int_{\partial D} \psi_0 \, ds = 0$ depends continuously on g. Therefore, as above, the corresponding single-layer potential u_0 depends continuously on g in the maximum norm. Finally, $u := u_0 - \int_{\partial D} u_0 \, ds / |\partial D|$ yields a solution vanishing in the integral mean on the boundary, and it depends continuously on g. □

6.5 Nonsmooth Boundaries

Despite the fact that the integral equation method of this chapter provides an elegant approach to constructively prove the existence of solutions for the boundary value problems of potential theory we do not want to disguise its major drawback: the relatively strong regularity assumption on the boundary to be of class C^2. It is possible to slightly weaken the regularity and allow *Lyapunov boundaries* instead of C^2 boundaries and still remain within the framework of compact operators. The boundary is said to satisfy a Lyapunov condition if at each point $x \in \partial D$ the normal vector ν exists and there are positive constants L and α such that for the angle $\vartheta(x, y)$ between the normal vectors at x and y the estimate $\vartheta(x, y) \leq L|x - y|^\alpha$ holds for all $x, y \in \partial D$. For the treatment of the Dirichlet and Neumann problem for Lyapunov boundaries, which does not differ essentially from that for C^2 boundaries, we refer to [169].

However, the situation changes considerably if the boundary is allowed to have edges and corners. This effects the form of the integral equations and the compactness of the integral operators as we will demonstrate by considering the interior Dirichlet problem in a two-dimensional domain D with corners. We assume that the boundary ∂D is piecewise twice differentiable, i.e., ∂D consists of a finite number of closed arcs $\Gamma_1, \dots, \Gamma_p$ that are all of class C^2 and that intersect only at the corners $x_1, \dots, x_p$. At the corners the normal vector is discontinuous (see Fig. 6.1 for a domain with three corners).

For simplicity, we restrict our analysis to boundaries that are straight lines in a neighborhood of each of the corners. In particular, this includes the case where ∂D is a polygon. The interior angle at the corner x_i we denote by γ_i and assume that $0 < \gamma_i < 2\pi$, $i = 1, \dots, p$, i.e., we exclude cusps. For a boundary with corners, the continuity of the double-layer potential with continuous density as stated in Theorem 6.18 remains valid, but at the corners the jump relation (6.28) has to be modified into the form

$$v_\pm(x_i) = \int_{\partial D} \varphi(y) \frac{\partial \Phi(x_i, y)}{\partial \nu(y)} \, ds(y) \pm \frac{1}{2} \, \delta_i^\mp \varphi(x_i), \quad i = 1, \dots, p, \tag{6.50}$$

where $\delta_i^+ = \gamma_i/\pi$ and $\delta_i^- = 2 - \gamma_i/\pi$. It is a matter of straightforward application of Green's theorem as in Example 6.17 to verify (6.50) for constant densities. For arbitrary continuous densities, the result can be obtained from the C^2 case of Theorem 6.18 by a superposition of two double-layer potentials on two C^2 curves intersecting at the corner with the density φ equal to zero on the parts of the two curves lying outside ∂D.

Trying to find the solution to the interior Dirichlet problem in the form of a double-layer potential with continuous density φ as in Theorem 6.22 reduces the boundary value problem to solving the integral equation $\varphi - \widetilde{K}\varphi = -2f$, where the operator $\widetilde{K} : C(\partial D) \to C(\partial D)$ is given by

$$(\widetilde{K}\varphi)(x) := \begin{cases} (K\varphi)(x), & x \neq x_i, \ i = 1, \dots, p, \\ (K\varphi)(x) + \left(\dfrac{\gamma_i}{\pi} - 1\right)\varphi(x_i), & x = x_i, \ i = 1, \dots, p. \end{cases}$$

Note that for $\varphi \in C(\partial D)$, in general, $K\varphi$ is not continuous at the corners. However $\widetilde{K}\varphi$ is continuous, since it is the sum $\widetilde{K}\varphi = v_+ + v_-$ of the continuous boundary values of the double-layer potential v.

By Problem 6.1 the kernel

$$k(x, y) := \frac{\nu(y) \cdot \{x - y\}}{\pi |x - y|^2}$$

of the integral operator K is continuous on $\Gamma_i \times \Gamma_i$ for $i = 1, \dots, p$. Singularities of the kernel occur when x and y approach a corner on the two different arcs intersecting at the corner.

For $n \in \mathbb{N}$ we use the continuous cutoff function h introduced in the proof of Theorem 2.29 to define the operators $K_n : C(\partial D) \to C(\partial D)$ by

$$(K_n\varphi)(x) := \int_{\partial D} h(n|x - y|) k(x, y) \varphi(y) \, ds(y), \quad x \in \partial D.$$

For each $n \in \mathbb{N}$ the operator K_n is compact, since its kernel is continuous on $\partial D \times \Gamma_i$ for $i = 1, \dots, p$, i.e., we can interpret K_n as the sum of p integral operators with continuous kernels on $\partial D \times \Gamma_i$ by subdividing the integral over ∂D into a sum of integrals over the arcs Γ_i for $i = 1, \dots, p$.

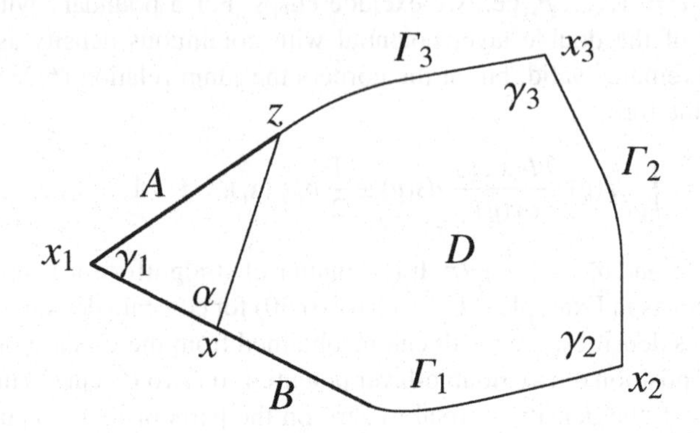

Fig. 6.1 Domain with corners

Now consider $\widetilde{K}_n := \widetilde{K} - K_n$ and assume that n is large enough that for each $x \in \partial D$ the disk $B[x; 1/n] = \{y \in \mathbb{R}^2 : |x - y| \leq 1/n\}$ intersects only either one or, in the vicinity of the corners, two of the arcs Γ_i. By our assumption on the nature of the corners we can assume n is large enough that in the second case the intersection consists of two straight lines A and B (see Fig. 6.1). Let

$$M := \max_{i=1,\dots,p} \max_{x,y \in \Gamma_i} |k(x, y)|.$$

Then, by projection onto the tangent line, for the first case we can estimate

$$|(\widetilde{K}_n \varphi)(x)| \leq M \|\varphi\|_\infty \int_{\partial D \cap B[x;1/n]} ds(y) \leq M \|\varphi\|_\infty \frac{4}{n}.$$

In the second case, we first note that for $x \in B \setminus \{x_i\}$, by Green's theorem (6.3) applied in the triangle with the corners at x and at the endpoints x_i and z of A we have

$$\int_A \left| \frac{\partial \Phi(x, y)}{\partial \nu(y)} \right| ds(y) = \left| \int_A \frac{\partial \Phi(x, y)}{\partial \nu(y)} \, ds(y) \right| = \frac{\alpha(x)}{2\pi},$$

where $\alpha(x)$ denotes the angle of this triangle at the corner x (see Fig. 6.1). Elementary triangle geometry shows that $\alpha(x) + \gamma_i \leq \pi$, where, without loss of generality, we have assumed that $\gamma_i < \pi$. Therefore, since for $x \in B \setminus \{x_i\}$ we have $k(x, y) = 0$ for all $y \in B \setminus \{x_i\}$, we obtain

$$|(\widetilde{K}_n \varphi)(x)| \leq 2 \|\varphi\|_\infty \left| \int_A \frac{\partial \Phi(x, y)}{\partial \nu(y)} \, ds(y) \right| \leq \frac{\alpha(x)}{\pi} \|\varphi\|_\infty \leq \left(1 - \frac{\gamma_i}{\pi}\right) \|\varphi\|_\infty.$$

Finally, for the corner x_i at the intersection of A and B we have

$$(\widetilde{K}_n\varphi)(x_i) = \left(\frac{\gamma_i}{\pi} - 1\right)\varphi(x_i),$$

since $k(x_i, y) = 0$ for all $y \in (A \cup B) \setminus \{x_i\}$. Combining these results we observe that we can choose n large enough that $\|\widetilde{K}_n\|_\infty \leq q$ where

$$q := \max_{i=1,\ldots,p} \left|1 - \frac{\gamma_i}{\pi}\right| < 1.$$

Hence, we have a decomposition $I - \widetilde{K} = I - \widetilde{K}_n - K_n$, where $I - \widetilde{K}_n$ has a bounded inverse by the Neumann series Theorem 2.14 and where K_n is compact. It is left to the reader to carry over the proof for injectivity of the operator $I - \widetilde{K}$ from Theorem 6.21 to the case of a boundary with corners. (For the application of Green's integral theorem analogous to the proof of Theorem 6.13 one needs to apply the dominated convergence theorem for Lebesgue integration; for a similar situation see [32, Lemma 3.10].) Then existence of a solution to the inhomogeneous equation $\varphi - \widetilde{K}\varphi = -2f$ follows by Corollary 3.6.

This idea of decomposing the integral operator into a compact operator and a bounded operator with norm less than one reflecting the behavior at the corners goes back to Radon [201] and can be extended to the general two-dimensional case and to three dimensions. For details we refer to Cryer [37], Král [131], and Wendland [246]. For a more comprehensive study of boundary value problems in domains with corners we refer to Grisvard [73]. For integral equations of the second kind in Lipschitz domains we refer to Verchota [241].

Finally, we wish to mention that the integral equations for the Dirichlet and Neumann problems can also be treated in the space $L^2(\partial D)$ allowing boundary data in $L^2(\partial D)$. This requires the boundary conditions to be understood in a weak sense, which we want to illustrate by again considering the interior Dirichlet problem. We say that a harmonic function u in D assumes the boundary values $f \in L^2(\partial D)$ in the L^2 sense if

$$\lim_{h\to+0} \int_{\partial D} [u(x - h\nu(x)) - f(x)]^2 ds(x) = 0.$$

To establish uniqueness under this weaker boundary condition, we choose parallel surfaces $\partial D_h := \{x - h\nu(x) : x \in \partial D\}$ to ∂D with $h > 0$ sufficiently small. Then, following Miranda [171], for

$$J(h) := \int_{\partial D_h} u^2 ds, \quad h > 0,$$

we can write

$$J(h) = \int_{\partial D} \left\{1 + 2hH(x) + h^2K(x)\right\} [u(x - h\nu(x))]^2 ds(x)$$

and differentiate to obtain

$$\frac{1}{2}\frac{dJ}{dh} = -\int_{\partial D_h} u\,\frac{\partial u}{\partial \nu}\,ds + \int_{\partial D} \{H(x) + hK(x)\}\,[u(x - h\nu(x))]^2 ds(x).$$

Hence, using Green's theorem (6.1), we have

$$\frac{1}{2}\frac{dJ}{dh} = -\int_{D_h} |\operatorname{grad} u|^2 dx + \int_{\partial D} \{H + hK\}\,[u(\cdot - h\nu)]^2 ds, \tag{6.51}$$

where D_h denotes the interior of the parallel surface ∂D_h. Now let u vanish on the boundary ∂D in the L^2 sense and assume that $\operatorname{grad} u \neq 0$ in D. Then there exists some closed ball B contained in D such that $I := \int_B |\operatorname{grad} u|^2 dx > 0$, and from (6.51) we deduce that $dJ/dh \leq -I$ for all $0 < h \leq h_0$ and some sufficiently small $h_0 > 0$. Since J is continuous on $[0, h_0]$, is continuously differentiable on $(0, h_0]$, and satisfies $J(0) = 0$, we see that $J(h) \leq -Ih$ for all $0 < h \leq h_0$. This is a contradiction to $J(h) \geq 0$ for all $h > 0$. Therefore u must be constant in D, and from $J(0) = 0$ we obtain $u = 0$ in D.

Using the fact that, due to Theorem 6.6, on the parallel surfaces ∂D_h for $h > 0$ there is more regularity of u, a different approach to establishing uniqueness under weaker assumptions was suggested by Calderón [23]. It is based on representing u in terms of the double-layer operator K_h on ∂D_h and establishing $\|K_h - K\|_{L^2(\partial D)} \to 0$ as $h \to 0$.

To prove existence of a solution for boundary conditions in the L^2 sense via the surface potential approach, it is necessary to extend the jump relations of Theorem 6.15, 6.18, 6.19, and 6.20 from $C(\partial D)$ onto $L^2(\partial D)$. This can be achieved quite elegantly through the use of Lax's Theorem 4.13 as worked out by Kersten [120]. In particular, for the double-layer potential v with density $\varphi \in L^2(\partial D)$, the jump relation (6.28) has to be replaced by

$$\lim_{h\to+0} \int_{\partial D} [2v(x \pm h\nu(x)) - (K\varphi)(x) \mp \varphi(x)]^2 ds(x) = 0. \tag{6.52}$$

From this, we see that the double-layer potential with density $\varphi \in L^2(\partial D)$ solves the Dirichlet problem with boundary values $f \in L^2(\partial D)$ provided the density solves the integral equation $\varphi - K\varphi = -2f$ in the space $L^2(\partial D)$. Noting that integral operators with weakly singular kernels are compact from $L^2(\partial D)$ into $L^2(\partial D)$ (see Problem 4.5), for existence and uniqueness of a solution to this integral equation we need to establish that the homogeneous equation admits only the trivial solution. From Theorem 6.21 we know that the operator $I - K$ has a trivial nullspace in $C(\partial D)$. Therefore we can apply Theorem 4.20 to obtain that $I - K$ also has a trivial nullspace in $L^2(\partial D)$. Corresponding to (6.52), for later use we also note the jump relation

$$\lim_{h\to+0} \int_{\partial D} \left[2\,\frac{\partial u}{\partial \nu}\,(x \pm h\nu(x)) - (K'\varphi)(x) \pm \varphi(x)\right]^2 ds(x) = 0 \tag{6.53}$$

for the single-layer potential u with density $\varphi \in L^2(\partial D)$.

We will come back to the potential theoretic boundary value problems in the next two chapters. In Section 7.5 we will solve the interior Dirichlet and Neumann problems in two dimensions by integral equations of the first kind in a Hölder space setting. Then in Section 8.3, again in two dimensions, we will solve the integral equations of the first and second kind in Sobolev spaces leading to weak solutions of the boundary value problems.

For integral equation methods for boundary value problems for the Helmholtz equation $\Delta u + \kappa^2 u = 0$, i.e., for acoustic and electromagnetic scattering problems, we refer to [31, 32].

Problems

6.1. Use a regular 2π-periodic parameterization $\partial D = \{x(t) = (x_1(t), x_2(t)) : 0 \le t \le 2\pi\}$ with counterclockwise orientation for the boundary curve to transform the integral equation (6.40) of the interior two-dimensional Dirichlet problem into the form

$$\widetilde{\varphi}(t) - \int_0^{2\pi} k(t,\tau)\widetilde{\varphi}(\tau)\,d\tau = -2\widetilde{f}(t), \quad 0 \le t \le 2\pi,$$

where $\widetilde{\varphi}(t) := \varphi(x(t))$, $\widetilde{f}(t) := f(x(t))$ and the kernel is given by

$$k(t,\tau) = \begin{cases} \dfrac{1}{\pi}\,\dfrac{[x'(\tau)]^\perp \cdot \{x(t) - x(\tau)\}}{|x(t) - x(\tau)|^2}, & t \ne \tau, \\[2ex] \dfrac{1}{2\pi}\,\dfrac{[x'(t)]^\perp \cdot x''(t)}{|x'(t)|^2}, & t = \tau, \end{cases}$$

where $[x']^\perp := (x_2', -x_1')$. Show that this kernel is continuous provided ∂D is of class C^2.

6.2. Show that for an ellipse with parametric representation $x(t) = (a\cos t, b\sin t)$, the kernel k of Problem 6.1 is given by

$$k(t,\tau) = -\frac{ab}{\pi}\,\frac{1}{a^2 + b^2 - (a^2 - b^2)\cos(t + \tau)}.$$

6.3. Extend Theorem 6.21 to domains D with non-connected boundaries and, in particular, show that $\dim N(I - K) = p$, where p denotes the number of bounded components of $\mathbb{R}^m \setminus \bar{D}$. For the interior Dirichlet problem in $\mathbb{R}^3$ establish existence of a solution through a modification of the integral equation (6.40) analogous to (6.41) by adding a point source in each of the bounded components of $\mathbb{R}^3 \setminus \bar{D}$. (For the two-dimensional case we refer to Problem 18.1.)

6.4. Let $D \subset \mathbb{R}^2$ be of class C^2 and strictly convex in the sense that the curvature of the boundary ∂D is strictly positive. Show that there exists a constant $0 < \delta < 1$ such that

$$\int_{\partial D} \left| \frac{\partial \Phi(x_1, y)}{\partial \nu(y)} - \frac{\partial \Phi(x_2, y)}{\partial \nu(y)} \right| ds(y) \le 1 - \delta$$

for all $x_1, x_2 \in \partial D$.
Hint: Use Example 6.17, Problem 6.1, and the property that

$$\frac{\partial \Phi(x, y)}{\partial \nu(y)} = \frac{\nu(y) \cdot \{x - y\}}{2\pi |x - y|^2}$$

is negative on $\partial D \times \partial D$ to verify that

$$\left| \int_\Gamma \left\{ \frac{\partial \Phi(x_1, y)}{\partial \nu(y)} - \frac{\partial \Phi(x_2, y)}{\partial \nu(y)} \right\} ds(y) \right| \le \frac{1}{2} - a|\partial D|,$$

for each Jordan measurable subset $\Gamma \subset \partial D$, where

$$a := \min_{x,y \in \partial D} \left| \frac{\partial \Phi(x, y)}{\partial \nu(y)} \right| > 0.$$

6.5. In 1870 Neumann [184] gave the first rigorous proof for the existence of a solution to the two-dimensional interior Dirichlet problem in a strictly convex domain of class C^2. By completely elementary means he established that the successive approximations

$$\varphi_{n+1} := \frac{1}{2}\,\varphi_n + \frac{1}{2}\,K\varphi_n - f, \quad n = 0, 1, 2, \ldots,$$

with arbitrary $\varphi_0 \in C(\partial D)$ converge uniformly to the unique solution φ of the integral equation $\varphi - K\varphi = -2f$. In functional analytic terms his proof amounted to showing that the operator L given by $L := \frac{1}{2}(I + K)$ is a contraction with respect to the norm

$$\|\varphi\| := |\sup_{z \in \partial D} \varphi(z) - \inf_{z \in \partial D} \varphi(z)| + \alpha \sup_{z \in \partial D} |\varphi(z)|,$$

where $\alpha > 0$ is appropriately chosen. This norm is equivalent to the maximum norm. Derive the above results for yourself.
Hint: Use Problem 6.4 to show that $\|L\| \le (2 - \delta + \alpha)/2$ by writing

$$(L\varphi)(x) = \int_{\partial D} [\varphi(y) - \varphi(x)]\, \frac{\partial \Phi(x, y)}{\partial \nu(y)}\, ds(y)$$

and

$$\begin{aligned}(L\varphi)(x_1) - (L\varphi)(x_2) &= \frac{1}{2}\,[\varphi(x_1) - \varphi(x)] - \frac{1}{2}\,[\varphi(x_2) - \varphi(x)] \\ &\quad + \int_{\partial D} [\varphi(y) - \varphi(x)] \left\{ \frac{\partial \Phi(x_1, y)}{\partial \nu(y)} - \frac{\partial \Phi(x_2, y)}{\partial \nu(y)} \right\} ds(y),\end{aligned}$$

where $x \in \partial D$ is chosen such that $\varphi(x) = \left\{\sup_{z \in \partial D} \varphi(z) + \inf_{z \in \partial D} \varphi(z)\right\}/2$ (see [117]).

Chapter 7
Singular Boundary Integral Equations

In this chapter we will consider one-dimensional singular integral equations involving Cauchy principal values that arise from boundary value problems for holomorphic functions in the classical Hölder space setting. The investigations of these integral equations with Cauchy kernels by Gakhov, Muskhelishvili, Vekua, and others have had a great impact on the further development of the general theory of singular integral equations. For our introduction to integral equations they will provide an application of the general idea of regularizing singular operators as described in Chapter 5. We assume the reader is acquainted with basic complex analysis.

For a comprehensive study of singular integral equations with Cauchy kernels we refer to Gakhov [59], Meister [167], Mikhlin and Prössdorf [170], Muskhelishvili [180], Prössdorf [197], Vekua [240], and Wendland [248].

A central piece of our analysis will be provided by the Sokhotski–Plemelj jump and regularity relations for the Cauchy integral. These will also be used at the end of the chapter to derive further regularity properties for single- and double-layer potentials in Hölder spaces and an existence analysis for the Dirichlet and Neumann problem for the Laplace equation via integral equations of the first kind.

7.1 Hölder Continuity

For our investigation of singular integrals we first need to introduce the concept of Hölder continuity.

Definition 7.1. A real- or complex- or vector-valued function φ defined on a set $G \subset \mathbb{R}^m$ is called *uniformly Hölder continuous* with *Hölder exponent* $0 < \alpha \le 1$ if there exists a constant C such that

$$|\varphi(x) - \varphi(y)| \le C|x - y|^\alpha$$

R. Kress, *Linear Integral Equations*, Applied Mathematical Sciences 82,
DOI 10.1007/978-1-4614-9593-2_7,

for all $x, y \in G$. Here, for a vector-valued function on the left-hand side the bar indicates the Euclidean norm (as on the right-hand side). By $C^{0,\alpha}(G)$ we denote the linear space of all functions defined on G that are bounded and uniformly Hölder continuous with exponent α. The space $C^{0,\alpha}(G)$ is called a *Hölder space*.

Note that each uniformly Hölder continuous function is uniformly continuous, but the converse is not true. We illustrate by two examples that uniformly Hölder continuous functions live between continuous and differentiable functions. The function $\varphi : [0, 1/2] \to \mathbb{R}$, given by $\varphi(x) = 1/\ln x$ for $x \in (0, 1/2]$ and $\varphi(0) = 0$, is uniformly continuous but not uniformly Hölder continuous. The function $\psi : [0, 1] \to \mathbb{R}$, given by $\psi(x) = \sqrt{x}$, is uniformly Hölder continuous with exponent $1/2$ but not continuously differentiable on $[0, 1]$. In general, by the mean value theorem, a continuously differentiable function on a convex set with bounded derivatives is uniformly Hölder continuous with exponent 1.

Theorem 7.2. *The Hölder space $C^{0,\alpha}(G)$ is a Banach space with the norm*

$$\|\varphi\|_\alpha := \sup_{x\in G} |\varphi(x)| + \sup_{\substack{x,y\in G\\ x\neq y}} \frac{|\varphi(x) - \varphi(y)|}{|x - y|^\alpha}.$$

Proof. It is clear that

$$|\varphi|_\alpha := \sup_{\substack{x,y\in G\\ x\neq y}} \frac{|\varphi(x) - \varphi(y)|}{|x - y|^\alpha}$$

defines a semi-norm on $C^{0,\alpha}(G)$, i.e., it satisfies all norm axioms with the exception of the definiteness (N2). Then $\|\cdot\|_\alpha = \|\cdot\|_\infty + |\cdot|_\alpha$ is a norm, since $\|\varphi\|_\infty := \sup_{x\in G} |\varphi(x)|$ defines a norm. Convergence in the supremum norm $\|\cdot\|_\infty$ is equivalent to uniform convergence on G. If G is compact, the supremum norm and the maximum norm on $C(G)$ coincide.

It remains to show that $C^{0,\alpha}(G)$ is complete. Let (φ_n) be a Cauchy sequence in $C^{0,\alpha}(G)$. Then obviously (φ_n) also is a Cauchy sequence with respect to the supremum norm and, by the sufficiency of the Cauchy criterion for uniform convergence, there exists a function $\varphi \in C(G)$ such that $\|\varphi_n - \varphi\|_\infty \to 0$, $n \to \infty$. Because (φ_n) is a Cauchy sequence in $C^{0,\alpha}(G)$, given $\varepsilon > 0$, there exists $N(\varepsilon) \in \mathbb{N}$ such that $|\varphi_n - \varphi_k|_\alpha < \varepsilon$ for all $n, k \geq N(\varepsilon)$, i.e.,

$$|\{\varphi_n(x) - \varphi_k(x)\} - \{\varphi_n(y) - \varphi_k(y)\}| < \varepsilon |x - y|^\alpha$$

for all $n, k \geq N(\varepsilon)$ and all $x, y \in G$. Since $\varphi_n \to \varphi$, $n \to \infty$, uniformly on G, by letting $k \to \infty$ we have

$$|\{\varphi_n(x) - \varphi(x)\} - \{\varphi_n(y) - \varphi(y)\}| \leq \varepsilon |x - y|^\alpha$$

for all $n \geq N(\varepsilon)$ and all $x, y \in G$. From this we conclude that $\varphi \in C^{0,\alpha}(G)$ and $|\varphi_n - \varphi|_\alpha \leq \varepsilon$ for all $n \geq N(\varepsilon)$, which implies $\|\varphi_n - \varphi\|_\alpha \to 0$, $n \to \infty$. □

Note that the product of two uniformly Hölder continuous functions φ and ψ is again uniformly Hölder continuous with

$$\|\varphi\psi\|_\alpha \leq \|\varphi\|_\infty\|\psi\|_\infty + \|\varphi\|_\infty|\psi|_\alpha + \|\psi\|_\infty|\varphi|_\alpha \leq \|\varphi\|_\alpha\|\psi\|_\alpha.$$

By the following technical lemma we illustrate that Hölder continuity is a local property.

Lemma 7.3. *Assume that the function φ satisfies $|\varphi(x)| \leq M$ for all $x \in G$ and*

$$|\varphi(x) - \varphi(y)| \leq C|x - y|^\alpha$$

for all $x, y \in G$ with $|x - y| \leq a$ and some constants a, C, M, and $0 < \alpha \leq 1$. Then $\varphi \in C^{0,\alpha}(G)$ with

$$\|\varphi\|_\alpha \leq M + \max\left(C, \frac{2M}{a^\alpha}\right).$$

Proof. If $|x - y| > a$ then

$$|\varphi(x) - \varphi(y)| \leq 2M \leq 2M\left(\frac{|x - y|}{a}\right)^\alpha,$$

whence $|\varphi|_\alpha \leq \max(C, 2M/a^\alpha)$ follows. □

In particular, from Lemma 7.3, we see that for $\alpha < \beta$ each function $\varphi \in C^{0,\beta}(G)$ is also contained in $C^{0,\alpha}(G)$. For this imbedding we have the following compactness property.

Theorem 7.4. *Let $0 < \alpha < \beta \leq 1$ and let G be compact. Then the imbedding operators*

$$I^\beta : C^{0,\beta}(G) \to C(G)$$

and

$$I^{\alpha,\beta} : C^{0,\beta}(G) \to C^{0,\alpha}(G)$$

are compact.

Proof. Let U be a bounded set in $C^{0,\beta}(G)$, i.e., $\|\varphi\|_\beta \leq C$ for all $\varphi \in U$ and some positive constant C. Then we have

$$|\varphi(x)| \leq C$$

for all $x \in G$ and

$$|\varphi(x) - \varphi(y)| \leq C|x - y|^\beta \tag{7.1}$$

for all $x, y \in G$ and $\varphi \in U$, i.e., U is bounded and equicontinuous. Therefore, by the Arzelà–Ascoli Theorem 1.18, the set U is relatively compact in $C(G)$, which implies that the imbedding operator $I^\beta : C^{0,\beta}(G) \to C(G)$ is compact.

It remains to verify that U is relatively compact in $C^{0,\alpha}(G)$. From (7.1) we deduce that

$$\begin{aligned}
&|\{\varphi(x) - \psi(x)\} - \{\varphi(y) - \psi(y)\}| \\
&= |\{\varphi(x) - \psi(x)\} - \{\varphi(y) - \psi(y)\}|^{\alpha/\beta} |\{\varphi(x) - \psi(x)\} - \{\varphi(y) - \psi(y)\}|^{1-\alpha/\beta} \\
&\leq (2C)^{\alpha/\beta} |x - y|^{\alpha} (2\|\varphi - \psi\|_\infty)^{1-\alpha/\beta}
\end{aligned}$$

for all $\varphi, \psi \in U$ and all $x, y \in G$. Therefore

$$|\varphi - \psi|_\alpha \leq (2C)^{\alpha/\beta} 2^{1-\alpha/\beta} \|\varphi - \psi\|_\infty^{1-\alpha/\beta}$$

for all $\varphi, \psi \in U$, and from this we can conclude that each sequence taken from U and converging in $C(G)$ also converges in $C^{0,\alpha}(G)$. This completes the proof. □

The following theorem illustrates how the above tools can be used to establish compactness of boundary integral operators in the Hölder space setting and is adopted from [31].

Theorem 7.5. *For $0 < \alpha < 1$ and $\partial D \in C^2$, the boundary integral operators K and K' from potential theory as defined by (6.35) and (6.36) are compact operators from $C^{0,\alpha}(\partial D)$ into itself.*

Proof. We abbreviate

$$k(x, y) := 2\,\frac{\partial \Phi(x, y)}{\partial \nu(y)}, \quad x, y \in \partial D,\ x \neq y,$$

and from Lemma 6.16 we observe that there exists a constant M_1 depending on ∂D such that

$$|k(x, y)| \leq \frac{M_1}{|x - y|^{m-2}}, \quad x \neq y. \tag{7.2}$$

Using the mean value theorem and the estimates (6.25) and (6.26), from the decompositions

$$\frac{\omega_m}{2}\,\{k(x_1, y) - k(x_2, y)\} = \left\{ \frac{1}{|x_1 - y|^m} - \frac{1}{|x_2 - y|^m} \right\} \nu(y) \cdot \{x_2 - y\} + \frac{\nu(y) \cdot \{x_1 - x_2\}}{|x_1 - y|^m}$$

and

$$\nu(y) \cdot \{x_1 - x_2\} = \nu(x_1) \cdot \{x_1 - x_2\} + \{\nu(y) - \nu(x_1)\} \cdot \{x_1 - x_2\}$$

we find that

$$|k(x_1, y) - k(x_2, y)| \leq M_2 \left\{ \frac{|x_1 - x_2|}{|x_1 - y|^{m-1}} + \frac{|x_1 - x_2|^2}{|x_1 - y|^m} \right\} \tag{7.3}$$

for all $x_1, x_2, y \in \partial D$ with $2|x_1 - x_2| \leq |x_1 - y|$ and some constant M_2 depending on ∂D.

As in the proof of Theorem 2.30 we choose $0 < R < 1$ such that $\partial D(x; R) := \partial D \cap B[x; R]$ can be projected bijectively onto the tangent plane at x for each $x \in \partial D$. Now let $x_1, x_2 \in \partial D$ be such that $0 < |x_1 - x_2| < R/4$ and for

$$r := 4|x_1 - x_2|$$

set $\partial D(x; r) := \partial D \cap B[x; r]$. Then, using (7.2), with the aid of $\partial D(x_1, r) \subset \partial D(x_2, 2r)$ we find

$$\left| \int_{\partial D(x_1, r)} \{k(x_1, y) - k(x_2, y)\}\varphi(y)\, ds(y) \right|$$

$$\leq M_1 \|\varphi\|_\infty \left\{ \int_{\partial D(x_1, r)} \frac{ds(y)}{|x_1 - y|^{m-2}} + \int_{\partial D(x_2, 2r)} \frac{ds(y)}{|x_2 - y|^{m-2}} \right\}$$

$$\leq C_1 \|\varphi\|_\infty |x_1 - x_2|$$

for all $\varphi \in C(\partial D)$ and some constant C_1 depending on ∂D. From (7.3) we obtain

$$\left| \int_{\partial D(x_1, R) \setminus D(x_1, r)} \{k(x_1, y) - k(x_2, y)\}\varphi(y)\, ds(y) \right|$$

$$\leq M_2 \|\varphi\|_\infty \left\{ |x_1 - x_2| \int_{r/4}^{R} \frac{d\rho}{\rho} + |x_1 - x_2|^2 \int_{r/4}^{R} \frac{d\rho}{\rho^2} \right\}$$

$$\leq C_2 \|\varphi\|_\infty |x_1 - x_2|^\alpha$$

for all $\varphi \in C(\partial D)$ and some constant C_2 depending on ∂D and α. Here we have used that

$$|x_1 - x_2| \ln \frac{1}{|x_1 - x_2|} \leq \frac{1}{1 - \alpha}\, |x_1 - x_2|^\alpha$$

for $|x_1 - x_2| < 1$. Finally, making again use of (7.3) we can estimate

$$\left| \int_{\partial D \setminus D(x_1, R)} \{k(x_1, y) - k(x_2, y)\}\varphi(y)\, ds(y) \right|$$

$$\leq M_2 \|\varphi\|_\infty \left\{ |x_1 - x_2| \int_{\partial D \setminus D(x_1, R)} \frac{ds(y)}{|x_1 - y|^{m-1}} + |x_1 - x_2|^2 \int_{\partial D \setminus D(x_1, R)} \frac{ds(y)}{|x_1 - y|^{m}} \right\}$$

$$\leq C_3 \|\varphi\|_\infty |x_1 - x_2|$$

for all $\varphi \in C(\partial D)$ and some constant C_3 depending on ∂D. Summing up the last three inequalities we obtain that

$$|(K\varphi)(x_1) - (K\varphi)(x_2)| \leq C_4 \|\varphi\|_\infty |x_1 - x_2|^\alpha$$

for all $x_1 . x_2 \in \partial D$ with $|x_1 - x_2| < R/4$ and some constant C_4 depending on ∂D and α. Now Lemma 7.3 together with the boundedness of $K : C(\partial D) \to C(\partial D)$

implies that $\|K\varphi\|_\alpha \le C\|\varphi\|_\infty$ for all $\varphi \in C(\partial D)$ and some constant C. Hence, $K : C(\partial D) \to C^{0,\alpha}(\partial D)$ is bounded and the statement of the theorem follows from Theorems 2.21 and 7.4. The proof for the operator K' is analogous. □

Since the Definition 7.1 of Hölder continuity also covers vector fields, for a set $G \subset \mathbb{R}^m$, we can introduce the Hölder space $C^{1,\alpha}(G)$ of uniformly Hölder continuously differentiable functions as the space of differentiable functions φ for which $\operatorname{grad}\varphi$ (or the surface gradient $\operatorname{Grad}\varphi$ in the case $G = \partial D$) belongs to $C^{0,\alpha}(G)$. With the norm

$$\|\varphi\|_{1,\alpha} := \|\varphi\|_\infty + \|\operatorname{grad}\varphi\|_{0,\alpha}$$

the Hölder space $C^{1,\alpha}(G)$ is again a Banach space and we also have an imbedding theorem corresponding to Theorem 7.4.

Theorem 7.6. *For $0 < \alpha < 1$, the boundary integral operator K defined by (6.35) is a compact operator from $C^{1,\alpha}(\partial D)$ into itself.*

Proof. This can be proven with techniques similar to those used in the proof of the previous Theorem 7.5 (see [31]). □

Note, that for C^2 boundaries ∂D we cannot expect Theorem 7.6 to also hold for the operator K' since the derivative of $K'\varphi$ contains the derivative of the normal vector ν which is only continuous.

7.2 The Cauchy Integral Operator

Now let D be a bounded and simply connected domain in the complex plane. We denote its boundary by $\Gamma := \partial D$ and assume it to be of class C^2. The normal vector ν is directed into the exterior of D. For complex integration along the contour Γ we assume that the direction of integration is counterclockwise. We confine our analysis to this basic configuration and note that it can be extended, for example, to multiply connected domains with less regular boundaries.

The *Cauchy integral*

$$f(z) := \frac{1}{2\pi i}\int_\Gamma \frac{\varphi(\zeta)}{\zeta - z}\,d\zeta, \quad z \in \mathbb{C}\setminus\Gamma,$$

with density $\varphi \in C(\Gamma)$ defines a function that is holomorphic in $\mathbb{C}\setminus\bar{D}$ and in D. Obviously, for z varying in an open domain not intersecting with Γ, the integrand is continuous with respect to ζ and continuously differentiable with respect to z. Therefore, we can differentiate under the integral to verify the Cauchy–Riemann equations for f. Occasionally, we will call a function that is holomorphic in $\mathbb{C}\setminus\bar{D}$ and in D *sectionally holomorphic*.

As in the case of the single- and double-layer potentials of Chapter 6 we are interested in the behavior of the Cauchy integral for points on the boundary, where the integral becomes singular. By integrating the equation $(\ln\ln x)' = (x\ln x)^{-1}$ between $0 \le x \le 1$ we observe that the Cauchy integral for points on the boundary, in general, will not exist if the density is merely continuous. Therefore we assume the density to be uniformly Hölder continuous. Note that the introduction of Hölder continuity by Definition 7.1 also covers functions on subsets of the complex plane by identifying $\mathbb{R}^2$ and $\mathbb{C}$. For the remainder of this chapter we will always assume that $\alpha \in (0,1)$ for the Hölder exponent, despite the fact that part of our results remain valid for $\alpha = 1$.

Theorem 7.7. *For densities $\varphi \in C^{0,\alpha}(\Gamma)$, $0 < \alpha < 1$, the Cauchy integral exists as a* Cauchy principal value

$$\frac{1}{2\pi i}\int_\Gamma \frac{\varphi(\zeta)}{\zeta - z}\,d\zeta = \frac{1}{2\pi i}\lim_{\rho\to 0}\int_{\Gamma\setminus\Gamma(z;\rho)} \frac{\varphi(\zeta)}{\zeta - z}\,d\zeta$$

for all $z \in \Gamma$. Here, $\Gamma(z;\rho) := \{\zeta \in \Gamma : |\zeta - z| \le \rho\}$.

Proof. Let $z \in \Gamma$ and set $H(z;\rho) := \{\zeta \in D : |\zeta - z| = \rho\}$ with counterclockwise orientation. Then, by Cauchy's integral theorem,

$$\int_{\Gamma\setminus\Gamma(z;\rho)} \frac{d\zeta}{\zeta - z} = \int_{H(z;\rho)} \frac{d\zeta}{\zeta - z}\,.$$

Writing $\zeta = z + \rho e^{i\vartheta}$, $d\zeta = i\rho e^{i\vartheta} d\vartheta$, we find

$$\lim_{\rho\to 0}\int_{H(z;\rho)} \frac{d\zeta}{\zeta - z} = \lim_{\rho\to 0}\int_{H(z;\rho)} i\,d\vartheta = i\pi,$$

since Γ is of class C^2. Hence

$$\frac{1}{2\pi i}\int_\Gamma \frac{d\zeta}{\zeta - z} = \frac{1}{2}\,,\quad z \in \Gamma, \tag{7.4}$$

in the sense of a Cauchy principal value.

The normal vector ν is continuous on Γ. Therefore, as in the proof of Theorem 2.30, we can choose $R \in (0,1]$ such that the scalar product of the normal vectors satisfies

$$\nu(z)\cdot\nu(\zeta) \ge \frac{1}{2} \tag{7.5}$$

for all $z, \zeta \in \Gamma$ with $|z - \zeta| \le R$. Furthermore, we can assume that R is small enough such that $\Gamma(z;R)$ is connected for each $z \in \Gamma$. Then the condition (7.5) implies that $\Gamma(z;R)$ can be bijectively projected onto the tangent line to Γ at the point z. The line elements $ds(\zeta)$ on Γ and $d\sigma$ on the tangent are related by

$$ds(\zeta) = \frac{d\sigma}{\nu(z)\cdot\nu(\zeta)} \le 2\,d\sigma.$$

Hence,

$$\int_{\Gamma(z;R)} \left| \frac{\varphi(\zeta) - \varphi(z)}{\zeta - z} \right| ds \le 2|\varphi|_\alpha \int_{-R}^{R} |\sigma|^{\alpha-1} \, d\sigma = \frac{4R^\alpha}{\alpha} \, |\varphi|_\alpha$$

and

$$\int_{\Gamma \setminus \Gamma(z;R)} \left| \frac{\varphi(\zeta) - \varphi(z)}{\zeta - z} \right| ds \le |\varphi|_\alpha \int_{\Gamma \setminus \Gamma(z;R)} |\zeta - z|^{\alpha-1} \, ds \le R^{\alpha-1} |\Gamma| \, |\varphi|_\alpha.$$

Therefore

$$\int_\Gamma \frac{\varphi(\zeta) - \varphi(z)}{\zeta - z} \, d\zeta$$

exists as an improper integral, and by the decomposition

$$\int_\Gamma \frac{\varphi(\zeta)}{\zeta - z} \, d\zeta = \int_\Gamma \frac{\varphi(\zeta) - \varphi(z)}{\zeta - z} \, d\zeta + \varphi(z) \int_\Gamma \frac{d\zeta}{\zeta - z}$$

the proof is completed. □

After establishing the existence of the Cauchy integral for points on the boundary, we want to show that the function f defined by the Cauchy integral is uniformly Hölder continuous in $\bar{D}$ and in $\mathbb{C} \setminus D$. For this we choose $h > 0$ small enough such that in the parallel strip

$$D_h := \{w + \eta h \nu(w) : w \in \Gamma, \ \eta \in [-1, 1]\}$$

each point z is uniquely representable through projection onto Γ in the form $z = w + \eta h \nu(w)$ with $w \in \Gamma$ and $\eta \in [-1, 1]$ (see p. 84). Then we show that the function $g : D_h \to \mathbb{C}$ defined by

$$g(z) := \frac{1}{2\pi i} \int_\Gamma \frac{\varphi(\zeta) - \varphi(w)}{\zeta - z} \, d\zeta, \quad z \in D_h,$$

is uniformly Hölder continuous with

$$\|g\|_\alpha \le C \|\varphi\|_\alpha, \tag{7.6}$$

where C is some constant independent of φ.

We begin by noting that we can choose h sufficiently small such that

$$|w_1 - w_2| \le 2|z_1 - z_2| \tag{7.7}$$

for each pair $z_1 = z_1(\eta_1) = w_1 + \eta_1 h \nu(w_1)$, $z_2 = z_2(\eta_2) = w_2 + \eta_2 h \nu(w_2)$. Since Γ is of class C^2, the normal ν is continuously differentiable. Therefore, by the mean value theorem, there exists a constant $M > 0$ such that $|\nu(w_1) - \nu(w_2)| \le M|w_1 - w_2|$ for all $w_1, w_2 \in \Gamma$. Then, from

$$|w_1 - w_2| \le |z_1(\eta) - z_2(\eta)| + h|\nu(w_1) - \nu(w_2)|, \quad \eta \in [-1, 1],$$

we observe that $|w_1 - w_2| \le 2|z_1(\eta) - z_2(\eta)|$ for all $\eta \in [-1, 1]$, if we restrict $h \le 1/2M$. Now (7.7) follows from the observation that for all $\eta_1, \eta_2 \in [-1, 1]$ by elementary geometry we have

$$\min\{|z_1(1) - z_2(1)|, |z_1(-1) - z_2(-1)|\} \le |z_1(\eta_1) - z_2(\eta_2)|.$$

We can use (7.7) to estimate $|\varphi(\zeta) - \varphi(w)| \le |\varphi|_\alpha |\zeta - w|^\alpha \le 2^\alpha |\varphi|_\alpha |\zeta - z|^\alpha$ and then, as in the proof of Theorem 7.7, by splitting the integral into the two parts over $\Gamma(w; R)$ and $\Gamma \setminus \Gamma(w; R)$, we obtain

$$|g(z)| \le C_0 |\varphi|_\alpha \tag{7.8}$$

for all $z \in D_h$ and some constant C_0.

Now let $z_1, z_2 \in D_{h_0}$ with $0 < |z_1 - z_2| \le R/4$ and set $\rho := 4|z_1 - z_2|$. Then we can estimate

$$\begin{aligned}
&\left| \int_{\Gamma(w_1;\rho)} \left\{ \frac{\varphi(\zeta) - \varphi(w_1)}{\zeta - z_1} - \frac{\varphi(\zeta) - \varphi(w_2)}{\zeta - z_2} \right\} d\zeta \right| \\
&\le 2^\alpha |\varphi|_\alpha \left\{ \int_{\Gamma(w_1;\rho)} |\zeta - z_1|^{\alpha-1}\, ds + \int_{\Gamma(w_2;2\rho)} |\zeta - z_2|^{\alpha-1}\, ds \right\} \\
&\le C_1 |z_1 - z_2|^\alpha |\varphi|_\alpha,
\end{aligned} \tag{7.9}$$

where C_1 is some constant depending on α. Here we have made use of the fact that $\Gamma(w_1; \rho) \subset \Gamma(w_2; 2\rho)$. For $|\zeta - w_1| \ge \rho = 4|z_1 - z_2|$ we have

$$|\zeta - z_1| \le |\zeta - z_2| + |z_2 - z_1| \le |\zeta - z_2| + \frac{1}{4}\, |\zeta - w_1| \le |\zeta - z_2| + \frac{1}{2}\, |\zeta - z_1|.$$

Hence, $|\zeta - z_1| \le 2|\zeta - z_2|$ for $|\zeta - w_1| \ge \rho$, and this inequality can now be used to obtain

$$\begin{aligned}
&\left| \int_{\Gamma \setminus \Gamma(w_1;\rho)} \{\varphi(\zeta) - \varphi(w_2)\} \left\{ \frac{1}{\zeta - z_1} - \frac{1}{\zeta - z_2} \right\} d\zeta \right| \\
&\le |z_1 - z_2|\, |\varphi|_\alpha \int_{\Gamma \setminus \Gamma(w_1;\rho)} \frac{|\zeta - w_2|^\alpha}{|\zeta - z_1|\, |\zeta - z_2|}\, ds \\
&\le C_2' |z_1 - z_2|\, |\varphi|_\alpha \left\{ \int_\rho^R \frac{d\sigma}{\sigma^{2-\alpha}} + \frac{|\Gamma|}{R^{2-\alpha}} \right\} \le C_2 |z_1 - z_2|^\alpha |\varphi|_\alpha,
\end{aligned} \tag{7.10}$$

where C_2 and C_2' are some constants depending on α and Γ. Note that for this estimate we need the restriction $\alpha < 1$, and we have split the integral into the two parts over $\Gamma(w_1; R) \setminus \Gamma(w_1; \rho)$ and $\Gamma \setminus \Gamma(w_1; R)$. Finally, since from (7.7) and the proof of Theorem 7.7 it is obvious that

$$\int_{\Gamma \setminus \Gamma(w_1;\rho)} \frac{d\zeta}{\zeta - z_1}$$

is bounded by 2π, we have the estimate

$$\left|\{\varphi(w_1) - \varphi(w_2)\} \int_{\Gamma\setminus\Gamma(w_1;\rho)} \frac{d\zeta}{\zeta - z_1}\right| \le C_3|z_1 - z_2|^\alpha|\varphi|_\alpha \tag{7.11}$$

for some constant C_3. Combining (7.9)–(7.11), we obtain

$$2\pi\,|g(z_1) - g(z_2)| \le (C_1 + C_2 + C_3)|z_1 - z_2|^\alpha|\varphi|_\alpha \tag{7.12}$$

for all $z_1, z_2 \in D_h$ with $|z_1 - z_2| \le R/4$. Now the desired result (7.6) follows from (7.8) and (7.12) with the aid of Lemma 7.3.

By Cauchy's integral theorem and (7.4) we have

$$\frac{1}{2\pi i}\int_\Gamma \frac{d\zeta}{\zeta - z} = \begin{cases} 1, & z \in D, \\ \frac{1}{2}, & z \in \Gamma, \\ 0, & z \in \mathbb{C}\setminus\bar{D}. \end{cases}$$

Therefore, from the decomposition

$$f(z) = g(z) + \frac{\varphi(w)}{2\pi i}\int_\Gamma \frac{d\zeta}{\zeta - z}, \quad z \in D_h,$$

we observe that $f \in C^{0,\alpha}(D_h \cap \bar{D})$ and $f \in C^{0,\alpha}(D_h \cap (\mathbb{C}\setminus D))$ with boundary values

$$f_\pm(z) = \frac{1}{2\pi i}\int_\Gamma \frac{\varphi(\zeta)}{\zeta - z}\,d\zeta \mp \frac{1}{2}\,\varphi(z), \quad z \in \Gamma.$$

Hölder continuity of f in $\bar{D}$ and $\mathbb{C}\setminus D$ then follows by again employing Lemma 7.3. We summarize our results in the following theorem.

Theorem 7.8 (Sokhotski–Plemelj). *For densities $\varphi \in C^{0,\alpha}(\Gamma)$ the holomorphic function f defined by the Cauchy integral*

$$f(z) := \frac{1}{2\pi i}\int_\Gamma \frac{\varphi(\zeta)}{\zeta - z}\,d\zeta, \quad z \in \mathbb{C}\setminus\Gamma, \tag{7.13}$$

can be uniformly Hölder continuously extended from D into $\bar{D}$ and from $\mathbb{C}\setminus\bar{D}$ into $\mathbb{C}\setminus D$ with limiting values

$$f_\pm(z) = \frac{1}{2\pi i}\int_\Gamma \frac{\varphi(\zeta)}{\zeta - z}\,d\zeta \mp \frac{1}{2}\,\varphi(z), \quad z \in \Gamma, \tag{7.14}$$

where $f_\pm(z) = \lim_{h\to+0} f(z \pm h\nu(z))$. Furthermore, we have the inequalities

$$\|f\|_{\alpha,\bar{D}} \le C\|\varphi\|_\alpha \quad \text{and} \quad \|f\|_{\alpha,\mathbb{C}\setminus D} \le C\|\varphi\|_\alpha$$

for some constant C depending on α and Γ.

The formula (7.14) was first derived by Sokhotski [224] in his doctoral thesis in 1873. A sufficiently rigorous proof was given by Plemelj [191] in 1908.

Corollary 7.9. *The* Cauchy integral operator $A : C^{0,\alpha}(\Gamma) \to C^{0,\alpha}(\Gamma)$, *defined by*

$$(A\varphi)(z) := \frac{1}{\pi i} \int_\Gamma \frac{\varphi(\zeta)}{\zeta - z} \, d\zeta, \quad z \in \Gamma,$$

is bounded.

Proof. We write the Sokhotski–Plemelj formula (7.14) in the short form

$$f_\pm = \frac{1}{2} A\varphi \mp \frac{1}{2} \varphi \tag{7.15}$$

and obtain the boundedness of A from Theorem 7.8. □

As a first application of the Sokhotski–Plemelj Theorem 7.8 we solve the problem of finding a sectionally holomorphic function with a given jump along the contour Γ.

Theorem 7.10. *Let $\varphi \in C^{0,\alpha}(\Gamma)$. Then there exists a unique function f that is holomorphic in D and $\mathbb{C} \setminus \bar{D}$, which can be extended continuously from D into $\bar{D}$ and from $\mathbb{C} \setminus \bar{D}$ into $\mathbb{C} \setminus D$ satisfying the boundary condition*

$$f_- - f_+ = \varphi \quad \text{on } \Gamma,$$

and for which $f(z) \to 0$, $z \to \infty$, uniformly for all directions. This function is given by (7.13).

Proof. By Theorem 7.8 the function f given by (7.13) has the required properties. To establish uniqueness, from $f_- - f_+ = 0$ on Γ, by Morera's theorem, we conclude that f is holomorphic everywhere in $\mathbb{C}$, i.e., f is an entire function. Then from $f(z) \to 0$, $z \to \infty$, by Liouville's theorem it follows that $f = 0$ in $\mathbb{C}$. From this proof we see that the general sectionally holomorphic function f satisfying $f_- - f_+ = \varphi$ on Γ is obtained from (7.13) by adding an arbitrary entire function. □

As a second application of Theorem 7.8 we state necessary and sufficient conditions for the existence of a holomorphic function in D or in $\mathbb{C} \setminus \bar{D}$ with given boundary values.

Theorem 7.11. *For a given function $\varphi \in C^{0,\alpha}(\Gamma)$, there exists a function f that is holomorphic in D and continuous in $\bar{D}$ with boundary values $f = \varphi$ on Γ if and only if φ is a solution of the homogeneous integral equation of the second kind*

$$\varphi - A\varphi = 0.$$

The solution is given by (7.13).

Proof. Let f be holomorphic with $f = \varphi$ on Γ. Then by Cauchy's integral formula we have

$$f(z) = \frac{1}{2\pi i}\int_\Gamma \frac{f(\zeta)}{\zeta - z}\,d\zeta = \frac{1}{2\pi i}\int_\Gamma \frac{\varphi(\zeta)}{\zeta - z}\,d\zeta, \quad z \in D,$$

and from Theorem 7.8 it follows that $2\varphi = 2f_- = A\varphi + \varphi$ on Γ, and therefore $\varphi - A\varphi = 0$. Conversely, if φ is a solution of $\varphi - A\varphi = 0$, then again by Theorem 7.8 the function f defined by (7.13) has boundary values $2f_- = A\varphi + \varphi = 2\varphi$ on Γ. □

Obviously, for the corresponding exterior problem in $\mathbb{C} \setminus \bar{D}$ with $f(z) \to 0$ as $z \to \infty$ we have the homogeneous integral equation $\varphi + A\varphi = 0$. From this and Theorem 7.11 we observe that the operator A provides an example for nullspaces $N(I - A)$ and $N(I + A)$ with infinite dimension, since there exist infinitely many linearly independent holomorphic functions. This, in particular, implies that A is not compact, and it also means that $I - A$ and $I + A$ cannot be regularized.

We now can prove a property of the Cauchy integral operator A, which is of central importance in our study of singular integral equations with Cauchy kernels, since it will allow the construction of regularizers. Note that the following theorem is a further indication that the Cauchy integral operator A is not compact.

Theorem 7.12. *The Cauchy integral operator A satisfies $A^2 = I$.*

Proof. For $\varphi \in C^{0,\alpha}(\Gamma)$ we define f by (7.13). Then, by Theorem 7.11, we have $f_+ + Af_+ = 0$ and $f_- - Af_- = 0$. Hence, using (7.15), we derive

$$A^2\varphi = A(f_- + f_+) = f_- - f_+ = \varphi,$$

and this is the desired result. □

Theorem 7.13. *The operators A and $-A$ are adjoint with respect to the dual system $\langle C^{0,\alpha}(\Gamma), C^{0,\alpha}(\Gamma)\rangle$ with the non-degenerate bilinear form*

$$\langle \varphi, \psi \rangle := \int_\Gamma \varphi(z)\psi(z)\,dz, \quad \varphi, \psi \in C^{0,\alpha}(\Gamma).$$

Proof. It is left as an exercise to show that $\langle\cdot\,,\cdot\rangle$ is non-degenerate on $C^{0,\alpha}(\Gamma)$. For $\varphi, \psi \in C^{0,\alpha}(\Gamma)$ define

$$f(z) := \frac{1}{2\pi i}\int_\Gamma \frac{\varphi(\zeta)}{\zeta - z}\,d\zeta, \quad g(z) := \frac{1}{2\pi i}\int_\Gamma \frac{\psi(\zeta)}{\zeta - z}\,d\zeta, \quad z \in \mathbb{C} \setminus \Gamma.$$

Then, by Theorem 7.8 and Cauchy's integral theorem we find that

$$\langle A\varphi, \psi\rangle + \langle \varphi, A\psi\rangle = 2\langle f_-, g_-\rangle - 2\langle f_+, g_+\rangle = -2\int_{|z|=R} f(z)g(z)\,dz \to 0, \quad R \to \infty,$$

since $f(z),\ g(z) = O(1/|z|),\ z \to \infty$. □

Note that Theorem 7.13 justifies interchanging the order of integration in

$$\int_\Gamma \left(\int_\Gamma \frac{\varphi(\zeta)}{\zeta - z}\, d\zeta \right) \psi(z)\, dz = \int_\Gamma \varphi(\zeta) \left(\int_\Gamma \frac{\psi(z)}{\zeta - z}\, dz \right) d\zeta.$$

Example 7.14. Let Γ be the unit circle. Then, substituting $z = e^{it}$, $\zeta = e^{i\tau}$, we find

$$\frac{d\zeta}{\zeta - z} = \frac{1}{2}\left(\cot \frac{\tau - t}{2} + i\right) d\tau,$$

i.e., the operator A can be expressed in the form

$$(A\varphi)\left(e^{it}\right) = \frac{1}{2\pi i} \int_0^{2\pi} \left\{\cot \frac{\tau - t}{2} + i\right\} \varphi\left(e^{i\tau}\right) d\tau, \quad t \in [0, 2\pi],$$

with the integral to be understood as a Cauchy principal value. Consider the integral equation of the first kind

$$\frac{1}{2\pi} \int_0^{2\pi} \cot \frac{\tau - t}{2}\, \widetilde{\varphi}(\tau)\, d\tau = \widetilde{\psi}(t), \quad t \in [0, 2\pi], \tag{7.16}$$

in the space $C^{0,\alpha}[0, 2\pi]$. Since

$$\int_0^{2\pi} \cot \frac{t}{2}\, dt = 0,$$

by integrating (7.16) with respect to t and interchanging the order of integration (use Theorem 7.13 for $\psi(z) = 1/z$ and transform the integrals), we find that

$$\int_0^{2\pi} \widetilde{\psi}(t)\, dt = 0 \tag{7.17}$$

is a necessary condition for the solvability of (7.16). It is also sufficient, because by making use of $A^2 = I$ we see that when (7.17) is satisfied the general solution of (7.16) is given by

$$\widetilde{\varphi}(t) = -\frac{1}{2\pi} \int_0^{2\pi} \cot \frac{\tau - t}{2}\, \widetilde{\psi}(\tau)\, d\tau + c, \quad t \in [0, 2\pi], \tag{7.18}$$

where c is an arbitrary constant. □

The formulas (7.16) and (7.18) represent the *inversion formulas* of Hilbert. The kernel in (7.16) and (7.18) is a called *Hilbert kernel*, since singular equations with this kernel were first encountered by Hilbert [95]. Splitting the integral equation of Theorem 7.11 into its real and imaginary part, we readily find that the Hilbert inversion formulas relate the real and imaginary part of holomorphic functions in the unit disk.

7.3 The Riemann Problem

The following boundary value problem was first formulated by Riemann in his inaugural dissertation. Because a first attempt toward a solution was made by Hilbert [95] in 1904 through the use of integral equations, what we will denote as the Riemann problem in the literature is sometimes also called the Hilbert problem.

Riemann Problem. *Find a function f that is holomorphic in D and in $\mathbb{C} \setminus \bar{D}$, which can be extended uniformly Hölder continuously from D into $\bar{D}$ and from $\mathbb{C} \setminus \bar{D}$ into $\mathbb{C} \setminus D$ satisfying the boundary condition*

$$f_- = gf_+ + h \quad \text{on } \Gamma \tag{7.19}$$

and

$$f(z) \to 0, \quad z \to \infty, \tag{7.20}$$

uniformly for all directions. Here, g and h are given uniformly Hölder continuous functions on Γ.

Before we give a solution to this Riemann problem for the case where g vanishes nowhere on Γ, we want to indicate its connection to integral equations.

Theorem 7.15. *Let $a, b, h \in C^{0,\alpha}(\Gamma)$. Then the Cauchy integral (7.13) maps solutions $\varphi \in C^{0,\alpha}(\Gamma)$ of the integral equation*

$$a\varphi + bA\varphi = h \tag{7.21}$$

linearly and bijectively onto solutions f of the Riemann problem with the boundary condition

$$(a + b)f_- = (a - b)f_+ + h \quad \text{on } \Gamma. \tag{7.22}$$

Proof. By Theorems 7.8 and 7.10, the Cauchy integral maps $C^{0,\alpha}(\Gamma)$ linearly and bijectively onto the linear space of sectionally holomorphic functions that vanish at infinity and can be extended uniformly Hölder continuously with Hölder exponent α from D into $\bar{D}$ and from $\mathbb{C} \setminus \bar{D}$ into $\mathbb{C} \setminus D$. From (7.15) and Theorem 7.11 we have the relation

$$a\varphi + bA\varphi = a(f_- - f_+) + b(f_- + f_+) = (a + b)f_- - (a - b)f_+$$

between the density φ and the function f defined by the Cauchy integral. This ends the proof. □

Definition 7.16. Let g be a complex-valued and nowhere vanishing function defined on the contour Γ. The *index* of the function g is the integer $\operatorname{ind} g$ given by the increment of its argument along the contour in a counterclockwise direction divided by 2π, i.e.,

$$\operatorname{ind} g := \frac{1}{2\pi} \int_\Gamma d \arg g.$$

The index can also be expressed by the logarithm of g through

$$\operatorname{ind} g = \frac{1}{2\pi i} \int_{\Gamma} d \ln g.$$

Now, with the aid of the Sokhotski–Plemelj jump relations, we explicitly solve the homogeneous Riemann problem.

Theorem 7.17. *Let $g \in C^{0,\alpha}(\Gamma)$ be a nowhere vanishing function with $\operatorname{ind} g = \kappa$. Then there exists a unique sectionally holomorphic function f_0 satisfying the homogeneous boundary condition*

$$f_{0-} = g f_{0+} \quad \text{on } \Gamma$$

and

$$\lim_{z \to \infty} z^{\kappa} f_0(z) = 1$$

uniformly for all directions. It is called the canonical *solution to the homogeneous Riemann problem, has the property that $f_0(z) \neq 0$ for all $z \in \mathbb{C}$, and is uniformly Hölder continuous up to the boundary.*

Proof. Choose a point $a \in D$ and define $G \in C^{0,\alpha}(\Gamma)$ by

$$G(z) := (z - a)^{-\kappa} g(z), \quad z \in \Gamma.$$

Because G has index $\operatorname{ind} G = 0$, we can reduce the homogeneous boundary condition $F_- = G F_+$ by taking logarithms to obtain $\ln F_- - \ln F_+ = \ln G$, where we may take any branch of the logarithm to arrive at a single-valued function $\ln G \in C^{0,\alpha}(\Gamma)$. According to Theorem 7.10 we set

$$\psi(z) := \frac{1}{2\pi i} \int_{\Gamma} \frac{\ln G(\zeta)}{\zeta - z} \, d\zeta, \quad z \in \mathbb{C} \setminus \Gamma,$$

and $F(z) := e^{\psi(z)}$, $z \in \mathbb{C} \setminus \Gamma$. Then, from $\psi_- - \psi_+ = \ln G$ on Γ we derive $F_- = G F_+$ on Γ, and since $\psi(z) \to 0$, $z \to \infty$, we have $F(z) \to 1$, $z \to \infty$. Now the function f_0, defined by,

$$f_0(z) := \begin{cases} F(z), & z \in D, \\ (z - a)^{-\kappa} F(z), & z \in \mathbb{C} \setminus \bar{D}, \end{cases}$$

has the required properties.

Let f be any other sectionally holomorphic function satisfying the homogeneous boundary condition $f_- = g f_+$ on Γ. Then the quotient $q := f / f_0$ is sectionally holomorphic (observe $f_0(z) \neq 0$ for all $z \in \mathbb{C}$) satisfying $q_- = q_+$ on Γ. Hence, by Morera's theorem, q is an entire function. From $z^{\kappa} f_0(z) \to 1$ and $z^{\kappa} f(z) \to 1$ for $z \to \infty$, it follows that $q(z) \to 1$ for $z \to \infty$, and Liouville's theorem implies $q = 1$ on $\mathbb{C}$. □

From the proof we note that the general sectionally holomorphic function f satisfying the homogeneous boundary condition $f_- = gf_+$ on Γ is obtained from the canonical solution f_0 by multiplying it by an arbitrary entire function.

Theorem 7.18. *Under the assumptions of Theorem 7.17, the homogeneous Riemann problem admits* $\max(\kappa, 0)$ *linearly independent solutions.*

Proof. For p an entire function, the product $f := pf_0$ vanishes at infinity if and only if

$$\lim_{z\to\infty} z^{-\kappa} p(z) = 0.$$

Hence, p must be a polynomial of degree less than or equal to $\kappa - 1$ if $\kappa > 0$ and it must be zero if $\kappa \le 0$. □

By the procedure used in Theorem 7.17 it is also possible to treat the inhomogeneous Riemann problem. But here we will proceed differently and use Theorem 5.6 to obtain results on the existence of solutions to the inhomogeneous integral equation (7.21), which then by the equivalence stated in Theorem 7.15 imply results on the existence of solutions to the inhomogeneous Riemann problem (see Problem 7.1).

7.4 Integral Equations with Cauchy Kernel

For $0 < \alpha, \beta \le 1$ by $C^{0,\beta,\alpha}(\Gamma \times \Gamma)$ we denote the set of all functions k defined on $\Gamma \times \Gamma$ satisfying

$$|k(z_1, \zeta_1) - k(z_2, \zeta_2)| \le M(|z_1 - z_2|^\beta + |\zeta_1 - \zeta_2|^\alpha) \tag{7.23}$$

for all $z_1, z_2, \zeta_1, \zeta_2 \in \Gamma$ and some constant M depending on k. Let $0 < \alpha < \beta \le 1$ and let $a \in C^{0,\alpha}(\Gamma)$ and $k \in C^{0,\beta,\alpha}(\Gamma \times \Gamma)$. Then the operator $K : C^{0,\alpha}(\Gamma) \to C^{0,\alpha}(\Gamma)$, defined by

$$(K\varphi)(z) := a(z)\varphi(z) + \frac{1}{\pi i} \int_\Gamma \frac{k(z,\zeta)}{\zeta - z}\, \varphi(\zeta)\, d\zeta, \quad z \in \Gamma,$$

is called a *singular integral operator* with a *Cauchy kernel*. The operator K^0 : $C^{0,\alpha}(\Gamma) \to C^{0,\alpha}(\Gamma)$, given by

$$(K^0\varphi)(z) := a(z)\varphi(z) + \frac{b(z)}{\pi i} \int_\Gamma \frac{\varphi(\zeta)}{\zeta - z}\, d\zeta, \quad z \in \Gamma,$$

where $b(z) := k(z, z)$, $z \in \Gamma$, is called the *dominant part* of the operator K. For the coefficients a and b of the dominant part we will assume that $a^2 - b^2$ vanishes nowhere on Γ. The dominant part can be written in the short form $K^0 = aI + bA$ and is bounded by Corollary 7.9. The splitting of the operator K into its dominant part K^0 and the remainder $K - K^0$ is justified by the following theorem.

Theorem 7.19. *The singular integral operator K is bounded and the difference between K and its dominant part K^0 is compact from $C^{0,\alpha}(\Gamma)$ into $C^{0,\alpha}(\Gamma)$.*

Proof. We choose $0 < \gamma < \alpha$ and show that the difference $H := K - K^0$ is bounded from $C^{0,\gamma}(\Gamma)$ into $C^{0,\alpha}(\Gamma)$. Then the assertion follows from Theorems 2.21 and 7.4.

We write

$$(H\varphi)(z) = \frac{1}{\pi i}\int_\Gamma h(z,\zeta)\varphi(\zeta)\,d\zeta, \quad z \in \Gamma,$$

with kernel

$$h(z,\zeta) := \frac{k(z,\zeta) - k(z,z)}{\zeta - z}, \quad z \neq \zeta. \tag{7.24}$$

Since $k \in C^{0,\beta,\alpha}(\Gamma \times \Gamma)$, there exists a constant M such that

$$|h(z,\zeta)| \leq M|z - \zeta|^{\alpha - 1}, \quad z \neq \zeta. \tag{7.25}$$

Hence, the kernel h is weakly singular and therefore we have

$$\|H\varphi\|_\infty \leq C_0\|\varphi\|_\infty \tag{7.26}$$

for some constant C_0.

To establish Hölder continuity, we proceed analogously to the proof of Theorem 7.8. Let $z_1, z_2 \in \Gamma$ with $0 < |z_1 - z_2| \leq R/4$ and set $\rho := 4|z_1 - z_2|$. Then, using (7.25), we can estimate

$$\begin{aligned} &\left|\int_{\Gamma(z_1;\rho)} \{h(z_1,\zeta) - h(z_2,\zeta)\}\varphi(\zeta)\,d\zeta\right| \\ &\leq M\|\varphi\|_\infty \int_{\Gamma(z_1;\rho)} \{|z_1 - \zeta|^{\alpha-1} + |z_2 - \zeta|^{\alpha-1}\}\,ds \\ &\leq C_1'\|\varphi\|_\infty \int_0^{2\rho} \sigma^{\alpha-1} d\sigma \leq C_1\|\varphi\|_\infty |z_1 - z_2|^\alpha \end{aligned} \tag{7.27}$$

for some constants C_1 and C_1' depending on M and α. Using (7.23), we can estimate

$$\begin{aligned} &\left|\int_{\Gamma\setminus\Gamma(z_1;\rho)} \frac{k(z_1,\zeta) - k(z_2,\zeta)}{\zeta - z_1}\varphi(\zeta)\,d\zeta\right| \\ &\leq M\|\varphi\|_\infty |z_1 - z_2|^\beta \int_{\Gamma\setminus\Gamma(z_1;\rho)} \frac{ds}{|\zeta - z_1|} \\ &\leq C_2'\|\varphi\|_\infty |z_1 - z_2|^\beta \left\{\int_{\rho/2}^R \frac{d\sigma}{\sigma} + \frac{|\Gamma|}{R}\right\} \\ &\leq C_2''\|\varphi\|_\infty |z_1 - z_2|^\beta \left\{\ln\frac{1}{\rho} + 1\right\} \leq C_2\|\varphi\|_\infty |z_1 - z_2|^\alpha \end{aligned} \tag{7.28}$$

for some constants C_2, C_2', and C_2'' depending on M, α, β, and Γ. Because for $\zeta \in \Gamma$ with $|\zeta - z_1| \geq \rho$ we have that $|\zeta - z_1| \leq 2|\zeta - z_2|$, we derive the further estimate

$$\begin{aligned}
&\left| \int_{\Gamma \setminus \Gamma(z_1;\rho)} \{k(z_2, \zeta) - k(z_2, z_2)\} \left\{ \frac{1}{\zeta - z_1} - \frac{1}{\zeta - z_2} \right\} \varphi(\zeta)\, d\zeta \right| \\
&\leq M \|\varphi\|_\infty |z_1 - z_2| \int_{\Gamma \setminus \Gamma(z_1;\rho)} \frac{ds}{|\zeta - z_1|\, |\zeta - z_2|^{1-\alpha}} \\
&\leq C_3' \|\varphi\|_\infty |z_1 - z_2| \left\{ \int_{\rho/2}^{R} \sigma^{\alpha - 2} d\sigma + \frac{|\Gamma|}{R^{2-\alpha}} \right\} \\
&\leq C_3 \|\varphi\|_\infty |z_1 - z_2|^\alpha
\end{aligned} \tag{7.29}$$

for some constants C_3 and C_3' depending on M, α, and Γ. Finally, we write

$$\int_{\Gamma \setminus \Gamma(z_1;\rho)} \frac{\varphi(\zeta)}{\zeta - z_1}\, d\zeta = \int_{\Gamma \setminus \Gamma(z_1;\rho)} \frac{\varphi(\zeta) - \varphi(z_1)}{\zeta - z_1}\, d\zeta + \varphi(z_1) \int_{\Gamma \setminus \Gamma(z_1;\rho)} \frac{d\zeta}{\zeta - z_1}$$

and note from the proof of Theorem 7.7 that the second integral on the right-hand side is bounded by 2π. For the first integral we have

$$\left| \int_{\Gamma \setminus \Gamma(z_1;\rho)} \frac{\varphi(\zeta) - \varphi(z_1)}{\zeta - z_1}\, d\zeta \right| \leq |\varphi|_\gamma \int_\Gamma \frac{ds}{|\zeta - z_1|^{1-\gamma}} \leq C_4' |\varphi|_\gamma$$

for some constant C_4' depending on Γ and γ. Hence, using again (7.23), we obtain

$$\left| \{k(z_1, z_1) - k(z_2, z_2)\} \int_{\Gamma \setminus \Gamma(z_1;\rho)} \frac{\varphi(\zeta)}{\zeta - z_1}\, d\zeta \right| \leq C_4 \|\varphi\|_\gamma |z_1 - z_2|^\alpha \tag{7.30}$$

for some constant C_4 depending on M, α and Γ. Combining (7.27)–(7.30), we find

$$\pi\, |(H\varphi)(z_1) - (H\varphi)(z_2)| \leq (C_1 + C_2 + C_3 + C_4) \|\varphi\|_\gamma |z_1 - z_2|^\alpha \tag{7.31}$$

for all $z_1, z_2 \in \Gamma$ with $|z_1 - z_2| \leq R/4$. Now the desired result

$$\|H\varphi\|_\alpha \leq C \|\varphi\|_\gamma$$

follows from (7.26) and (7.31) with the aid of Lemma 7.3. □

Theorem 7.20. *The operators $K^0 = aI + bA$ and $K^{0\prime} := aI - Ab$ are adjoint with respect to the dual system $\langle C^{0,\alpha}(\Gamma), C^{0,\alpha}(\Gamma) \rangle$. They both have finite-dimensional nullspaces and the index of the operator K^0 is given by*

$$\operatorname{ind} K^0 = \operatorname{ind} \frac{a - b}{a + b}\,.$$

Proof. That K^0 and $K^{0\prime}$ are adjoint is easily derived as a consequence of Theorem 7.13. From Theorems 7.15 and 7.18 we obtain

$$\dim N(K^0) = \max(\kappa, 0), \tag{7.32}$$

where

$$\kappa := \operatorname{ind} \frac{a-b}{a+b}\,.$$

Analogous to Theorem 7.15, the homogeneous adjoint equation

$$a\psi - Ab\psi = 0 \tag{7.33}$$

is equivalent to the Riemann problem with homogeneous boundary condition

$$(a-b)f_- = (a+b)f_+ \quad \text{on } \Gamma. \tag{7.34}$$

Let ψ solve the integral equation (7.33). Then $\varphi := b\psi$ solves $a\varphi - bA\varphi = 0$ and therefore, by Theorem 7.15, the sectionally holomorphic function f defined by (7.13) satisfies the boundary condition (7.34). Conversely, let f be a solution to the Riemann problem with boundary condition (7.34). Then for the function $\psi := 2(a+b)^{-1}f_-$ we have $f_- + f_+ = a\psi$ and $f_- - f_+ = b\psi$. Hence, by Theorem 7.11 and its analogue for $\mathbb{C} \setminus \bar{D}$, we find $a\psi - Ab\psi = f_- + f_+ - A(f_- - f_+) = 0$.

With the aid of $\operatorname{ind}[(a+b)^{-1}(a-b)] = -\operatorname{ind}[(a-b)^{-1}(a+b)]$, Theorem 7.18 now yields

$$\dim N(K^{0\prime}) = \max(-\kappa, 0) \tag{7.35}$$

and $\operatorname{ind} K^0 = \dim N(K^0) - \dim N(K^{0\prime}) = \kappa$ follows by combining (7.32) and (7.35). □

From now on, for symmetry, we assume that $k \in C^{0,\beta,\beta}(\Gamma \times \Gamma)$ with $0 < \alpha < \beta \le 1$. The singular integral operators with Cauchy kernel

$$(K\varphi)(z) := a(z)\varphi(z) + \frac{1}{\pi i}\int_\Gamma \frac{k(z,\zeta)}{\zeta - z}\,\varphi(\zeta)\,d\zeta, \quad z \in \Gamma,$$

and

$$(K'\psi)(z) := a(z)\psi(z) - \frac{1}{\pi i}\int_\Gamma \frac{k(\zeta,z)}{\zeta - z}\,\psi(\zeta)\,d\zeta, \quad z \in \Gamma,$$

are adjoint. This follows by writing $K = K^0 + H$ and $K' = K^{0\prime} + H'$, where H and H' have weakly singular kernels $h(z,\zeta)$ and $h(\zeta,z)$, respectively, with h given by (7.24). For weakly singular kernels, as in Theorem 4.7, the order of integration may be interchanged.

Theorem 7.21. *The operator $(a^2 - b^2)^{-1}K'$ is a regularizer of K.*

Proof. First, we observe that for each $c \in C^{0,\alpha}(\Gamma)$ the difference $Ac - cA$ is compact from $C^{0,\alpha}(\Gamma)$ into $C^{0,\alpha}(\Gamma)$. This follows from Theorem 7.19, applied to the kernel

$c(\zeta)$ contained in $C^{0,\beta,\alpha}(\Gamma \times \Gamma)$. Actually, the need for the compactness of the commutator $Ac - cA$ was our reason for allowing different exponents α and β in Theorem 7.19. Now, using $A^2 = I$, we derive

$$K^{0\prime} K^0 = (aI - Ab)(aI + bA) = (a^2 - b^2)I + M$$

where $M := abA - Aba + (b^2A - Ab^2)A$ is compact, since A is bounded by Corollary 7.9. Hence, $(a^2 - b^2)^{-1}K^{0\prime}$ is a left regularizer of K^0. Analogously,

$$K^0 K^{0\prime} = (aI + bA)(aI - Ab) = (a^2 - b^2)I + \widetilde{M}$$

where $\widetilde{M} := bAa - aAb$ is compact, i.e., $K^{0\prime}(a^2 - b^2)^{-1}$ is a right regularizer of K^0. Again, the difference $(a^2 - b^2)^{-1}K^{0\prime} - K^{0\prime}(a^2 - b^2)^{-1}$ is compact. Therefore, $(a^2 - b^2)^{-1}K^{0\prime}$ is a regularizer of K^0. Now the assertion of the theorem follows from the fact that $K = K^0 + H$ and $K = K^{0\prime} + H'$ with compact H and H'. □

We now are in a position to derive the classical Noether theorems from our general theory on regularization in dual systems of Chapter 5.

Theorem 7.22 (First Noether Theorem). *The singular integral operator with Cauchy kernel has a finite-dimensional nullspace.*

Proof. This follows from Theorems 5.6 and 7.21. □

Theorem 7.23 (Second Noether Theorem). *The index of the singular integral operator K with Cauchy kernel is given by*

$$\operatorname{ind} K = \operatorname{ind} \frac{a - b}{a + b} .$$

Proof. This follows from Theorems 7.20 and 7.21 and Corollary 5.12. □

Theorem 7.24 (Third Noether Theorem). *The inhomogeneous singular integral equation*

$$K\varphi = h$$

is solvable if and only if

$$\langle h, \psi \rangle = 0$$

is satisfied for all $\psi \in N(K')$.

Proof. This follows from Theorems 5.6 and 7.21. □

Theorem 7.25. *Let $0 < \alpha < \beta \le 1$, $a \in C^{0,\alpha}(\Gamma)$, and $k \in C^{0,\beta,\beta}(\Gamma \times \Gamma)$ and assume that $a^2 - b^2$ vanishes nowhere on Γ, where $b(z) := k(z, z)$, $z \in \Gamma$. Then, in the Hölder space $C^{0,\alpha}(\Gamma)$, the number of linearly independent solutions of the homogeneous singular integral equation*

$$a(z)\varphi(z) + \frac{1}{\pi i} \int_\Gamma \frac{k(z,\zeta)}{\zeta - z}\, \varphi(\zeta)\, d\zeta = 0, \quad z \in \Gamma,$$

and of its adjoint equation

$$a(z)\psi(z) - \frac{1}{\pi i}\int_\Gamma \frac{k(\zeta, z)}{\zeta - z}\,\psi(\zeta)\,d\zeta = 0, \quad z \in \Gamma,$$

are both finite, and their difference is given by the index

$$\frac{1}{2\pi i}\int_\Gamma d\ln\frac{a-b}{a+b}\,.$$

The inhomogeneous singular integral equation

$$a(z)\varphi(z) + \frac{1}{\pi i}\int_\Gamma \frac{k(z, \zeta)}{\zeta - z}\,\varphi(\zeta)\,d\zeta = h(z), \quad z \in \Gamma,$$

is solvable if and only if

$$\int_\Gamma h(z)\psi(z)\,dz = 0$$

is satisfied for all solutions ψ of the homogeneous adjoint equation.

Proof. This is a reformulation of the three preceding theorems. □

For equations of the first kind, i.e., equations with $a = 0$, we note the following corollary.

Corollary 7.26. *Singular integral operators of the first kind with Cauchy kernel have index zero. In particular, injective singular integral operators of the first kind are bijective with bounded inverse.*

Proof. The index zero follows from Theorem 7.23. The bounded inverse for an injective operator of the first kind is a consequence of the Riesz Theorem 3.4 together with the fact that A provides an equivalent regularizer. □

With the aid of Example 7.14, choosing for Γ the unit circle, we obtain the following corollary, which contains the equations for which Noether [185] proved the theorems named after him.

Corollary 7.27. *Let a and k be real-valued and 2π-periodic and assume that $a^2 + b^2$ is strictly positive, where $b(t) := k(t, t)$ for $t \in [0, 2\pi]$. Then for the singular integral equation with Hilbert kernel*

$$a(t)\varphi(t) - \frac{1}{2\pi}\int_0^{2\pi} k(t, \tau)\cot\frac{\tau - t}{2}\,\varphi(\tau)\,d\tau = h(t), \quad t \in [0, 2\pi],$$

and its adjoint equation

$$a(t)\psi(t) + \frac{1}{2\pi}\int_0^{2\pi} k(\tau, t)\cot\frac{\tau - t}{2}\,\psi(\tau)\,d\tau = g(t), \quad t \in [0, 2\pi],$$

the three Noether theorems are valid, i.e., the numbers of linearly independent solutions of the homogeneous equation and the homogeneous adjoint equation are both finite, and their difference is given by the index

$$\frac{1}{\pi}\int_0^{2\pi} d\arctan\frac{b}{a}\,.$$

The inhomogeneous equation is solvable if and only if

$$\int_0^{2\pi} h(t)\psi(t)\,dt = 0$$

is satisfied for all solutions ψ of the homogeneous adjoint equation.

Proof. After transforming them onto the unit circle Γ by setting $z = e^{it}$, $\zeta = e^{i\tau}$, the two equations read

$$\widetilde{a}(z)\widetilde{\varphi}(z) - \frac{1}{\pi}\int_\Gamma \frac{\widetilde{k}(z,\zeta)}{\zeta - z}\,\widetilde{\varphi}(\zeta)\,d\zeta = \widetilde{h}(z), \quad z \in \Gamma,$$

and

$$\widetilde{a}(z)\widetilde{\psi}(z) + \frac{1}{\pi}\int_\Gamma \frac{\widetilde{k}(\zeta,z)}{\zeta - z}\,\widetilde{\psi}(\zeta)\,d\zeta = \widetilde{g}(z), \quad z \in \Gamma,$$

for $\widetilde{\varphi}(e^{it}) := \varphi(t)$ and $\widetilde{\psi}(e^{it}) := \psi(t)/e^{it}$. We have set $\widetilde{k}(e^{it}, e^{i\tau}) := (e^{it} + e^{i\tau})k(t,\tau)/2e^{i\tau}$ and $\widetilde{a}(e^{it}) := a(t)$, $\widetilde{h}(e^{it}) := h(t)$, and $\widetilde{g}(e^{it}) := g(t)/e^{it}$. Hence, the corollary is established through transforming the Noether theorems for the pair of equations on the unit circle back into the equations on $[0, 2\pi]$. For the index we compute

$$\ln\frac{a+ib}{a-ib}\bigg|_0^{2\pi} = 2\ln(a+ib)\bigg|_0^{2\pi} = 2i\arctan\frac{b}{a}\bigg|_0^{2\pi},$$

and the proof is finished. □

Concluding this section, we wish to mention that the above integral equations can also be treated in various other normed function spaces. In this context we make use of Lax's Theorem 4.13 to establish the boundedness of the Cauchy integral operator A in $L^2(\Gamma)$.

Theorem 7.28. *The Cauchy integral operator A is bounded from $L^2(\Gamma)$ into $L^2(\Gamma)$.*

Proof. We want to apply Theorem 4.13 in the positive dual system $(C^{0,\alpha}(\Gamma), C^{0,\alpha}(\Gamma))$ generated by the scalar product

$$(\varphi, \psi) := \int_\Gamma \varphi\bar{\psi}\,ds, \quad \varphi, \psi \in C^{0,\alpha}(\Gamma).$$

Therefore, we need the adjoint A^* of the Cauchy integral operator A with respect to this scalar product. Complex integration and contour integration along Γ are related by

$$dz = e^{i\gamma(z)}\, ds(z) \tag{7.36}$$

for all $z \in \Gamma$, where $\gamma(z)$ denotes the angle between the tangent to Γ at the point z and the real axis. Using Theorem 7.13 and (7.36), we see that

$$\int_\Gamma A\varphi\, \bar{\psi}\, ds = -\int_\Gamma \varphi\chi\, A(\overline{\psi\chi})\, ds,$$

where $\chi(z) := e^{i\gamma(z)}$, $z \in \Gamma$. Hence, the adjoint A^* is given by

$$A^*\psi = -\overline{\chi A(\overline{\psi\chi})},$$

and $A^* : C^{0,\alpha}(\Gamma) \to C^{0,\alpha}(\Gamma)$ is bounded, since $A : C^{0,\alpha}(\Gamma) \to C^{0,\alpha}(\Gamma)$ is bounded. Now the statement follows from Theorem 4.13 and the denseness of $C^{0,\alpha}(\Gamma)$ in $L^2(\Gamma)$. □

In particular, as a consequence of Theorem 7.28, the property $A^2 = I$ of Theorem 7.12 remains valid in $L^2(\Gamma)$ and can be used to construct regularizers as in Theorem 7.21.

7.5 Cauchy Integral and Logarithmic Potential

We now want to use the close connection between holomorphic and two-dimensional harmonic functions to derive regularity results on the logarithmic single- and double-layer potential from the Sokhotski–Plemelj theorem.

In the Cauchy integral for fixed $z \in \mathbb{C}$ we substitute $\zeta - z = re^{i\vartheta}$, where $r = |\zeta - z|$ and $\vartheta = \arg(\zeta - z)$. Then

$$\int_\Gamma \frac{d\zeta}{\zeta - z} = \int_\Gamma d\ln(\zeta - z) = \int_\Gamma d(\ln r + i\vartheta) = \int_\Gamma \left(\frac{\partial \ln r}{\partial s} + i\,\frac{\partial \vartheta}{\partial s} \right) ds.$$

By the Cauchy–Riemann equations, applied to $\ln(\zeta - z) = \ln r + i\vartheta$, we have

$$\frac{\partial \vartheta}{\partial s} = \frac{\partial \ln r}{\partial \nu} = -\frac{\partial}{\partial \nu} \ln \frac{1}{r}\,.$$

Consequently,

$$\begin{aligned} \frac{1}{2\pi i} \int_\Gamma \frac{\varphi(\zeta)}{\zeta - z}\, d\zeta = &-\frac{1}{2\pi} \int_\Gamma \varphi(\zeta)\, \frac{\partial}{\partial \nu(\zeta)} \ln \frac{1}{|\zeta - z|}\, ds(\zeta) \\ &-\frac{1}{2\pi i} \int_\Gamma \varphi(\zeta)\, \frac{\partial}{\partial s(\zeta)} \ln \frac{1}{|\zeta - z|}\, ds(\zeta). \end{aligned} \tag{7.37}$$

This formula indicates that for real-valued densities the real part of the Cauchy integral coincides with the logarithmic double-layer potential

$$v(z) := \frac{1}{2\pi} \int_\Gamma \varphi(\zeta) \frac{\partial}{\partial \nu(\zeta)} \ln \frac{1}{|\zeta - z|} \, ds(\zeta), \quad z \in \mathbb{R}^2 \setminus \Gamma.$$

Note that we use the same symbol z for the complex number $z = x_1 + ix_2$ and the vector $z = (x_1, x_2) \in \mathbb{R}^2$. Hence, from Theorem 7.8 and the Sokhotski–Plemelj formula (7.14), we derive the following theorem on the regularity of the double-layer potential with a Hölder continuous density. Recall that in Theorem 6.18 the density is only assumed to be continuous.

Theorem 7.29. *The logarithmic double-layer potential v with Hölder continuous density φ can be extended uniformly Hölder continuously from D into $\bar{D}$ and from $\mathbb{R}^2 \setminus \bar{D}$ into $\mathbb{R}^2 \setminus D$ with limiting values*

$$v_\pm(z) = \frac{1}{2\pi} \int_\Gamma \varphi(\zeta) \frac{\partial}{\partial \nu(\zeta)} \ln \frac{1}{|\zeta - z|} \, ds(\zeta) \pm \frac{1}{2} \varphi(z), \quad z \in \Gamma. \tag{7.38}$$

Furthermore, we have the inequalities

$$\|v\|_{\alpha,\bar{D}} \le C \|\varphi\|_\alpha \quad \textit{and} \quad \|v\|_{\alpha,\mathbb{R}^2 \setminus D} \le C \|\varphi\|_\alpha$$

for some constant C depending on α and Γ.

Recall the relation (7.36) between complex integration and contour integration along Γ. Since Γ is of class C^2, the angle γ is continuously differentiable with respect to z. Therefore, from Theorem 7.8 we immediately deduce that

$$f(z) := \frac{1}{2\pi} \int_\Gamma \frac{\varphi(\zeta)}{\zeta - z} \, ds(\zeta), \quad z \in \mathbb{C} \setminus \Gamma, \tag{7.39}$$

is holomorphic in D and in $\mathbb{C} \setminus \bar{D}$ and that it can be extended uniformly Hölder continuously from D into $\bar{D}$ and from $\mathbb{C} \setminus \bar{D}$ into $\mathbb{C} \setminus D$ with limiting values

$$f_\pm(z) := \frac{1}{2\pi} \int_\Gamma \frac{\varphi(\zeta)}{\zeta - z} \, ds(\zeta) \mp \frac{i}{2} e^{-i\gamma(z)} \varphi(z), \quad z \in \Gamma. \tag{7.40}$$

The gradient of the logarithmic single-layer potential

$$u(z) := \frac{1}{2\pi} \int_\Gamma \varphi(\zeta) \ln \frac{1}{|\zeta - z|} \, ds(\zeta), \quad z \in \mathbb{R}^2 \setminus \Gamma,$$

is given by

$$\operatorname{grad} u(z) = \frac{1}{2\pi} \int_\Gamma \varphi(\zeta) \frac{\zeta - z}{|\zeta - z|^2} \, ds(\zeta), \quad z \in \mathbb{R}^2 \setminus \Gamma.$$

We observe that the gradient of the single-layer potential u corresponds to the complex conjugate of the holomorphic function f given by the Cauchy integral (7.39).

Therefore, observing the fact that the complex conjugate of $ie^{-i\gamma(z)}$ corresponds to the normal $\nu(z)$ at $z \in \Gamma$, we can again use Theorem 7.8 to prove the following theorem.

Theorem 7.30. *The first derivatives of the logarithmic single-layer potential u with Hölder continuous density φ can be extended uniformly Hölder continuously from D into $\bar{D}$ and from $\mathbb{R}^2 \setminus \bar{D}$ into $\mathbb{R}^2 \setminus D$ with limiting values*

$$\operatorname{grad} u_{\pm}(z) = \frac{1}{2\pi} \int_\Gamma \varphi(\zeta) \operatorname{grad} \ln \frac{1}{|\zeta - z|} \, ds(\zeta) \mp \frac{1}{2} \, \nu(z)\varphi(z), \quad z \in \Gamma. \tag{7.41}$$

Furthermore, we have the inequalities

$$\| \operatorname{grad} u\|_{\alpha,\bar{D}} \le C\|\varphi\|_\alpha \quad \text{and} \quad \| \operatorname{grad} u\|_{\alpha,\mathbb{R}^2\setminus D} \le C\|\varphi\|_\alpha$$

for some constant C depending on α and Γ.

Recall that by $C^{1,\alpha}(\Gamma)$ we denote the normed space of all functions defined on Γ that have a uniformly Hölder continuous first derivative furnished with the norm

$$\|\varphi\|_{1,\alpha} := \|\varphi\|_\infty + \|\varphi'\|_{0,\alpha}$$

with the prime indicating differentiation with respect to arc length.

Corollary 7.31. *The single-layer operator $S : C^{0,\alpha}(\Gamma) \to C^{1,\alpha}(\Gamma)$ given by*

$$(S\varphi)(z) := \frac{1}{\pi} \int_\Gamma \varphi(\zeta) \ln \frac{1}{|\zeta - z|} \, ds(\zeta), \quad z \in \Gamma, \tag{7.42}$$

is bounded.

Consider once again the logarithmic double-layer potential v and assume that $\varphi \in C^{1,\alpha}(\Gamma)$. For $a = (a_1, a_2) \in \mathbb{R}^2$ we denote $a^\perp = (a_2, -a_1)$. Then, for $r = |\zeta - z|$, we write

$$\frac{\partial}{\partial \nu(\zeta)} \ln \frac{1}{r} = \nu(\zeta) \cdot \operatorname{grad}_\zeta \ln \frac{1}{r} = -\nu(\zeta) \cdot \operatorname{grad}_z \ln \frac{1}{r} = -\operatorname{div}_z \left(\ln \frac{1}{r} \, [\tau(\zeta)]^\perp \right)$$

where we have made use of the relation $\nu = \tau^\perp$ between the tangent vector τ and the normal vector ν. From this, using the vector identity

$$\operatorname{grad} \operatorname{div} w = \Delta w + [\operatorname{grad} \operatorname{div} w^\perp]^\perp,$$

the identity $[w^\perp]^\perp = -w$, and $\Delta \ln 1/r = 0$, we see that

$$\operatorname{grad}_z \frac{\partial}{\partial \nu(\zeta)} \ln \frac{1}{r} = \left[\operatorname{grad}_z \left(\tau(\zeta) \cdot \operatorname{grad}_z \ln \frac{1}{r} \right) \right]^\perp .$$

Then, by partial integration, for the gradient of the double-layer potential we obtain that

$$\begin{aligned}\operatorname{grad} v(z) &= \frac{-1}{2\pi}\left[\operatorname{grad}\int_\Gamma \tau(\zeta)\cdot\operatorname{grad}_\zeta \ln\frac{1}{|\zeta - z|}\,\varphi(\zeta)\,ds(\zeta)\right]^\perp\\ &= \frac{1}{2\pi}\left[\operatorname{grad}\int_\Gamma \frac{d\varphi}{ds}(\zeta)\ln\frac{1}{|\zeta - z|}\,ds(\zeta)\right]^\perp, \quad z\in\mathbb{R}^2\setminus\Gamma.\end{aligned}$$

Therefore, from Theorem 7.30 we derive the following result which is due to Maue [164].

Theorem 7.32. *The first derivatives of the logarithmic double-layer potential v with density $\varphi \in C^{1,\alpha}(\Gamma)$ can be extended uniformly Hölder continuously from D into $\bar{D}$ and from $\mathbb{R}^2 \setminus \bar{D}$ into $\mathbb{R}^2 \setminus D$. The normal derivative is given by*

$$\frac{\partial v_\pm}{\partial \nu}(z) = \frac{1}{2\pi}\frac{d}{ds(z)}\int_\Gamma \frac{d\varphi}{ds}(\zeta)\ln\frac{1}{|\zeta - z|}\,ds(\zeta), \quad z\in\Gamma, \tag{7.43}$$

and the tangential derivative by

$$\frac{\partial v_\pm}{\partial s}(z) = -\frac{1}{2\pi}\int_\Gamma \frac{d\varphi}{ds}(\zeta)\frac{\partial}{\partial\nu(z)}\ln\frac{1}{|\zeta - z|}\,ds(\zeta) \pm \frac{1}{2}\frac{d\varphi}{ds}(z), \quad z\in\Gamma. \tag{7.44}$$

Furthermore, we have the inequalities

$$\|\operatorname{grad} v\|_{\alpha,\bar{D}} \le C\|\varphi\|_{1,\alpha} \quad \textit{and} \quad \|\operatorname{grad} v\|_{\alpha,\mathbb{R}^2\setminus D} \le C\|\varphi\|_{1,\alpha}$$

for some constant C depending on α and Γ.

Corollary 7.33. *The operator T, defined by the normal derivative of the logarithmic double-layer potential*

$$(T\varphi)(z) := \frac{1}{\pi}\frac{\partial}{\partial\nu(z)}\int_\Gamma \varphi(\zeta)\frac{\partial}{\partial\nu(\zeta)}\ln\frac{1}{|\zeta - z|}\,ds(\zeta), \quad z\in\Gamma, \tag{7.45}$$

can be expressed by the single-layer integral operator S in the form

$$T\varphi = \frac{d}{ds}\,S\,\frac{d\varphi}{ds}. \tag{7.46}$$

It is a bounded operator from $C^{1,\alpha}(\Gamma)$ into $C^{0,\alpha}(\Gamma)$.

Since the singularity of the operator T is stronger than the Cauchy type singularity it is called a *hypersingular operator*.

We wish to mention that these results on the regularity of single- and double-layer potentials with uniformly Hölder continuous densities can be extended to the three-dimensional case by techniques similar to those used in the proofs of Theorems 7.5, 7.8 and 7.19 (see [31]).

For the logarithmic single-layer potential with L^2 densities we have the following continuity property.

Theorem 7.34. *The logarithmic single-layer potential with L^2 density is continuous in all of $\mathbb{R}^2$.*

Proof. This can be shown analogously to Theorems 2.29 and 2.30 by cutting off the singularity and estimating with the aid of the Cauchy–Schwarz inequality.

7.6 Boundary Integral Equations in Hölder Spaces

Armed with Theorems 7.5, 7.6, 7.29, 7.30 and 7.32 together with their corollaries we now revisit the boundary integral equations from potential theory in $\mathbb{R}^m$, $m = 2, 3$, to make a few additions to the results of Section 6.4 in the classical Hölder space setting. In the sequel we always assume for the exponent that $0 < \alpha < 1$ and that $D \subset \mathbb{R}^m$ is a bounded domain with connected boundary $\partial D \in C^2$. For convenience we recall the definition of the four main boundary integral operators given by the single- and double-layer operators

$$(S\varphi)(x) := 2 \int_{\partial D} \Phi(x, y)\varphi(y)\, ds(y)$$

and

$$(K\varphi)(x) := 2 \int_{\partial D} \frac{\partial \Phi(x, y)}{\partial \nu(y)}\, \varphi(y)\, ds(y)$$

and the normal derivative operators

$$(K'\varphi)(x) := 2 \int_{\partial D} \frac{\partial \Phi(x, y)}{\partial \nu(x)}\, \varphi(y)\, ds(y)$$

and

$$(T\varphi)(x) := 2\, \frac{\partial}{\partial \nu(x)} \int_{\partial D} \frac{\partial \Phi(x, y)}{\partial \nu(y)}\, \varphi(y)\, ds(y)$$

for $x \in \partial D$. From Section 6.4 we know that K and K' are adjoint with respect to the L^2 bilinear form and, clearly, S is self-adjoint, i.e., $\langle S\varphi, \psi\rangle = \langle \varphi, S\psi\rangle$ for all $\varphi, \psi \in C(\partial D)$. To derive further properties of the boundary integral operators, let u and v denote the double-layer potentials with densities φ and ψ in $C^{1,\alpha}(\partial D)$, respectively. Then by the jump relation of Theorem 6.18, Green's Theorem 6.3 and the asymptotic behavior (6.20) we find that

$$\int_{\partial D} T\varphi\,\psi\, ds = 2 \int_{\partial D} \frac{\partial u}{\partial \nu}\, (v_+ - v_-)\, ds = 2 \int_{\partial D} (u_+ - u_-)\, \frac{\partial v}{\partial \nu}\, ds = \int_{\partial D} \varphi T\psi\, ds,$$

that is, T also is self-adjoint. Now, in addition, let w denote the single-layer potential with density $\varphi \in C(\partial D)$. Then, using the jump relations from Theorems 6.15, 6.18

and 6.19, we find

$$\int_{\partial D} S\varphi T\psi \, ds = 4\int_{\partial D} w\,\frac{\partial v}{\partial \nu}\, ds = 4\int_{\partial D} v_-\,\frac{\partial w_-}{\partial \nu}\, ds = \int_{\partial D} (K-I)\psi(K'+I)\varphi\, ds,$$

whence

$$\int_{\partial D} \varphi\, ST\psi\, ds = \int_{\partial D} \varphi(K^2 - I)\psi\, ds$$

follows for all $\varphi \in C(\partial D)$ and $\psi \in C^{1,\alpha}(\partial D)$. Thus, we have proven the relation

$$ST = K^2 - I \tag{7.47}$$

and similarly it can be shown that the adjoint relation

$$TS = K'^2 - I \tag{7.48}$$

is also valid.

Theorem 7.35. *The unique solution u to the interior Dirichlet problem with boundary value $f \in C^{1,\alpha}(\partial D)$ belongs to $C^{1,\alpha}(D)$ and depends continuously on f, i.e.,*

$$\|\operatorname{grad} u\|_{\alpha,D} \le C\|f\|_{1,\alpha,\partial D} \tag{7.49}$$

with some constant C depending on D and α. The Dirichlet-to-Neumann operator *A mapping f onto the normal derivative $\partial u/\partial\nu$ on ∂D is a bounded operator $A : C^{1,\alpha}(\partial D) \to C^{0,\alpha}(\partial D)$.*

Proof. By Theorem 7.6 the operator K is compact from $C^{1,\alpha}(\partial D)$ into itself. Hence, for the unique solution φ of the integral equation $\varphi - K\varphi = -2f$ from Theorem 6.22 we have that $\varphi \in C^{1,\alpha}(\partial D)$ and $\|\varphi\|_{1,\alpha} \le c\|f\|_{1,\alpha}$ with some constant c. Now $u \in C^{1,\alpha}(D)$ and the boundedness (7.49) follow from Theorem 7.32. The statement on the Dirichlet-to-Neumann operator is a consequence of $A = -T(I-K)^{-1}$ and the mapping properties of T and K. □

Theorem 7.36. *For $g \in C^{0,\alpha}(\partial D)$ satisfying the solvability condition $\int_{\partial D} g\, ds = 0$, the unique solution u to the interior Neumann problem subject to the normalization condition $\int_{\partial D} u\, ds = 0$ belongs to $C^{1,\alpha}(D)$ and depends continuously on g, i.e.,*

$$\|\operatorname{grad} u\|_{\alpha,D} \le C\|g\|_{\alpha,\partial D} \tag{7.50}$$

with some constant C depending on D and α. The Neumann-to-Dirichlet operator *B mapping g onto the trace $u|_{\partial D}$ is a bounded operator $B : C_0^{1,\alpha}(\partial D) \to C_0^{0,\alpha}(\partial D)$ where the subscripts indicate the subspaces of functions with mean value zero.*

Proof. We refer back to the proof of Theorem 6.30. Since K' is compact from $C_0^{\alpha}(\partial D)$ into itself and $I + K'$ has a trivial nullspace in $C_0(\partial D)$, for the unique solution $\psi \in C_0(\partial D)$ of the integral equation $\psi + K'\psi = 2g$ from Theorem 6.26 we have that $\psi \in C_0^{\alpha}(\partial D)$ and $\|\psi\|_{\alpha} \le c\|g\|_{\alpha}$ with some constant c. Now $u \in C^{1,\alpha}(D)$ and the boundedness (7.50) follow from Theorem 7.30. The statement on the Neumann-to-Dirichlet operator is a consequence of $B = S(I + K')^{-1}$ and the mapping properties of S and K'. □

Note that, by construction, the operators $A : C_0^{0,\alpha}(\partial D) \to C_0^{1,\alpha}(\partial D)$ and $B : C_0^{1,\alpha}(\partial D) \to C_0^{0,\alpha}(\partial D)$ are inverse to each other.

Theorem 7.37. *For a harmonic function $u \in C^2(D) \cap C^{1,\alpha}(\bar{D})$ the boundary value and the normal derivative satisfy*

$$\begin{pmatrix} u \\ \partial u/\partial \nu \end{pmatrix} = \begin{pmatrix} -K & S \\ -T & K' \end{pmatrix} \begin{pmatrix} u \\ \partial u/\partial \nu \end{pmatrix}, \tag{7.51}$$

i.e., the operator in (7.51) is a projection operator in the product space of the boundary values and the normal derivatives of harmonic functions. This projection operator is known as the Calderón projection.

Proof. This follows from Green's representation formula (6.4) using the jump relations. □

Note that by inserting a single- or a double-layer potential into (7.51) the relations (7.47) and (7.48) can again be obtained.

Obviously, given the Dirichlet data f, the second equation in (7.51) leads to

$$g - K'g = -Tf \tag{7.52}$$

as an integral equation of the second kind for the unknown Neumann data g. The operator in (7.52) is the adjoint of the operator in the equation $\varphi - K\varphi = -2f$ from the double-layer potential approach (see Theorem 6.22). Therefore, (7.52) is uniquely solvable.

Conversely, given the Neumann data g, the first equation in (7.51) leads to

$$f + Kf = Sg \tag{7.53}$$

as an integral equation of the second kind for the unknown Dirichlet data. The operator in this equation is the adjoint of the operator in the equation $\varphi + K'\varphi = 2g$ from the single-layer potential approach (see Theorem 6.26). The solvability condition from the Fredholm alternative is satisfied for (7.53). For the solution of the homogeneous adjoint equation ψ_0 from Theorem 6.21 we have that $S\psi_0 =$ constant and therefore $\langle \psi_0, Sg \rangle = \langle S\psi_0, g \rangle = 0$ provided g satisfies the solvability condition for the interior Neumann problem.

The integral equations (7.52) and (7.53) obtained from Green's representation theorem are also known as the *direct approach* to boundary integral equations for the Dirichlet and Neumann problems. As an advantage of this approach we note that (7.52) and (7.53) are equations for the physically interesting unknown Neumann or Dirichlet data, respectively, which in the potential approach that we have considered so far need an additional evaluation of a potential. However, there is a conservation of computational costs since the equations (7.52) and (7.53) require evaluations of potentials for their right-hand sides. We also note that, if these equations are used without the parallel discussion of the integral equations from the potential approach, for a complete existence analysis one still has to show that solutions to the integral equations (7.52) and (7.53) indeed lead to solutions of the boundary value problems.

7.7 Boundary Integral Equations of the First Kind

We will conclude this chapter by considering integral equations of the first kind for the solution of the Dirichlet and Neumann problems. Given $f \in C^{1,\alpha}(\partial D)$, the single-layer potential

$$u(x) := \int_{\partial D} \Phi(x,y)\varphi(y)\,ds(y), \quad x \in \mathbb{R}^m, \tag{7.54}$$

solves the interior Dirichlet problem in D provided the density $\varphi \in C^{0,\alpha}(\partial D)$ is a solution of the integral equation

$$S\varphi = 2f. \tag{7.55}$$

In two dimensions, where (7.55) is known as Symm's [228] integral equation, due to the logarithmic fundamental solution the operator S, in general, is not injective. For example, from Lemma 8.23 we can conclude that $S1 = 0$ when ∂D is the unit circle. The following theorem that we adopt from [126] ensures injectivity of S under a geometric assumption.

Theorem 7.38. *In two dimensions, assume there exist $z_0 \in D$ such that $|x - z_0| \neq 1$ for all $x \in \Gamma$. Then the single-layer operator $S : C(\partial D) \to C(\partial D)$ is injective.*

Proof. By continuity, from $|x - z_0| \neq 1$ for all $x \in \partial D$ we have that either $|x - z_0| < 1$ for all $x \in \partial D$ or $|x - z_0| > 1$ for all $x \in \partial D$. We consider the case where $|x - z_0| < 1$ for all $x \in \partial D$ and choose a neighborhood U of z_0 such that $U \subset D$ and $|x - z| < 1$ for all $x \in \partial D$ and all $z \in U$. For a solution $\varphi \in C(\partial D)$ of $S\varphi = 0$ the single-layer potential u defined by (7.54) satisfies $u = 0$ on ∂D and by Theorem 6.15 is continuous in $\mathbb{R}^2$. The maximum-minimum principle (Theorem 6.9) implies that $u = 0$ in D.

We will show that $u = 0$ also in $\mathbb{R}^2 \setminus D$. To this end we prove that $\int_{\partial D} \varphi\,ds = 0$ and assume without loss of generality that $\int_{\partial D} \varphi\,ds \geq 0$. We consider the harmonic function

$$v(x) := u(x) + \frac{1}{2\pi} \ln|x - z| \int_{\partial D} \varphi\,ds, \quad x \in \mathbb{R}^2 \setminus D,$$

for some $z \in U$. Then

$$v(x) = \frac{1}{2\pi} \ln|x - z| \int_{\partial D} \varphi\,ds \leq 0, \quad x \in \partial D,$$

since $|x-z| < 1$ for $x \in \partial D$ and $z \in U$. Furthermore, elementary estimates show that

$$v(x) = \frac{1}{2\pi\,|x|^2}\, x \cdot \int_{\partial D} (y-z)\,\varphi(y)\,ds(y) + O\left(\frac{1}{|x|^2}\right), \quad x \to \infty. \tag{7.56}$$

Consequently $v(x) \to 0, x \to \infty$, and the maximum-minimum principle implies that $v(x) \le 0$ for all $x \in \mathbb{R}^2 \setminus D$. Thus we can conclude from (7.56) that

$$e \cdot \int_{\partial D} (y-z)\,\varphi(y)\,ds(y) \le 0$$

for all $e \in \mathbb{R}^2$ with $|e| = 1$ which in turn yields

$$\int_{\partial D} y\,\varphi(y)\,ds(y) = z \int_{\partial D} \varphi(y)\,ds(y).$$

Since this holds for all $z \in U$ we finally conclude that $\int_{\partial D} \varphi\,ds = 0$.

Now we have $u = v$ and consequently $u(x) \to 0, x \to \infty$. Therefore, in view of $u = 0$ on ∂D, by the uniqueness for the exterior Dirichlet problem (Theorem 6.12) we obtain that that $u = 0$ in $\mathbb{R}^2 \setminus D$ and consequently $u = 0$ in $\mathbb{R}^2$. Hence $\varphi = 0$ follows by the jump relation of Theorem 6.19.

The case where $|x - z_0| > 1$ for all $x \in \partial D$ is treated analogously. □

Theorem 7.39. *In three dimensions the single-layer operator $S : C(\partial D) \to C(\partial D)$ is injective.*

Proof. For a solution $\varphi \in C(\partial D)$ to $S\varphi = 0$ the single-layer potential (7.54) solves both the homogeneous interior Dirichlet problem in D and the homogeneous exterior Dirichlet problem in $\mathbb{R}^3 \setminus \bar{D}$. Therefore $u = 0$ in $\mathbb{R}^3$ by Theorem 6.12 whence $\varphi = 0$ follows by the jump relation of Theorem 6.19. □

Theorem 7.40. *For $f \in C^{1,\alpha}(\partial D)$ there exists a unique solution of the integral equation of the first kind (7.55) if $m = 3$ or if $m = 2$ and the geometric condition of Theorem 7.38 is satisfied.*

Proof. By Theorems 7.5 and 7.6 the operators $K' : C^{0,\alpha}(\partial D) \to C^{0,\alpha}(\partial D)$ and $K : C^{1,\alpha}(\partial D) \to C^{1,\alpha}(\partial D)$ are compact and by Corollaries 7.31 and 7.33 the operators $S : C^{0,\alpha}(\partial D) \to C^{1,\alpha}(\partial D)$ and $T : C^{1,\alpha}(\partial D) \to C^{0,\alpha}(\partial D)$ are bounded. Therefore by (7.47) and (7.48) the self-adjoint operator T is a regularizer of the self-adjoint operator S (and vice versa). Now Theorem 5.6 implies that S is a Fredholm operator with index zero and the statement of the theorem follows from the two uniqueness results of Theorems 7.38 and 7.39. □

To avoid the geometric condition in the uniqueness result of Theorem 7.38 for Symm's integral equation we describe a modified single-layer approach for the Dirichlet problem in $\mathbb{R}^2$. For notational brevity we introduce the mean value operator M by

$$M : \varphi \mapsto \frac{1}{|\partial D|} \int_{\partial D} \varphi\,ds. \tag{7.57}$$

Then the modified single-layer potential

$$u(x) := \int_{\partial D} \Phi(x, y)[\varphi(y) - M\varphi]\, ds(y) + M\varphi, \quad x \in \mathbb{R}^2 \setminus \partial D, \tag{7.58}$$

solves the interior (and the exterior) Dirichlet problem with boundary condition $u = f$ on ∂D provided the density $\varphi \in C^{0,\alpha}(\partial D)$ solves the integral equation

$$S_0\varphi = 2f \tag{7.59}$$

with the operator

$$S_0 := S - SM + M. \tag{7.60}$$

Note that our modification of the logarithmic single-layer potential in (7.58) ensures boundedness of u at infinity for the exterior problem. As we will see it is also crucial for the interior problem since it ensures injectivity of the modified operator S_0.

Theorem 7.41. *The modified single-layer operator* $S_0 : C^{0,\alpha}(\Gamma) \to C^{1,\alpha}(\Gamma)$ *is bijective.*

Proof. Clearly, S_0 is a compact perturbation of S since the difference $S_0 - S$ is bounded and has finite-dimensional range. Therefore the operator T remains a regularizer of S_0. Hence by Theorem 5.6 is suffices to show that S_0 is injective. Let $\varphi \in C(\partial D)$ be a solution to $S_0\varphi = 0$. Then the single-layer potential u given by (7.58) solves the homogeneous interior and exterior Dirichlet problem. Hence, by the uniqueness Theorem 6.12, we have $u = 0$ in D and in $\mathbb{R}^2 \setminus D$. From the behavior at infinity we deduce $\int_{\partial D} \varphi\, ds = 0$, and then the jump relation of Theorem 6.19 implies $\varphi = 0$ on ∂D. □

The single-layer potential (7.54) with the solution φ of the integral equation of the first kind (7.55) also provides the solution to the exterior Dirichlet problem in three dimensions. In the two-dimensional case the solution to the exterior Dirichlet problem is obtained by the modified single-layer potential and the solution of (7.59).

We now turn to the Neumann problem. Given $g \in C^{0,\alpha}(\partial D)$, the double-layer potential

$$v(x) := \int_{\partial D} \frac{\partial \Phi(x, y)}{\partial \nu(y)}\, \varphi(y)\, ds(y), \quad x \in \mathbb{R}^m \setminus \partial D, \tag{7.61}$$

solves the interior Neumann problem in D provided the density $\varphi \in C^{1,\alpha}(\partial D)$ is a solution of the integral equation

$$T\varphi = 2g. \tag{7.62}$$

Theorem 7.42. *The hypersingular operator* $T : C^{1,\alpha}(\partial D) \to C^{0,\alpha}(\partial D)$ *has nullspace* $N(T) = \operatorname{span}\{1\}$.

Proof. For a solution $\varphi \in C^{1,\alpha}(\partial D)$ to $T\varphi = 0$ the double-layer potential (7.61) solves both the homogeneous interior Neumann problem in D and the homogeneous exterior Neumann problem in $\mathbb{R}^m \setminus \bar{D}$. Therefore $v = 0$ in $\mathbb{R}^m \setminus \bar{D}$ and $v =$ constant in D by Theorem 6.13 whence $N(T) \subset \text{span}\{1\}$ follows by the jump relation of Theorem 6.18. The proof is completed by the observation that $T1 = 0$ by Example 6.17. □

Theorem 7.43. *For each* $g \in C^{0,\alpha}(\partial D)$ *satisfying the solvability condition* $\int_{\partial D} g\, ds = 0$ *for the interior Neumann problem there exists a solution of the integral equation of the first kind* (7.62)*. Two solutions can differ only by a constant.*

Proof. In the proof of Theorem 7.40 we already noted that the self-adjoint operator S is a regularizer of the self-adjoint operator T. Thus the statement of the theorem again follows from Theorem 5.6. □

We note that in two dimensions, with the aid of either Theorem 7.40 or Theorem 7.41, the hypersingular integral equation (7.62) can also be solved using the relation (7.46) (see also Problem 7.4).

The double-layer potential (7.61) with the solution φ of (7.62) also provides the solution to the exterior Neumann problem in the two-dimensional case. Recall from Theorem 6.29 that the condition $\int_{\partial D} g\, ds = 0$ is also necessary for the exterior Neumann problem. In three dimensions, in order to obtain the correct asymptotics at infinity, we add the single-layer potential u with density given by the natural charge ψ_0 from Theorem 6.21 to the double-layer potential (7.61), noting that $\partial u_+/\partial \nu = -\psi_0$ on ∂D, see (6.38). The modified double-layer potential

$$v(x) := \int_{\partial D} \frac{\partial \Phi(x, y)}{\partial \nu(y)}\, \varphi(y)\, ds(y) - \int_{\partial D} g\, ds \int_{\partial D} \Phi(x, y)\psi_0(y)\, ds(y), \quad x \in \mathbb{R}^3 \setminus \partial D,$$

solves the exterior Neumann problem in three dimensions provided the density $\varphi \in C^{1,\alpha}(\partial D)$ solves the integral equation

$$T\varphi = 2g - 2\psi_0 \int_{\partial D} g\, ds. \tag{7.63}$$

Since $\int_{\partial D} \psi_0\, ds = 1$ the right-hand side of (7.63) is in the range of T.

We conclude this section by noting that we can also obtain boundary integral equations of the first kind in terms of the operator S and T from the Calderón projection in the direct approach. These two equations coincide with (7.52) and (7.53) with the roles of the known and unknown functions reversed. We leave their discussion as an exercise to the reader.

7.8 Logarithmic Single-Layer Potential on an Arc

In the final section of this chapter we wish to indicate how the modifications of the integral equations of the first kind of Section 7.7 arising from the Dirichlet or Neumann problem for the exterior of an arc can be reduced to the case of a closed boundary contour by using a cosine substitution as suggested by Multhopp [178].

For brevity we confine ourselves to the case of the single-layer potential for the Dirichlet problem. For the Neumann problem we refer to Mönch [172].

Assume that $\Gamma \subset \mathbb{R}^2$ is an arc of class C^3, i.e.,

$$\Gamma = \{x(s) : s \in [-1, 1]\},$$

where $x : [-1, 1] \to \mathbb{R}^2$ is an injective and three times continuously differentiable function with $x'(s) \neq 0$ for all $s \in [-1, 1]$. We consider the Dirichlet problem for the Laplace equation in the exterior of the arc Γ: Given a function $f \in C^{1,\alpha}(\Gamma)$, find a bounded solution $u \in C^2(\mathbb{R}^2 \setminus \Gamma) \cap C(\mathbb{R}^2)$ to the Laplace equation

$$\Delta u = 0 \quad \text{in } \mathbb{R}^2 \setminus \Gamma \tag{7.64}$$

satisfying the Dirichlet boundary condition

$$u = f \quad \text{on } \Gamma. \tag{7.65}$$

Note that besides continuity we do not explicitly assume any condition for the behavior of the solution at the two endpoints $z_1 := x(1)$ and $z_{-1} := x(-1)$ of Γ.

Theorem 7.44. *The Dirichlet problem for the exterior of an arc has at most one solution.*

Proof. The statement follows analogously to the proof of Theorem 6.12 from the maximum-minimum principle and the boundedness condition at infinity. □

As in the case of a closed boundary curve (compare (7.58)) we will establish existence by seeking the solution in the form

$$u(x) := \frac{1}{2\pi} \int_\Gamma [\psi(y) - M\psi] \ln \frac{1}{|y - x|}\, ds(y) + M\psi, \quad x \in \mathbb{R}^2 \setminus \Gamma, \tag{7.66}$$

of a single-layer potential. Here, analogous to (7.57) we have abbreviated

$$M\psi := \frac{1}{|\Gamma|} \int_\Gamma \psi\, ds.$$

For the case of the arc Γ the density ψ is assumed to be of the form

$$\psi(x) = \frac{\tilde{\psi}(x)}{\sqrt{|x - z_1|\,|x - z_{-1}|}}, \quad x \in \Gamma \setminus \{z_1, z_{-1}\}, \tag{7.67}$$

where $\tilde{\psi} \in C^{0,\alpha}(\Gamma)$. Note that the integrals in (7.66) exist as improper integrals with respect to the two endpoints of Γ.

Recall the cut-off function h from the proof of Theorem 2.29 and, for $n \in \mathbb{N}$, denote by u_n the single-layer potential (7.66) with the density $\psi - M\psi$ replaced by χ_n, where $\chi_n(x) := h(n|x - z_1|)h(n|x - z_{-1}|)[\psi(x) - M\psi]$. Then, by interpreting Γ as part of a closed contour $\tilde{\Gamma}$ and setting $\chi_n = 0$ on $\tilde{\Gamma} \setminus \Gamma$, from Theorem 6.15 we conclude that u_n is continuous in all of $\mathbb{R}^2$. For sufficiently large n we have

$$\int_{\Gamma(z_1;1/n)} \frac{1}{|y-z_1|^{3/4}}\, ds(y) \le \frac{C_1}{n^{1/4}}$$

for some constant C_1. Therefore, using Hölder's inequality for integrals and (7.67), we can estimate

$$\int_{\Gamma(z_1;1/n)} \left| [\psi(y) - M\psi] \ln \frac{1}{|y-x|} \right| ds(y) \le \frac{C}{n^{1/6}} \left(\int_{\Gamma} \left| \ln \frac{1}{|y-x|} \right|^3 ds(y) \right)^{1/3}$$

for sufficiently large n and some constant C. Analogous to the logarithmic single-layer potential the integral on the right-hand side is continuous and therefore is bounded on compact subsets of $\mathbb{R}^2$. Therefore we can conclude that

$$\int_{\Gamma(z_1;1/n)} \left| [\psi(y) - M\psi] \ln \frac{1}{|y-x|} \right| ds(y) \to 0, \quad n \to \infty,$$

uniformly on compact subsets of $\mathbb{R}^2$. From this and the corresponding property for the integral over a neighborhood of the other endpoint z_{-1}, it follows that u is the uniform limit of a sequence of continuous functions and therefore is continuous in all of $\mathbb{R}^2$ (see also [91], p. 276). Therefore, as in the case of a closed boundary curve (compare (7.59)), the single-layer potential (7.66) solves the Dirichlet problem (7.64)–(7.65) provided the density satisfies the integral equation

$$\frac{1}{2\pi} \int_{\Gamma} [\psi(y) - M\psi] \ln \frac{1}{|y-x|}\, ds(y) + M\psi = f(x), \quad x \in \Gamma. \tag{7.68}$$

From the uniqueness Theorem 7.44 and the jump relations of Theorem 6.19 we can conclude that the integral equation (7.68) has at most one solution.

Following Multhopp [178] we substitute $s = \cos t$ into the parameterization $\Gamma = \{x(s) : s \in [-1, 1]\}$ to transform the integral equation (7.68) into the parameterized form

$$\frac{1}{4\pi} \int_0^{\pi} \left\{ -\ln(4[\cos t - \cos \tau]^2) + p(t,\tau) \right\} \varphi(\tau)\, d\tau = g(t), \quad t \in [0, \pi]. \tag{7.69}$$

Here we have set

$$\varphi(t) := |\sin t\, x'(\cos t)|\, \psi(x(\cos t)), \quad g(t) := f(x(\cos t)), \quad t \in [0, \pi],$$

and the kernel p is given by

$$p(t,\tau) := \ln \frac{4[\cos t - \cos \tau]^2}{|x(\cos t) - x(\cos \tau)|^2} + \frac{4\pi}{|\Gamma|}$$

$$-\frac{1}{|\Gamma|} \int_0^{\pi} \ln \frac{1}{|x(\cos t) - x(\cos \sigma)|^2}\, |x'(\cos \sigma)| \sin \sigma\, d\sigma$$

for $t \neq \tau$. (Compare Problem 7.2 and use (7.70).) Solving the integral equation (7.68) for the function ψ on Γ is equivalent to solving the integral equation (7.69) for the even 2π-periodic function φ.

Writing

$$\frac{x(s) - x(\sigma)}{s - \sigma} = \int_0^1 x'[s + \lambda(\sigma - s)]\, d\lambda,$$

it can be seen that the kernel p can be extended as a twice continuously differentiable function on $\mathbb{R}^2$ that is even and 2π-periodic with respect to both variables. Hence, in view of Theorem 7.4 the integral operator

$$(L\varphi)(t) := \frac{1}{4\pi} \int_0^\pi \{p(t, \tau) - 4\}\varphi(\tau)\, d\tau, \quad t \in [0, \pi],$$

is a compact operator from the Hölder space $C^{0,\alpha}_{2\pi,e} \subset C^{0,\alpha}(\mathbb{R})$ of even 2π-periodic Hölder continuous functions into the Hölder space of even 2π-periodic Hölder continuously differentiable functions $C^{1,\alpha}_{2\pi,e} \subset C^{1,\alpha}(\mathbb{R})$. From the identity

$$\ln(4[\cos t - \cos \tau]^2) = \ln\left(4 \sin^2 \frac{t - \tau}{2}\right) + \ln\left(4 \sin^2 \frac{t + \tau}{2}\right) \tag{7.70}$$

it follows that for even and 2π-periodic functions φ and g the integral equation

$$\frac{1}{4\pi} \int_0^\pi \left\{- \ln(4[\cos t - \cos \tau]^2) + 4\right\} \varphi(\tau)\, d\tau = g(t), \quad t \in [0, \pi],$$

is equivalent to

$$\frac{1}{4\pi} \int_0^{2\pi} \left\{- \ln\left(4 \sin^2 \frac{t - \tau}{2}\right) + 2\right\} \varphi(\tau)\, d\tau = g(t), \quad t \in [0, 2\pi].$$

From Problem 7.2, by choosing $x(t) = (\cos t, \sin t)$ as the parameterization of the unit circle, it can be seen that the latter equation corresponds to the parameterized form of the single-layer integral equation (7.59) for the case of the unit disk D. Hence, as a consequence of Theorem 7.41, the integral operator

$$(L_0\varphi)(t) := \frac{1}{4\pi} \int_0^\pi \left\{- \ln(4[\cos t - \cos \tau]^2) + 4\right\} \varphi(\tau)\, d\tau, \quad t \in [0, \pi],$$

is a bounded and bijective operator from $C^{0,\alpha}_{2\pi,e}$ into $C^{1,\alpha}_{2\pi,e}$ with a bounded inverse. Therefore, writing (7.69) in the form $L_0\varphi + L\varphi = g$, from Corollary 3.6 and the injectivity for (7.68) we have the following theorem.

Theorem 7.45. *For each $f \in C^{1,\alpha}(\Gamma)$ the single-layer integral equation (7.68) has a unique solution ψ of the form (7.67) with $\widetilde{\psi} \in C^{0,\alpha}(\Gamma)$.*

From this we finally conclude the following result.

Theorem 7.46. *The Dirichlet problem (7.64)–(7.65) for the exterior of an arc is uniquely solvable.*

From the form (7.67) of the density of the single-layer potential we expect that the gradient of the solution develops singularities at the endpoints of the arc. For a closer study of the nature of these singularities we refer to Grisvard [73].

Problems

7.1. Let $g \in C^{0,\alpha}(\Gamma)$ be a nowhere vanishing function. Show that the inhomogeneous Riemann problem with boundary conditions $f_- = gf_+ + h$ on Γ is solvable if and only if the condition $\int_\Gamma h\widetilde{f}_- \, dz = 0$ is satisfied for all solutions $\widetilde{f}$ to the homogeneous adjoint Riemann problem with boundary condition $\widetilde{f}_+ = g\widetilde{f}_-$.

7.2. Use a regular 2π-periodic parameterization $\Gamma = \{x(t) : 0 \le t \le 2\pi\}$ with counterclockwise orientation for the boundary curve to transform the integral equation of the first kind (7.59) for the Dirichlet problem into the form

$$\frac{1}{4\pi}\int_0^{2\pi}\left\{-\ln\left(4\sin^2\frac{t-\tau}{2}\right) + p(t,\tau)\right\}\widetilde{\psi}(\tau)\,d\tau = \widetilde{f}(t), \quad 0 \le t \le 2\pi,$$

where $\widetilde{\psi}(t) := |x'(t)|\,\psi(x(t))$, $\widetilde{f}(t) := f(x(t))$, and where the kernel is given by

$$p(t,\tau) := \ln\frac{4\sin^2\frac{t-\tau}{2}}{|x(t)-x(\tau)|^2} - \frac{1}{|\Gamma|}\int_0^{2\pi}\ln\frac{1}{|x(t)-x(\sigma)|^2}\,|x'(\sigma)|\,d\sigma + \frac{4\pi}{|\Gamma|}, \quad t \ne \tau.$$

Show that the kernel p is continuously differentiable provided Γ is of class C^2. Transform the integral equation of the first kind (7.62) for the Neumann problem into the form

$$\frac{1}{4\pi}\int_0^{2\pi}\left\{\cot\frac{\tau-t}{2} + q(t,\tau)\right\}\widetilde{\varphi}'(\tau)\,d\tau = \widetilde{g}(t), \quad 0 \le t \le 2\pi,$$

where $\widetilde{\varphi}(t) := \varphi(x(t))$, $\widetilde{g}(t) := |x'(t)|\,g(x(t))$, and where the kernel is given by

$$q(t,\tau) := 2\,\frac{x'(t)\cdot[x(\tau)-x(t)]}{|x(t)-x(\tau)|^2} - \cot\frac{\tau-t}{2}, \quad t \ne \tau.$$

Show that the kernel q is continuous provided Γ is of class C^2.
Hint: Use the integrals of Lemma 8.23 from the next chapter for $m = 0$.

7.3. Formulate and prove Noether's theorems for the singular integral equation with Cauchy kernel where the complex integration is replaced by contour integration.
Hint: Use (7.36).

7.4. Show that in two dimensions the bounded operator $R : C^{1,\alpha}(\partial D) \to C^{0,\alpha}(\partial D)$ given by $Rf := f' + Mf$ in terms of the derivative f' with respect to arc length and the mean value operator (7.57) is injective. Therefore Symm's integral equation (7.55) is equivalent to $RSf = 2Rf$. Use this to conclude that RS differs from the singular operator with Cauchy kernel

$$k(z,\zeta) = (\zeta - z)\,\frac{\partial}{\partial s(z)}\ln\frac{1}{|\zeta - z|}, \quad z \ne \zeta,$$

only by a compact operator. Show that

$$k(z,\zeta) == \frac{\mathrm{Re}\{\overline{h(z)}\,(\zeta - z)\}}{|\zeta - z|^2}\,(\zeta - z) = \frac{1}{2}\,h(z) + \frac{1}{2}\,\overline{h(z)}\,\frac{(\zeta - z)^2}{|\zeta - z|^2}, \quad z \neq \zeta,$$

where $\tau = (\tau_1, \tau_2)$ is the unit tangent vector to ∂D and $h(z) := \tau_1(z) + i\tau_2(z)$. Use a parametric representation of ∂D to conclude that

$$\frac{(\zeta - z)^2}{|\zeta - z|^2}, \quad z \neq \zeta,$$

can be extended as a continuously differentiable function onto $\partial D \times \partial D$ and therefore k belongs to $C^{0,1,1}(\partial D \times \partial D)$. Consequently one can apply Noether's theorems in the form of Problem 7.3 to obtain an alternative proof for the existence result of Theorem 7.40. The same approach can be used for the modified single-layer approach of Theorem 7.41 which is equivalent to $RS_0 f = 2Rf$. The idea to reduce the logarithmic single-layer potential integral equation by differentiation to a Cauchy type integral equation is due to Fichera [53] and in a modified version to Hsiao and MacCamy [102].

7.5. Show that for each real-valued function $\varphi \in C^{0,\alpha}(\Gamma)$ there exists a function f that is holomorphic in D, which can be extended continuously from D into $\bar{D}$ and that has real part $\mathrm{Re}\, f = \varphi$ on the boundary Γ. Show that two functions with this property can differ only by a purely imaginary constant.
Hint: Solve a Dirichlet problem for the real part of f and use (7.37) and Problem 7.4.

Chapter 8
Sobolev Spaces

In this chapter we study the concept of weak solutions to boundary value problems for harmonic functions. We shall extend the classical theory of boundary integral equations as described in the two previous chapters from the spaces of continuous or Hölder continuous functions to appropriate Sobolev spaces. For the sake of brevity we will confine ourselves to interior boundary value problems in two dimensions.

We wish to mention that our introduction of Sobolev spaces of periodic functions and Sobolev spaces on a closed contour differs from the usual approach in two regards. Firstly, instead of using the Fourier transform we only need Fourier series, and secondly, the Sobolev spaces in the contour case are defined via a global rather than a local procedure. The motivation for our approach is the hope that this analysis might make the basic ideas on these Sobolev spaces more easily accessible. We also want to emphasize that we do not rely heavily on Lebesgue integration as a necessary prerequisite. For our purpose, it will be sufficient to understand the Hilbert space $L^2[0, 2\pi]$ as the completion of the space $C[0, 2\pi]$ of continuous functions with respect to the mean square norm.

8.1 The Sobolev Space $H^p[0, 2\pi]$

As the basis of our presentation of Sobolev spaces we begin with a brief review of the classical Fourier series expansion. For a function $\varphi \in L^2[0, 2\pi]$ the series

$$\sum_{m=-\infty}^{\infty} \hat{\varphi}_m e^{imt}, \tag{8.1}$$

where

$$\hat{\varphi}_m := \frac{1}{2\pi} \int_0^{2\pi} \varphi(t) e^{-imt} dt,$$

is called the *Fourier series* of φ, its coefficients $\hat{\varphi}_m$ are called the *Fourier coefficients* of φ. On $L^2[0, 2\pi]$, as usual, the mean square norm is introduced by the scalar

R. Kress, *Linear Integral Equations*, Applied Mathematical Sciences 82,
DOI 10.1007/978-1-4614-9593-2_8, © Springer Science+Business Media New York 2014

product

$$(\varphi, \psi) := \int_0^{2\pi} \varphi(t)\overline{\psi(t)}\, dt.$$

We denote by f_m the trigonometric monomials

$$f_m(t) := e^{imt}$$

for $t \in \mathbb{R}$ and $m \in \mathbb{Z}$. Then the set $\{f_m : m \in \mathbb{Z}\}$ is an orthogonal system. By the Weierstrass approximation theorem (see [40]), the trigonometric polynomials are dense with respect to the maximum norm in the space of 2π-periodic continuous functions, and $C[0, 2\pi]$ is dense in $L^2[0, 2\pi]$ in the mean square norm. Therefore, by Theorem 1.28, the orthogonal system is complete and the Fourier series (8.1) converges in the mean square norm. Because of the orthonormality factor $\|f_m\|_2^2 = 2\pi$, Parseval's equality assumes the form

$$\sum_{m=-\infty}^{\infty} |\hat{\varphi}_m|^2 = \frac{1}{2\pi} \int_0^{2\pi} |\varphi(t)|^2 dt = \frac{1}{2\pi} \|\varphi\|_2^2. \tag{8.2}$$

We will now define subspaces $H^p[0, 2\pi]$ of $L^2[0, 2\pi]$ by requiring for their elements φ a certain decay of the Fourier coefficients $\hat{\varphi}_m$ as $|m| \to \infty$.

Definition 8.1. Let $0 \leq p < \infty$. By $H^p[0, 2\pi]$ we denote the space of all functions $\varphi \in L^2[0, 2\pi]$ with the property

$$\sum_{m=-\infty}^{\infty} (1 + m^2)^p |\hat{\varphi}_m|^2 < \infty$$

for the Fourier coefficients $\hat{\varphi}_m$ of φ. The space $H^p[0, 2\pi]$ is called a *Sobolev space*. Frequently we will abbreviate $H^p = H^p[0, 2\pi]$. Note that $H^0[0, 2\pi]$ coincides with $L^2[0, 2\pi]$.

Theorem 8.2. *The Sobolev space $H^p[0, 2\pi]$ is a Hilbert space with the scalar product defined by*

$$(\varphi, \psi)_p := \sum_{m=-\infty}^{\infty} (1 + m^2)^p \hat{\varphi}_m \overline{\hat{\psi}_m}$$

for $\varphi, \psi \in H^p[0, 2\pi]$ with Fourier coefficients $\hat{\varphi}_m$ and $\hat{\psi}_m$, respectively. Note that the norm on $H^p[0, 2\pi]$ is given by

$$\|\varphi\|_p = \left\{ \sum_{m=-\infty}^{\infty} (1 + m^2)^p |\hat{\varphi}_m|^2 \right\}^{1/2}.$$

The trigonometric polynomials are dense in $H^p[0, 2\pi]$.

Proof. We leave it as an exercise to verify that H^p is a linear space and that $(\cdot, \cdot)_p$ is a scalar product. That $(\cdot, \cdot)_p$ is well defined can be concluded from the

Cauchy–Schwarz inequality

$$\left|\sum_{m=-\infty}^{\infty}(1+m^2)^p\hat{\varphi}_m\overline{\hat{\psi}_m}\right|^2 \le \sum_{m=-\infty}^{\infty}(1+m^2)^p|\hat{\varphi}_m|^2 \sum_{m=-\infty}^{\infty}(1+m^2)^p|\hat{\psi}_m|^2.$$

To prove that H^p is complete, let (φ_n) be a Cauchy sequence, i.e., given $\varepsilon > 0$, there exists $N(\varepsilon) \in \mathbb{N}$ such that $\|\varphi_n - \varphi_k\|_p < \varepsilon$ for all $n, k \ge N(\varepsilon)$, or in terms of the Fourier coefficients $\hat{\varphi}_{m,n}$ of φ_n,

$$\sum_{m=-\infty}^{\infty}(1+m^2)^p|\hat{\varphi}_{m,n} - \hat{\varphi}_{m,k}|^2 < \varepsilon^2$$

for all $n, k \ge N(\varepsilon)$. From this we observe that

$$\sum_{m=-M_1}^{M_2}(1+m^2)^p|\hat{\varphi}_{m,n} - \hat{\varphi}_{m,k}|^2 < \varepsilon^2 \tag{8.3}$$

for all $M_1, M_2 \in \mathbb{N}$ and all $n, k \ge N(\varepsilon)$. Therefore, since $\mathbb{C}$ is complete, there exists a sequence $(\hat{\varphi}_m)$ in $\mathbb{C}$ such that $\hat{\varphi}_{m,n} \to \hat{\varphi}_m$, $n \to \infty$, for each $m \in \mathbb{Z}$. Passing to the limit $k \to \infty$ in (8.3) now yields

$$\sum_{m=-M_1}^{M_2}(1+m^2)^p|\hat{\varphi}_{m,n} - \hat{\varphi}_m|^2 \le \varepsilon^2$$

for all $M_1, M_2 \in \mathbb{N}$ and all $n \ge N(\varepsilon)$. Hence

$$\sum_{m=-\infty}^{\infty}(1+m^2)^p|\hat{\varphi}_{m,n} - \hat{\varphi}_m|^2 \le \varepsilon^2$$

for all $n \ge N(\varepsilon)$. From this we conclude that

$$\varphi := \sum_{m=-\infty}^{\infty}\hat{\varphi}_m f_m$$

defines a function $\varphi \in H^p$ with $\|\varphi - \varphi_n\|_p \to 0$, $n \to \infty$. Hence, H^p is complete.

Let $\varphi \in H^p$ with Fourier coefficients $\hat{\varphi}_m$. Then for the partial sums

$$\varphi_n := \sum_{m=-n}^{n}\hat{\varphi}_m f_m$$

of the Fourier series we have that

$$\|\varphi - \varphi_n\|_p^2 = \sum_{|m|=n+1}^{\infty}(1+m^2)^p|\hat{\varphi}_m|^2 \to 0, \quad n \to \infty,$$

since the series $\sum_{m=-\infty}^{\infty}(1+m^2)^p|\hat{\varphi}_m|^2$ converges. Therefore, the trigonometric polynomials are dense in H^p. □

Theorem 8.3. *If $q > p$ then $H^q[0,2\pi]$ is dense in $H^p[0,2\pi]$ with compact imbedding from $H^q[0,2\pi]$ into $H^p[0,2\pi]$.*

Proof. From $(1+m^2)^p \le (1+m^2)^q$ for $m \in \mathbb{Z}$ it follows that $H^q \subset H^p$ with bounded imbedding

$$\|\varphi\|_p \le \|\varphi\|_q \tag{8.4}$$

for all $\varphi \in H^q$. Then the denseness of H^q in H^p is a consequence of the denseness of the trigonometric polynomials in H^p.

We denote the imbedding operator by $I : H^q \to H^p$. For $n \in \mathbb{N}$ we define finite-dimensional bounded operators $I_n : H^q \to H^p$ by setting

$$I_n\varphi := \sum_{m=-n}^{n} \hat{\varphi}_m f_m$$

for $\varphi \in H^q$ with Fourier coefficients $\hat{\varphi}_m$. Then

$$\begin{aligned}\|(I_n - I)\varphi\|_p^2 &= \sum_{|m|=n+1}^{\infty} (1+m^2)^p|\hat{\varphi}_m|^2\\ &\le \frac{1}{(1+n^2)^{q-p}} \sum_{|m|=n+1}^{\infty} (1+m^2)^q|\hat{\varphi}_m|^2\\ &\le \frac{1}{(1+n^2)^{q-p}}\|\varphi\|_q^2.\end{aligned}$$

Hence, compactness of the imbedding follows from Theorem 2.22. □

Theorem 8.4. *Let $p > 1/2$. Then the Fourier series for $\varphi \in H^p[0,2\pi]$ converges absolutely and uniformly. Its limit is continuous and 2π-periodic and coincides with φ almost everywhere. This imbedding of the Sobolev space $H^p[0,2\pi]$ in the space $C_{2\pi}$ of 2π-periodic continuous functions is compact.*

Proof. For the Fourier series of $\varphi \in H^p[0,2\pi]$, by the Cauchy–Schwarz inequality, we conclude that

$$\left\{\sum_{m=-\infty}^{\infty} \left|\hat{\varphi}_m e^{imt}\right|\right\}^2 \le \sum_{m=-\infty}^{\infty} \frac{1}{(1+m^2)^p} \sum_{m=-\infty}^{\infty} (1+m^2)^p|\hat{\varphi}_m|^2$$

for all $t \in \mathbb{R}$. Since the series

$$\sum_{m=-\infty}^{\infty} \frac{1}{(1+m^2)^p}$$

converges for $p > 1/2$, this estimate establishes the absolute and uniform convergence of the Fourier series. Its limit is continuous and 2π-periodic because it is the uniform limit of a sequence of continuous and 2π-periodic functions. It must coincide with φ almost everywhere, because we already have convergence of the Fourier series to φ in the mean square norm, which is weaker than the maximum norm.

The above estimate also shows that

$$\|\varphi\|_\infty \leq C\|\varphi\|_p$$

for all $\varphi \in H^p[0, 2\pi]$ and some constant C depending on p. Therefore, the imbedding from $H^p[0, 2\pi]$ into $C_{2\pi}$ is bounded, and consequently, by Theorems 2.21 and 8.3, is compact for $p > 1/2$. □

For the introduction of the Sobolev spaces $H^p(\Gamma)$ for a closed contour Γ in $\mathbb{R}^2$ we shall need additional norms that are equivalent to $\|\cdot\|_p$. We will denote them by $\|\cdot\|_{p,[p]}$, where $[p]$ is the largest integer less than or equal to p. For their definition we have to distinguish three cases. For $k \in \mathbb{N}$, by $C^k_{2\pi}$ we will denote the space of k times continuously differentiable 2π-periodic functions from $\mathbb{R}$ into $\mathbb{C}$.

Theorem 8.5. *For $k \in \mathbb{N}$ we have $C^k_{2\pi} \subset H^k[0, 2\pi]$ and on $C^k_{2\pi}$ the norm $\|\cdot\|_k$ is equivalent to*

$$\|\varphi\|_{k,k} := \left(\int_0^{2\pi} \left\{|\varphi(t)|^2 + |\varphi^{(k)}(t)|^2\right\} dt\right)^{1/2}. \tag{8.5}$$

Proof. Using

$$\int_0^{2\pi} \varphi^{(k)}(t)e^{-imt}dt = (im)^k \int_0^{2\pi} \varphi(t)e^{-imt}dt \tag{8.6}$$

and Parseval's equality (8.2) for the Fourier coefficients of φ and $\varphi^{(k)}$ we obtain

$$\|\varphi\|^2_{k,k} = 2\pi \sum_{m=-\infty}^{\infty} (1 + m^{2k})|\hat{\varphi}_m|^2.$$

Now the inequalities

$$(1 + m^{2k}) \leq (1 + m^2)^k \leq (2m^2)^k \leq 2^k(1 + m^{2k}), \quad m \in \mathbb{Z},$$

yield the equivalence of the norms $\|\cdot\|_k$ and $\|\cdot\|_{k,k}$. □

Theorem 8.5 shows that for $k \in \mathbb{N}$ the Sobolev spaces $H^k[0, 2\pi]$ are isomorphic to the classical Sobolev spaces given by the completion of the space of k times continuously differentiable 2π-periodic functions with respect to the norm $\|\cdot\|_{k,k}$.

The norms considered in the following theorem and its corollaries are known as *Sobolev–Slobodeckij norms*. To some extent they resemble Hölder norms.

Theorem 8.6. *For $0 < p < 1$ on $C^1_{2\pi}$ the norm $\|\cdot\|_p$ is equivalent to*

$$\|\varphi\|_{p,0} := \left\{ \int_0^{2\pi} |\varphi(t)|^2 dt + \int_0^{2\pi} \int_0^{2\pi} \frac{|\varphi(t) - \varphi(\tau)|^2}{\left|\sin \dfrac{t-\tau}{2}\right|^{2p+1}} \, d\tau dt \right\}^{1/2}. \tag{8.7}$$

Proof. We first observe that the norm $\|\cdot\|_{p,0}$ corresponds to the scalar product

$$(\varphi, \psi) := \int_0^{2\pi} \varphi(t)\overline{\psi(t)}\, dt + \int_0^{2\pi} \int_0^{2\pi} \frac{\{\varphi(t) - \varphi(\tau)\}\overline{\{\psi(t) - \psi(\tau)\}}}{\left|\sin \dfrac{t-\tau}{2}\right|^{2p+1}} \, d\tau dt$$

on $C^1_{2\pi}$. Note that the second term on the right-hand side is well defined, since by the mean value theorem continuously differentiable functions are uniformly Hölder continuous with Hölder exponent 1. Straightforward integration shows that the trigonometric monomials f_m are orthogonal with respect to this scalar product, i.e.,

$$(f_m, f_k) = (2\pi + \gamma_m)\delta_{mk},$$

where

$$\gamma_m := 16\pi \int_0^{\pi} \frac{\sin^2 \dfrac{m}{2} t}{\sin^{2p+1} \dfrac{t}{2}} \, dt,$$

$\delta_{mk} = 1$ for $k = m$, and $\delta_{mk} = 0$ for $k \neq m$. Using $2t/\pi < \sin t < t$ for $0 < t < \pi/2$, we can estimate

$$2^{2p+5}\pi \int_0^{\pi} \frac{\sin^2 \dfrac{m}{2} t}{t^{2p+1}} \, dt < \gamma_m < 16\pi^{2p+2} \int_0^{\pi} \frac{\sin^2 \dfrac{m}{2} t}{t^{2p+1}} \, dt.$$

For $m > 0$, we substitute $mt = \tau$ and use

$$0 < \int_0^{\pi} \frac{\sin^2 \dfrac{\tau}{2}}{\tau^{2p+1}} \, d\tau \le \int_0^{m\pi} \frac{\sin^2 \dfrac{\tau}{2}}{\tau^{2p+1}} \, d\tau < \int_0^{\infty} \frac{\sin^2 \dfrac{\tau}{2}}{\tau^{2p+1}} \, d\tau < \infty$$

to obtain

$$c_0 |m|^{2p} < \gamma_m < c_1 |m|^{2p} \tag{8.8}$$

for all $m \neq 0$ with some constants c_0 and c_1 depending on p. Using the estimate $|e^{imt} - e^{im\tau}| \le |m(t-\tau)|$, from Parseval equality (8.2) for $\varphi \in C^1_{2\pi}$ and the Cauchy–Schwarz inequality, proceeding as in the proof of Theorem 8.4 we see that the series

$$\varphi(t) - \varphi(\tau) = \sum_{m=-\infty}^{\infty} \hat{\varphi}_m \left\{ e^{imt} - e^{im\tau} \right\}$$

is absolutely and uniformly convergent for all $t, \tau \in [0, 2\pi]$. Therefore we can integrate termwise to derive

$$\int_0^{2\pi} \int_0^{2\pi} \frac{|\varphi(t) - \varphi(\tau)|^2}{\left|\sin \dfrac{t-\tau}{2}\right|^{2p+1}} \, d\tau dt = \sum_{m=-\infty}^{\infty} \gamma_m |\hat{\varphi}_m|^2.$$

Now the statement of the theorem follows from the inequalities (8.8) and

$$(1 + |m|^{2p}) \le 2|m|^{2p} \le 2(1 + m^2)^p \le 2^{p+1}(1 + |m|^{2p}), \quad m \in \mathbb{Z},$$

together with Parseval's equality (8.2). □

Corollary 8.7. *When $p = k + q$, where $k \in \mathbb{N}$ and $0 < q < 1$, then on $C_{2\pi}^{k+1}$ the norm $\|\cdot\|_p$ is equivalent to*

$$\|\varphi\|_{p,k} := \left\{ \|\varphi\|_0^2 + \|\varphi^{(k)}\|_{q,0}^2 \right\}^{1/2}. \tag{8.9}$$

Proof. Combine the proofs of Theorems 8.5 and 8.6 (see Problem 8.1). □

Corollary 8.8. *For a nonnegative integer k let $f \in C_{2\pi}^k$ and assume that $0 \le p \le k$. Then for all $\varphi \in H^p[0, 2\pi]$ the product $f\varphi$ belongs to $H^p[0, 2\pi]$ and*

$$\|f\varphi\|_p \le C \left\{ \|f\|_\infty + \|f^{(k)}\|_\infty \right\} \|\varphi\|_p$$

for some constant C depending on p.

Proof. Use the equivalent norms of Theorems 8.5 and 8.6 and Corollary 8.7 (see Problem 8.1). □

Definition 8.9. For $0 \le p < \infty$ by $H^{-p}[0, 2\pi]$ we denote the *dual space* of $H^p[0, 2\pi]$, i.e., the space of *bounded linear functionals* on $H^p[0, 2\pi]$.

The space H^{-p} is characterized by the following theorem.

Theorem 8.10. *For $F \in H^{-p}[0, 2\pi]$ the norm is given by*

$$\|F\|_p = \left\{ \sum_{m=-\infty}^{\infty} (1 + m^2)^{-p} |\hat{F}_m|^2 \right\}^{1/2},$$

where $\hat{F}_m = F(f_m)$. Conversely, to each sequence $(\hat{F}_m)$ in $\mathbb{C}$ satisfying

$$\sum_{m=-\infty}^{\infty} (1 + m^2)^{-p} |\hat{F}_m|^2 < \infty$$

there exists a bounded linear functional $F \in H^{-p}[0, 2\pi]$ with $F(f_m) = \hat{F}_m$.

Proof. Assume the sequence $(\hat{F}_m)$ satisfies the required inequality and define a functional $F : H^p[0, 2\pi] \to \mathbb{C}$ by

$$F(\varphi) := \sum_{m=-\infty}^{\infty} \hat{\varphi}_m \hat{F}_m$$

for $\varphi \in H^p[0, 2\pi]$ with Fourier coefficients $\hat{\varphi}_m$. Then from the Cauchy–Schwarz inequality we have

$$|F(\varphi)|^2 \le \sum_{m=-\infty}^{\infty} (1 + m^2)^{-p} |\hat{F}_m|^2 \sum_{m=-\infty}^{\infty} (1 + m^2)^p |\hat{\varphi}_m|^2.$$

Hence, F is well defined and bounded with

$$\|F\|_p \le \left\{ \sum_{m=-\infty}^{\infty} (1 + m^2)^{-p} |\hat{F}_m|^2 \right\}^{1/2}.$$

On the other hand, for $n \in \mathbb{N}$ the function $\varphi_n := \sum_{m=-n}^{n} (1 + m^2)^{-p} \overline{\hat{F}_m} f_m$ has norm

$$\|\varphi_n\|_p = \left\{ \sum_{m=-n}^{n} (1 + m^2)^{-p} |\hat{F}_m|^2 \right\}^{1/2}.$$

Therefore

$$\|F\|_p \ge \frac{|F(\varphi_n)|}{\|\varphi_n\|_p} = \left\{ \sum_{m=-n}^{n} (1 + m^2)^{-p} |\hat{F}_m|^2 \right\}^{1/2}.$$

Hence,

$$\left\{ \sum_{m=-\infty}^{\infty} (1 + m^2)^{-p} |\hat{F}_m|^2 \right\}^{1/2} \le \|F\|_p,$$

and this completes the proof. □

Theorem 8.11. *For each function $g \in L^2[0, 2\pi]$ the sesquilinear* duality pairing

$$G(\varphi) := \frac{1}{2\pi} \int_0^{2\pi} \varphi(t) \overline{g(t)}\, dt, \quad \varphi \in H^p[0, 2\pi], \tag{8.10}$$

defines a linear functional $G \in H^{-p}[0, 2\pi]$. In this sense, $L^2[0, 2\pi]$ is a subspace of each dual space $H^{-p}[0, 2\pi]$, and the trigonometric polynomials are dense in $H^{-p}[0, 2\pi]$.

Proof. Denote by $\hat{g}_m$ the Fourier coefficients of g. Then, since $G(f_m) = \overline{\hat{g}_m}$, by the second part of Theorem 8.10 we have $G \in H^{-p}$. Let $F \in H^{-p}$ with $F(f_m) = \hat{F}_m$ and define a sequence (F_n) in H^{-p} by

$$F_n(\varphi) := \frac{1}{2\pi} \int_0^{2\pi} \varphi(t)\overline{g_n(t)}\, dt,$$

where g_n is the trigonometric polynomial $g_n := \sum_{m=-n}^{n} \overline{\hat{F}_m} f_m$. Then we have

$$\|F - F_n\|_p^2 = \sum_{|m|=n+1}^{\infty} (1 + m^2)^{-p} |\hat{F}_m|^2 \to 0, \quad n \to \infty,$$

analogous to Theorem 8.2. □

Obviously, H^{-p} becomes a Hilbert space by appropriately extending the definition of the scalar product from Theorem 8.2 to negative p. For $p = 0$ the duality map described in Theorem 8.11 is bijective with $\|G\|_0 = \|g\|_0$. Therefore, we can identify H^{-0} and H^0 and obtain a *Hilbert scale* of Hilbert spaces H^p for all real p with compact imbedding from H^q into H^p for $q > p$.

The Sobolev spaces H^p frequently are called *interpolation spaces* because of the interpolating properties indicated by the following two theorems.

Theorem 8.12. *Let $p < q$ and $r = \lambda p + (1 - \lambda)q$ with $0 < \lambda < 1$. Then*

$$\|\varphi\|_r \le \|\varphi\|_p^{\lambda} \, \|\varphi\|_q^{1-\lambda}$$

for all $\varphi \in H^q[0, 2\pi]$.

Proof. With the aid of Hölder's inequality we estimate

$$\|\varphi\|_r^2 = \sum_{m=-\infty}^{\infty} \{(1 + m^2)^p |\hat{\varphi}_m|^2\}^{\lambda} \{(1 + m^2)^q |\hat{\varphi}_m|^2\}^{1-\lambda}$$

$$\le \left\{ \sum_{m=-\infty}^{\infty} (1 + m^2)^p |\hat{\varphi}_m|^2 \right\}^{\lambda} \left\{ \sum_{m=-\infty}^{\infty} (1 + m^2)^q |\hat{\varphi}_m|^2 \right\}^{1-\lambda} = \|\varphi\|_p^{2\lambda} \, \|\varphi\|_q^{2-2\lambda},$$

and the proof is complete. □

For real s consider the operator D^s transferring the trigonometric monomials $f_m(t) = e^{imt}$ to

$$D^s f_m := (1 + m^2)^s f_m.$$

For each $p \in \mathbb{R}$ it clearly defines a bounded linear operator from H^{p+2s} onto H^p with

$$\|D^s \varphi\|_p = \|\varphi\|_{p+2s} \tag{8.11}$$

for all $\varphi \in H^{p+2s}$. Obviously, $D^{s+t} = D^s D^t$ for all $s, t \in \mathbb{R}$. In particular, D^{-s} is the inverse of D^s. For $s = 1$ the operator $D = D^1$ corresponds to the second derivative $D\varphi = \varphi - \varphi''$ if φ is twice continuously differentiable. Therefore, we may interpret

D as a generalization of this derivative for elements of the Sobolev spaces and the operators D^s as *intermediate derivatives*. With the aid of D^s we can transform the scalar products by

$$(\varphi, \psi)_{p+s} = (D^{s/2}\varphi, D^{s/2}\psi)_p \tag{8.12}$$

for all $\varphi, \psi \in H^{p+s}$ and

$$(\varphi, \psi)_{p+s} = (\varphi, D^s\psi)_p \tag{8.13}$$

for all $\varphi \in H^{p+s}$ and $\psi \in H^{p+2s}$.

Noting that $Df_m = (1 + m^2)f_m$, i.e., the f_m are eigenfunctions of the operator D with eigenvalues $1 + m^2$, we indicate obvious possibilities for extending this procedure to introduce interpolation spaces generated by other operators.

Theorem 8.13. *Let $p, q \in \mathbb{R}$, $r > 0$, and let A be a linear operator mapping H^p boundedly into H^q and H^{p+r} boundedly into H^{q+r}, i.e.,*

$$\|A\varphi\|_q \le C_1\|\varphi\|_p$$

for all $\varphi \in H^p$ and

$$\|A\varphi\|_{q+r} \le C_2\|\varphi\|_{p+r}$$

for all $\varphi \in H^{p+r}$ and some constants C_1 and C_2. Then

$$\|A\varphi\|_{q+\lambda r} \le C_1^{1-\lambda}C_2^{\lambda}\|\varphi\|_{p+\lambda r}$$

for all $0 < \lambda < 1$ and all $\varphi \in H^{p+\lambda r}$, i.e., A maps $H^{p+\lambda r}$ boundedly into $H^{q+\lambda r}$ for $0 < \lambda < 1$.

Proof. By $A^* : H^q \to H^p$ we denote the adjoint of $A : H^p \to H^q$, i.e.,

$$(A\varphi, \psi)_q = (\varphi, A^*\psi)_p \tag{8.14}$$

for all $\varphi \in H^p$ and $\psi \in H^q$. By Theorem 4.11 we have

$$\|A^*\psi\|_p \le C_1\|\psi\|_q$$

for all $\psi \in H^q$. From this, making use of (8.11), we observe that the operator

$$\widetilde{A} := D^{-r/2}A^*D^{r/2}$$

is bounded from H^{q+r} into H^{p+r} with

$$\|\widetilde{A}\psi\|_{p+r} \le C_1\|\psi\|_{q+r}$$

for all $\psi \in H^{q+r}$.

For all $\varphi \in H^{p+r}$ and $\psi \in H^{q+r}$, with the aid of (8.13) and (8.14), we deduce

$$(A\varphi, \psi)_{q+r/2} = (A\varphi, D^{r/2}\psi)_q = (\varphi, A^*D^{r/2}\psi)_p = (\varphi, D^{-r/2}A^*D^{r/2}\psi)_{p+r/2}.$$

Hence, the bounded operators $A : H^{p+r} \to H^{q+r}$ and $\widetilde{A} : H^{q+r} \to H^{p+r}$ are adjoint with respect to the positive dual systems (H^{p+r}, H^{p+r}) and (H^{q+r}, H^{q+r}) generated by the $(p+r/2)$– and $(q+r/2)$–scalar products. Therefore, applying Lax's Theorem 4.13 and using the denseness of H^{p+r} in $H^{p+r/2}$, we conclude that A is bounded from $H^{p+r/2}$ into $H^{q+r/2}$ with

$$\|A\varphi\|_{q+r/2} \le C_1^{1/2} C_2^{1/2} \|\varphi\|_{p+r/2}$$

for all $\varphi \in H^{p+r/2}$.

Repeating this argument, we see that

$$\|A\varphi\|_{q+\lambda r} \le C_1^{1-\lambda} C_2^{\lambda} \|\varphi\|_{p+\lambda r}$$

for all $\varphi \in H^{p+\lambda r}$ and all rational binary numbers $0 < \lambda < 1$. For arbitrary $0 < \lambda < 1$ we can choose a monotonic decreasing sequence (λ_n) of binary rationals such that $\lambda_n \to \lambda$, $n \to \infty$. Then, using (8.4), we have

$$\|A\varphi\|_{q+\lambda r} \le \|A\varphi\|_{q+\lambda_n r} \le C_1^{1-\lambda_n} C_2^{\lambda_n} \|\varphi\|_{p+\lambda_n r}$$

for all $\varphi \in H^{p+r}$ and $n \in \mathbb{N}$. Passing to the limit $n \to \infty$ we see that

$$\|A\varphi\|_{q+\lambda r} \le C_1^{1-\lambda} C_2^{\lambda} \|\varphi\|_{p+\lambda r}$$

for all $\varphi \in H^{p+r}$, and because H^{p+r} is dense in $H^{p+\lambda r}$, for all $\varphi \in H^{p+\lambda r}$. The idea to use Lax's theorem to prove this interpolation theorem is due to Kirsch [122]. □

8.2 The Sobolev Space $H^p(\Gamma)$

Let Γ be the boundary of a simply connected bounded domain $D \subset \mathbb{R}^2$ of class C^k, $k \in \mathbb{N}$. With the aid of a regular and k-times continuously differentiable 2π-periodic parametric representation

$$\Gamma = \{z(t) : t \in [0, 2\pi)\}$$

for $0 \le p \le k$ we can define the Sobolev space $H^p(\Gamma)$ as the space of all functions $\varphi \in L^2(\Gamma)$ with the property that $\varphi \circ z \in H^p[0, 2\pi]$. By $\varphi \circ z$, as usual, we denote the 2π-periodic function given by $(\varphi \circ z)(t) := \varphi(z(t))$, $t \in \mathbb{R}$. The scalar product and norm on $H^p(\Gamma)$ are defined through the scalar product on $H^p[0, 2\pi]$ by

$$(\varphi, \psi)_{H^p(\Gamma)} := (\varphi \circ z, \psi \circ z)_{H^p[0,2\pi]}.$$

Without loss of generality we have restricted the parameter domain to be the interval $[0, 2\pi)$. However, we must allow the possibility of different regular parametric representations for the boundary curve Γ. Therefore, we have to convince ourselves that our definition is invariant with respect to the parameterization.

Theorem 8.14. *Assume that z and $\tilde{z}$ are two different regular 2π-periodic parametric representations for the boundary Γ of a simply connected bounded domain $D \subset \mathbb{R}^2$ of class C^k, $k \in \mathbb{N}$, i.e., $\Gamma = \{z(t) : t \in [0, 2\pi)\}$ and $\Gamma = \{\tilde{z}(t) : t \in [0, 2\pi)\}$. Then, for $0 \leq p \leq k$ the Sobolev spaces*

$$H^p(\Gamma) := \left\{\varphi \in L^2(\Gamma) : \varphi \circ z \in H^p[0, 2\pi]\right\}$$

with scalar product

$$(\varphi, \psi)_{H^p(\Gamma)} := (\varphi \circ z, \psi \circ z)_{H^p[0,2\pi]}$$

and

$$\widetilde{H}^p(\Gamma) := \left\{\varphi \in L^2(\Gamma) : \varphi \circ \tilde{z} \in H^p[0, 2\pi]\right\}$$

with scalar product

$$(\varphi, \psi)_{\widetilde{H}^p(\Gamma)} := (\varphi \circ \tilde{z}, \psi \circ \tilde{z})_{H^p[0,2\pi]}$$

are isomorphic.

Proof. Assume that both parameterizations have the same orientation. (The case of two opposite parameterizations is treated analogously.) Because of the 2π-periodicity, without loss of generality, we may assume that $z(0) = \tilde{z}(0)$. Denote by $z^{-1} : \Gamma \to [0, 2\pi)$ the inverse of $z : [0, 2\pi) \to \Gamma$. Then the regularity of the parameterization implies that the mapping $f : [0, 2\pi) \to [0, 2\pi)$ given by $f := z^{-1} \circ \tilde{z}$ is bijective and k-times continuously differentiable with $f(0) = 0$. By setting $f(t + 2\pi) := f(t) + 2\pi$ it can be extended as a k-times continuously differentiable function on $\mathbb{R}$.

To establish the statement of the theorem, obviously, it suffices to show that for all $\varphi \in H^p[0, 2\pi]$ we have $\varphi \circ f \in H^p[0, 2\pi]$ with

$$\|\varphi \circ f\|_p \leq C\|\varphi\|_p, \tag{8.15}$$

where C is some constant depending on k and p. For this we make use of the equivalent norms of Theorems 8.5 and 8.6 and Corollary 8.7. If $p \in \mathbb{N}$, we use the norm given by (8.5). Because of Theorem 8.5 there exist positive constants c_1 and c_2 such that

$$c_1\|\varphi\|_{j,j}^2 \leq \|\varphi\|_j^2 \leq c_2\|\varphi\|_{j,j}^2$$

for all $\varphi \in C_{2\pi}^p$ and all $j = 1, \ldots, p$. Then, using the chain rule and the Cauchy–Schwarz inequality, we can estimate

$$\|\varphi \circ f\|_p^2 \leq c_2 \int_0^{2\pi} \left\{|\varphi(f(t))|^2 + \left|\frac{d^p}{dt^p}\,\varphi(f(t))\right|^2\right\} dt \leq c_2 M \sum_{j=0}^{p} \int_0^{2\pi} \left|\varphi^{(j)}(f(t))\right|^2 dt,$$

where M is some constant depending on p that contains bounds on the derivatives of f up to order p. Substituting $s = f(t)$ on the right-hand side and using (8.4), we get the further estimates

$$\|\varphi \circ f\|_p^2 \le c_2 M\gamma \sum_{j=0}^{p} \int_0^{2\pi} \left|\varphi^{(j)}(s)\right|^2 ds \le c_2 M\gamma \sum_{j=1}^{p} \|\varphi\|_{j,j}^2 \le \frac{c_2}{c_1} M\gamma p \, \|\varphi\|_p^2.$$

Here, we have set $\gamma := \|1/f'\|_\infty$. This completes the proof of (8.15) for $p \in \mathbb{N}$, since $C_{2\pi}^p$ is dense in $H^p[0,\pi]$.

If $0 < p < 1$, we use the norm given by (8.7). From l'Hôpital's rule and the property $f(t+2\pi) := f(t) + 2\pi$ we can conclude that the function

$$b(t,\tau) := \begin{cases} \left| \dfrac{\sin \dfrac{f(t)-f(\tau)}{2}}{\sin \dfrac{t-\tau}{2}} \right|, & \dfrac{t-\tau}{2\pi} \notin \mathbb{Z}, \\[2ex] |f'(t)|, & \dfrac{t-\tau}{2\pi} \in \mathbb{Z}, \end{cases}$$

is continuous and 2π-periodic with respect to both variables. Hence, b is bounded and we can estimate

$$\frac{|\varphi(f(t)) - \varphi(f(\tau))|^2}{\left|\sin \dfrac{t-\tau}{2}\right|^{2p+1}} \le B^{2p+1} \frac{|\varphi(f(t)) - \varphi(f(\tau))|^2}{\left|\sin \dfrac{f(t)-f(\tau)}{2}\right|^{2p+1}},$$

where B is a bound on b. From this, substituting $s = f(t)$ and $\sigma = f(\tau)$, we obtain

$$\int_0^{2\pi}\int_0^{2\pi} \frac{|\varphi(f(t)) - \varphi(f(\tau))|^2}{\left|\sin \dfrac{t-\tau}{2}\right|^{2p+1}} \, d\tau dt \le \gamma^2 B^{2p+1} \int_0^{2\pi}\int_0^{2\pi} \frac{|\varphi(s) - \varphi(\sigma)|^2}{\left|\sin \dfrac{s-\sigma}{2}\right|^{2p+1}} \, d\sigma ds.$$

With this inequality, (8.15) follows from the equivalence of the norms $\|\cdot\|_p$ and $\|\cdot\|_{p,0}$ on $C_{2\pi}^1$ and the denseness of $C_{2\pi}^1$ in $H^p[0,2\pi]$.

Finally, the case of an arbitrary noninteger positive p is settled by combining the two previous cases with the aid of the norm given by (8.9). □

The classical Sobolev space $H^1(D)$ for a bounded domain $D \subset \mathbb{R}^2$ with boundary ∂D of class C^1 is defined as the completion of the space $C^1(\bar{D})$ of continuously differentiable functions with respect to the norm

$$\|u\|_{H^1(D)} := \left(\int_D \left\{ |u(x)|^2 + |\operatorname{grad} u(x)|^2 \right\} dx \right)^{1/2}.$$

For a detailed description of this space, in particular, its equivalent definition through the concept of weak derivatives, and its extension to the Sobolev spaces $H^k(D)$ of arbitrary order $k \in \mathbb{N}$ we refer to Adams and Fournier [2], Gilbarg and Trudinger [63], McLean [165], and Treves [236]. Since each Cauchy sequence with respect to

$\|\cdot\|_{H^1(D)}$ is also a Cauchy sequence with respect to $\|\cdot\|_{L^2(D)}$, we may interpret $H^1(D)$ as a subspace of $L^2(D)$. Obviously the gradient can be extended from $C^1(\bar{D})$ as a bounded linear operator from $H^1(D)$ into $L^2(D)$ (see Problem 2.1).

We want to illustrate the connection between Sobolev spaces on the domain D and Sobolev spaces on its boundary Γ through the simplest case of the trace theorem. For functions defined on the closure $\bar{D}$ their values on the boundary are clearly defined and the restriction of the function to the boundary Γ is called the *trace*. The operator mapping a function onto its trace is called the *trace operator*.

As a first step, we consider continuously differentiable functions u in the strip $\mathbb{R}\times[0,1]$ that are 2π-periodic with respect to the first variable, i.e., $u(t+2\pi,\cdot) = u(t,\cdot)$ for all $t \in \mathbb{R}$. By Q we denote the rectangle $Q := [0, 2\pi) \times [0, 1]$. For $0 \le \eta \le 1$ we consider the Fourier coefficients

$$\hat{u}_m(\eta) := \frac{1}{2\pi} \int_0^{2\pi} u(t,\eta) e^{-imt} dt.$$

By Parseval's equality we have

$$\sum_{m=-\infty}^{\infty} |\hat{u}_m(\eta)|^2 = \frac{1}{2\pi} \int_0^{2\pi} |u(t,\eta)|^2 dt, \quad 0 \le \eta \le 1.$$

Because the $\hat{u}_m$ and u are continuous, by Dini's theorem (see Problem 8.2), the series is uniformly convergent. Hence, we can integrate term by term to obtain

$$\sum_{m=-\infty}^{\infty} \int_0^1 |\hat{u}_m(\eta)|^2 d\eta = \frac{1}{2\pi} \|u\|_{L^2(Q)}^2. \tag{8.16}$$

Similarly, from

$$\hat{u}_m'(\eta) = \frac{1}{2\pi} \int_0^{2\pi} \frac{\partial u}{\partial \eta}(t,\eta) e^{-imt} dt$$

and

$$im\hat{u}_m(\eta) = \frac{1}{2\pi} \int_0^{2\pi} \frac{\partial u}{\partial t}(t,\eta) e^{-imt} dt$$

we see that

$$\sum_{m=-\infty}^{\infty} \int_0^1 |\hat{u}_m'(\eta)|^2 d\eta = \frac{1}{2\pi} \left\| \frac{\partial u}{\partial \eta} \right\|_{L^2(Q)}^2 \tag{8.17}$$

and

$$\sum_{m=-\infty}^{\infty} \int_0^1 m^2 |\hat{u}_m(\eta)|^2 d\eta = \frac{1}{2\pi} \left\| \frac{\partial u}{\partial t} \right\|_{L^2(Q)}^2. \tag{8.18}$$

In the second step we shall show that

$$\|u(\cdot,0)\|_{H^{1/2}[0,2\pi]}^2 \le \frac{1}{\pi} \|u\|_{H^1(Q)}^2 \tag{8.19}$$

for all u with $u(\cdot\,,1) = 0$. The latter property implies that $\hat{u}_m(1) = 0$ for all $m \in \mathbb{Z}$. From this, by the Cauchy–Schwarz inequality, we obtain that

$$|\hat{u}_m(0)|^2 = 2\,\mathrm{Re}\int_1^0 \hat{u}_m'(\eta)\overline{\hat{u}_m(\eta)}\,d\eta \le 2\left\{\int_0^1 |\hat{u}_m'(\eta)|^2 d\eta\right\}^{1/2}\left\{\int_0^1 |\hat{u}_m(\eta)|^2 d\eta\right\}^{1/2}$$

for all $m \in \mathbb{Z}$ and consequently, using again the Cauchy–Schwarz inequality, we can estimate

$$\begin{aligned}
\|u(\cdot\,,0)\|^2_{H^{1/2}[0,2\pi]} &= \sum_{m=-\infty}^{\infty} (1+m^2)^{1/2} |\hat{u}_m(0)|^2 \\
&\le 2\left\{\sum_{m=-\infty}^{\infty}\int_0^1 |\hat{u}_m'(\eta)|^2 d\eta\right\}^{1/2}\left\{\sum_{m=-\infty}^{\infty}(1+m^2)\int_0^1 |\hat{u}_m(\eta)|^2 d\eta\right\}^{1/2} \\
&= \frac{1}{\pi}\left\|\frac{\partial u}{\partial \eta}\right\|_{L^2(Q)}\left\{\|u\|^2_{L^2(Q)} + \left\|\frac{\partial u}{\partial t}\right\|^2_{L^2(Q)}\right\}^{1/2} \le \frac{1}{\pi}\,\|u\|^2_{H^1(Q)}.
\end{aligned}$$

We are now ready to establish the following theorem.

Theorem 8.15. *Let D be a simply connected bounded domain in $\mathbb{R}^2$ with boundary Γ of class C^2. Then there exists a positive constant C such that*

$$\|u\|_{H^{1/2}(\Gamma)} \le C\|u\|_{H^1(D)}$$

for all $u \in C^1(\bar{D})$.

Proof. As in the proof of Theorem 7.8, we choose a parallel strip

$$D_h := \{x + \eta h\nu(x) : x \in \Gamma,\ \eta \in [0,1]\}$$

with some $h > 0$ such that each $y \in D_h$ is uniquely representable through projection onto Γ in the form $y = x + \eta h\nu(x)$ with $x \in \Gamma$ and $\eta \in [0,1]$. Here, for convenience, deviating from our general rule, we assume the normal ν is directed into D. By Γ_h we denote the interior boundary $\Gamma_h := \{y = x + h\nu(x) : x \in \Gamma\}$ of D_h. Then, through a regular 2π-periodic parametric representation

$$\Gamma = \{z(t) : 0 \le t < 2\pi\}$$

of the contour Γ we have a parameterization

$$\zeta(t,\eta) = z(t) + \eta h\nu(z(t)), \quad 0 \le t < 2\pi, \quad 0 \le \eta \le 1, \tag{8.20}$$

of the strip D_h. In particular, ζ provides a bijective and continuously differentiable mapping from Q onto D_h with a continuously differentiable inverse. Thus, applying the chain rule, for all $u \in C^1(D_h)$ with $u = 0$ on Γ_h from (8.19) we conclude that

$$\|u\|_{H^{1/2}(\Gamma)} = \|u \circ z\|_{H^{1/2}[0,2\pi]} \le \frac{1}{\sqrt{\pi}}\, \|u \circ \zeta\|_{H^1(Q)} \le c\|u\|_{H^1(D_h)},$$

where c is some constant containing a bound on the first derivatives of the mapping ζ and its inverse.

Finally, to extend the estimate to arbitrary $u \in C^1(\bar{D})$, we choose a cut-off function $g \in C^1(\bar{D})$ given by $g(y) = 0$ for $y \notin D_h$ and $g(y) = f(\eta)$ for $y \in D_h$ with $y = x + \eta h \nu(x)$. Here, f is defined by

$$f(\eta) := (1 - \eta)^2(1 + 3\eta)$$

and satisfies $f(0) = f'(0) = 1$ and $f(1) = f'(1) = 0$. The property $f'(0) = 1$ will not be used in this proof but will be needed later, in the proof of Theorem 8.18. Then, applying the product rule, we obtain

$$\|u\|_{H^{1/2}(\Gamma)} = \|gu\|_{H^{1/2}(\Gamma)} \le c\|gu\|_{H^1(D)} \le C\|u\|_{H^1(D)}$$

for all $u \in C^1(\bar{D})$ and some constant C containing a bound on g and its first derivatives. □

Corollary 8.16. *The trace operator can be uniquely extended as a continuous operator $\gamma : H^1(D) \to H^{1/2}(\Gamma)$.*

Proof. This follows from Theorem 8.15 and Problem 2.1. □

We proceed by showing that the trace operator γ of Corollary 8.16 has a bounded right inverse.

Theorem 8.17. *There exists a bounded linear operator $\beta : H^{1/2}(\Gamma) \to H^1(D)$ such that $\gamma\beta = I$.*

Proof. We use the notations introduced in the proof of Theorem 8.15 and first consider the strip $\mathbb{R} \times [0, 1]$. For $m \in \mathbb{Z}$ we introduce functions $w_m : [0, 1] \to \mathbb{R}$ by

$$w_m(\eta) := (1 - \eta)^{2|m|}, \quad 0 \le \eta \le 1.$$

Elementary calculations yield

$$\int_0^1 [w_m(t)]^2 dt = \frac{1}{4|m| + 1} \tag{8.21}$$

and

$$\int_0^1 [w_m'(t)]^2 dt = \frac{4m^2}{4|m| - 1}\,. \tag{8.22}$$

Let ψ be a trigonometric polynomial with Fourier coefficients $\hat{\psi}_m$ (of which only finitely many are different from zero). We define a function v on $\mathbb{R} \times [0, 1]$ by

$$v(t, \eta) := \sum_{m=-\infty}^{\infty} \hat{\psi}_m w_m(\eta) e^{imt}$$

with the properties $v(\cdot\,, 0) = \psi$ and $v(\cdot\,, 1) = \partial_\eta v(\cdot\,, 1) = 0$. With the aid of (8.21), (8.22) and the orthogonality of the trigonometric monomials, we obtain

$$\|v\|^2_{H^1(Q)} = 2\pi \sum_{m=-\infty}^{\infty} \left\{ \frac{1 + m^2}{4|m| + 1} + \frac{4m^2}{4|m| - 1} \right\} |\hat{\psi}_m|^2$$

whence the estimate

$$\|v\|_{H^1(Q)} \le c\, \|\psi\|_{H^{1/2}[0,2\pi]} \tag{8.23}$$

follows for some constant c independent of ψ.

In terms of the bijective mapping $\zeta : Q \to D_h$ given by (8.20) we define a function $u \in C^1(\bar{D})$ by $u = v \circ \zeta^{-1}$ in D_h and $u = 0$ in $D \setminus D_h$. Since $\gamma u = \psi \circ z^{-1}$, on the subspace $U \subset H^{1/2}(\Gamma)$ given by the images $\psi \circ z^{-1}$ of all trigonometric polynomials ψ we now define a linear operator $\beta : U \to H^1(D)$ by setting

$$\beta(\psi \circ z^{-1}) := u.$$

By (8.23) it is bounded and clearly it satisfies $\gamma\beta = I$. Now the statement follows from the denseness of the trigonometric polynomials in $H^{1/2}[0, 2\pi]$ and Problem 2.1. □

In the following theorem, by $|\cdot|_{L^1(\Gamma)}$ we will denote the semi-norm

$$|u|_{L^1(\Gamma)} := \left| \int_\Gamma u\, ds \right|.$$

Theorem 8.18. *Let D be as in Theorem 8.15. Then there exists a positive constant C such that*

$$\|u\|^2_{L^2(D)} \le C \left\{ |u|^2_{L^1(\Gamma)} + \|\operatorname{grad} u\|^2_{L^2(D)} \right\}$$

and

$$\|u\|^2_{L^2(D)} \le C \left\{ \|u\|^2_{L^2(\Gamma)} + \|\operatorname{grad} u\|^2_{L^2(D)} \right\}$$

for all $u \in H^1(D)$.

Proof. Since $C^1(\bar{D})$ is dense in $H^1(D)$, it suffices to establish the inequalities for all $u \in C^1(\bar{D})$. In addition, obviously, we only need to verify the first inequality.

We use the notations introduced in the proof of Theorem 8.15 and consider a function u in the strip $\mathbb{R} \times [0, 1]$. From

$$\int_0^{2\pi} u(t, \eta)\, dt - \int_0^{2\pi} u(t, 0)\, dt = \int_0^{2\pi} \int_0^{\eta} \frac{\partial u}{\partial \xi}(t, \xi)\, d\xi dt,$$

using the Cauchy–Schwarz inequality, we derive

$$\left|\int_0^{2\pi} u(t,\eta)\,dt\right|^2 \le 2\left\{\left|\int_0^{2\pi} u(t,0)\,dt\right|^2 + 2\pi\left\|\frac{\partial u}{\partial \eta}\right\|^2_{L^2(Q)}\right\}.$$

Integrating this inequality with respect to η we obtain

$$\int_0^1\left|\int_0^{2\pi} u(t,\eta)\,dt\right|^2 d\eta \le C_1\left\{\left|\int_0^{2\pi} u(t,0)\,dt\right|^2 + \|\,\mathrm{grad}\,u\|^2_{L^2(Q)}\right\}$$

for some constant C_1. Combining this with (8.16) and (8.18), we see that

$$\|u\|^2_{L^2(Q)} \le 2\pi\int_0^1 |\hat{u}_0(\eta)|^2 d\eta + \left\|\frac{\partial u}{\partial t}\right\|^2_{L^2(Q)}$$

$$\le C_2\left\{\left|\int_0^{2\pi} u(t,0)\,dt\right|^2 + \|\,\mathrm{grad}\,u\|^2_{L^2(Q)}\right\}$$

for some constant C_2. Substituting back into the domain D as in the proof of Theorem 8.15, we see that there exists a constant C_3 such that

$$\|u\|^2_{L^2(D_h)} \le C_3\left\{|u|^2_{L^1(\Gamma)} + \|\,\mathrm{grad}\,u\|^2_{L^2(D)}\right\}. \tag{8.24}$$

For this transformation, without loss of generality we have assumed that the parameter t in the representation of the boundary Γ is given through the arc length multiplied by $2\pi/|\Gamma|$.

Now let $u \in C^1(\bar{D})$ satisfy $u = 0$ and $\mathrm{grad}\,u = 0$ on the boundary Γ. Then we can extend u to a continuously differentiable function on $\mathbb{R}^2$ by setting $u = 0$ in the exterior of D. Choose R large enough such that D is contained in the disk with radius R and center at the origin. Then from

$$u(x) = \int_{-R}^{x_1} \frac{\partial u}{\partial x_1}\,dx_1,$$

with the aid of the Cauchy–Schwarz inequality, we see that

$$|u(x)|^2 \le 2R\int_{-R}^{R}\left|\frac{\partial u}{\partial x_1}\right|^2 dx_1.$$

Integrating this inequality and using $u = 0$ outside of D yields

$$\|u\|^2_{L^2(D)} \le 4R^2\|\,\mathrm{grad}\,u\|^2_{L^2(D)} \tag{8.25}$$

for all $u \in C^1(\bar{D})$ with $u = 0$ and $\mathrm{grad}\,u = 0$ on Γ.

Finally, for arbitrary $u \in C^1(\bar{D})$ we use the function g introduced in the proof of Theorem 8.15 to decompose

$$u = gu + (1 - g)u$$

and then apply (8.24) to the first term and (8.25) to the second term on the right-hand side. Observing that

$$\|\operatorname{grad}(gu)\|_{L^2(D)} \leq C_4 \left\{\|\operatorname{grad} u\|_{L^2(D)} + \|u\|_{L^2(D_h)}\right\}$$

for some constant C_4 depending on g ends the proof. □

Corollary 8.19. *On $H^1(D)$ the norm $\|\cdot\|_{H^1(D)}$ is equivalent to each of the two norms*

$$\|u\| := \left(|u|^2_{L^1(\Gamma)} + \|\operatorname{grad} u\|^2_{L^2(D)}\right)^{1/2}$$

and

$$\|u\| := \left(\|u\|^2_{L^2(\Gamma)} + \|\operatorname{grad} u\|^2_{L^2(D)}\right)^{1/2}.$$

Proof. This follows from Theorems 8.3, 8.15, and 8.18.

8.3 Weak Solutions to Boundary Value Problems

We will demonstrate how the Sobolev spaces $H^{1/2}(\Gamma)$ and $H^{-1/2}(\Gamma)$ occur in a natural way through the solution of boundary value problems in a weak formulation. For a function $u \in H^1(D)$, the integral

$$\int_D |\operatorname{grad} u|^2 dx$$

is called the *Dirichlet integral*. For a harmonic function u the Dirichlet integral represents the energy of the potential u. Therefore, it is natural to attempt to develop an approach for the solution of boundary value problems for harmonic functions in which it is required that the solutions have finite energy, i.e., they belong to the Sobolev space $H^1(D)$. Since the functions contained in $H^1(D)$, in general, are not twice continuously differentiable in D and since they do not attain boundary values or normal derivatives on the boundary in the classical sense, we have to extend the classical formulation of the boundary value problems by generalized versions. Our procedure will be quite typical in the sense that we use specific properties of classical solutions to formulate what we mean by a weak solution.

For each solution $u \in C^2(D) \cap C^1(\bar{D})$ to the Laplace equation $\Delta u = 0$ in D, by the first Green's Theorem 6.3 we have

$$\int_D \operatorname{grad} u \cdot \operatorname{grad} v \, dx = 0 \qquad (8.26)$$

for all $v \in C^1(\bar{D})$ with $v = 0$ on Γ. Since the integral in this equation is well defined for all $u, v \in H^1(D)$, we call a function $u \in H^1(D)$ a *weak solution* to the Laplace equation in D if (8.26) is satisfied for all $v \in H_0^1(D)$, where $H_0^1(D)$ denotes the subspace of all functions $v \in H^1(D)$ satisfying $v = 0$ on Γ in the sense of the trace operator.

Weak Interior Dirichlet Problem. *For a given $f \in H^{1/2}(\Gamma)$, find a weak solution $u \in H^1(D)$ to the Laplace equation in D such that*

$$u = f \quad \text{on } \Gamma \tag{8.27}$$

in the sense of the trace operator γ of Corollary 8.16.

For each solution $u \in C^2(D) \cap C^1(\bar{D})$ to the Neumann problem $\Delta u = 0$ in D with $\partial u/\partial \nu = g$ on Γ by the first Green's theorem we have

$$\int_D \operatorname{grad} u \cdot \operatorname{grad} v \, dx = \int_\Gamma g v \, ds$$

for all $v \in C^1(\bar{D})$. The following lemma extends this property to weak solutions of the Laplace equation by introducing a weak formulation of the normal derivative.

Lemma 8.20. *Let $u \in H^1(D)$ be a weak solution to the Laplace equation in D. Then there exists a unique $g \in H^{-1/2}(\Gamma)$ such that*

$$\int_D \operatorname{grad} u \cdot \operatorname{grad} v \, dx = \int_\Gamma g \, \gamma v \, ds \tag{8.28}$$

for all $v \in H^1(D)$. The integral in (8.28) has to be understood in the sense of the bilinear duality pairing

$$\int_\Gamma g \varphi \, ds := g(\varphi) \tag{8.29}$$

for $\varphi \in H^{1/2}(\Gamma)$ and g satisfies

$$\|g\|_{H^{-1/2}(\Gamma)} \le C \, \|u\|_{H^1(D)} \tag{8.30}$$

for some constant C depending only on D.

Proof. In terms of the right inverse β of Theorem 8.17, we define $g \in H^{-1/2}(\Gamma)$ by

$$g(\varphi) := \int_D \operatorname{grad} u \cdot \operatorname{grad} \beta\varphi \, dx \tag{8.31}$$

for $\varphi \in H^{1/2}(\Gamma)$. Indeed g is bounded because of

$$|g(\varphi)| \le \|u\|_{H^1(D)} \|\beta\varphi\|_{H^1(D)} \le \|\beta\| \, \|u\|_{H^1(D)} \|\varphi\|_{H^{1/2}(\Gamma)}.$$

From this inequality the estimate (8.30) follows.

In order to show that g satisfies (8.28), for $v \in H^1(D)$ we consider $v_0 = v - \beta\gamma v$. Then $v_0 \in H^1_0(D)$ and consequently

$$\int_D \operatorname{grad} u \cdot \operatorname{grad} v_0 \, dx = 0,$$

since u is a weak solution to the Laplace equation. Therefore, in view of the definition (8.31) we obtain

$$\int_D \operatorname{grad} u \cdot \operatorname{grad} v \, dx = \int_D \operatorname{grad} u \cdot \operatorname{grad}(v_0 + \beta\gamma v) \, dx = g(\gamma v)$$

as required. To establish uniqueness, we observe that $g(\gamma v) = 0$ for all $v \in H^1(D)$, by Theorem 8.17, implies that $g(\varphi) = 0$ for all $\varphi \in H^{1/2}(\Gamma)$ and consequently $g = 0$. □

We particularly note that if the weak normal derivative of a weak solution to the Laplace equation in the sense of Lemma 8.20 is used, Green's theorem remains valid.

We are now in the position to introduce the following weak formulation of the interior Neumann problem.

Weak Interior Neumann Problem. *For a given $g \in H^{-1/2}(\Gamma)$, find a weak solution $u \in H^1(D)$ to the Laplace equation in D such that*

$$\frac{\partial u}{\partial \nu} = g \quad \text{on } \Gamma \tag{8.32}$$

with the weak normal derivative in the sense of Lemma 8.20.

Theorem 8.21. *The weak interior Dirichlet problem has at most one solution. Two solutions to the weak interior Neumann problem can differ only by a constant.*

Proof. The difference $u := u_1 - u_2$ between two solutions to the Dirichlet problem is a weak solution with homogeneous boundary condition $u = 0$ on Γ. Then we may insert $v = u$ in (8.26) and obtain $\|\operatorname{grad} u\|_{L^2(D)} = 0$. Since $u = 0$ on Γ, from Corollary 8.19 we see that $u = 0$ in D. As a consequence of (8.28) for two solutions of the Neumann problem the difference $u := u_1 - u_2 - c$ again satisfies $\|\operatorname{grad} u\|_{L^2(D)} = 0$ for all constants c. We choose c such that $\int_\Gamma u \, ds = 0$. Then from Corollary 8.19 we obtain $u = 0$ in D. □

To establish the existence of weak solutions we will proceed analogously to the classical case and try to find the solution in the form of a logarithmic single- or double-layer potential with densities in $H^{-1/2}(\Gamma)$ or $H^{1/2}(\Gamma)$. Therefore, we first need to investigate the properties of these potentials in our Sobolev space setting. This will be achieved through making use of corresponding properties in the spaces of uniformly Hölder continuous functions. For the remainder of this section, by α we will always denote a Hölder exponent with $0 < \alpha < 1$.

Theorem 8.22. *The operators $K : H^{1/2}(\Gamma) \to H^{1/2}(\Gamma)$ defined by the logarithmic double-layer potential*

$$(K\varphi)(x) := \frac{1}{\pi} \int_\Gamma \varphi(y) \frac{\partial}{\partial\nu(y)} \ln \frac{1}{|x-y|}\, ds(y), \quad x \in \Gamma,$$

and $K' : H^{-1/2}(\Gamma) \to H^{-1/2}(\Gamma)$ defined by the formal normal derivative of the logarithmic single-layer potential

$$(K'\psi)(x) := \frac{1}{\pi} \int_\Gamma \psi(y) \frac{\partial}{\partial\nu(x)} \ln \frac{1}{|x-y|}\, ds(y), \quad x \in \Gamma,$$

are compact and adjoint in the dual system $(H^{1/2}(\Gamma), H^{-1/2}(\Gamma))_{L^2(\Gamma)}$, i.e.,

$$(K\varphi, \psi)_{L^2(\Gamma)} = (\varphi, K'\psi)_{L^2(\Gamma)} \tag{8.33}$$

for all $\varphi \in H^{1/2}(\Gamma)$ and $\psi \in H^{-1/2}(\Gamma)$. Here, again, we use the notation $(f, g)_{L^2(\Gamma)} := g(f)$ for $f \in H^{1/2}(\Gamma)$ and $g \in H^{-1/2}(\Gamma)$.

Proof. We shall show that K is bounded from $H^{1/2}(\Gamma)$ into $H^1(\Gamma)$. Then the assertion of the compactness of K follows from Theorems 2.21 and 8.3.

For functions $\varphi \in C^{1,\alpha}(\Gamma)$ from Theorem 7.32, we know that the gradient of the double-layer potential v with density φ can be expressed in terms of the derivatives of the single-layer potential w with density $d\varphi/ds$. In particular, by (7.44) the tangential derivative is given by

$$\frac{\partial v_-}{\partial s}(x) = -\frac{1}{2}\frac{d\varphi}{ds}(x) - \frac{1}{2\pi}\int_\Gamma \frac{d\varphi}{ds}(y) \frac{\partial}{\partial\nu(x)} \ln \frac{1}{|x-y|}\, ds(y), \quad x \in \Gamma.$$

From this, with the aid of the jump relations of Theorem 7.29 for double-layer potentials with uniformly Hölder continuous densities, we conclude that

$$\frac{dK\varphi}{ds}(x) = -\frac{1}{\pi}\int_\Gamma \frac{d\varphi}{ds}(y) \frac{\partial}{\partial\nu(x)} \ln \frac{1}{|x-y|}\, ds(y), \quad x \in \Gamma.$$

Performing a partial integration yields

$$\frac{dK\varphi}{ds}(x) = \frac{1}{\pi}\int_\Gamma \{\varphi(y) - \varphi(x)\} \frac{\partial}{\partial s(y)} \frac{\partial}{\partial\nu(x)} \ln \frac{1}{|x-y|}\, ds(y), \quad x \in \Gamma,$$

where the integral has to be understood in the sense of a Cauchy principal value. Now we use a 2π-periodic parameterization $\Gamma = \{x(t) : t \in [0, 2\pi)\}$ of Γ to transform the integral into

$$\frac{dK\varphi}{ds}(x(t)) = \int_0^{2\pi} \frac{k(t,\tau)}{\sin\dfrac{t-\tau}{2}} \{\varphi(x(\tau)) - \varphi(x(t))\}\, d\tau, \quad t \in [0, 2\pi],$$

where k is a bounded weakly singular kernel. Thus, by the Cauchy–Schwarz inequality, we can estimate

$$\int_0^{2\pi} \left| \frac{dK\varphi}{ds}(x(t)) \right|^2 dt \le M \int_0^{2\pi} \int_0^{2\pi} \frac{|\varphi(x(\tau)) - \varphi(x(t))|^2}{\sin^2 \frac{t-\tau}{2}} \, d\tau dt \le M \|\varphi \circ x\|^2_{1/2,0}$$

for some constant M. Hence, by Theorem 8.6,

$$\left\| \frac{dK\varphi}{ds} \right\|_{L^2(\Gamma)} \le c\|\varphi\|_{H^{1/2}(\Gamma)}$$

for some constant c. Since K has continuous kernel (see Problem 6.1), it is bounded from $L^2(\Gamma)$ into $L^2(\Gamma)$. Therefore we conclude

$$\|K\varphi\|_{H^1(\Gamma)} \le C\|\varphi\|_{H^{1/2}(\Gamma)}$$

for all $\varphi \in C^{1,\alpha}(\Gamma)$ and some constant C. This completes the proof of the boundedness of K from $H^{1/2}(\Gamma)$ into $H^1(\Gamma)$, since $C^{1,\alpha}(\Gamma)$ is dense in $H^{1/2}(\Gamma)$ by Theorem 8.2.

For $\varphi, \psi \in C^{1,\alpha}(\Gamma)$, clearly (8.33) holds and we can estimate

$$|(\varphi, K'\psi)_{L^2(\Gamma)}| = |(K\varphi, \psi)_{L^2(\Gamma)}| \le \|K\varphi\|_{H^1(\Gamma)} \|\psi\|_{H^{-1}(\Gamma)} \le C\|\varphi\|_{H^{1/2}(\Gamma)} \|\psi\|_{H^{-1}(\Gamma)}.$$

From this it follows that

$$\|K'\psi\|_{H^{-1/2}(\Gamma)} \le C\|\psi\|_{H^{-1}(\Gamma)}$$

for all $\psi \in C^{1,\alpha}(\Gamma)$ and, because $C^{1,\alpha}(\Gamma)$ is dense in $H^{-1}(\Gamma)$, also for all $\psi \in H^{-1}(\Gamma)$. Therefore K' is bounded from $H^{-1}(\Gamma)$ into $H^{-1/2}(\Gamma)$ and the statement on the compactness of K' again follows from Theorems 2.21 and 8.3.

Finally, the validity of (8.33) for all $\varphi \in H^{1/2}(\Gamma)$ and $\psi \in H^{-1/2}(\Gamma)$ follows from the boundedness of K and K' by the denseness of $C^{1,\alpha}(\Gamma)$ in $H^{1/2}(\Gamma)$ and in $H^{-1/2}(\Gamma)$. □

Lemma 8.23. *For the trigonometric monomials we have the integrals*

$$\frac{1}{2\pi} \int_0^{2\pi} \ln\left(4 \sin^2 \frac{t}{2}\right) e^{imt} dt = \begin{cases} 0, & m = 0, \\ -\dfrac{1}{|m|}, & m = \pm 1, \pm 2, \dots. \end{cases}$$

Proof. Integrating the geometric sum

$$1 + 2\sum_{k=1}^{m-1} e^{ikt} + e^{imt} = i\,(1 - e^{imt}) \cot \frac{t}{2}, \quad 0 < t < 2\pi, \tag{8.34}$$

we obtain

$$\int_0^{2\pi} e^{imt} \cot \frac{t}{2}\, dt = 2\pi i, \quad m = 1, 2, \ldots, \tag{8.35}$$

in the sense of Cauchy principal values. Integrating the identity

$$\frac{d}{dt}\left\{\left[e^{imt} - 1\right] \ln\left(4 \sin^2 \frac{t}{2}\right)\right\} = im\, e^{imt} \ln\left(4 \sin^2 \frac{t}{2}\right) + \left[e^{imt} - 1\right] \cot \frac{t}{2}$$

yields

$$\int_0^{2\pi} e^{imt} \ln\left(4 \sin^2 \frac{t}{2}\right) dt = -\frac{1}{im} \int_0^{2\pi} e^{imt} \cot \frac{t}{2}\, dt = -\frac{2\pi}{m}$$

for $m = 1, 2, \ldots$, and the statement is proven for $m \neq 0$. To evaluate the integral for $m = 0$ we set

$$I := \int_0^{2\pi} \ln\left(4 \sin^2 \frac{t}{2}\right) dt.$$

Then

$$2I = \int_0^{2\pi} \ln\left(4 \sin^2 \frac{t}{2}\right) dt + \int_0^{2\pi} \ln\left(4 \cos^2 \frac{t}{2}\right) dt$$

$$= \int_0^{2\pi} \ln(4 \sin^2 t)\, dt = \frac{1}{2} \int_0^{4\pi} \ln\left(4 \sin^2 \frac{t}{2}\right) dt = I,$$

and therefore $I = 0$. □

By Problem 7.2 the operator S_0 in the following theorem corresponds to the modified single-layer operator of Theorem 7.41 for the case of the unit circle.

Theorem 8.24. *The operator* $\widetilde{S}_0 : H^p[0, 2\pi] \to H^{p+1}[0, 2\pi]$, *defined by*

$$(\widetilde{S}_0\varphi)(t) := \frac{1}{2\pi} \int_0^{2\pi} \left\{\ln\left(4 \sin^2 \frac{t - \tau}{2}\right) - 2\right\} \varphi(\tau)\, d\tau, \quad t \in [0, 2\pi], \tag{8.36}$$

is bounded and has a bounded inverse for all $p \in \mathbb{R}$. *(For negative* p *the integral in (8.36) has to be understood as the duality pairing.)*

Proof. From Lemma 8.23, for $f_m(t) = e^{imt}$, we have

$$\widetilde{S}_0 f_m = \beta_m f_m, \quad m \in \mathbb{Z},$$

where $\beta_m = -1/|m|$ for $m \neq 0$ and $\beta_0 = -2$. This implies the boundedness of $\widetilde{S}_0 : H^p[0, 2\pi] \to H^{p+1}[0, 2\pi]$ and its invertibility with the inverse operator $\widetilde{S}_0^{-1} : H^{p+1}[0, 2\pi] \to H^p[0, 2\pi]$ given by

$$\widetilde{S}_0^{-1} f_m = \frac{1}{\beta_m} f_m, \quad m \in \mathbb{Z},$$

which again is bounded. □

Theorem 8.25. *The operators* $S : H^{-1/2}(\Gamma) \to H^{1/2}(\Gamma)$ *defined by the single-layer potential*

$$(S\varphi)(x) := \frac{1}{\pi} \int_\Gamma \varphi(y) \ln \frac{1}{|x-y|}\, ds(y), \quad x \in \Gamma,$$

and $T : H^{1/2}(\Gamma) \to H^{-1/2}(\Gamma)$ *defined by the normal derivative of the double-layer potential*

$$(T\psi)(x) := \frac{1}{\pi} \frac{\partial}{\partial \nu(x)} \int_\Gamma \psi(y) \frac{\partial}{\partial \nu(y)} \ln \frac{1}{|x-y|}\, ds(y), \quad x \in \Gamma,$$

are bounded. The operator S *is self-adjoint with respect to the dual systems* $(H^{-1/2}(\Gamma), H^{1/2}(\Gamma))_{L^2(\Gamma)}$ *and* $(H^{1/2}(\Gamma), H^{-1/2}(\Gamma))_{L^2(\Gamma)}$, *i.e.,*

$$(S\varphi, \psi)_{L^2(\Gamma)} = (\varphi, S\psi)_{L^2(\Gamma)} \tag{8.37}$$

for all $\varphi, \psi \in H^{-1/2}(\Gamma)$. *The operator* T *is self-adjoint with respect to the dual systems* $(H^{1/2}(\Gamma), H^{-1/2}(\Gamma))_{L^2(\Gamma)}$ *and* $(H^{-1/2}(\Gamma), H^{1/2}(\Gamma))_{L^2(\Gamma)}$, *i.e.,*

$$(T\varphi, \psi)_{L^2(\Gamma)} = (\varphi, T\psi)_{L^2(\Gamma)} \tag{8.38}$$

for all $\varphi, \psi \in H^{1/2}(\Gamma)$.

Proof. We use a 2π-periodic parameterization $\Gamma = \{x(t) : t \in [0, 2\pi)\}$ of Γ. By Theorem 8.14, we may assume that the parameter t is given by the arc length on Γ multiplied by $2\pi/|\Gamma|$. Then, for $\varphi \in C^{0,\alpha}(\Gamma)$, we can write (see Problem 7.2)

$$(S\varphi)(x(t)) = \frac{|\Gamma|}{4\pi^2} \int_0^{2\pi} \left\{ -\ln\left(4\sin^2 \frac{t-\tau}{2}\right) + p(t,\tau) \right\} \varphi(x(\tau))\, d\tau$$

for $t \in [0, 2\pi]$, where p is 2π-periodic and continuously differentiable. By the previous Theorem 8.24, the leading term with the logarithmic singularity represents a bounded operator from $H^{-1/2}(\Gamma)$ into $H^{1/2}(\Gamma)$. Because the kernel function p is continuously differentiable, using the Cauchy–Schwarz inequality and an integration by parts, the second term on the right is seen to be bounded from $H^0(\Gamma)$ into $H^1(\Gamma)$ and from $H^{-1}(\Gamma)$ into $H^0(\Gamma)$. Hence, by the interpolation Theorem 8.13, it is also bounded from $H^{-1/2}(\Gamma)$ into $H^{1/2}(\Gamma)$.

For $\varphi, \psi \in C^{1,\alpha}(\Gamma)$ we use the relation (7.46) between S and T and a partial integration to write

$$(T\varphi, \psi)_{L^2(\Gamma)} = \left(\frac{d}{ds} S \frac{d\varphi}{ds}, \psi \right)_{L^2(\Gamma)} = -\left(S \frac{d\varphi}{ds}, \frac{d\psi}{ds} \right)_{L^2(\Gamma)}. \tag{8.39}$$

Then we can estimate

$$|(T\varphi,\psi)|_{L^2(\Gamma)} \le \left\| S\,\frac{d\varphi}{ds}\right\|_{H^{1/2}(\Gamma)} \left\|\frac{d\psi}{ds}\right\|_{H^{-1/2}(\Gamma)}$$

$$\le C\left\|\frac{d\varphi}{ds}\right\|_{H^{-1/2}(\Gamma)} \left\|\frac{d\psi}{ds}\right\|_{H^{-1/2}(\Gamma)}$$

$$\le C\|\varphi\|_{H^{1/2}(\Gamma)}\,\|\psi\|_{H^{1/2}(\Gamma)}$$

for some constant C. Since $C^{1,\alpha}(\Gamma)$ is dense in $H^{1/2}(\Gamma)$, this implies that T is bounded from $H^{1/2}(\Gamma)$ into $H^{-1/2}(\Gamma)$.

The self-adjointness property (8.37) follows as in the proof of Theorem 8.22 by extending it from the case of smooth functions. Finally, combining (8.37) and (8.39) yields (8.38). □

Theorem 8.26. *The logarithmic single-layer potential defines a bounded linear operator from $H^{-1/2}(\Gamma)$ into $H^1(D)$. The logarithmic double-layer potential defines a bounded linear operator from $H^{1/2}(\Gamma)$ into $H^1(D)$.*

Proof. Let u be the single-layer potential with density $\varphi \in C^{0,\alpha}(\Gamma)$. Then, by Green's theorem and the jump relations of Theorem 7.30, we have

$$\int_D |\operatorname{grad} u|^2 dx = \int_\Gamma \bar{u}\,\frac{\partial u}{\partial \nu}\,ds = \frac{1}{4}\int_\Gamma S\bar{\varphi}\,(\varphi + K'\varphi)\,ds.$$

Therefore, by Theorems 8.22 and 8.25, with a constant c we can estimate

$$\|\operatorname{grad} u\|^2_{L^2(D)} \le \frac{1}{4}\,\|S\bar{\varphi}\|_{H^{1/2}(\Gamma)}\,\|\varphi + K'\varphi\|_{H^{-1/2}(\Gamma)} \le c\,\|\varphi\|^2_{H^{-1/2}(\Gamma)}.$$

Since S is bounded from $L^2(\Gamma)$ into $L^2(\Gamma)$ (see Problem 5.5), using Theorem 8.18, we see that

$$\|u\|_{H^1(D)} \le C\|\varphi\|_{H^{-1/2}(\Gamma)}$$

for some constant C. Now the statement on the single-layer potential follows from the denseness of $C^{0,\alpha}(\Gamma)$ in $H^{-1/2}(\Gamma)$.

The case of the double-layer potential v with density ψ is treated analogously through the relation

$$\int_D |\operatorname{grad} v|^2 dx = \frac{1}{4}\int_\Gamma T\bar{\varphi}\,(K\varphi - \varphi)\,ds,$$

which again follows from Green's theorem and the jump relations. Note that K is bounded from $L^2(\Gamma)$ into $L^2(\Gamma)$ since it has a continuous kernel. □

We are now prepared for the solution of the weak Dirichlet and Neumann problem via boundary integral equations.

Theorem 8.27. *The weak Dirichlet problem has a unique solution. The mapping taking the given boundary data $f \in H^{1/2}(\Gamma)$ into the solution $u \in H^1(D)$ is bounded.*

Proof. Analogous to the classical treatment we try to find the solution in the form of a double-layer potential

$$u(x) = \frac{1}{2\pi} \int_\Gamma \varphi(y) \frac{\partial}{\partial \nu(y)} \ln \frac{1}{|x-y|} \, ds(y), \quad x \in D. \tag{8.40}$$

If $\varphi \in C^{1,\alpha}(\Gamma)$, by Theorem 7.32 and Green's theorem, we have $u \in C^1(\bar{D})$ and

$$\int_D \operatorname{grad} u \cdot \operatorname{grad} v \, dx = \frac{1}{2} \int_\Gamma T\varphi \, v \, ds \tag{8.41}$$

for all $v \in C^1(\bar{D})$ and

$$2u = K\varphi - \varphi \quad \text{on } \Gamma.$$

In the first equation, by Theorems 8.25 and 8.26 and the trace theorem, both sides are continuous bilinear mappings from $H^{1/2}(\Gamma) \times H^1(D) \to \mathbb{C}$. Hence, the double-layer potential with density $\varphi \in H^{1/2}(\Gamma)$ satisfies (8.26) for all $v \in H_0^1(D)$, that is, it is a weak solution to the Laplace equation. In the second equation both sides are again bounded from $H^{1/2}(\Gamma)$ into $H^{1/2}(\Gamma)$ by Theorems 8.22 and 8.26 and Corollary 8.16. Therefore u is a weak solution to the Dirichlet problem provided the density φ solves the integral equation

$$\varphi - K\varphi = -2f.$$

Let $\varphi \in H^{1/2}(\Gamma)$ be a solution to the homogeneous equation $\varphi - K\varphi = 0$. Then, since K has continuous kernel (see Problem 6.1), we have $\varphi \in C(\Gamma)$ and from the classical Theorem 6.21 we deduce that $N(I - K) = \{0\}$ in $H^{1/2}(\Gamma)$. Now existence of a solution in $H^{1/2}(\Gamma)$ to the inhomogeneous equation follows from the Riesz theory by Theorems 3.4 and 8.22.

The statement on the continuous dependence of the solution on the given boundary data is a consequence of the boundedness of the inverse of $I - K$ (Theorem 3.4) and Theorem 8.26. □

Corollary 8.28. *The Dirichlet-to-Neumann operator A taking the boundary values of a harmonic function in $H^1(D)$ into its normal derivative is bounded from $H^{1/2}(\Gamma)$ into $H^{-1/2}(\Gamma)$.*

Proof. From Theorem 8.27 and Lemma 8.20. □

Since each weak solution $u \in H^1(D)$ to the Laplace equation can be viewed as the solution of a weak Dirichlet problem, from the representation of this solution as the double-layer potential (8.40) we can conclude that weak solutions to the Laplace equation are twice continuously differentiable in D, i.e., they are automatically classical solutions. The weak formulation only affects the behavior of the solutions at the boundary Γ.

Theorem 8.29. *The weak interior Neumann problem is solvable if and only if*

$$\int_\Gamma g\,ds = 0.$$

The operator mapping the given boundary data $g \in H^{-1/2}(\Gamma)$ into the unique solution $u \in H^1(D)$ satisfying the additional condition $\int_\Gamma u\,ds = 0$ is bounded.

Proof. By Theorem 7.30, the single-layer potential

$$u(x) = \frac{1}{2\pi}\int_\Gamma \psi(y)\ln\frac{1}{|x-y|}\,ds(y), \quad x \in D,$$

with density $\psi \in C^{0,\alpha}(\Gamma)$ belongs to $C^1(\bar{D})$. It satisfies

$$\int_D \operatorname{grad} u \cdot \operatorname{grad} v\,dx = \frac{1}{2}\int_\Gamma (\psi + K'\psi)\,v\,ds$$

for all $v \in C^1(\bar{D})$. As in the previous proof, both sides of this equation are continuous from $H^{-1/2}(\Gamma) \times H^1(D) \to \mathbb{C}$ by Theorems 8.22 and 8.26 and the trace theorem. Hence, the single-layer potential u with density $\psi \in H^{-1/2}(\Gamma)$ is a weak solution to the Neumann problem provided ψ solves the integral equation

$$\psi + K'\psi = 2g.$$

As in the proof of Theorem 8.27 from Theorem 6.21 we conclude that $N(I+K) = \operatorname{span}\{1\}$ in $H^{1/2}(\Gamma)$. Therefore, by Theorem 8.22 and the Fredholm alternative, the inhomogeneous equation $\psi + K'\psi = 2g$ is solvable in $H^{-1/2}(\Gamma)$ provided g satisfies the solvability condition $\int_\Gamma g\,ds = 0$. The necessity of the latter condition for the existence of a weak solution to the Neumann problem follows from (8.28) for $v = 1$.

The continuous dependence of the solution on the given boundary data is a consequence of Theorem 8.26 analogous to the classical case of Theorem 6.30. □

Corollary 8.30. *The Neumann-to-Dirichlet operator B taking the normal derivative of a harmonic function in $H^1(D)$ into its boundary values is bounded from $H_0^{-1/2}(\Gamma)$ into $H_0^{1/2}(\Gamma)$. Here, the subscripts indicate the subspaces with mean value zero.*

Proof. From Theorem 8.29 and Corollary 8.16. □

As in Section 7.7, it is also possible to obtain the weak solvability to the Dirichlet and Neumann problem through integral equations of the first kind

$$S_0\psi = 2f$$

and

$$T\varphi = 2g$$

respectively. Here, S_0 denotes the modified single-layer operator as introduced by (7.60). By denseness and continuity arguments it can be seen that the relations (7.47) and (7.48) carry over into the Sobolev space setting. Therefore, analogues of Theorems 7.41 and 7.43 can be proven with the aid of Theorem 5.6. Alternatively, bijectivity of $S_0 : H^{-1/2}(\Gamma) \to H^{1/2}(\Gamma)$ can also be obtained by Corollary 3.5 and Theorem 8.24 (see also Problem 8.3) and the hypersingular integral equation with the operator T can also be solved using the relation (7.46).

For the solution of the single-layer potential integral equation of Section 7.8 for an arc in a Sobolev space setting we refer to [143].

Since weak solutions to the Laplace equation in D are classical solutions (see p. 167) and Green's theorem remains valid for them (see p. 161), from the proof of Green's representation formula Theorem 6.5 it can be seen that this formula remains valid for harmonic functions in $H^1(D)$. The jump-relations are also preserved in the Sobolev space setting (we have used them already in the proofs of Theorems 8.27 and 8.29). Therefore the Calderón projection of Theorem 7.37 is also valid for harmonic functions in $H^1(D)$ with the projection operator acting on $H^{1/2}(\Gamma) \times H^{-1/2}(\Gamma)$. This opens up the option of also considering the direct approach for the boundary integral equations in the Sobolev space setting as in Section 7.6. To avoid repetitions, we leave this as an exercise to the reader.

Our presentation clearly demonstrates that the results on weak solutions heavily rely on the classical analysis on single- and double-layer potentials with uniformly Hölder continuous densities. The mapping properties of the integral operators in the Sobolev spaces essentially follow from the corresponding properties in the Hölder spaces through applications of functional analytic tools, in particular, denseness and continuation arguments. For a proof of the mapping property of the single-layer operator from Theorem 8.25 with the aid of Lax's Theorem 4.13, we refer to [32].

The major advantage of the weak approach to the boundary integral equations, besides the fact that it allows less regular boundary data, stems from the fact that Sobolev spaces are Hilbert spaces rather than just normed spaces. This, as we have seen, adds some elegance to the analysis.

Another advantage of the Sobolev space setting is the possibility to extend the analysis to less regular boundaries such as *Lipschitz domains*. These domains are defined analogously to domains of class C^1 on p. 30 with the continuous differentiability replaced by Lipschitz continuity, i.e., Hölder continuity with Hölder exponent 1. In particular, Lipschitz domains can have corners. For extensions of the mapping properties of single- and double layer potentials of this section to Lipschitz domains and its implications for the existence analysis of the boundary integral equations we refer to McLean [165].

For a more detailed study of boundary integral equations in Sobolev space settings in the framework of *pseudodifferential operators* (see Taylor [230, 231] and Treves [237]) including the extensions of the analysis to exterior problems and to three-dimensional problems we refer to Hsiao and Wendland [104] and the literature therein. For extensions of the approach based on Lax's Theorem 4.13 we refer to Kirsch [121, 123].

Problems

8.1. Work out the proofs of Corollaries 8.7 and 8.8.

8.2. Prove Dini's theorem: Let $D \subset \mathbb{R}^m$ be compact and let (φ_n) be a nondecreasing sequence of continuous real-valued functions on D converging pointwise to a continuous function φ. Then (φ_n) converges uniformly on D to φ.

8.3. Let $K : [0, 2\pi] \times [0, 2\pi] \to \mathbb{R}$ be n-times continuously differentiable. Show that the integral operator A with kernel K is compact from $H^p[0, 2\pi]$ into $H^q[0, 2\pi]$ for all $|p|, |q| \leq n$.

8.4. Formulate and prove the analogue of the Noether theorems in the form of Corollary 7.26 in the Sobolev spaces $H^p[0, 2\pi]$.
Hint: Use the integrals (8.35).

8.5. Use the separation of variables solutions of the Laplace equation in polar coordinates to verify the regularity and mapping properties of the weak solution to the Dirichlet and Neumann problem for the case where D is the unit disk.

Chapter 9
The Heat Equation

The temperature distribution u in a homogeneous and isotropic heat conducting medium with conductivity k, heat capacity c, and mass density ρ satisfies the partial differential equation

$$\frac{\partial u}{\partial t} = \kappa \Delta u$$

where $\kappa = k/c\rho$. This is called the *equation of heat conduction* or, shortly, the *heat equation;* it was first derived by Fourier. Simultaneously, the heat equation also occurs in the description of diffusion processes. The heat equation is the standard example for a *parabolic* differential equation. In this chapter we want to indicate the application of Volterra-type integral equations of the second kind for the solution of initial boundary value problems for the heat equation. Without loss of generality we assume the constant $\kappa = 1$. For a more comprehensive study of integral equations of the second kind for the heat equation we refer to Cannon [25], Friedman [58], and Pogorzelski [193].

9.1 Initial Boundary Value Problem: Uniqueness

The mathematical modeling of the heat conduction in a medium with a given temperature distribution at some initial time and a given temperature distribution on the boundary of the medium for all times leads to the following initial boundary value problem for the heat equation.

Initial Boundary Value Problem. *Let $D \subset \mathbb{R}^m$, $m = 1, 2, 3$, be a bounded domain with boundary $\Gamma := \partial D$ and let T denote a positive real number. Find a function $u \in C(\bar{D} \times [0, T])$ that is twice continuously differentiable with respect to the space variable x and continuously differentiable with respect to the time variable t in $D \times (0, T]$ and that satisfies the heat equation*

$$\frac{\partial u}{\partial t} = \Delta u \quad \text{in } D \times (0, T], \tag{9.1}$$

R. Kress, *Linear Integral Equations*, Applied Mathematical Sciences 82,
DOI 10.1007/978-1-4614-9593-2_9, © Springer Science+Business Media New York 2014

the initial condition

$$u(\cdot\,,0) = w \quad \textit{in } D, \tag{9.2}$$

and the Dirichlet boundary condition

$$u = f \quad \textit{on } \Gamma \times [0,T]. \tag{9.3}$$

Here, $w \in C(\bar{D})$ and $f \in C(\Gamma \times [0,T])$ are given continuous functions subject to the compatibility condition

$$w = f(\cdot\,,0) \quad \textit{on } \Gamma. \tag{9.4}$$

Analogous to the case of the Laplace equation as studied in the previous three chapters, instead of the Dirichlet boundary condition also a Neumann boundary condition

$$\frac{\partial u}{\partial \nu} = g \quad \text{on } \Gamma \times [0,T]$$

can be considered. Here, as in Chapters 6–8, we denote by ν the unit normal to Γ directed into the exterior of D and the given function $g \in C(\Gamma \times [0,T])$ must satisfy the compatibility condition

$$\frac{\partial w}{\partial \nu} = g(\cdot\,,0) \quad \text{on } \Gamma.$$

For our short introduction to the use of Volterra type integral equations for initial boundary value problems for the heat equation for brevity we choose to only consider the Dirichlet boundary condition. However, for the Neumann boundary condition in the one-dimensional case see Problem 9.5.

Our aim is to establish that this initial boundary value problem (9.1)–(9.3) has a unique solution that depends continuously on the given initial and boundary data. As in the case of the Dirichlet problem for the Laplace equation, uniqueness and continuous dependence follow from a maximum-minimum principle.

Theorem 9.1 (Weak Maximum-Minimum Principle). *Assume that the function $u \in C(\bar{D} \times [0,T])$ is twice continuously differentiable with respect to the space variable and continuously differentiable with respect to the time variable and solves the heat equation in $D \times (0,T]$. Then u attains both its maximum and its minimum on the* parabolic boundary

$$B := \{(x,0) : x \in D\} \cup \{(x,t) : x \in \Gamma,\ t \in [0,T]\}.$$

Proof. We confine ourselves to the proof for the maximum. Define a continuous function v on $\bar{D} \times [0,T]$ by

$$v(x,t) := u(x,t) + \varepsilon(T-t)$$

with $\varepsilon > 0$. Assume that v attains its maximum value in the *parabolic interior* $D \times (0, T]$, i.e., there exists a point $(x, t) \in D \times (0, T]$ such that

$$v(x,t) = \max_{(y,\tau)\in\bar{D}\times[0,T]} v(y,\tau).$$

Then the necessary conditions for a maximum

$$\frac{\partial^2 v}{\partial x_i^2}(x,t) \le 0, \quad i = 1,\ldots,m,$$

and

$$\frac{\partial v}{\partial t}(x,t) \ge 0$$

must be fulfilled. Hence, we have

$$\Delta v(x,t) - \frac{\partial v}{\partial t}(x,t) \le 0,$$

which is a contradiction to

$$\Delta v - \frac{\partial v}{\partial t} = \varepsilon > 0$$

throughout $D \times (0, T]$. Therefore, v attains its maximum on the parabolic boundary B. Then we can conclude that

$$u(x,t) \le v(x,t) \le \max_{(y,\tau)\in B} v(y,\tau) \le \max_{(y,\tau)\in B} u(y,\tau) + \varepsilon T$$

for all $(x, t) \in \bar{D} \times [0, T]$. Because ε can be chosen arbitrarily, it follows that

$$u(x,t) \le \max_{(y,\tau)\in B} u(y,\tau),$$

which ends the proof. □

We wish to mention that analogous to the maximum-minimum principle Theorem 6.9 for harmonic functions, there is also a strong maximum-minimum principle, stating that, under the conditions of Theorem 9.1, a solution to the heat equation cannot assume its maximum or minimum in the parabolic interior unless it is constant. For a proof in the case $m = 1$ we refer to Cannon [25].

The maximum-minimum principle implies uniqueness and continuous dependence for the initial boundary value problem as formulated in the following two theorems.

Theorem 9.2. *The initial boundary value problem (9.1)–(9.3) for the heat equation has at most one solution.*

Theorem 9.3. *The solution to the initial boundary value problem (9.1)–(9.3) for the heat equation depends continuously in the maximum norm on the given initial and boundary data.*

In the statement of Theorem 9.3 we have tacitly assumed that we will be able to establish existence of the solution.

9.2 Heat Potentials

The function

$$G(x,t;y,\tau) := \frac{1}{\sqrt{4\pi(t-\tau)}^{\,m}} \exp\left\{-\frac{|x-y|^2}{4(t-\tau)}\right\}, \qquad t > \tau, \tag{9.5}$$

is called *fundamental solution* of the heat equation. Straightforward differentiation shows that G satisfies the heat equation with respect to the variables x and t. With the aid of this fundamental solution, as in the analysis of Laplace's equation in Chapter 6, we will define so-called *heat potentials*. Then we shall use these potentials to solve the initial boundary value problem. For the remainder of this chapter, in the case $m \geq 2$, we will always assume that the bounded domain $D \subset \mathbb{R}^m$ is of class C^2. We will take care of the initial condition through the Poisson integral introduced in the following theorem.

Theorem 9.4. *Let w be a continuous function on $\mathbb{R}^m$ with compact support, i.e., w vanishes outside some compact set. Then the* Poisson integral

$$u(x,t) := \frac{1}{\sqrt{4\pi t}^{\,m}} \int_{\mathbb{R}^m} \exp\left\{-\frac{|x-y|^2}{4t}\right\} w(y)\,dy \tag{9.6}$$

defines an infinitely differentiable solution of the heat equation in $\mathbb{R}^m \times (0,\infty)$. It can be continuously extended into $\mathbb{R}^m \times [0,\infty)$ with initial values

$$u(\cdot\,,0) = w. \tag{9.7}$$

Proof. Obviously, for all $x \in \mathbb{R}^m$ and $t \in (0,\infty)$ differentiation and integration can be interchanged to show that the Poisson integral is infinitely differentiable and satisfies the heat equation. Substituting $y = x + 2\sqrt{t}\,z$, we obtain

$$\frac{1}{\sqrt{4\pi t}^{\,m}} \int_{\mathbb{R}^m} \exp\left\{-\frac{|x-y|^2}{4t}\right\} w(y)\,dy = \frac{1}{\sqrt{\pi}^{\,m}} \int_{\mathbb{R}^m} w(x + 2\sqrt{t}\,z)\, e^{-|z|^2}\,dz.$$

Since w has compact support, it is bounded and uniformly continuous. Let M be a bound for w. Then for $\varepsilon > 0$ we choose R such that

$$\frac{1}{\sqrt{\pi}^{\,m}} \int_{|z|\geq R} e^{-|z|^2}\,dz < \frac{\varepsilon}{4M}\,.$$

Because w is uniformly continuous there exists $\delta > 0$ such that $|w(x) - w(y)| < \varepsilon/2$ for all x and y with $|x-y| < \delta$. Setting $\eta = \delta^2/4R^2$ and making use of the integral

(see Problem 9.1)

$$\int_{\mathbb{R}^m} e^{-|z|^2} dz = \sqrt{\pi}^{\,m} \tag{9.8}$$

and the fact that for $|z| \le R$ and $t < \eta$ we have $2\sqrt{t}\,|z| < 2\sqrt{\eta}\,R = \delta$, we deduce that

$$\left| \frac{1}{\sqrt{4\pi t}^{\,m}} \int_{\mathbb{R}^m} \exp\left\{ -\frac{|x-y|^2}{4t} \right\} w(y)\,dy - w(x) \right|$$

$$= \left| \frac{1}{\sqrt{\pi}^{\,m}} \int_{\mathbb{R}^m} \left\{ w(x + 2\sqrt{t}\,z) - w(x) \right\} e^{-|z|^2} dz \right|$$

$$< \frac{\varepsilon}{2\sqrt{\pi}^{\,m}} \int_{|z|\le R} e^{-|z|^2} dz + \frac{2M}{\sqrt{\pi}^{\,m}} \int_{|z|\ge R} e^{-|z|^2} dz < \varepsilon$$

for all $x \in \mathbb{R}^m$ and all $t < \eta$. This implies continuity of u at $t = 0$ and (9.7). □

In order to deal with the boundary condition we will need *surface heat potentials*. Analogous to the terminology used for harmonic functions we define single- and double-layer potentials. For a function $\varphi \in C(\Gamma \times [0, T])$ the *single-layer heat potential* is given by

$$u(x,t) := \int_0^t \int_\Gamma G(x,t;y,\tau)\varphi(y,\tau)\,ds(y)d\tau \tag{9.9}$$

and the *double-layer heat potential* by

$$u(x,t) := \int_0^t \int_\Gamma \frac{\partial G(x,t;y,\tau)}{\partial \nu(y)}\,\varphi(y,\tau)\,ds(y)d\tau. \tag{9.10}$$

These potentials are well defined for $x \in D$ and $t \in (0, T]$ with the time integral to be understood as an improper integral with respect to the upper limit. For the double-layer potential we assume the unit normal vector ν to the boundary surface Γ to be directed into the exterior of D.

In the case of one space dimension for an interval $D := (a, b)$ the two potentials have the form

$$u(x,t) := \int_0^t \frac{1}{\sqrt{4\pi(t-\tau)}} \left[\exp\left\{ -\frac{(x-a)^2}{4(t-\tau)} \right\} + \exp\left\{ -\frac{(x-b)^2}{4(t-\tau)} \right\} \right] \varphi(b,\tau)\,d\tau \tag{9.11}$$

and

$$\begin{aligned} u(x,t) := &\int_0^t \frac{a-x}{4\sqrt{\pi}\,\sqrt{(t-\tau)}^{\,3}} \exp\left\{ -\frac{(x-a)^2}{4(t-\tau)} \right\} \varphi(a,\tau)\,d\tau \\ &+ \int_0^t \frac{x-b}{4\sqrt{\pi}\,\sqrt{(t-\tau)}^{\,3}} \exp\left\{ -\frac{(x-b)^2}{4(t-\tau)} \right\} \varphi(b,\tau)\,d\tau. \end{aligned} \tag{9.12}$$

Standard analysis again shows that both the single- and double-layer heat potentials are infinitely differentiable solutions to the heat equation in $D \times (0, T]$. In addition they can be continuously extended into $D \times [0, T]$ with initial values $u(\cdot\,, 0) = 0$ in D. For the discussion of the boundary values we will confine our analysis to the double-layer potential.

Theorem 9.5. *The double-layer heat potential with continuous density φ can be continuously extended from $D \times (0, T]$ into $\bar{D} \times (0, T]$ with limiting values*

$$u(x, t) = \int_0^t \int_\Gamma \frac{\partial G(x, t; y, \tau)}{\partial \nu(y)}\, \varphi(y, \tau)\, ds(y) d\tau - \frac{1}{2}\, \varphi(x, t) \tag{9.13}$$

for $x \in \Gamma$ and $t \in (0, T]$. Here the time integral exists as improper integral.

Proof. We carry out the proof for the cases $m = 2, 3$, and leave it to the reader to work out the details for the simpler case $m = 1$ (see Problem 9.3). For $x \in D$ we may interchange the integrations over Γ and $[0, t]$, since the integrand is continuous on $\Gamma \times [0, t]$ with value zero at the upper limit $\tau = t$. Then we substitute for the time integral

$$\sigma = \frac{|x - y|}{2\sqrt{t - \tau}}$$

to obtain

$$\begin{aligned} u(x, t) &= \int_0^t \frac{1}{\sqrt{4\pi(t - \tau)}^{\,m}} \int_\Gamma \frac{\nu(y) \cdot (x - y)}{2(t - \tau)} \exp\left\{-\frac{|x - y|^2}{4(t - \tau)}\right\} \varphi(y, \tau)\, ds(y) d\tau \\ &= \frac{1}{\sqrt{\pi}^{\,m}} \int_\Gamma \frac{\nu(y) \cdot (x - y)}{|x - y|^m} \int_{|x-y|/2\sqrt{t}}^\infty \sigma^{m-1} e^{-\sigma^2} \varphi\left(y, t - \frac{|x - y|^2}{4\sigma^2}\right) d\sigma ds(y). \end{aligned}$$

Therefore, we can view the double-layer heat potential as a harmonic double-layer potential with the density

$$\begin{aligned} \psi(x, y, t) &:= \frac{1}{\sqrt{\pi}^{\,m}} \int_{|x-y|/2\sqrt{t}}^\infty \sigma^{m-1} e^{-\sigma^2} \varphi\left(y, t - \frac{|x - y|^2}{4\sigma^2}\right) d\sigma \\ &= |x - y|^m \int_0^t \frac{1}{\sqrt{4\pi(t - \tau)}^{\,m}}\, \frac{1}{2(t - \tau)} \exp\left\{-\frac{|x - y|^2}{4(t - \tau)}\right\} \varphi(y, \tau)\, d\tau \end{aligned}$$

depending on the time t as a parameter. First, we will show that ψ is continuous on $\mathbb{R}^m \times \Gamma \times (0, T]$ with

$$\lim_{x \to y} \psi(x, y, t) = \gamma_m \varphi(y, t) \tag{9.14}$$

for all $y \in \Gamma$ and $t \in (0, T]$, where

$$\gamma_m := \frac{1}{\sqrt{\pi}^{\,m}} \int_0^\infty s^{m-1} e^{-s^2} ds$$

and the limit holds uniformly on Γ and compact subintervals of $(0, T]$. Clearly the function ψ is continuous for all $x \neq y$ and all $t \in (0, T]$. We establish the limit (9.14) by splitting

$$\psi(x, y, t) = \frac{1}{\sqrt{\pi}^m} \int_{|x-y|/2\sqrt{t}}^{\sqrt{|x-y|}} \sigma^{m-1} e^{-\sigma^2} \varphi\left(y, t - \frac{|x-y|^2}{4\sigma^2}\right) d\sigma$$

$$+ \frac{1}{\sqrt{\pi}^m} \int_{\sqrt{|x-y|}}^{\infty} \sigma^{m-1} e^{-\sigma^2} \left\{\varphi\left(y, t - \frac{|x-y|^2}{4\sigma^2}\right) - \varphi(y, t)\right\} d\sigma$$

$$+ \varphi(y, t) \frac{1}{\sqrt{\pi}^m} \int_{\sqrt{|x-y|}}^{\infty} \sigma^{m-1} e^{-\sigma^2} d\sigma =: I_1 + I_2 + I_3,$$

with obvious definitions of I_1, I_2 and I_3. Clearly we have

$$\lim_{x \to y} I_1(x, y, t) = 0$$

uniformly on Γ and on compact subintervals of $(0, T]$ and

$$\lim_{x \to y} I_3(x, y, t) = \gamma_m \varphi(y, t)$$

uniformly on $\Gamma \times [0, T]$. Since φ is uniformly continuous, for any $\varepsilon > 0$ there exists $\delta > 0$ such that

$$|\varphi(y, t_1) - \varphi(y, t_2)| < \varepsilon$$

for all $y \in \Gamma$ and all $t_1, t_2 \in [0, T]$ with $|t_1 - t_2| < \delta$. Then for all $|x - y| < 4\delta$ and all $\sigma \geq \sqrt{|x - y|}$ we have

$$\frac{|x-y|^2}{4\sigma^2} \leq \frac{|x-y|}{4} < \delta$$

and therefore

$$\left|\varphi\left(y, t - \frac{|x-y|^2}{4\sigma^2}\right) - \varphi(y, t)\right| < \varepsilon.$$

Hence,

$$|I_2(x, y, t)| < \frac{\varepsilon}{\sqrt{\pi}^m} \int_{\sqrt{|x-y|}}^{\infty} \sigma^{m-1} e^{-\sigma^2} d\sigma \leq \gamma_m \varepsilon$$

and thus

$$\lim_{x \to y} I_2(x, y, t) = 0$$

uniformly on $\Gamma \times [0, T]$. Now, using $\gamma_2 = 1/2\pi$ and $\gamma_3 = 1/4\pi$, from Theorem 6.18 we can deduce that

$$\lim_{h \to 0} u(x - h\nu(x), t) = \int_0^t \int_\Gamma \frac{\partial G(x, t; y, \tau)}{\partial \nu(y)} \varphi(y, \tau)\, ds(y) d\tau - \frac{1}{2} \varphi(x, t) \tag{9.15}$$

with uniform convergence on Γ and on compact subintervals of $(0, T]$. Note that we need a slightly generalized version of Theorem 6.18 where the density is allowed to also depend on x and a parameter t. It is left to the reader to go over the proof of Theorem 6.18 to see that it can be extended to this more general case.

Finally, the statement on the continuity of the double-layer heat potential on $\bar{D} \times (0, T]$ follows from the fact that the right-hand side of (9.15) is continuous on $\Gamma \times [0, T]$ by the following Theorem 9.6. □

Consider the integral operator $H : C(\Gamma \times [0, T]) \to C(\Gamma \times [0, T])$ defined by

$$(H\varphi)(x, t) := 2 \int_0^t \int_\Gamma \frac{\partial G(x, t; y, \tau)}{\partial \nu(y)} \varphi(y, \tau)\, ds(y) d\tau \tag{9.16}$$

for $x \in \Gamma$ and $t \in (0, T]$ with improper integral over $[0, t]$. For its kernel

$$2\, \frac{\partial G(x, t; y, \tau)}{\partial \nu(y)} = \frac{1}{\sqrt{4\pi(t-\tau)}^{\,m}} \, \frac{\nu(y) \cdot (x-y)}{t-\tau} \exp\left\{-\frac{|x-y|^2}{4(t-\tau)}\right\},$$

using Lemma 6.16, we can estimate

$$2\left|\frac{\partial G(x, t; y, \tau)}{\partial \nu(y)}\right| \le \frac{L}{\sqrt{t-\tau}^{\,m}} \, \frac{|x-y|^2}{t-\tau} \exp\left\{-\frac{|x-y|^2}{4(t-\tau)}\right\}, \qquad t > \tau,$$

with some constant L. Applying the inequality (see Problem 9.2)

$$s^\beta e^{-s} \le \beta^\beta e^{-\beta}, \tag{9.17}$$

which is valid for all $0 < s, \beta < \infty$, for the special case

$$s = \frac{|x-y|^2}{4(t-\tau)}$$

and $\beta = 1 + m/2 - \alpha$, we derive the further estimate

$$2\left|\frac{\partial G(x, t; y, \tau)}{\partial \nu(y)}\right| \le \frac{M}{(t-\tau)^\alpha |x-y|^{m-2\alpha}}, \qquad t > \tau, \quad x \neq y, \tag{9.18}$$

for all $0 < \alpha < 1 + m/2$ and some constant M depending on L and α. Hence, choosing $1/2 < \alpha < 1$, the kernel of H is seen to be weakly singular with respect to both the integrals over Γ and over time. Therefore, proceeding as in the proofs of Theorems 2.29 and 2.30, we can establish the following result.

Theorem 9.6. *The double-layer heat potential operator H is a compact operator from $C(\Gamma \times [0, T])$ into $C(\Gamma \times [0, T])$.*

Summarizing, the double-layer heat potential with continuous density is continuous in $\bar{D} \times [0, T]$ with the exception of possible discontinuities on the boundary Γ for $t = 0$. From (9.18) we observe that $(H\varphi)(\cdot\,, 0) = 0$. Therefore we can expect continuity in $\bar{D} \times [0, T]$ only if $\varphi(\cdot\,, 0) = 0$ on Γ. And indeed, in this case we can extend the density φ continuously onto $\Gamma \times (-\infty, T]$ by setting $\varphi(\cdot\,, t) = 0$ for $t < 0$. Then, as a consequence of Theorem 9.5, we can state the following corollary.

Corollary 9.7. *The double-layer heat potential is continuous on $\bar{D} \times [0, T]$ provided the continuous density satisfies $\varphi(\cdot\,, 0) = 0$.*

9.3 Initial Boundary Value Problem: Existence

We now return to the initial boundary value problem. First we reduce it to the special case of a homogeneous initial condition. To this end, consider

$$u(x, t) := v(x) + \frac{1}{\sqrt{4\pi t}^{\,m}} \int_{\mathbb{R}^m} \exp\left\{-\frac{|x - y|^2}{4t}\right\} \{w(y) - v(y)\}\, dy. \tag{9.19}$$

Here, v denotes the unique solution to the Dirichlet problem for Laplace's equation in D with boundary condition $v = f(\cdot\,, 0)$ on Γ (see Theorem 6.23). Because of the compatibility condition (9.4), the function $w - v$ can be continuously extended into $\mathbb{R}^m$ by setting it to zero outside D. Then, from Theorems 6.6 and 9.4, we deduce that (9.19) defines an infinitely differentiable solution to the heat equation in $D \times (0, T]$ that is continuous in $\bar{D} \times [0, T]$ and satisfies the initial condition $u(\cdot\,, 0) = w$ in D. Hence, by superposition, it suffices to treat the special case of the initial boundary value problem with initial condition

$$u(\cdot\,, 0) = 0 \quad \text{in } D \tag{9.20}$$

and boundary condition

$$u = f \quad \text{on } \Gamma \times [0, T], \tag{9.21}$$

where f satisfies the compatibility condition

$$f(\cdot\,, 0) = 0 \quad \text{on } \Gamma. \tag{9.22}$$

In order to deal with the boundary condition (9.21) we seek the solution in the form of a double-layer heat potential. Again we will leave the case where $m = 1$ as an exercise for the reader (see Problem 9.4).

Theorem 9.8. *The double-layer heat potential (9.10) solves the initial boundary value problem (9.20)–(9.22) provided that the continuous density φ for all $x \in \Gamma$ and $t \in (0, T]$ solves the integral equation*

$$\varphi(x, t) - 2\int_0^t \int_\Gamma \frac{\partial G(x, t; y, \tau)}{\partial \nu(y)}\, \varphi(y, \tau)\, ds(y) d\tau = -2f(x, t). \tag{9.23}$$

Proof. This follows from Theorem 9.5 and Corollary 9.7. The compatibility condition (9.22) for f ensures that $\varphi(\cdot\,,0) = 0$ for solutions to (9.23). □

By Theorem 9.6 and the Riesz theory Theorem 3.4, the inhomogeneous integral equation $\varphi - H\varphi = -2f$ is solvable if the corresponding homogeneous integral equation $\varphi - H\varphi = 0$ has only the trivial solution. From the weak singularity (9.18) we see that for the integral operator H the integration over Γ corresponds to a compact operator. Therefore we can estimate for the maximum norm with respect to the space variable by

$$\|(H\varphi)(\cdot\,,t)\|_{\infty,\Gamma} \le C \int_0^t \frac{1}{(t-\tau)^\alpha}\, \|\varphi(\cdot\,,\tau)\|_{\infty,\Gamma}\, d\tau \tag{9.24}$$

for all $0 < t \le T$ and some constant C depending on Γ and α. Note that (see Problem 9.1)

$$\begin{aligned} \int_\sigma^t \frac{d\tau}{(t-\tau)^\alpha(\tau-\sigma)^\beta} &= \frac{1}{(t-\sigma)^{\alpha+\beta-1}} \int_0^1 \frac{ds}{s^\beta(1-s)^\alpha} \\ &\le \frac{1}{(t-\sigma)^{\alpha+\beta-1}} \int_0^1 \frac{ds}{[s(1-s)]^\alpha} \end{aligned} \tag{9.25}$$

for $\beta \le \alpha < 1$. Then, from (9.24), by induction and by interchanging the integrals in

$$\int_0^t \int_0^\tau \frac{d\sigma}{(t-\tau)^\alpha(\tau-\sigma)^\beta}\, d\tau = \int_0^t \int_\sigma^t \frac{d\tau}{(t-\tau)^\alpha(\tau-\sigma)^\beta}\, d\sigma,$$

we find that

$$\|(H^k\varphi)(\cdot\,,t)\|_{\infty,\Gamma} \le C^k I^{k-1} \int_0^t \frac{1}{(t-\tau)^{k(\alpha-1)+1}}\, \|\varphi(\cdot\,,\tau)\|_{\infty,\Gamma}\, d\tau$$

for all $k \in \mathbb{N}$ and all $0 < t \le T$, where

$$I := \int_0^1 \frac{ds}{[s(1-s)]^\alpha}\,.$$

Hence there exists an integer k such that for $\widetilde{H} := H^k$ we have an estimate of the form

$$\|(\widetilde{H}\varphi)(\cdot\,,t)\|_{\infty,\Gamma} \le M \int_0^t \|\varphi(\cdot\,,\tau)\|_{\infty,\Gamma}\, d\tau \tag{9.26}$$

for all $0 \le t \le T$ and some constant M.

Now, let φ be a solution to the homogeneous equation $\varphi - H\varphi = 0$. Then, by iteration, φ also solves $\varphi - \widetilde{H}\varphi = 0$. Proceeding as in the proof of Theorem 3.10, i.e., as for Volterra equations with a continuous kernel, from (9.26) we can derive that

$$\|\varphi(\cdot\,,t)\|_{\infty,\Gamma} \le \|\varphi\|_\infty \frac{M^n t^n}{n!}, \quad 0 \le t \le T,$$

for all $n \in \mathbb{N}$. Hence $\|\varphi(\cdot, t)\|_{\infty,\Gamma} = 0$ for all $t \in [0, T]$, i.e., $\varphi = 0$. Thus we have established the following existence theorem.

Theorem 9.9. *The initial boundary value problem for the heat equation has a unique solution.*

We wish to mention that, by using a single-layer heat potential, the analysis of this chapter can be carried over to an initial boundary value problem where the Dirichlet boundary condition is replaced by a Neumann boundary condition. For the one-dimensional case, see Problem 9.5.

The idea to proceed along the same line in the case of the heat equation as for the Laplace equation is due to Holmgren [100, 101] and Gevrey [62]. The first rigorous existence proof was given by Müntz [179] using successive approximations for the integral equation in two dimensions. Integral equations of the first kind for the heat equation have been investigated by Arnold and Noon [7], Costabel [34], Costabel and Saranen [35] and Hsiao and Saranen [103] in Sobolev spaces, and by Baderko [13] in Hölder spaces.

Problems

9.1. Prove the integrals (9.8) and (9.25).

9.2. Prove the inequality (9.17).

9.3. Carry out the proof of Theorem 9.5 in the one-dimensional case.

9.4. Show that the double-layer heat potential (9.12) solves the initial boundary value problem for the heat equation in $(a, b) \times (0, T]$ with homogeneous initial condition $u(\cdot, 0) = 0$ and Dirichlet boundary conditions

$$u(a, \cdot) = f(a, \cdot) \quad \text{and} \quad u(b, \cdot) = f(b, \cdot)$$

(with compatibility condition $f(a, 0) = f(b, 0) = 0$) provided the densities satisfy the system of Volterra integral equations

$$\varphi(a, t) - \int_0^t h(t, \tau)\varphi(b, \tau)\, d\tau = -2f(a, t),$$

$$\varphi(b, t) - \int_0^t h(t, \tau)\varphi(a, \tau)\, d\tau = -2f(b, t)$$

for $0 \le t \le T$ with kernel

$$h(t, \tau) := \frac{a - b}{2\sqrt{\pi}}\, \frac{1}{\sqrt{(t - \tau)}^{\,3}} \exp\left\{-\frac{(a - b)^2}{4(t - \tau)}\right\}, \quad 0 \le \tau < t.$$

Establish existence and uniqueness for this system.

9.5. Show that the single-layer heat potential (9.11) solves the initial boundary value problem for the heat equation in $(a, b) \times (0, T]$ with homogeneous initial condition $u(\cdot, 0) = 0$ and Neumann boundary conditions

$$-\frac{\partial u(a, \cdot)}{\partial x} = g(a, \cdot) \quad \text{and} \quad \frac{\partial u(b, \cdot)}{\partial x} = g(b, \cdot)$$

(with compatibility condition $g(a, 0) = g(b, 0) = 0$) provided the densities satisfy the system of Volterra integral equations

$$\varphi(a,t) + \int_0^t h(t,\tau)\varphi(b,\tau)\,d\tau = 2g(a,t),$$

$$\varphi(b,t) + \int_0^t h(t,\tau)\varphi(a,\tau)\,d\tau = 2g(b,t)$$

for $0 \le t \le T$ with the kernel h given as in Problem 9.4. Establish existence and uniqueness for this system.

Chapter 10
Operator Approximations

In subsequent chapters we will study the numerical solution of integral equations. It is our intention to provide the basic tools for the investigation of approximate solution methods and their error analysis. We do not aim at a complete review of all the various numerical methods that have been developed in the literature. However, we will develop some of the principal ideas and illustrate them with a few instructive examples.

A fundamental concept for approximately solving an operator equation

$$A\varphi = f$$

with a bounded linear operator $A : X \to Y$ mapping a Banach space X into a Banach space Y is to replace it by an equation

$$A_n\varphi_n = f_n$$

with an approximating sequence of bounded linear operators $A_n : X \to Y$ and an approximating sequence $f_n \to f$, $n \to \infty$. For practical purposes, the approximating equations will be chosen so that they can be reduced to solving a finite-dimensional linear system. In this chapter we will provide an error analysis for such general approximation schemes. In particular, we will derive convergence results and error estimates for the cases where we have either norm or pointwise convergence of the sequence $A_n \to A$, $n \to \infty$. For the latter case, we will present the concept of collectively compact operators.

10.1 Approximations via Norm Convergence

Theorem 10.1. *Let X and Y be Banach spaces and let $A : X \to Y$ be a bounded linear operator with a bounded inverse $A^{-1} : Y \to X$, i.e., an isomorphism. Assume the sequence $A_n : X \to Y$ of bounded linear operators to be norm convergent*

R. Kress, *Linear Integral Equations*, Applied Mathematical Sciences 82,
DOI 10.1007/978-1-4614-9593-2_10,

$\|A_n - A\| \to 0$, $n \to \infty$. *Then for sufficiently large n, more precisely for all n with*

$$\|A^{-1}(A_n - A)\| < 1,$$

the inverse operators $A_n^{-1} : Y \to X$ *exist and are bounded by*

$$\|A_n^{-1}\| \le \frac{\|A^{-1}\|}{1 - \|A^{-1}(A_n - A)\|}\,. \tag{10.1}$$

For the solutions of the equations

$$A\varphi = f \quad \text{and} \quad A_n\varphi_n = f_n$$

we have the error estimate

$$\|\varphi_n - \varphi\| \le \frac{\|A^{-1}\|}{1 - \|A^{-1}(A_n - A)\|}\,\{\|(A_n - A)\varphi\| + \|f_n - f\|\}\,.$$

Proof. If $\|A^{-1}(A_n - A)\| < 1$, then by the Neumann series Theorem 2.14, the inverse $[I - A^{-1}(A - A_n)]^{-1}$ of $I - A^{-1}(A - A_n) = A^{-1}A_n$ exists and is bounded by

$$\|\,[I - A^{-1}(A - A_n)]^{-1}\| \le \frac{1}{1 - \|A^{-1}(A_n - A)\|}\,.$$

But then $[I - A^{-1}(A - A_n)]^{-1}A^{-1}$ is the inverse of A_n and bounded by (10.1). The error estimate follows from

$$A_n(\varphi_n - \varphi) = f_n - f + (A - A_n)\varphi,$$

and the proof is complete. □

In Theorem 10.1 from the unique solvability of the original equation we conclude unique solvability of the approximating equation provided the approximation is sufficiently close in the operator norm. The converse situation is described in the following theorem.

Theorem 10.2. *Assume there exists some* $N \in \mathbb{N}$ *such that for all* $n \ge N$ *the inverse operators* $A_n^{-1} : Y \to X$ *exist and are uniformly bounded. Then the inverse operator* $A^{-1} : Y \to X$ *exists and is bounded by*

$$\|A^{-1}\| \le \frac{\|A_n^{-1}\|}{1 - \|A_n^{-1}(A_n - A)\|} \tag{10.2}$$

for all n with $\|A_n^{-1}(A_n - A)\| < 1$. *For the solutions of the equations*

$$A\varphi = f \quad \text{and} \quad A_n\varphi_n = f_n$$

we have the error estimate

$$\|\varphi_n - \varphi\| \leq \frac{\|A_n^{-1}\|}{1 - \|A_n^{-1}(A_n - A)\|} \{\|(A_n - A)\varphi_n\| + \|f_n - f\|\}.$$

Proof. This follows from Theorem 10.1 by interchanging the roles of A and A_n. □

Note that Theorem 10.2 provides an error bound that, in principle, can be evaluated, because it involves A_n^{-1} and φ_n but neither A^{-1} nor φ. The error bound of Theorem 10.1 shows that the accuracy of the approximate solution depends on how well $A_n\varphi$ approximates $A\varphi$ for the exact solution as expressed through the following corollary.

Corollary 10.3. *Under the assumptions of Theorem 10.1 we have the error estimate*

$$\|\varphi_n - \varphi\| \leq C\,\{\|(A_n - A)\varphi\| + \|f_n - f\|\} \tag{10.3}$$

for all sufficiently large n and some constant C.

Proof. This is an immediate consequence of Theorem 10.1. □

In Chapter 11 we will apply the error analysis of Theorems 10.1 and 10.2 to the approximation of integral equations of the second kind by replacing the kernels through so-called degenerate kernels.

10.2 Uniform Boundedness Principle

To develop a similar analysis for the case where the sequence (A_n) is merely pointwise convergent, i.e., $A_n\varphi \to A\varphi$, $n \to \infty$, for all φ, we will have to bridge the gap between norm and pointwise convergence. This goal will be achieved through compactness properties and one of the necessary tools will be provided by the *principle of uniform boundedness* that we will develop in this section.

Analogous to $\mathbb{R}^m$, in Banach spaces we have the following properties of nested balls.

Lemma 10.4. *Let X be a Banach space and let (B_n) be a sequence of closed balls with radii r_n and the properties $B_{n+1} \subset B_n$ for $n = 1, 2, \ldots$ and $\lim_{n\to\infty} r_n = 0$. Then there exists a unique $\varphi \in X$ with $\varphi \in \bigcap_{n=1}^{\infty} B_n$.*

Proof. We denote the centers of the balls B_n by φ_n. Then for $m \geq n$ we have

$$\|\varphi_m - \varphi_n\| \leq r_n \to 0, \quad n \to \infty, \tag{10.4}$$

since $\varphi_m \in B_m \subset B_n$. Hence (φ_n) is a Cauchy sequence and consequently $\varphi_n \to \varphi$, $n \to \infty$, for some $\varphi \in X$. Passing to the limit $m \to \infty$ in (10.4) yields $\|\varphi - \varphi_n\| \leq r_n$, that is, $\varphi \in B_n$ for all $n \in \mathbb{N}$.

Assume that φ, ψ are two elements satisfying $\varphi, \psi \in \bigcap_{n=1}^{\infty} B_n$. Then

$$\|\varphi - \psi\| \leq \|\varphi - \varphi_n\| + \|\varphi_n - \psi\| \leq 2r_n \to 0, \quad n \to \infty,$$

that is, $\varphi = \psi$. □

Theorem 10.5 (Baire). *Let X be a Banach space and let (U_n) be a sequence of closed subsets of X such that $\bigcup_{n=1}^{\infty} U_n$ contains an open ball. Then for some $n \in \mathbb{N}$ the set U_n contains an open ball.*

Proof. By assumption there exists an open ball $V \subset \bigcup_{n=1}^{\infty} U_n$. We assume that none of the sets U_n contains an open ball. Then by induction we construct sequences (r_n) in $\mathbb{R}$ and (φ_n) in X with the properties

$$0 < r_n \leq \frac{1}{n}, \qquad B_n := B[\varphi_n; r_n] \subset V, \quad B_n \subset X \setminus U_n, \quad B_n \subset B_{n-1}$$

for $n = 1, 2, \ldots$. We start the sequence by choosing $B_0 = B[\varphi_0; r_0] \subset V$ arbitrary and assume that for some $n \geq 1$ the partial sequences $r_0, \ldots, r_n$ and $\varphi_0, \ldots, \varphi_n$ with the required properties are constructed. Then the set $(X \setminus U_{n+1}) \cap B(\varphi_n; r_n)$ is open and not empty, since otherwise $B(\varphi_n; r_n) \subset U_{n+1}$ would contradict our assumption that U_{n+1} does not contain an open ball. Therefore there exist r_{n+1} with $0 < r_{n+1} \leq 1/(n+1)$ and $\varphi_{n+1} \in X$ such that

$$B_{n+1} = B[\varphi_{n+1}; r_{n+1}] \subset (X \setminus U_{n+1}) \cap B_n.$$

The sequence (B_n) satisfies the assumptions of Lemma 10.4. Hence there exists some $\varphi \in X$ such that $\varphi \in \bigcap_{n=1}^{\infty} B_n$. Then on one hand we have that

$$\varphi \in \bigcap_{n=1}^{\infty} B_n \subset \bigcap_{n=1}^{\infty} (X \setminus U_n) = X \setminus \bigcup_{n=1}^{\infty} U_n,$$

that is, $\varphi \notin \bigcup_{n=1}^{\infty} U_n$. One the other hand we also have that

$$\varphi \in \bigcap_{n=1}^{\infty} B_n \subset V \subset \bigcup_{n=1}^{\infty} U_n,$$

i.e., our assumption that none of the sets U_n contains an open ball leads to a contradiction and the theorem is proven. □

Theorem 10.6 (Uniform boundedness principle). *Let the set $\mathcal{A} := \{A : X \to Y\}$ of bounded linear operators A mapping a Banach space X into a normed space Y be pointwise bounded, i.e., for each $\varphi \in X$ there exists a positive number C_φ depending on φ such that $\|A\varphi\| \leq C_\varphi$ for all $A \in \mathcal{A}$. Then the set $\mathcal{A}$ is uniformly bounded, i.e., there exists some constant C such that $\|A\| \leq C$ for all $A \in \mathcal{A}$.*

Proof. We define

$$U_n := \{\varphi \in X : \|A\varphi\| \leq n, \ A \in \mathcal{A}\}.$$

Because of the continuity of the operators A the sets U_n are closed. Clearly, for each $\varphi \in X$ we have $\varphi \in U_n$ for all $n \geq C_\varphi$ and therefore $X = \bigcup_{n=1}^{\infty} U_n$. By Baire's Theorem 10.5 there exist $n \in \mathbb{N}$ and an open ball $B(\varphi_0; r)$ with radius $r > 0$ centered at some $\varphi_0 \in X$ such that $B(\varphi_0; r) \subset U_n$, that is, $\|A\varphi\| \leq n$ for all $\varphi \in B(\varphi_0; r)$ and all $A \in \mathcal{A}$.

Finally, for all $\psi \in X$ with $\|\psi\| \leq 1$ and all $A \in \mathcal{A}$, this implies that

$$\|A\psi\| = \frac{1}{r}\, \|A(r\psi + \varphi_0) - A\varphi_0\| \leq \frac{2n}{r}\,,$$

since $r\psi + \varphi_0 \in B(\varphi_0; r)$. Therefore $\|A\| \leq 2n/r$ for all $A \in \mathcal{A}$. □

As a simple consequence of the uniform boundedness principle we note that the limit operator of a pointwise convergent sequence of bounded linear operators again is bounded.

Corollary 10.7. *Let X be a Banach space, Y be a normed space, and assume that the bounded linear operators $A_n : X \to Y$ converge pointwise with limit operator $A : X \to Y$. Then the convergence is uniform on compact subsets U of X, i.e.,*

$$\sup_{\varphi \in U} \|A_n\varphi - A\varphi\| \to 0, \quad n \to \infty.$$

Proof. For $\varepsilon > 0$ consider the open balls $B(\varphi; r) = \{\psi \in X : \|\psi - \varphi\| < r\}$ with center $\varphi \in X$ and radius $r = \varepsilon/3C$, where C is a bound on the operators A_n. Clearly

$$U \subset \bigcup_{\varphi \in U} B(\varphi; r)$$

forms an open covering of U. Since U is compact, there exists a finite subcovering

$$U \subset \bigcup_{j=1}^{m} B(\varphi_j; r).$$

Pointwise convergence of (A_n) guarantees the existence of an integer $N(\varepsilon)$ such that

$$\|A_n\varphi_j - A\varphi_j\| < \frac{\varepsilon}{3}$$

for all $n \geq N(\varepsilon)$ and all $j = 1, \ldots, m$. Now let $\varphi \in U$ be arbitrary. Then φ is contained in some ball $B(\varphi_j; r)$ with center φ_j. Hence for all $n \geq N(\varepsilon)$ we have

$$\begin{aligned}\|A_n\varphi - A\varphi\| &\leq \|A_n\varphi - A_n\varphi_j\| + \|A_n\varphi_j - A\varphi_j\| + \|A\varphi_j - A\varphi\| \\ &\leq \|A_n\|\,\|\varphi - \varphi_j\| + \frac{\varepsilon}{3} + \|A\|\,\|\varphi_j - \varphi\| \leq 2Cr + \frac{\varepsilon}{3} = \varepsilon.\end{aligned}$$

Therefore the convergence is uniform on U. □

Although the *Banach open mapping theorem* does not play a central role in this book, for completeness of our presentation of the functional analysis for bounded linear operators we include the following proof since with Baire's theorem its main ingredient is already available to us.

Theorem 10.8 (Banach open mapping theorem). *Let X and Y be Banach spaces and let $A : X \to Y$ be a bijective bounded linear operator. Then its inverse $A^{-1} : Y \to X$ is also bounded, i.e., A is an isomorphism.*

Proof. **1.** We begin by showing that there exists a closed ball centered at some $\psi \in Y$ with radius ρ such that

$$B[\psi;\rho] \subset \overline{A(B[0;1])}. \tag{10.5}$$

To this end, for $f \in Y$ we choose $m \in \mathbb{N}$ satisfying $m \geq \|A^{-1}f\|$. Then we have that

$$f = AA^{-1}f \in A(B[0;m]) \subset \overline{A(B[0;m])},$$

whence

$$Y = \bigcup_{m=1}^{\infty} \overline{A(B[0;m])}$$

follows. Baire's Theorem 10.5 now implies that there exist $m \in \mathbb{N}$ and an open ball with radius $2\rho_m$ centered at some $\psi_m \in Y$ with $B(\psi_m;2\rho_m) \subset \overline{A(B[0;m])}$ and consequently

$$B[\psi_m;\rho_m] \subset \overline{A(B[0;m])}.$$

Setting $\psi := \psi_m/m$ and $\rho := \rho_m/m$ for $f \in B[\psi;\rho]$ we have that $mf \in B[\psi_m;\rho_m]$. Therefore there exists a sequence (φ_n) in X with $\|\varphi_n\| \leq m$ and $A\varphi_n \to mf$ as $n \to \infty$, that is, (10.5) is proven.

2. In the next step we shift the center of the ball and show that

$$B[0;r\rho] \subset \overline{A(B[0;r])} \tag{10.6}$$

for all $r > 0$. Let $f \in B[0;r\rho]$ and set

$$f_1 = \psi + \frac{f}{r}\,, \quad f_2 = \psi - \frac{f}{r}\,.$$

Then $f_1, f_2 \in B[\psi;\rho] \subset \overline{A(B[0;1])}$ as a consequence of (10.5). Hence, for $j = 1, 2$, there exist sequences $(\varphi_{n,j})$ in X satisfying $\|\varphi_{n,j}\| \leq 1$ and $f_j = \lim_{n\to\infty} A\varphi_{n,j}$. This implies

$$f = \frac{r}{2}\,(f_1 - f_2) = \lim_{n\to\infty} A\,\frac{r}{2}\,(\varphi_{n,1} - \varphi_{n,2})$$

and consequently $f \in \overline{A(B[0;r])}$.

3. Now we complete the proof of the theorem by showing that

$$\|A^{-1}f\| \leq 2 \tag{10.7}$$

for all $f \in Y$ with $\|f\| \leq \rho$. To this end by induction we construct a sequence (f_n) in Y with the properties

$$f_n \in A\left(B[0; 1/2^{n-1}]\right)$$

and

$$\left\| f - \sum_{k=1}^{n} f_k \right\| \leq \frac{\rho}{2^n}$$

for all $n \in \mathbb{N}$. By (10.6) we have that $f \in \overline{A(B[0; 1])}$ and therefore there exists an $f_1 \in A(B[0; 1])$ with $\|f - f_1\| \leq \rho/2$. Assume we have constructed the sequence up to the n-th term. Then again by (10.6) we have

$$f - \sum_{k=1}^{n} f_k \in \overline{A\,(B[0; 1/2^n])}$$

and consequently there exists an $f_{n+1} \in A(B[0; 1/2^n])$ with

$$\left\| f - \sum_{k=1}^{n+1} f_k \right\| \leq \frac{\rho}{2^{n+1}}$$

as the next element of the sequence. Now clearly the partial sums $g_n := \sum_{k=1}^{n} f_k$ converge to f. Because of $f_k \in A\left(B[0; 1/2^{k-1}]\right)$ we have $\|A^{-1}f_k\| \leq 1/2^{k-1}$, and consequently absolute convergence

$$\sum_{k=1}^{\infty} \|A^{-1}f_k\| \leq \sum_{k=1}^{\infty} \frac{1}{2^{k-1}} = 2.$$

By Problem 1.5 the latter implies the existence of $\varphi \in X$ with the properties

$$\varphi = \sum_{k=1}^{\infty} A^{-1} f_k = \lim_{n\to\infty} A^{-1} g_n$$

and $\|\varphi\| \leq 2$. The continuity of A finally yields

$$A\varphi = A \lim_{n\to\infty} A^{-1} g_n = \lim_{n\to\infty} AA^{-1} g_n = \lim_{n\to\infty} g_n = f$$

and $\|A^{-1}f\| = \|\varphi\| \leq 2$, that is, (10.7) is proven. □

We leave it as an exercise for the reader to show that a function $A : U \subset X \to Y$ mapping a subset U of a normed space into a normed space Y is continuous if and only if for each open set $V \subset A(U)$ the pre-image

$$A^{-1}(V) := \{\varphi \in U : A\varphi \in V\}$$

is open. Analogous to the concept of a continuous function an open function, or an open mapping, is defined as a function for which the images of open sets are open. Consequently, the previous theorem on the bounded inverse implies that a bijective bounded linear operator between Banach spaces is an open map and this explains why the theorem is known as the open mapping theorem.

10.3 Collectively Compact Operators

Motivated through Corollary 10.7 and following Anselone [5], we introduce the concept of collectively compact operators.

Definition 10.9. A set $\mathcal{A} = \{A : X \to Y\}$ of linear operators mapping a normed space X into a normed space Y is called *collectively compact* if for each bounded set $U \subset X$ the image set $\mathcal{A}(U) = \{A\varphi : \varphi \in U,\ A \in \mathcal{A}\}$ is relatively compact.

Clearly, every operator in a collectively compact set is compact. Each finite set of compact operators is collectively compact. A sequence (A_n) is called collectively compact when the corresponding set is. Pointwise convergence $A_n \to A$, $n \to \infty$, of a collectively compact sequence implies compactness of the limit operator, since

$$A(U) \subset \overline{\{A_n\varphi : \varphi \in U,\ n \in \mathbb{N}\}}.$$

Theorem 10.10. *Let X, Z be normed spaces and let Y be a Banach space. Let $\mathcal{A}$ be a collectively compact set of operators mapping X into Y and let $L_n : Y \to Z$ be a pointwise convergent sequence of bounded linear operators with limit operator $L : Y \to Z$. Then*

$$\|(L_n - L)A\| \to 0, \quad n \to \infty,$$

uniformly for all $A \in \mathcal{A}$, that is, $\sup_{A\in\mathcal{A}} \|(L_n - L)A\| \to 0$, $n \to \infty$.

Proof. The set $U := \{A\varphi : \|\varphi\| \le 1,\ A \in \mathcal{A}\}$ is relatively compact. By Corollary 10.7 the convergence $L_n\psi \to L\psi$, $n \to \infty$, is uniform for all $\psi \in U$. Hence, for every $\varepsilon > 0$ there exists an integer $N(\varepsilon)$ such that

$$\|(L_n - L)A\varphi\| < \varepsilon$$

for all $n \ge N(\varepsilon)$, all $\varphi \in X$ with $\|\varphi\| \le 1$, and all $A \in \mathcal{A}$. Therefore

$$\|(L_n - L)A\| \le \varepsilon$$

for all $n \ge N(\varepsilon)$ and all $A \in \mathcal{A}$. □

Corollary 10.11. *Let X be a Banach space and let $A_n : X \to X$ be a collectively compact and pointwise convergent sequence with limit operator $A : X \to X$. Then*

$$\|(A_n - A)A\| \to 0 \quad \textit{and} \quad \|(A_n - A)A_n\| \to 0, \quad n \to \infty.$$

10.4 Approximations via Pointwise Convergence

The following error analysis for equations of the second kind takes advantage of Corollary 10.11 and is due to Brakhage [19] and Anselone and Moore [6].

Theorem 10.12. *Let $A : X \to X$ be a compact linear operator in a Banach space X and let $I - A$ be injective. Assume the sequence $A_n : X \to X$ to be collectively compact and pointwise convergent $A_n\varphi \to A\varphi$, $n \to \infty$, for all $\varphi \in X$. Then for sufficiently large n, more precisely for all n with*

$$\|(I - A)^{-1}(A_n - A)A_n\| < 1,$$

the inverse operators $(I - A_n)^{-1} : X \to X$ exist and are bounded by

$$\|(I - A_n)^{-1}\| \leq \frac{1 + \|(I - A)^{-1}A_n\|}{1 - \|(I - A)^{-1}(A_n - A)A_n\|} \,. \tag{10.8}$$

For the solutions of the equations

$$\varphi - A\varphi = f \quad \text{and} \quad \varphi_n - A_n\varphi_n = f_n$$

we have the error estimate

$$\|\varphi_n - \varphi\| \leq \frac{1 + \|(I - A)^{-1}A_n\|}{1 - \|(I - A)^{-1}(A_n - A)A_n\|} \{\|(A_n - A)\varphi\| + \|f_n - f\|\}.$$

Proof. By the Riesz Theorem 3.4 the inverse operator $(I - A)^{-1} : X \to X$ exists and is bounded. The identity

$$(I - A)^{-1} = I + (I - A)^{-1}A$$

suggests

$$B_n := I + (I - A)^{-1}A_n$$

as an approximate inverse for $I - A_n$. Elementary calculations yield

$$B_n(I - A_n) = I - S_n, \tag{10.9}$$

where

$$S_n := (I - A)^{-1}(A_n - A)A_n.$$

From Corollary 10.11 we conclude that $\|S_n\| \to 0$, $n \to \infty$. For $\|S_n\| < 1$ the Neumann series Theorem 2.14 implies that $(I - S_n)^{-1}$ exists and is bounded by

$$\|(I - S_n)^{-1}\| \leq \frac{1}{1 - \|S_n\|} \,.$$

Now (10.9) implies first that $I - A_n$ is injective, and therefore, since A_n is compact, by Theorem 3.4 the inverse $(I - A_n)^{-1}$ exists. Then (10.9) also yields $(I - A_n)^{-1} =$

$(I - S_n)^{-1}B_n$, whence the estimate (10.8) follows. The error estimate follows from

$$(I - A_n)(\varphi_n - \varphi) = f_n - f + (A_n - A)\varphi,$$

and the proof is complete. □

Analogous to Theorem 10.1, in Theorem 10.12 from the unique solvability of the original equation we conclude unique solvability of the approximating equation provided the approximation is sufficiently close. The converse situation is described through the following theorem.

Theorem 10.13. *Assume there exists some $N \in \mathbb{N}$ such that for all $n \geq N$ the inverse operators $(I-A_n)^{-1}$ exist and are uniformly bounded. Then the inverse $(I-A)^{-1}$ exists and is bounded by*

$$\|(I - A)^{-1}\| \leq \frac{1 + \|(I - A_n)^{-1}A\|}{1 - \|(I - A_n)^{-1}(A_n - A)A\|} \tag{10.10}$$

for all n with

$$\|(I - A_n)^{-1}(A_n - A)A\| < 1.$$

For the solutions of the equations

$$\varphi - A\varphi = f \quad \text{and} \quad \varphi_n - A_n\varphi_n = f_n$$

we have the error estimate

$$\|\varphi_n - \varphi\| \leq \frac{1 + \|(I - A_n)^{-1}A\|}{1 - \|(I - A_n)^{-1}(A_n - A)A\|} \{\|(A_n - A)\varphi_n\| + \|f_n - f\|\}.$$

Proof. This follows from Theorem 10.12 by interchanging the roles of A and A_n. □

We note that Theorem 10.13 provides an error estimate that, in principle, can be evaluated, since it involves $(I - A_n)^{-1}$ and φ_n but neither $(I - A)^{-1}$ nor φ. From Theorem 10.12 we can conclude that the accuracy of the approximate solution essentially depends on how well $A_n\varphi$ approximates $A\varphi$ for the exact solution as expressed in the following corollary.

Corollary 10.14. *Under the assumptions of Theorem 10.12 we have the error estimate*

$$\|\varphi_n - \varphi\| \leq C\{\|(A_n - A)\varphi\| + \|f_n - f\|\} \tag{10.11}$$

for all sufficiently large n and some constant C.

Proof. This is an immediate consequence of Theorem 10.12. □

In Chapter 12 we will apply the error analysis of Theorems 10.12 and 10.13 to the approximate solution of integral equations of the second kind by using numerical quadratures.

10.5 Successive Approximations

We now briefly revisit the investigation of the convergence of successive approximations

$$\varphi_{n+1} := A\varphi_n + f$$

to solve the equation of the second kind $\varphi - A\varphi = f$, where $A : X \to X$ is a bounded linear operator in a Banach space X. So far we have used $\|A\| < 1$ as a sufficient condition for convergence (see Theorem 2.15). It is our aim now to show that, similar to the case of finite-dimensional linear equations, convergence of successive approximations can be characterized through the spectral radius of the operator A. For the notion of the resolvent set $\rho(A)$, the resolvent $R(\lambda; A) = (\lambda I - A)^{-1}$, the spectrum $\sigma(A)$, and the spectral radius $r(A)$ of a bounded linear operator A recall Definition 3.8.

Theorem 10.15. *Let $A : X \to X$ be a bounded linear operator mapping a Banach space X into itself. Then the Neumann series*

$$(\lambda I - A)^{-1} = \sum_{k=0}^{\infty} \lambda^{-k-1} A^k \tag{10.12}$$

converges in the operator norm for all $|\lambda| > r(A)$ and diverges for all $|\lambda| < r(A)$.

Proof. Provided the series (10.12) converges it defines a bounded linear operator $S(\lambda)$. As in the proof of Theorem 2.14, this operator $S(\lambda)$ can be seen to be the inverse of $\lambda I - A$, since $|\lambda|^{-n-1}\|A^n\| \to 0$, $n \to \infty$, is necessary for the convergence of the series (10.12).

Let λ_0 belong to the resolvent set $\rho(A)$ of A. Then, by Theorem 2.14, for all λ with

$$|\lambda - \lambda_0| \, \|R(\lambda_0; A)\| < 1$$

the series

$$T(\lambda) := \sum_{k=0}^{\infty} (\lambda_0 - \lambda)^k R(\lambda_0; A)^{k+1} \tag{10.13}$$

converges in the operator norm and defines a bounded linear operator with

$$T(\lambda) = R(\lambda_0; A)\,[I - (\lambda_0 - \lambda)R(\lambda_0; A)]^{-1} = [I - (\lambda_0 - \lambda)R(\lambda_0; A)]^{-1} R(\lambda_0; A).$$

Hence,

$$\begin{aligned}(\lambda I - A)T(\lambda) &= [\lambda_0 I - A - (\lambda_0 - \lambda)I]\, T(\lambda)\\ &= [I - (\lambda_0 - \lambda)R(\lambda_0; A)]\,(\lambda_0 I - A)T(\lambda) = I\end{aligned}$$

and similarly $T(\lambda)(\lambda I - A) = I$. Therefore $\lambda \in \rho(A)$ and $T(\lambda) = R(\lambda; A)$ for all λ with $|\lambda - \lambda_0| \, \|R(\lambda_0; A)\| < 1$. In particular, this implies that the resolvent set $\rho(A)$ is an open subset of $\mathbb{C}$.

For a bounded linear functional F on the space $L(X, X)$ of bounded linear operators in X by the continuity of F from (10.13) we observe that the Taylor series

$$F\left((\lambda I - A)^{-1}\right) = \sum_{k=0}^{\infty} (\lambda_0 - \lambda)^k F\left(R(\lambda_0; A)^{k+1}\right)$$

converges for all λ with $|\lambda - \lambda_0| \, \|R(\lambda_0; A)\| < 1$, i.e., the function

$$\lambda \mapsto F\left((\lambda I - A)^{-1}\right)$$

is an analytic function from $\rho(A)$ into $\mathbb{C}$. Therefore the integral

$$J_n(F) := \frac{1}{2\pi i} \int_{|\lambda|=r} \lambda^n F\left((\lambda I - A)^{-1}\right) d\lambda \tag{10.14}$$

is well defined for all $n \in \mathbb{N}$ and $r > r(A)$, and by Cauchy's integral theorem does not depend on r. If we choose $r > \|A\|$, by Theorem 2.14 the series (10.12) converges and we can insert it into (10.14) to obtain

$$J_n(F) = \sum_{k=0}^{\infty} \frac{1}{2\pi i} \int_{|\lambda|=r} \lambda^{n-k-1} F(A^k) \, d\lambda = F(A^n).$$

By the Corollary 2.11 of the Hahn–Banach theorem, for each $n \in \mathbb{N}$ there exists a bounded linear functional $F_n \in (L(X, X))^*$ with the properties $\|F_n\| = 1$ and $F_n(A^n) = \|A^n\|$. Therefore, estimating in (10.14), we obtain

$$\|A^n\| = F_n(A^n) = J_n(F_n) \leq \frac{r^n}{2\pi} \int_{|\lambda|=r} \|(\lambda I - A)^{-1}\| \, |d\lambda| \leq \gamma_r r^n \tag{10.15}$$

with a bound γ_r on the continuous function $\lambda \mapsto \lambda \|(\lambda I - A)^{-1}\|$ on the circle of radius r centered at the origin.

Now let $|\lambda| > r(A)$ and choose r such that $|\lambda| > r > r(A)$. Then from (10.15) it follows that

$$\left|\lambda^{-n}\right| \|A^n\| \leq \gamma_r \left(\frac{r}{|\lambda|}\right)^n$$

and therefore we arrived at a convergent geometric series as majorant for (10.12). Thus by Problem 1.5 the series (10.12) converges in the Banach space $L(X, X)$.

On the other hand, assume that the Neumann series converges for $|\lambda_0| < r(A)$. Then there exists a constant M such that $|\lambda_0|^{-k-1} \|A^k\| \leq M$ for all $k \in \mathbb{N}$. Hence, for all λ with $|\lambda| > |\lambda_0|$ the Neumann series has a convergent geometric series as a majorant, and thus it converges in the Banach space X (see Problem 1.5). Its limit represents the inverse of $\lambda I - A$. Hence all λ with $|\lambda| > |\lambda_0|$ belong to the resolvent set $\rho(A)$, which is a contradiction to $|\lambda_0| < r(A)$. □

Theorem 10.16. *Let $A : X \to X$ be a bounded linear operator in a Banach space X with spectral radius $r(A) < 1$. Then the successive approximations*

$$\varphi_{n+1} := A\varphi_n + f, \quad n = 0, 1, 2, \ldots,$$

converge for each $f \in X$ and each $\varphi_0 \in X$ to the unique solution of $\varphi - A\varphi = f$.

Proof. This follows, as in the proof of Theorem 2.15, from the convergence of the Neumann series for $\lambda = 1 > r(A)$. □

That the condition $r(A) < 1$ for convergence cannot be weakened is demonstrated by the following remark.

Remark 10.17. *Let A have spectral radius $r(A) > 1$. Then the successive approximations with $\varphi_0 = 0$ cannot converge for all $f \in X$.*

Proof. Assume the statement is false. Convergence of the successive approximations for all $f \in X$ is equivalent to pointwise convergence of the Neumann series for $\lambda = 1$. Hence, by the uniform boundedness principle Theorem 10.6, there exists a positive constant C such that

$$\left\| \sum_{k=0}^{n} A^k \right\| \le C$$

for all $n \in \mathbb{N}$. Then, by the triangle inequality, it follows that $\|A^n\| \le 2C$ for all $n \in \mathbb{N}$. Since $r(A) > 1$, there exists $\lambda_0 \in \sigma(A)$ with $|\lambda_0| > 1$. For the partial sums $S_n := \sum_{k=0}^{n} \lambda_0^{-k-1} A^k$ of the Neumann series we have

$$\|S_m - S_n\| \le \frac{2C|\lambda_0|^{-n-1}}{|\lambda_0| - 1}$$

for all $m > n$. Therefore (S_n) is a Cauchy sequence in the Banach space $L(X, X)$ (see Theorem 2.6). This implies convergence of the Neumann series for λ_0, which contradicts $\lambda_0 \in \sigma(A)$. □

If A is compact then the successive approximations, in general, cannot converge if $r(A) = 1$. By Theorem 3.9 all spectral values of a compact operator different from zero are eigenvalues. Hence, for $r(A) = 1$, there exists $f \in X$ with $f \neq 0$ and λ_0 with $|\lambda_0| = 1$ such that $Af = \lambda_0 f$. But then the successive approximations with $\varphi_0 = 0$ satisfy

$$\varphi_n = \sum_{k=0}^{n-1} \lambda_0^k f, \quad n = 1, 2, \ldots,$$

and therefore they diverge.

Usually, in practical calculations, some approximation of the operator A will be used. Therefore we have to establish that our results on convergence remain stable under small perturbations. This is done through the following two theorems.

Theorem 10.18. *Let $A : X \to X$ be a bounded linear operator in a Banach space X with spectral radius $r(A) < 1$ and let the sequence $A_m : X \to X$ of bounded linear operators be norm convergent $\|A_m - A\| \to 0$, $m \to \infty$. Then for all sufficiently large m the equation $\varphi - A_m\varphi = f$ can be solved by successive approximations.*

Proof. Choose $\lambda_0 \in (r(A), 1)$. Then $R(\lambda; A) = (\lambda I - A)^{-1}$ exists and is bounded for all $\lambda \in \mathbb{C}$ with $|\lambda| \geq \lambda_0$. Since $R(\lambda; A)$ is analytic and $\|R(\lambda; A)\| \to 0$, $|\lambda| \to \infty$ (this follows from the Neumann series expansion of the resolvent), we have

$$C := \sup_{|\lambda| \geq \lambda_0} \|(\lambda I - A)^{-1}\| < \infty.$$

Therefore, because $\|A_m - A\| \to 0$, $m \to \infty$, there exists an integer N such that $\|(\lambda I - A)^{-1}(A_m - A)\| < 1$ for all $m \geq N$ and all $|\lambda| \geq \lambda_0$. But then from Theorem 10.1, applied to $\lambda I - A$ and $\lambda I - A_m$, we deduce that $(\lambda I - A_m)^{-1}$ exists and is bounded for all $m \geq N$ and all $|\lambda| \geq \lambda_0$. Hence $r(A_m) < 1$ for all $m \geq N$ and the statement follows from Theorem 10.16. □

Similarly, based on Theorem 10.12, we can prove the following theorem.

Theorem 10.19. *Let $A : X \to X$ be a compact linear operator in a Banach space X with spectral radius $r(A) < 1$ and let the sequence $A_m : X \to X$ of collectively compact linear operators be pointwise convergent $A_m\varphi \to A\varphi$, $m \to \infty$, for all $\varphi \in X$. Then for all sufficiently large m the equation $\varphi - A_m\varphi = f$ can be solved by successive approximations.*

We conclude this section with two examples for the application of Theorem 10.16.

Theorem 10.20. *Volterra integral equations of the second kind with continuous or weakly singular kernels can be solved by successive approximations.*

Proof. Volterra integral operators have spectral radius zero. For continuous kernels this is a consequence of Theorems 3.9 and 3.10. The case of a weakly singular kernel can be dealt with by reducing it to a continuous kernel through iteration as in the proof of Theorem 9.9. □

Historically, the first existence proofs for Volterra integral equations of the second kind actually were obtained by directly establishing the convergence of successive approximations.

The following classical potential theoretic result goes back to Plemelj [192].

Theorem 10.21. *The integral operators $K, K' : C(\partial D) \to C(\partial D)$, defined by (6.35) and (6.36), have spectrum*

$$-1 \in \sigma(K) = \sigma(K') \subset [-1, 1).$$

Proof. By Theorem 3.9, the spectrum $\sigma(K')\setminus\{0\}$ of the compact operator K' consists only of eigenvalues. Let $\lambda \neq -1$ be an eigenvalue of K' with eigenfunction φ and define the single-layer potential u with density φ. Then, from the jump relations of Theorems 6.15 and 6.19, we obtain $u_+ = u_-$ and

$$\frac{\partial u_\pm}{\partial \nu} = \frac{1}{2}\,K'\varphi \mp \frac{1}{2}\,\varphi = \frac{1}{2}(\lambda \mp 1)\varphi$$

on ∂D. From this we see that

$$(\lambda + 1)\int_{\partial D} \bar{u}_+ \,\frac{\partial u_+}{\partial \nu}\,ds = (\lambda - 1)\int_{\partial D} \bar{u}_- \,\frac{\partial u_-}{\partial \nu}\,ds.$$

In addition, by the Gauss theorem Corollary 6.4, we deduce $\int_{\partial D} \varphi\,ds = 0$. Therefore, in view of (6.18) the single-layer potential u has the asymptotic behavior

$$u(x) = O\left(\frac{1}{|x|^{m-1}}\right), \quad \operatorname{grad} u(x) = O\left(\frac{1}{|x|^{m}}\right), \quad |x| \to \infty,$$

uniformly for all directions. Now we can apply Green's Theorem 6.3 to obtain

$$(1 + \lambda)\int_{\mathbb{R}^m\setminus D} |\operatorname{grad} u|^2 dx = (1 - \lambda)\int_D |\operatorname{grad} u|^2 dx. \tag{10.16}$$

Assume that $\int_{\mathbb{R}^m\setminus D} |\operatorname{grad} u|^2 dx = \int_D |\operatorname{grad} u|^2 dx = 0$. Then $\operatorname{grad} u = 0$ in $\mathbb{R}^m$ and from the jump relations we have $\varphi = 0$, which is a contradiction to the fact that φ is an eigenfunction. Hence, from (10.16) we conclude that λ is contained in $[-1, 1]$. From Theorem 6.21 we already know that -1 is an eigenvalue of K' whereas 1 is not an eigenvalue. Finally, using the fact that $\sigma(K')$ is real, the equality $\sigma(K) = \sigma(K')$ is a consequence of the Fredholm alternative for the adjoint operators K and K'. □

Corollary 10.22. *The successive approximations*

$$\varphi_{n+1} := \frac{1}{2}\,\varphi_n + \frac{1}{2}\,K\varphi_n - f, \quad n = 0, 1, 2, \ldots, \tag{10.17}$$

converge uniformly for all φ_0 and all f in $C(\partial D)$ to the unique solution φ of the integral equation $\varphi - K\varphi = -2f$ from Theorem 6.22 for the interior Dirichlet problem.

Proof. We apply Theorem 10.16 to the equation $\varphi - A\varphi = -f$ with the operator $A := 2^{-1}(I + K)$. Then we have

$$\sigma(A) = \{1/2 + \lambda/2 : \lambda \in \sigma(K)\}$$

and therefore $r(A) < 1$ by Theorem 10.21. □

Remark 10.23. *The successive approximations (10.17) also converge in $L^2(\partial D)$ and in the Sobolev space $H^{1/2}(\partial D)$.*

Proof. Applying Theorem 4.20 it can be seen that the spectrum of the operator K is the same for all three spaces $C(\partial D)$, $L^2(\partial D)$ and $H^{1/2}(\partial D)$. □

We note that the proof of Corollary 10.22 relies on the compactness of K since we made use of the spectral Theorem 3.9. As pointed out in Section 6.5, the double-layer potential operator K no longer remains compact for domains with corners or, more generally, for Lipschitz domains. For an alternative approach to establish convergence of Neumann's iteration method in Lipschitz domains we refer to Steinbach and Wendland [226] and further references therein.

We already mentioned in Problem 6.5 that Neumann used the successive approximation scheme of Corollary 10.22 to give the first rigorous existence proof for the Dirichlet problem in two-dimensional convex domains.

Problems

10.1. Prove the Banach–Steinhaus theorem: Let $A : X \to Y$ be a bounded linear operator and let $A_n : X \to Y$ be a sequence of bounded linear operators from a Banach space X into a normed space Y. For pointwise convergence $A_n\varphi \to A\varphi$, $n \to \infty$, for all $\varphi \in X$ it is necessary and sufficient that $\|A_n\| \le C$ for all $n \in \mathbb{N}$ and some constant C and that $A_n\varphi \to A\varphi$, $n \to \infty$, for all $\varphi \in U$, where U is some dense subset of X.

10.2. Show that a sequence $A_n : X \to Y$ of compact linear operators mapping a normed space X into a normed space Y is collectively compact if and only if for each bounded sequence (φ_n) in X the sequence $(A_n\varphi_n)$ is relatively compact in Y.

10.3. Let X and Y be Banach spaces, $S : X \to Y$ be a bounded linear operator with bounded inverse $S^{-1} : Y \to X$, and $A : X \to Y$ be compact. Formulate and prove extensions of Theorems 10.12 and 10.13 for the approximation of the equation

$$S\varphi - A\varphi = f$$

by equations

$$S_n\varphi_n - A_n\varphi_n = f_n,$$

where the $A_n : X \to Y$ are collectively compact with pointwise convergence $A_n \to A$, $n \to \infty$, the $S_n : X \to Y$ are bounded with pointwise convergence $S_n \to S$, $n \to \infty$, and with uniformly bounded inverses $S_n^{-1} : Y \to X$, and $f_n \to f$, $n \to \infty$.

10.4. Solve the Volterra integral equation

$$\varphi(x) - \int_0^x e^{x-y}\varphi(y)\,dy = f(x)$$

by successive approximations.

10.5. How can the integral equation from Theorem 6.26 for the interior Neumann problem be solved by successive approximations?

Chapter 11
Degenerate Kernel Approximation

In this chapter we will consider the approximate solution of integral equations of the second kind by replacing the kernels by *degenerate kernels*, i.e., by approximating a given kernel $K(x, y)$ through a sum of a finite number of products of functions of x alone by functions of y alone. In particular, we will describe the construction of appropriate degenerate kernels by interpolation of the given kernel and by orthonormal expansions. The corresponding error analysis will be settled by our results in Section 10.1. We also include a discussion of some basic facts on piecewise linear interpolation and trigonometric interpolation, which will be used in this and subsequent chapters.

The degenerate kernel approximation is the simplest method for numerically solving integral equations of the second kind as far as its understanding and error analysis are concerned. However, because the actual implementation requires the numerical evaluation of either a single or double integral for each matrix coefficient and for each right-hand side of the approximating linear system, in general, it cannot compete in efficiency with other methods, for example, with the Nyström method of Chapter 12 or the projection methods of Chapter 13.

For further studies on degenerate kernel methods, we refer to Atkinson [11], Baker [14], Fenyö and Stolle [52], Hackbusch [76], Kantorovic and Krylov [117] and Kanwal [118].

11.1 Degenerate Operators and Kernels

Let $\langle X, X\rangle$ be a dual system, generated by a bounded non-degenerate bilinear form, and let $A_n : X \to X$ be a bounded linear operator of the form

$$A_n\varphi = \sum_{j=1}^{n} \langle \varphi, b_j\rangle a_j, \tag{11.1}$$

R. Kress, *Linear Integral Equations*, Applied Mathematical Sciences 82,
DOI 10.1007/978-1-4614-9593-2_11,

where $a_1, \dots, a_n$ and $b_1, \dots, b_n$ are elements of X such that the $a_1, \dots, a_n$ are linearly independent. For the sake of notational clarity we restrict ourselves to the case where the index n for the operator A_n coincides with the number of elements $a_1, \dots, a_n$ and $b_1, \dots, b_n$. The solution of the equation of the second kind $\varphi_n - A_n\varphi_n = f$ with such a *degenerate operator* reduces to solving a finite-dimensional linear system.

Theorem 11.1. *Each solution of the equation*

$$\varphi_n - \sum_{j=1}^{n} \langle \varphi_n, b_j \rangle a_j = f \tag{11.2}$$

has the form

$$\varphi_n = f + \sum_{k=1}^{n} \gamma_k a_k, \tag{11.3}$$

where the coefficients $\gamma_1, \dots, \gamma_n$ satisfy the linear system

$$\gamma_j - \sum_{k=1}^{n} \langle a_k, b_j \rangle \gamma_k = \langle f, b_j \rangle, \quad j = 1, \dots, n. \tag{11.4}$$

Conversely, to each solution $\gamma_1, \dots, \gamma_n$ of the linear system (11.4) there corresponds a solution φ_n of (11.2) defined by (11.3).

Proof. Let φ_n be a solution of (11.2). Writing $\gamma_k := \langle \varphi_n, b_k \rangle$ for $k = 1, \dots, n$, and solving for φ_n we obtain the form (11.3). Taking the bilinear form of (11.3) with b_j we find that the coefficients $\gamma_1, \dots, \gamma_n$ must satisfy the linear system (11.4). Conversely, let $\gamma_1, \dots, \gamma_n$ be a solution to the linear system (11.4) and define φ_n by (11.3). Then

$$\varphi_n - \sum_{j=1}^{n} \langle \varphi_n, b_j \rangle a_j = f + \sum_{j=1}^{n} \gamma_j a_j - \sum_{j=1}^{n} \left\langle f + \sum_{k=1}^{n} \gamma_k a_k, b_j \right\rangle a_j$$

$$= f + \sum_{j=1}^{n} \left\{ \gamma_j - \sum_{k=1}^{n} \langle a_k, b_j \rangle \gamma_k - \langle f, b_j \rangle \right\} a_j = f,$$

and the proof is complete. □

For a given operator A we will use finite-dimensional approximations A_n of the form (11.1) to obtain approximate solutions of equations of the second kind by using Theorems 10.1 and 10.2 for the operators $I - A$ and $I - A_n$. We wish to mention that this approach via finite-dimensional approximations can also be used to establish the Fredholm alternative for the case of linear operators in Banach spaces, which can be split into the sum of an operator of the form (11.1) and an operator with norm less than one (see Problems 11.1 and 11.2). For integral operators with continuous or weakly singular kernels such an approximation can be achieved by Problem 2.3

and by using an approximation of weakly singular kernels by continuous kernels, as in the proof of Theorem 2.29.

In the case of the dual system $\langle C(G), C(G)\rangle$ introduced in Theorem 4.4 the finite-dimensional operator A_n has the form

$$(A_n\varphi)(x) = \int_G K_n(x, y)\varphi(y)\, dy, \quad x \in G,$$

of an integral operator with a *degenerate kernel*

$$K_n(x, y) = \sum_{j=1}^{n} a_j(x)b_j(y).$$

The solution of the integral equation of the second kind

$$\varphi_n(x) - \int_G \sum_{j=1}^{n} a_j(x)b_j(y)\varphi_n(y)\, dy = f(x), \quad x \in G, \tag{11.5}$$

with such a degenerate kernel is described through Theorem 11.1. We will use integral operators A_n with degenerate kernels K_n as approximations for integral operators A with continuous kernels K in the sense of Section 10.1. In view of Theorems 2.13, 10.1, and 10.2 we need approximations with the property that

$$\|A_n - A\|_\infty = \max_{x\in G} \int_G |K_n(x, y) - K(x, y)|\, dy$$

becomes small.

Here, we do not attempt to give a complete account of the various possibilities for degenerate kernel approximations. Instead, we will introduce the reader to the basic ideas by considering a few simple examples.

11.2 Interpolation

One important method to construct degenerate kernels approximating a given continuous kernel is by interpolation. For its investigation, and for use in subsequent chapters, we collect some basic facts on interpolation.

Theorem 11.2. *Let $U_n \subset C(G)$ be an n-dimensional subspace and let $x_1, \dots, x_n$ be n points in G such that U_n is* unisolvent *with respect to $x_1, \dots, x_n$, i.e., each function from U_n that vanishes in $x_1, \dots, x_n$ vanishes identically. Then, given n values $g_1, \dots, g_n$, there exists a uniquely determined function $u \in U_n$ with the interpolation property*

$$u(x_j) = g_j, \quad j = 1, \dots, n.$$

With the interpolation data given by the values $g_j = g(x_j)$, $j = 1, \dots, n$, of a function $g \in C(G)$ the mapping $g \mapsto u$ defines a bounded linear operator $P_n : C(G) \to U_n$ which is called interpolation operator.

Proof. Let $U_n = \operatorname{span}\{u_1, \dots, u_n\}$. Then the solution to the interpolation problem is given by

$$u = \sum_{k=1}^{n} \gamma_k u_k,$$

where the coefficients $\gamma_1, \dots, \gamma_n$ are determined by the uniquely solvable linear system

$$\sum_{k=1}^{n} \gamma_k u_k(x_j) = g_j, \quad j = 1, \dots, n.$$

Let $L_1, \dots, L_n$ denote the *Lagrange basis* for U_n, i.e., we have the interpolation property

$$L_k(x_j) = \delta_{jk}, \quad j, k = 1, \dots, n,$$

where δ_{jk} is the Kronecker symbol with $\delta_{jk} = 1$ for $k = j$, and $\delta_{jk} = 0$ for $k \neq j$. Then, from

$$P_n g = \sum_{k=1}^{n} g(x_k) L_k \tag{11.6}$$

we conclude the linearity and boundedness of P_n. □

From (11.6) we have that

$$\|P_n\|_\infty \le \max_{x \in G} \sum_{k=1}^{n} |L_k(x)|.$$

Now choose $z \in G$ such that

$$\sum_{k=1}^{n} |L_k(z)| = \max_{x \in G} \sum_{k=1}^{n} |L_k(x)|$$

and a function $f \in C(G)$ with $\|f\|_\infty = 1$ and

$$\sum_{k=1}^{n} f(x_k) L_k(z) = \sum_{k=1}^{n} |L_k(z)|.$$

Then

$$\|P_n\|_\infty \ge \|P_n f\|_\infty \ge |(P_n f)(z)| = \max_{x \in G} \sum_{k=1}^{n} |L_k(x)|,$$

and therefore, in terms of the Lagrange basis, we have the norm

$$\|P_n\|_\infty = \max_{x\in G} \sum_{k=1}^{n} |L_k(x)|. \tag{11.7}$$

For examples we confine ourselves to interpolation on a finite interval $[a, b]$ in $\mathbb{R}$ with $a < b$. First we consider linear splines, i.e., continuous piecewise linear functions. Let $x_j = a + jh$, $j = 0, \dots, n$, denote an equidistant subdivision with step size $h = (b - a)/n$ and let U_n be the space of continuous functions on $[a, b]$ whose restrictions on each of the subintervals $[x_{j-1}, x_j]$, $j = 1, \dots, n$, are linear. Existence and uniqueness for the corresponding linear spline interpolation are evident. Here, the Lagrange basis is given by the so-called *hat functions*

$$L_j(x) = \begin{cases} \dfrac{1}{h}\,(x - x_{j-1}), & x \in [x_{j-1}, x_j], \quad j \geq 1, \\[2ex] \dfrac{1}{h}\,(x_{j+1} - x), & x \in [x_j, x_{j+1}], \quad j \leq n-1, \\[2ex] 0, & \text{otherwise}. \end{cases} \tag{11.8}$$

Because the L_j are nonnegative, we have

$$\sum_{j=0}^{n} |L_j(x)| = \sum_{j=0}^{n} L_j(x) = 1$$

for all $x \in [a, b]$. Therefore, (11.7) implies that

$$\|P_n\|_\infty = 1 \tag{11.9}$$

for piecewise linear interpolation. For the interpolation of twice continuously differentiable functions g we have the following error estimate.

Theorem 11.3. *Let $g \in C^2[a, b]$. Then, for the error in piecewise linear interpolation we have the estimate*

$$\|P_n g - g\|_\infty \leq \frac{1}{8}\,h^2 \|g''\|_\infty. \tag{11.10}$$

Proof. Obviously, the maximum of $|P_n g - g|$ in $[x_j, x_{j+1}]$ is attained at an interior point ξ where

$$g'(\xi) = (P_n g)'(\xi) = \frac{1}{h}\,\{g(x_{j+1}) - g(x_j)\}.$$

Without loss of generality we may assume that $\xi - x_j \leq h/2$. Then, using Taylor's formula, we derive

$$(P_n g)(\xi) - g(\xi) = g(x_j) + (P_n g)'(\xi)(\xi - x_j) - g(\xi)$$

$$= g(x_j) - g(\xi) - (x_j - \xi)g'(\xi)$$

$$= \frac{1}{2}\,(x_j - \xi)^2 g''(\eta),$$

with some $\eta \in (x_j, \xi)$. Hence,

$$\max_{x_j \leq x \leq x_{j+1}} |(P_n g)(x) - g(x)| \leq \frac{1}{8}\, h^2 \|g''\|_\infty$$

and the estimate (11.10) follows. □

11.3 Trigonometric Interpolation

As we have seen in Chapters 6–8 the parameterization of boundary integral equations for two-dimensional boundary value problems leads to integral equations for periodic functions. For these, instead of a local approximation by piecewise linear interpolation (or higher-order spline interpolation), quite often it will be preferable to use a global approximation in the form of trigonometric interpolation, since this leads to a more rapid convergence. To support this preference for using trigonometric polynomial approximations we quote from Atkinson [11]: ... *the most efficient numerical methods for solving boundary integral equations on smooth planar boundaries are those based on trigonometric polynomial approximations, and such methods are sometimes called spectral methods. When calculations using piecewise polynomial approximations are compared with those using trigonometric polynomial approximations, the latter are almost always the more efficient.*

Let $t_j = j\pi/n$, $j = 0, \ldots, 2n-1$, be an equidistant subdivision of the interval $[0, 2\pi]$ with an even number of grid points. Then, given the values $g_0, \ldots, g_{2n-1}$, there exists a unique trigonometric polynomial of the form

$$u(t) = \frac{\alpha_0}{2} + \sum_{k=1}^{n-1} [\alpha_k \cos kt + \beta_k \sin kt] + \frac{\alpha_n}{2}\, \cos nt \tag{11.11}$$

with the interpolation property $u(t_j) = g_j$, $j = 0, \ldots, 2n-1$. Its coefficients are given by

$$\alpha_k = \frac{1}{n} \sum_{j=0}^{2n-1} g_j \cos kt_j, \quad k = 0, \ldots, n,$$

$$\beta_k = \frac{1}{n} \sum_{j=0}^{2n-1} g_j \sin kt_j, \quad k = 1, \ldots, n-1.$$

From this we deduce that the Lagrange basis for the trigonometric interpolation has the form

$$L_j(t) = \frac{1}{2n} \left\{ 1 + 2 \sum_{k=1}^{n-1} \cos k(t - t_j) + \cos n(t - t_j) \right\} \tag{11.12}$$

for $t \in [0, 2\pi]$ and $j = 0, \ldots, 2n-1$. Using the real part of the geometric sum (8.34), we can transform (11.12) into

$$L_j(t) = \frac{1}{2n} \sin n(t - t_j) \cot \frac{t - t_j}{2}, \quad t \neq t_j. \tag{11.13}$$

Theorem 11.4. *For $n \geq 2$, the trigonometric interpolation operator with $2n$ equidistant interpolation points has norm*

$$\|P_n\|_\infty < 3 + \frac{2}{\pi} \ln(2n). \tag{11.14}$$

Proof. The function $\lambda : [0, \pi] \to \mathbb{R}$ defined by

$$\lambda(t) := \sum_{j=0}^{2n-1} |L_j(t)|$$

is even and has period π/n. Therefore, in view of (11.7), we have

$$\|P_n\|_\infty = \max_{0 \leq t \leq \pi/2n} \lambda(t).$$

For $0 \leq t \leq \pi/2n$, using (11.12) and (11.13), we find

$$\lambda(t) \leq 3 + \frac{1}{2n} \sum_{j=2}^{2n-2} \left| \cot \frac{t - t_j}{2} \right|.$$

Since the cotangent function $s \mapsto \cot(s/2)$ is positive and monotonically decreasing on the interval $(0, \pi]$, for $0 \leq t \leq \pi/2n$ in view of $t_1 - t \geq \pi/2n$ we can estimate

$$\frac{1}{2n} \sum_{j=2}^{n} \left| \cot \frac{t - t_j}{2} \right| < \frac{1}{2\pi} \int_{t_1}^{t_n} \cot \frac{s - t}{2} \, ds < -\frac{1}{\pi} \ln \sin \frac{\pi}{4n} .$$

Analogously we have

$$\frac{1}{2n}\sum_{j=n+1}^{2n-2}\left|\cot\frac{t-t_j}{2}\right| < \frac{1}{2\pi}\int_{t_{n+1}}^{t_{2n-1}}\cot\frac{t-s}{2}\,ds < -\frac{1}{\pi}\ln\sin\frac{\pi}{4n}\,.$$

Now, using the inequality

$$\sin\frac{\pi}{4n} > \frac{1}{2n}$$

the statement of the lemma follows by piecing the above inequalities together. □

The estimate (11.14) gives the correct behavior for large n, because proceeding as in the proof of Theorem 11.4 we can show that

$$\|P_n\|_\infty \geq \lambda\left(\frac{\pi}{2n}\right) > \frac{1}{2n}\cot\frac{\pi}{4n} - \frac{2}{\pi}\ln\sin\frac{\pi}{4n} \tag{11.15}$$

for all $n \geq 2$. This estimate implies that $\|P_n\|_\infty \to \infty$ as $n \to \infty$. Therefore, from the uniform boundedness principle Theorem 10.6 it follows that there exists a continuous 2π-periodic function g for which the sequence $(P_n g)$ does not converge to g in $C[0, 2\pi]$. However, by the Banach–Steinhaus theorem (see Problem 10.1) and the denseness of the trigonometric polynomials in $C[0, 2\pi]$, the following lemma implies mean square convergence

$$\|P_n g - g\|_2 \to 0, \quad n \to \infty,$$

continuous 2π-periodic functions g.

Lemma 11.5. *The trigonometric interpolation operator satisfies*

$$\|P_n g\|_2 \leq \sqrt{3\pi}\,\|g\|_\infty \tag{11.16}$$

for all $n \in \mathbb{N}$ and all continuous 2π-periodic functions g.

Proof. Using (11.12), elementary integrations yield

$$\int_0^{2\pi} L_j(t)L_k(t)\,dt = \frac{\pi}{n}\,\delta_{jk} - (-1)^{j-k}\,\frac{\pi}{4n^2} \tag{11.17}$$

for $j, k = 0, \ldots, 2n-1$. Then, by the triangle inequality we can estimate

$$\|P_n g\|_2^2 \leq \|g\|_\infty^2 \sum_{j,k=0}^{2n-1}\left|\int_0^{2\pi} L_j(t)L_k(t)\,dt\right| \leq 3\pi\,\|g\|_\infty^2,$$

and (11.16) is proven. □

Provided the interpolated function g has additional regularity, we also have uniform convergence of the trigonometric interpolation polynomials as expressed by the following three theorems. For $0 < \alpha \leq 1$, by $C^{0,\alpha}_{2\pi} \subset C^{0,\alpha}(\mathbb{R})$ we denote the space of 2π-periodic Hölder continuous functions. Furthermore, for $m \in \mathbb{N}$ by $C^{m,\alpha}_{2\pi} \subset C^{m,\alpha}(\mathbb{R})$ we denote the space of 2π-periodic and m-times Hölder continuously differentiable functions g equipped with the norm

$$\|g\|_{m,\alpha} = \|g\|_\infty + \|g^{(m)}\|_{0,\alpha}.$$

The following theorem provides a uniform error estimate and implies uniform convergence for the trigonometric interpolation of Hölder continuous functions.

Theorem 11.6. *Let $m \in \mathbb{N} \cup \{0\}$ and $0 < \alpha \leq 1$. Then for the trigonometric interpolation we have*

$$\|P_n g - g\|_\infty \leq C \, \frac{\ln n}{n^{m+\alpha}} \, \|g\|_{m,\alpha} \tag{11.18}$$

for all $g \in C^{m,\alpha}_{2\pi}$ and some constant C depending on m and α.

Proof. For $g \in C^{m,\alpha}_{2\pi}$ let p_n be a best approximation of g with respect to trigonometric polynomials of the form (11.11) and the maximum norm (see Theorem 1.24). Then, by *Jackson's theorem* (see [40, 166]) there exists a constant c that does not depend on g, such that

$$\|p_n - g\|_\infty \leq \frac{c}{n^{m+\alpha}} \, \|g\|_{m,\alpha}. \tag{11.19}$$

Now, writing

$$P_n g - g = P_n(g - p_n) - (g - p_n),$$

the estimate (11.18) follows from (11.14) and (11.19). □

Theorem 11.6 can be extended to

$$\|P_n g - g\|_{\ell,\beta} \leq C \, \frac{\ln n}{n^{m-\ell+\alpha-\beta}} \, \|g\|_{m,\alpha} \tag{11.20}$$

for $g \in C^{m,\alpha}_{2\pi}$, $m, \ell \in \mathbb{N} \cup \{0\}$ with $\ell \leq m$, $0 < \beta \leq \alpha \leq 1$ and some constant C depending on m, ℓ, α, and β (see Prössdorf and Silbermann [199, 200]). For the trigonometric interpolation of 2π-periodic analytic functions we have the following error estimate (see [133]).

Theorem 11.7. *Let $g : \mathbb{R} \to \mathbb{R}$ be analytic and 2π-periodic. Then there exists a strip $D = \mathbb{R} \times (-s, s) \subset \mathbb{C}$ with $s > 0$ such that g can be extended to a holomorphic and 2π-periodic bounded function $g : D \to \mathbb{C}$. The error for the trigonometric interpolation can be estimated by*

$$\|P_n g - g\|_\infty \leq M \, \frac{\coth \frac{s}{2}}{\sinh ns}, \tag{11.21}$$

where M denotes a bound for the holomorphic function g on D.

Proof. Because $g : \mathbb{R} \to \mathbb{R}$ is analytic, at each point $t \in \mathbb{R}$ the Taylor expansion provides a holomorphic extension of g into some open disk in the complex plane with radius $r(t) > 0$ and center t. The extended function again has period 2π, since the coefficients of the Taylor series at t and at $t + 2\pi$ coincide for the 2π-periodic function $g : \mathbb{R} \to \mathbb{R}$. The disks corresponding to all points of the interval $[0, 2\pi]$ provide an open covering of $[0, 2\pi]$. Because $[0, 2\pi]$ is compact, a finite number of these disks suffices to cover $[0, 2\pi]$. Then we have an extension into a strip D with finite width $2s$ contained in the union of the finite number of disks. Without loss of generality we may assume that g is bounded on D.

Let $0 < \sigma < s$ be arbitrary. By Γ we denote the boundary of the rectangle $[-\pi/2n, 2\pi - \pi/2n] \times [-\sigma, \sigma]$ with counterclockwise orientation. A straightforward application of the residue theorem yields

$$\frac{1}{2\pi i} \int_\Gamma \frac{\cot \dfrac{\tau - t}{2}}{\sin n\tau}\, g(\tau)\, d\tau = \frac{2g(t)}{\sin nt} - \frac{1}{n} \sum_{j=0}^{2n-1} (-1)^j g(t_j) \cot \frac{t - t_j}{2}$$

for $-\pi/2n \le t < 2\pi - \pi/2n$, $t \neq t_k$, $k = 0, \ldots, 2n - 1$. Hence, in view of (11.13) we obtain

$$g(t) - (P_n g)(t) = \frac{\sin nt}{4\pi i} \int_\Gamma \frac{\cot \dfrac{\tau - t}{2}}{\sin n\tau}\, g(\tau)\, d\tau,$$

where we can obviously drop the restriction that t does not coincide with an interpolation point. From the periodicity of the integrand and since, by the Schwarz reflection principle, g enjoys the symmetry property $g(\overline{\tau}) = \overline{g(\tau)}$, we find the representation

$$g(t) - (P_n g)(t) = \frac{1}{2\pi} \sin nt \, \mathrm{Re} \left\{ \int_{i\sigma}^{i\sigma + 2\pi} \frac{i \cot \dfrac{\tau - t}{2}}{\sin n\tau}\, g(\tau)\, d\tau \right\} \tag{11.22}$$

with $0 < \sigma < s$. Because of

$$|\sin n\tau| \ge \sinh n\sigma \quad \text{and} \quad |\cot \tau/2| \le \coth \sigma/2$$

for $\mathrm{Im}\, \tau = \sigma$, the bound (11.21) now follows by estimating in (11.22) and then passing to the limit $\sigma \to s$. □

We can summarize Theorem 11.7 by the estimate

$$\|P_n g - g\|_\infty \le C e^{-ns} \tag{11.23}$$

for the trigonometric interpolation of periodic analytic functions, where C and s are some positive constants depending on g, i.e., the interpolation error decays at least exponentially.

We conclude this section on trigonometric interpolation with an error estimate in the Sobolev spaces $H^p[0, 2\pi]$. Note that by Theorem 8.4 each function $g \in H^p[0, 2\pi]$ is continuous if $p > 1/2$.

Theorem 11.8. *For the trigonometric interpolation we have*

$$\|P_n g - g\|_q \leq \frac{C}{n^{p-q}} \|g\|_p, \quad 0 \leq q \leq p, \ \frac{1}{2} < p, \tag{11.24}$$

for all $g \in H^p[0, 2\pi]$ *and some constant* C *depending on* p *and* q.

Proof. Consider the monomials $f_m(t) = e^{imt}$ and write $m = 2kn + \mu$, where k is an integer and $-n < \mu \leq n$. Since $f_m(t_j) = f_\mu(t_j)$ for $j = 0, \dots, 2n-1$, we have

$$P_n f_m = P_n f_\mu = \begin{cases} f_\mu, & \mu \neq n, \\ \dfrac{1}{2}(f_n + f_{-n}), & \mu = n. \end{cases}$$

Therefore the trigonometric interpolation polynomial is given by

$$P_n g = \sum_{k=-\infty}^{\infty} \left\{ \frac{1}{2} \hat{g}_{2kn+n}(f_n + f_{-n}) + \sum_{\mu=-n+1}^{n-1} \hat{g}_{2kn+\mu} f_\mu \right\}$$

in terms of the Fourier coefficients $\hat{g}_k$ of $g \in H^p[0, 2\pi]$. Hence,

$$\|g - P_n g\|_q^2 = S_1 + S_2 + S_3,$$

where

$$S_1 := \sum_{|m| \geq n+1} (1 + m^2)^q |\hat{g}_m|^2, \quad S_2 := \sum_{\mu=-n+1}^{n-1} (1 + \mu^2)^q \left| \sum_{k \neq 0} \hat{g}_{2kn+\mu} \right|^2,$$

$$S_3 := (1 + n^2)^q \left\{ \left| \hat{g}_n - \frac{1}{2} \sum_{k=-\infty}^{\infty} \hat{g}_{2kn+n} \right|^2 + \left| \hat{g}_{-n} - \frac{1}{2} \sum_{k=-\infty}^{\infty} \hat{g}_{2kn+n} \right|^2 \right\}.$$

Since $p \geq q$, we can estimate

$$S_1 \leq (1 + n^2)^{q-p} \sum_{|m| \geq n+1} (1 + m^2)^p |\hat{g}_m|^2 \leq c_1 n^{2q-2p} \|g\|_p^2$$

with some constant c_1 depending on p and q. For the second term, with the aid of the Cauchy–Schwarz inequality we find

$$S_2 \leq c_2 \sum_{\mu=-n+1}^{n-1} \frac{(1 + \mu^2)^q}{n^{2p}} \sum_{k \neq 0} |2kn + \mu|^{2p} |\hat{g}_{2kn+\mu}|^2$$

with some constant c_2 depending on p and q. Similarly we can estimate

$$S_3 \leq c_3 \, \frac{(1+n^2)^q}{n^{2p}} \sum_{k=-\infty}^{\infty} |2kn+n|^{2p} |\hat{g}_{2kn+n}|^2$$

with some constant c_3 depending on p and q. Since $q \geq 0$, we have

$$\frac{(1+\mu^2)^q}{n^{2p}} \leq 2^{2q} n^{2q-2p}, \quad |\mu| \leq n,$$

and thus we obtain

$$S_2 + S_3 \leq c_4 n^{2q-2p} \sum_{\mu=-n+1}^{n} \sum_{k=-\infty}^{\infty} |2kn+\mu|^{2p} |\hat{g}_{2kn+\mu}|^2 \leq c_5 n^{2q-2p} \|g\|_p^2$$

with some constants c_4 and c_5 depending on p and q. Summarizing the two estimates on S_1 and $S_2 + S_3$ concludes the proof. □

11.4 Degenerate Kernels via Interpolation

Now we return to integral equations of the second kind of the form

$$\varphi(x) - \int_a^b K(x,y)\varphi(y)\,dy = f(x), \quad a \leq x \leq b.$$

Recalling the interpolation Theorem 11.2, we approximate a given continuous kernel K by the kernel K_n interpolating K with respect to x, i.e., $K_n(\cdot\,,y) \in U_n$ and

$$K_n(x_j, y) = K(x_j, y), \quad j = 1, \dots, n,$$

for each y in $[a,b]$. Then we can write

$$K_n(x,y) = \sum_{j=1}^{n} L_j(x) K(x_j, y).$$

Hence, K_n is a degenerate kernel with $a_j = L_j$ and $b_j = K(x_j, \cdot)$. In this case the linear system (11.4) reads

$$\gamma_j - \sum_{k=1}^{n} \gamma_k \int_a^b K(x_j,y) L_k(y)\,dy = \int_a^b K(x_j,y) f(y)\,dy \tag{11.25}$$

for $j = 1, \ldots, n$, and the solution of the integral equation (11.5) is given by

$$\varphi_n = f + \sum_{k=1}^{n} \gamma_k L_k.$$

We illustrate the method by two examples, where we use piecewise linear interpolation and trigonometric interpolation as discussed in the two previous sections.

From Theorems 2.13 and 11.3, for the approximation of an integral operator A with a twice continuously differentiable kernel K by a degenerate kernel operator A_n via linear spline interpolation we conclude the estimate

$$\|A_n - A\|_\infty \le \frac{1}{8}\, h^2(b-a) \left\| \frac{\partial^2 K}{\partial x^2} \right\|_\infty . \tag{11.26}$$

Therefore, by Theorem 10.1, the corresponding approximation of the solution to the integral equation is of order $O(h^2)$. In principle, using Theorem 10.2, it is possible to derive computable error bounds based on the estimate (11.26). In general, these bounds will be difficult to evaluate in applications. However, in most practical problems it will be sufficient to judge the accuracy of the computed solution by refining the subdivision and then comparing the results for the fine and the coarse grid with the aid of the convergence order, i.e., in the case of linear splines with the aid of the approximation order $O(h^2)$.

Only in special cases the integrals for the coefficients and the right-hand side of the linear system (11.25) can be evaluated in closed form. Therefore the degenerate kernel method, in principle, is only a semi-discrete method. For a fully discrete method we have to incorporate a numerical evaluation of the integrals occurring in (11.25). As a general rule, these numerical quadratures should be performed so that the approximation order for the solution to the integral equation is maintained.

To be consistent with our approximations, we replace $K(x_j, \cdot)$ by its linear spline interpolation, i.e., we approximate

$$\int_a^b K(x_j, y) L_k(y)\, dy \approx \sum_{m=0}^{n} K(x_j, x_m) \int_a^b L_m(y) L_k(y)\, dy, \tag{11.27}$$

for $j, k = 0, \ldots, n$. Straightforward calculations yield the tridiagonal matrix

$$W = \frac{h}{6} \begin{pmatrix} 2 & 1 & & & & \\ 1 & 4 & 1 & & & \\ & 1 & 4 & 1 & & \\ & \cdot & \cdot & \cdot & \cdot & \cdot \\ & & & 1 & 4 & 1 \\ & & & & 1 & 2 \end{pmatrix}$$

for the weights $w_{mk} = \int_a^b L_m(y) L_k(y)\, dy$. For the right-hand side of (11.25) we simultaneously replace f and $K(x_j, \cdot)$ by its spline interpolations. This leads to the approximations

$$\int_a^b K(x_j, y)f(y)\,dy \approx \sum_{k,m=0}^{n} K(x_j, x_m)f(x_k) \int_a^b L_m(y)L_k(y)\,dy \tag{11.28}$$

for $j = 0, \ldots, n$. We now investigate the influence of these approximations on the error analysis. For this we interpret the solution of the system (11.25) with the approximate values for the coefficients and the right-hand sides as the solution $\widetilde{\varphi}_n$ of a further approximate equation

$$\widetilde{\varphi}_n - \widetilde{A}_n\widetilde{\varphi}_n = \widetilde{f}_n,$$

namely, of the degenerate kernel equation

$$\widetilde{\varphi}_n(x) - \int_a^b \widetilde{K}_n(x, y)\widetilde{\varphi}_n(y)\,dy = \widetilde{f}_n(x), \quad a \le x \le b,$$

with

$$\widetilde{K}_n(x, y) := \sum_{j,m=0}^{n} K(x_j, x_m)L_j(x)L_m(y) \quad \text{and} \quad \widetilde{f}_n(x) := \sum_{m=0}^{n} f(x_m)L_m(x).$$

Provided that the kernel K is twice continuously differentiable, writing

$$\begin{aligned} K(x, y) - \widetilde{K}_n(x, y) = {} & K(x, y) - [P_nK(\cdot\,, y)](x) \\ & +[P_n[K(\cdot\,, y) - [P_nK(\cdot\,,\cdot)](y)](x), \end{aligned} \tag{11.29}$$

by Theorem 11.3 and (11.9), we can estimate

$$|K(x, y) - \widetilde{K}_n(x, y)| \le \frac{1}{8}\, h^2 \left\{ \left\| \frac{\partial^2 K}{\partial x^2} \right\|_\infty + \left\| \frac{\partial^2 K}{\partial y^2} \right\|_\infty \right\}$$

for all $a \le x, y \le b$. Hence, for the integral operator $\widetilde{A}_n$ with kernel $\widetilde{K}_n$ we have $\|\widetilde{A}_n - A\|_\infty = O(h^2)$. When f is twice continuously differentiable, we also have $\|\tilde{f}_n - f\|_\infty = O(h^2)$. Now, Theorem 10.1 yields the error estimate $\|\widetilde{\varphi}_n - \varphi\| = O(h^2)$. Therefore we have the following theorem.

Theorem 11.9. *The degenerate kernel approximation via equidistant piecewise linear interpolation with the matrix entries and the right-hand sides of the approximating linear system (11.25) replaced by the quadratures (11.27) and (11.28) approximates the solution to the integral equation with order $O(h^2)$ if the kernel K and the right-hand side f are twice continuously differentiable.*

Example 11.10. Consider the integral equation

$$\varphi(x) - \frac{1}{2}\int_0^1 (x+1)e^{-xy}\varphi(y)\,dy = e^{-x} - \frac{1}{2} + \frac{1}{2}e^{-(x+1)}, \quad 0 \le x \le 1. \tag{11.30}$$

Obviously it has the solution $\varphi(x) = e^{-x}$. For its kernel we have

$$\max_{0\le x\le 1} \int_0^1 |K(x,y)|dy = \sup_{0<x\le 1} \frac{x+1}{2x}(1-e^{-x}) < 1.$$

Therefore, by Corollary 2.16, the solution is unique.

Table 11.1 shows the error between the approximate and the exact solution at the points $x = 0, 0.25, 0.5, 0.75$, and 1 for various n. It clearly exhibits the behavior $O(h^2)$ as predicted by Theorem 11.9. □

Table 11.1 Numerical results for Example 11.10

n	$x = 0$	$x = 0.25$	$x = 0.5$	$x = 0.75$	$x = 1$
4	0.004808	0.005430	0.006178	0.007128	0.008331
8	0.001199	0.001354	0.001541	0.001778	0.002078
16	0.000300	0.000338	0.000385	0.000444	0.000519
32	0.000075	0.000085	0.000096	0.000111	0.000130

From Theorem 11.6, for the approximation of an integral operator A with a 2π-periodic and m-times continuously differentiable kernel K by degenerate kernels via trigonometric interpolation we conclude the estimate

$$\|A_n - A\|_\infty \le O\left(\frac{1}{n^{m-\varepsilon}}\right) \tag{11.31}$$

for all $\varepsilon > 0$. If the kernel is analytic, then from Theorem 11.7 we have

$$\|A_n - A\|_\infty \le O\left(e^{-ns}\right) \tag{11.32}$$

for some $s > 0$. For the derivation of (11.32) we have to use the fact that for the kernel function, which is assumed to be analytic and periodic with respect to both variables, there exists a strip D with width $2s$ such that $K(\cdot\,,\tau)$ can be continued holomorphicly and uniformly bounded into D for all $\tau \in [0, 2\pi]$. We leave it to the reader to go over the argument given in the proof of Theorem 11.7 to verify this property. Now, by Theorem 10.1, the corresponding approximation of the solution to the integral equation is also of order $O\left(n^{\varepsilon-m}\right)$ or $O(e^{-ns})$, respectively.

Again we have to describe the numerical evaluation of the coefficients and the right-hand side of the linear system (11.25). We proceed analogously to the preceding example and approximate the matrix entries by replacing $K(t_j, \cdot)$ by its trigonometric interpolation polynomial, i.e.,

$$\int_0^{2\pi} K(t_j,\tau)L_k(\tau)\,d\tau \approx \sum_{m=0}^{2n-1} K(t_j,t_m)\int_0^{2\pi} L_m(\tau)L_k(\tau)\,d\tau \tag{11.33}$$

for $j, k = 0, \dots, 2n-1$ with the integrals on the right-hand side given by (11.17). For the right-hand side of (11.25) we simultaneously replace f and $K(t_j, \cdot)$ by its trigonometric interpolation. This leads to

$$\int_0^{2\pi} K(t_j, \tau) f(\tau)\, d\tau \approx \sum_{k,m=0}^{2n-1} K(t_j, t_m) f(t_k) \int_0^{2\pi} L_m(\tau) L_k(\tau)\, d\tau \tag{11.34}$$

for $j = 0, \dots, 2n-1$. Note that despite the global nature of the trigonometric interpolation and its Lagrange basis, due to the simple structure of the weights (11.17) in the quadrature rule, the approximate computation of the matrix elements and right-hand side is not too costly. In view of (11.18) and splitting analogous to (11.29), we remain with a total error $O(n^{\varepsilon - m})$ or $O(e^{-ns})$, respectively, for the approximate solution of the integral equation. Here we use that $\ln n\, e^{-ns} = O(e^{-n\sigma})$ for all $0 < \sigma < s$. Hence we have the following theorem.

Theorem 11.11. *The degenerate kernel approximation via trigonometric interpolation with the matrix entries and the right-hand sides of the approximating linear system (11.25) replaced by the quadratures (11.33) and (11.34) approximates the solution to the integral equation of order $O(n^{\varepsilon - m})$ or $O(e^{-ns})$, respectively, if the 2π-periodic kernel K and the 2π-periodic right-hand side f are m-times continuously differentiable or analytic, respectively.*

Example 11.12. Consider the integral equation

$$\varphi(t) + \frac{ab}{\pi} \int_0^{2\pi} \frac{\varphi(\tau)\, d\tau}{a^2 + b^2 - (a^2 - b^2)\cos(t + \tau)} = f(t), \quad 0 \le t \le 2\pi, \tag{11.35}$$

corresponding to the Dirichlet problem for the Laplace equation in the interior of an ellipse with semi-axis $a \ge b > 0$ (see Problem 6.2). We numerically want to solve the case where the unique solution is given by

$$\varphi(t) = e^{\cos t} \cos(\sin t) = \operatorname{Re} \sum_{m=0}^{\infty} \frac{1}{m!}\, e^{imt}, \quad 0 \le t \le 2\pi.$$

Then, the right-hand side becomes

$$f(t) = \varphi(t) + e^{c \cos t} \cos(c \sin t), \quad 0 \le t \le 2\pi,$$

where $c = (a - b)/(a + b)$ (compare also Problem 11.4).

Table 11.2 shows the error between the approximate and the exact solution at the three points $t = 0, \pi/2$, and π for various n and depending on the ratio b/a. It clearly shows the exponential decay of the error: doubling the number of grid points doubles the number of correct digits in the approximate solution. The rate of the exponential decay depends on the parameters a and b, which describe the location of the singularities of the integrands in the complex plane, i.e., they determine the value for the strip parameter s. □

Table 11.2 Numerical results for Example 11.12

	$2n$	$t = 0$	$t = \pi/2$	$t = \pi$
	4	-0.10752855	-0.03243176	0.03961310
a=1	8	-0.00231537	0.00059809	0.00045961
b=0.5	16	-0.00000044	0.00000002	-0.00000000
	32	0.00000000	0.00000000	0.00000000
	4	-0.56984945	-0.18357135	0.06022598
a=1	8	-0.14414257	-0.00368787	-0.00571394
b=0.2	16	-0.00602543	-0.00035953	-0.00045408
	32	-0.00000919	-0.00000055	-0.00000069

11.5 Degenerate Kernels via Expansions

A second possibility for constructing approximate degenerate kernels is by expansions, in particular, by orthonormal expansions. Let $(\cdot\,,\cdot)$ denote a scalar product on $C(G)$ and let $\{u_1, u_2, \dots\}$ be a complete orthonormal system (recall Theorem 1.28). Then a given continuous kernel K is expanded with respect to x for each fixed $y \in G$, i.e., $K(x, y)$ is approximated by the partial sum

$$K_n(x, y) := \sum_{j=1}^{n} u_j(x)(K(\cdot\,, y), u_j)$$

of the Fourier series. For this degenerate kernel the linear system (11.4) reads

$$\gamma_j - \sum_{k=1}^{n} \gamma_k \int_G u_k(y)(K(\cdot\,, y), u_j)\,dy = \int_G f(y)(K(\cdot\,, y), u_j)\,dy \qquad (11.36)$$

for $j = 1, \dots, n$. Usually, the scalar product will be given in terms of an integral. Therefore the system (11.36) requires a double integration for each coefficient and for each right-hand side. Since it will turn out that the degenerate kernel method via orthonormal expansion is closely related to the Galerkin method (see Section 13.6), we omit further discussion and a description of the numerical implementation.

For the orthonormal expansion degenerate kernel approximations, in principle, we can derive error estimates from Theorems 10.1 and 10.2 only with respect to the scalar product norm. In the subsequent analysis, for later theoretical use, we will describe how estimates in the maximum norm can be obtained under additional regularity assumptions on the kernel.

The *Chebyshev polynomials* of the first kind are defined by

$$T_n(z) := \cos(n \arccos z), \quad -1 \le z \le 1, \quad n = 0, 1, \dots. \qquad (11.37)$$

From $T_0(z) = 1$ and $T_1(z) = z$ and from the recursion formula

$$T_{n+1}(z) + T_{n-1}(z) = 2zT_n(z),$$

which follows from the cosine addition theorem, we observe that the T_n indeed are polynomials of degree n and therefore, in particular, well defined on the whole complex plane. A substitution $x = \cos t$ readily shows that

$$\int_{-1}^{1} \frac{T_n(x)T_m(x)}{\sqrt{1-x^2}}\, dx = \frac{\pi}{2}\,[\delta_{nm} + \delta_{n0}],$$

i.e., the T_n form an orthogonal system with respect to the scalar product

$$(\varphi, \psi) := \int_{-1}^{1} \frac{\varphi(x)\overline{\psi(x)}}{\sqrt{1-x^2}}\, dx \tag{11.38}$$

on $C[-1,1]$. By the same substitution, the denseness of the trigonometric polynomials in $C[0,2\pi]$ implies the denseness of $\operatorname{span}\{T_0, T_1, \ldots\}$ in $C[-1,1]$ with respect to the scalar product (11.38), i.e., the T_n form a complete orthogonal system.

Theorem 11.13. *Let $g : [-1,1] \to \mathbb{R}$ be analytic. Then there exists an ellipse E with foci at -1 and 1 such that g can be extended to a holomorphic and bounded function $g : D \to \mathbb{C}$, where D denotes the open interior of E. The orthonormal expansion with respect to the Chebyshev polynomials*

$$g = \frac{a_0}{2}\,T_0 + \sum_{n=1}^{\infty} a_n T_n \tag{11.39}$$

with coefficients

$$a_n = \frac{2}{\pi}\int_{-1}^{1} \frac{g(x)T_n(x)}{\sqrt{1-x^2}}\, dx$$

is uniformly convergent with the estimate

$$\left\| g - \frac{a_0}{2}\,T_0 - \sum_{m=1}^{n} a_m T_m \right\|_\infty \le \frac{2M}{R-1}\, R^{-n}.$$

Here, R is given through the semi-axis a and b of E by $R = a + b$ and M is a bound on g in D.

Proof. The existence of the holomorphic extension of the analytic function $g : [-1,1] \to \mathbb{R}$ into the open interior D of some ellipse with foci at -1 and 1 is shown analogously to the proof of Theorem 11.7.

The function

$$z = \frac{w}{2} + \frac{1}{2w}$$

maps the annulus $1/R < |w| < R$ of the w–plane onto the interior D of the ellipse E of the z–plane. For each holomorphic function g on D the function $\widetilde{g}$, defined by

$$\widetilde{g}(w) := 2g\left(\frac{w}{2} + \frac{1}{2w}\right),$$

is holomorpic in the annulus. Therefore it can be expanded into a Laurent series

$$\widetilde{g}(w) = \sum_{n=-\infty}^{\infty} a_n w^n$$

with coefficients

$$a_n = \frac{1}{\pi i} \int_{|w|=r} g\left(\frac{w}{2} + \frac{1}{2w}\right) \frac{dw}{w^{n+1}},$$

where $1/R < r < R$. By substituting $\widetilde{w} = 1/w$ and using Cauchy's integral theorem we find that $a_n = a_{-n}$ for all $n \in \mathbb{N}$. Hence,

$$\widetilde{g}(w) = a_0 + \sum_{n=1}^{\infty} a_n \left(w^n + \frac{1}{w^n}\right).$$

In particular, this Laurent expansion converges uniformly on the circle $|w| = 1$. Writing $w = e^{it}$, we derive

$$T_n\left(\frac{w}{2} + \frac{1}{2w}\right) = \frac{1}{2}\left(w^n + \frac{1}{w^n}\right)$$

first on the unit circle, and then, since both sides of the equation represent rational functions, for all $w \neq 0$. Inserting this into the previous series, we obtain the expansion (11.39) and its uniform convergence on $[-1, 1]$.

Estimating the coefficients of the Laurent expansion and then passing to the limit $r \to R$ yields

$$|a_n| \leq \frac{2M}{R^n}, \quad n = 0, 1, 2 \ldots,$$

with a bound M for g on D. Hence, using $|T_n(x)| \leq 1$ for all $-1 \leq x \leq 1$ and all $n \in \mathbb{N}$, the remainder can be estimated by

$$\left| \sum_{m=n+1}^{\infty} a_m T_m(x) \right| \leq \sum_{m=n+1}^{\infty} |a_m| \leq \frac{2M}{R-1} R^{-n},$$

which finishes the proof. □

From Theorem 11.13, analogous to the derivation of the estimate (11.32), for the approximation of an integral operator A with an analytic kernel by orthonormal expansion with respect to Chebyshev polynomials we find the estimate

$$\|A_n - A\|_\infty = O(R^{-n}) \tag{11.40}$$

with some constant $R > 1$ depending on the kernel of A.

In closing this chapter, we wish to mention the use of Taylor expansions as another method to construct degenerate kernels that may be useful in special cases (see Problem 11.3).

Problems

11.1. Let $\langle X, Y\rangle$ be a dual system with two normed spaces X and Y and let $A_n : X \to X$ and $B_n : Y \to Y$ be adjoint finite-dimensional operators of the form

$$A_n\varphi = \sum_{j=1}^{n} \langle \varphi, b_j\rangle a_j, \quad B_n\psi = \sum_{j=1}^{n} \langle a_j, \psi\rangle b_j$$

with linearly independent elements $a_1, \dots, a_n \in X$ and $b_1, \dots, b_n \in Y$. By reduction to linear systems as in Theorem 11.1, establish the Fredholm alternative for the operators $I - A_n$ and $I - B_n$.

11.2. Let $\langle X, Y\rangle$ be a dual system with two Banach spaces X and Y, let $A_n : X \to X$ and $B_n : Y \to Y$ be adjoint finite-dimensional operators as in Problem 11.1, and let $S : X \to X$ and $T : Y \to Y$ be adjoint operators with norm less than one. With the aid of Theorem 2.14 and Problem 11.1 establish the Fredholm alternative for the operators $I - A$ and $I - B$ where $A = A_n + S$ and $B = B_n + T$.
Hint: Transform the equations with the operators $I - A$ and $I - B$ equivalently into equations with $I - (I - S)^{-1}A_n$ and $I - B_n(I - T)^{-1}$.

11.3. Solve the integral equation (11.30) approximately through degenerate kernels by using the Taylor series for e^{xy}.

11.4. Show that the eigenvalues of the integral equation (11.35) are described by

$$\frac{ab}{\pi}\int_0^{2\pi} \frac{e^{in\tau}\,d\tau}{(a^2 + b^2) - (a^2 - b^2)\cos(t + \tau)} = \left(\frac{a - b}{a + b}\right)^n e^{-int}, \quad n = 0, 1, 2, \dots.$$

11.5. Show that, analogous to (11.40), we have $\|A_n - A\|_\infty = o\,(1/\sqrt{n})$ for the approximation of continuously differentiable kernels through degenerate kernels via Chebyshev polynomials.
Hint: Use Dini's theorem from Problem 8.2.

Chapter 12
Quadrature Methods

In this chapter we shall describe the *quadrature* or *Nyström method* for the approximate solution of integral equations of the second kind with continuous or weakly singular kernels. As we have pointed out in Chapter 11, the implementation of the degenerate kernel method, in general, requires some use of numerical quadrature. Therefore it is natural to try the application of numerical integration in a more direct approach to approximate integral operators by numerical integration operators. This will lead to a straightforward but widely applicable method for approximately solving equations of the second kind. The reason we placed the description of the quadrature method after the degenerate kernel method is only because its error analysis is more involved.

Through numerical examples we will illustrate that the Nyström method provides a highly efficient method for the approximate solution of the boundary integral equations of the second kind for two-dimensional boundary value problems as considered in Chapters 6–8. We also wish to point out that, despite the fact that our numerical examples for the application of the Nyström method are all for one-dimensional integral equations, this method can be applied to multi-dimensional integral equations provided appropriate numerical quadratures are available. The latter requirement puts some limits on the application to the weakly singular integral equations arising for boundary value problems in more than two space dimensions.

12.1 Numerical Integration

We start with a brief account of the basics of numerical quadrature. In general, a quadrature formula is a numerical method for approximating an integral of the form

$$Q(g) := \int_G w(x)g(x)\,dx, \tag{12.1}$$

R. Kress, *Linear Integral Equations*, Applied Mathematical Sciences 82,
DOI 10.1007/978-1-4614-9593-2_12,

where w is some *weight function* and $g \in C(G)$. Throughout this chapter $G \subset \mathbb{R}^m$ will be compact and Jordan measurable. We consider only quadrature rules of the form

$$Q_n(g) := \sum_{j=1}^{n} \alpha_j^{(n)} g(x_j^{(n)})$$

with *quadrature points* $x_1^{(n)}, \dots, x_n^{(n)}$ contained in G and real *quadrature weights* $\alpha_1^{(n)}, \dots, \alpha_n^{(n)}$. For the sake of notational clarity we restrict ourselves to the case where the index n of the quadrature rule coincides with the number of quadrature points. Occasionally, we also will write $x_1, \dots, x_n$ instead of $x_1^{(n)}, \dots, x_n^{(n)}$ and $\alpha_1, \dots, \alpha_n$ instead of $\alpha_1^{(n)}, \dots, \alpha_n^{(n)}$. The basic numerical integrations are *interpolatory quadratures*. They are constructed by replacing the integrand g by an interpolation with respect to the quadrature points $x_1, \dots, x_n$, usually a polynomial, a trigonometric polynomial, or a spline, and then integrating the interpolating function instead of g. Clearly, polynomial interpolation quadratures of degree $n-1$ on an interval $[a, b]$ with n quadrature points integrate polynomials of degree less than n exactly. The classical Newton–Cotes rules are a special case of these interpolatory quadratures with polynomial interpolation on an equidistantly spaced subdivision of the interval $[a, b]$. Since the Newton–Cotes rules have unsatisfactory convergence behavior as the degree of the interpolation increases, it is more practical to use so-called *composite rules*. These are obtained by subdividing the interval of integration and then applying a fixed rule with low interpolation order to each subinterval. The most frequently used quadrature rules of this type are the *composite trapezoidal rule* and the *composite Simpson's rule*. For convenience we will discuss their error analysis, which we base on the representation of the remainder terms by means of *Peano kernels*. Let $x_j = a + jh$, $j = 0, \dots, n$, be an equidistant subdivision with step size $h = (b-a)/n$.

Theorem 12.1. *Let $g \in C^2[a, b]$. Then the remainder*

$$R_T(g) := \int_a^b g(x)\,dx - h\left[\frac{1}{2}\,g(x_0) + g(x_1) + \cdots + g(x_{n-1}) + \frac{1}{2}\,g(x_n)\right]$$

for the composite trapezoidal rule can be estimated by

$$|R_T(g)| \le \frac{1}{12}\,h^2(b-a)\,\|g''\|_\infty.$$

Proof. Define the Peano kernel for the trapezoidal rule by

$$K_T(x) := \frac{1}{2}\,(x - x_{j-1})(x_j - x), \quad x_{j-1} \le x \le x_j,$$

for $j = 1, \dots, n$. Then, straightforward partial integrations yield

$$\int_a^b K_T(x) g''(x)\,dx = -R_T(g).$$

Now the estimate follows from

$$\int_a^b K_T(x)\,dx = \frac{h^2}{12}\,(b-a)$$

and the observation that K_T is nonnegative on $[a,b]$. □

Theorem 12.2. *Let $g \in C^4[a,b]$ and let n be even. Then the remainder*

$$R_S(g) := \int_a^b g(x)\,dx - \frac{h}{3}\,[g(x_0) + 4g(x_1) + 2g(x_2) + 4g(x_3) + 2g(x_4)$$

$$+ \cdots + 2g(x_{n-2}) + 4g(x_{n-1}) + g(x_n)]$$

for the composite Simpson's rule can be estimated by

$$|R_S(g)| \le \frac{1}{180}\,h^4(b-a)\,\|g^{(4)}\|_\infty.$$

Proof. With the Peano kernel of Simpson's rule given by

$$K_S(x) := \begin{cases} \dfrac{h}{18}\,(x - x_{2j-2})^3 - \dfrac{1}{24}\,(x - x_{2j-2})^4, & x_{2j-2} \le x \le x_{2j-1}, \\[2ex] \dfrac{h}{18}\,(x_{2j} - x)^3 - \dfrac{1}{24}\,(x_{2j} - x)^4, & x_{2j-1} \le x \le x_{2j}, \end{cases}$$

for $j = 1, \dots, n/2$, via

$$\int_a^b K_S(x) g^{(4)}(x)\,dx = -R_S(g),$$

the proof is analogous to the proof of Theorem 12.1. □

More satisfying than Newton–Cotes quadratures are Gauss quadratures. These are polynomial interpolation quadratures with n quadrature points that are chosen such that polynomials of degree less than $2n$ are integrated exactly. For the integral (12.1) over an interval $[a,b]$ with integrable positive weight function w the quadrature points of the Gauss quadrature are given by the n distinct zeros in (a,b) of the polynomial of order n that is orthogonal to all polynomials of degree less than n with respect to the scalar product $(f,g) := \int_a^b wfg\,dx$. The *Legendre polynomial* L_n of degree n is defined by

$$L_n(x) := \frac{1}{2^n n!}\,\frac{d^n}{dx^n}\,(x^2-1)^n.$$

If $m < n$, by repeated partial integration we see that

$$\int_{-1}^{1} x^m \frac{d^n}{dx^n}(x^2-1)^n dx = 0.$$

Therefore L_n is orthogonal to all polynomials of degree less than n with respect to the L^2 scalar product on $[-1, 1]$. Hence, the quadrature points of the Gauss–Legendre quadrature are given by the n zeros $-1 < \xi_1 < \xi_2 < \cdots < \xi_n < 1$ of L_n and the corresponding quadrature weights can be computed as

$$\alpha_j := \frac{2(1-\xi_j^2)}{[nL_{n-1}(\xi_j)]^2}, \quad j = 1, \dots, n.$$

We note that these weights are positive. For a more detailed analysis of Gauss quadratures we refer to Davis and Rabinowitz [41] and Engels [47].

Definition 12.3. A sequence (Q_n) of quadrature formulas is called *convergent* if $Q_n(g) \to Q(g)$, $n \to \infty$, for all $g \in C(G)$, i.e., if the sequence of linear functionals (Q_n) converges pointwise to the integral Q.

With the aid of the uniform boundedness principle we can state the following necessary and sufficient conditions for convergence.

Theorem 12.4 (Szegö). *The quadrature formulas (Q_n) converge if and only if $Q_n(g) \to Q(g)$, $n \to \infty$, for all g in some dense subset $U \subset C(G)$ and*

$$\sup_{n \in \mathbb{N}} \sum_{j=1}^{n} |\alpha_j^{(n)}| < \infty.$$

Proof. It is left as an exercise to show that

$$\|Q_n\|_\infty = \sum_{j=1}^{n} |\alpha_j^{(n)}|.$$

Now the statement follows by the Banach–Steinhaus theorem (see Problem 10.1). □

Corollary 12.5 (Steklov). *Assume that $Q_n(1) \to Q(1)$, $n \to \infty$, and that the quadrature weights are all nonnegative. Then the quadrature formulas (Q_n) converge if and only if $Q_n(g) \to Q(g)$, $n \to \infty$, for all g in some dense subset $U \subset C(G)$.*

Proof. This follows from

$$\sum_{j=1}^{n} |\alpha_j^{(n)}| = \sum_{j=1}^{n} \alpha_j^{(n)} = Q_n(1) \to Q(1), \quad n \to \infty,$$

and the preceding theorem. □

In particular, from Theorems 12.1 and 12.2 and Corollary 12.5, we observe that the composite trapezoidal and the composite Simpson's rule are convergent. Also the sequence of Gauss–Legendre quadratures is convergent because of their positive weights.

We conclude our remarks on numerical quadrature by describing an error estimate for the integration of periodic analytic functions due to Davis [39] (see also Problem 12.2). Note that for periodic functions the composite trapezoidal rule coincides with the composite rectangular rule.

Theorem 12.6. *Let g be as in Theorem 11.7. Then the error*

$$R_T(g) := \frac{1}{2\pi}\int_0^{2\pi} g(t)\,dt - \frac{1}{2n}\sum_{j=0}^{2n-1} g\left(\frac{j\pi}{n}\right)$$

for the composite trapezoidal rule can be estimated by

$$|R_T(g)| \le M\,(\coth ns - 1).$$

Proof. Using the integrals (8.35), we integrate the remainder term (11.22) for the trigonometric interpolation to obtain

$$R_T(g) = \frac{1}{2\pi}\,\mathrm{Re}\left\{\int_{i\sigma}^{i\sigma+2\pi}(1 - i\cot n\tau)\,g(\tau)\,d\tau\right\}$$

for all $0 < \sigma < s$. This can also be shown directly through the residue theorem. Now the estimate follows from $|1 - i\cot n\tau| \le \coth n\sigma - 1$ for $\mathrm{Im}\,\tau = \sigma$ and passing to the limit $\sigma \to s$. □

We can summarize Theorem 12.6 by the estimate

$$|R_T(g)| \le Ce^{-2ns} \tag{12.2}$$

for the composite trapezoidal rule for periodic analytic functions, where C and s are some positive constants depending on g. The improvement by the factor 2 in the exponent as compared with (11.23) reflects the fact that the trapezoidal rule integrates trigonometric polynomials not only of degree less than or equal to n but also of degree less than or equal to $2n - 1$ exactly.

For m-times continuously differentiable 2π-periodic functions g the Euler–MacLaurin expansion yields the estimate

$$|R_T(g)| \le \frac{C}{n^m}\,\|g^{(m)}\|_\infty \tag{12.3}$$

for some constant C depending on m (see [142]).

Obviously, numerical quadrature of 2π-periodic functions is equivalent to numerical quadrature of contour integrals over the unit circle, or more generally over closed curves in $\mathbb{R}^2$ that are homeomorphic to the unit circle. As an analogue in $\mathbb{R}^3$

we briefly describe a numerical quadrature scheme for the integration of analytic functions over closed surfaces Γ which are homeomorphic to the unit sphere $\mathbb{S}^2$. Clearly, it suffices to describe the quadrature for integrating over the unit sphere. For this we recall the quadrature points ξ_j and the weights α_j of the Gauss–Legendre quadrature and choose a set of points x_{jk} on the unit sphere given in spherical coordinates by

$$x_{jk} := (\sin\theta_j \cos\varphi_k, \sin\theta_j \sin\varphi_k, \cos\theta_j)$$

for $j = 1, \dots, n$ and $k = 0, \dots, 2n-1$ where $\theta_j := \arccos \xi_j$ and $\varphi_k = \pi k/n$. Then, writing the integral over the unit sphere as a double integral

$$\int_{\mathbb{S}^2} g\,ds = \int_0^{2\pi}\int_0^{\pi} g(\sin\theta\cos\varphi, \sin\theta\sin\varphi, \cos\theta)\sin\theta\,d\theta d\varphi$$

$$= \int_0^{2\pi}\int_{-1}^{1} g\left(\sqrt{1-\xi^2}\cos\varphi, \sqrt{1-\xi^2}\sin\varphi, \xi\right) d\xi d\varphi,$$

we apply the Gauss–Legendre quadrature to the integration over ξ and the trapezoidal rule to the integration over φ to obtain the so-called Gauss trapezoidal product rule

$$\int_{\mathbb{S}^2} g\,ds \approx \frac{\pi}{n}\sum_{j=1}^{n}\sum_{k=0}^{2n-1} \alpha_j g(x_{jk}) \tag{12.4}$$

for the numerical integration over the unit sphere. Wienert [249] has shown that for analytic functions g there exist positive constants C and σ depending on g such that the quadrature error in (12.4) can be estimated by $Ce^{-n\sigma}$, i.e, as for the composite trapezoidal rule we have exponential convergence as $n \to \infty$.

For a comprehensive study of numerical integration, we refer the reader to Davis and Rabinowitz [41] and Engels [47].

12.2 Nyström's Method

We choose a convergent sequence (Q_n) of quadrature formulas for the integral $Q(g) = \int_G g(x)\,dx$ and approximate the integral operator

$$(A\varphi)(x) := \int_G K(x, y)\varphi(y)\,dy, \quad x \in G, \tag{12.5}$$

with continuous kernel K, as introduced in Theorem 2.13, by a sequence of numerical integration operators

$$(A_n\varphi)(x) := \sum_{k=1}^{n} \alpha_k^{(n)} K(x, x_k^{(n)})\varphi(x_k^{(n)}), \quad x \in G. \tag{12.6}$$

Then the solution to the integral equation of the second kind

$$\varphi - A\varphi = f$$

is approximated by the solution of

$$\varphi_n - A_n\varphi_n = f,$$

which reduces to solving a finite-dimensional linear system.

Theorem 12.7. *Let* φ_n *be a solution of*

$$\varphi_n(x) - \sum_{k=1}^{n} \alpha_k K(x, x_k)\varphi_n(x_k) = f(x), \quad x \in G. \tag{12.7}$$

Then the values $\varphi_j^{(n)} = \varphi_n(x_j)$, $j = 1, \dots, n$, *at the quadrature points satisfy the linear system*

$$\varphi_j^{(n)} - \sum_{k=1}^{n} \alpha_k K(x_j, x_k)\varphi_k^{(n)} = f(x_j), \quad j = 1, \dots, n. \tag{12.8}$$

Conversely, let $\varphi_j^{(n)}$, $j = 1, \dots, n$, *be a solution of the system (12.8). Then the function* φ_n *defined by*

$$\varphi_n(x) := f(x) + \sum_{k=1}^{n} \alpha_k K(x, x_k)\varphi_k^{(n)}, \quad x \in G, \tag{12.9}$$

solves equation (12.7).

Proof. The first statement is trivial. For a solution $\varphi_j^{(n)}$, $j = 1, \dots, n$, of the system (12.8), the function φ_n defined by (12.9) has values

$$\varphi_n(x_j) = f(x_j) + \sum_{k=1}^{n} \alpha_k K(x_j, x_k)\varphi_k^{(n)} = \varphi_j^{(n)}, \quad j = 1, \dots, n.$$

Inserting this into (12.9), we see that φ_n satisfies (12.7). □

The formula (12.9) may be viewed as a natural interpolation of the values $\varphi_j^{(n)}$, $j = 1, \dots, n$, at the quadrature points to obtain the approximating function φ_n and goes back to Nyström [187].

The error analysis will be based on the following result.

Theorem 12.8. *Assume the quadrature formulas* (Q_n) *are convergent. Then the sequence* (A_n) *is collectively compact and pointwise convergent (i.e.,* $A_n\varphi \to A\varphi$, $n \to \infty$, *for all* $\varphi \in C(G)$*), but not norm convergent.*

Proof. Because the quadrature formulas (Q_n) are assumed to be convergent, by Theorem 12.4, there exists a constant C such that the weights satisfy

$$\sum_{k=1}^{n} |\alpha_k^{(n)}| \leq C$$

for all $n \in \mathbb{N}$. Then we can estimate

$$\|A_n\varphi\|_\infty \leq C \max_{x,y\in G} |K(x,y)| \, \|\varphi\|_\infty \tag{12.10}$$

and

$$|(A_n\varphi)(x_1) - (A_n\varphi)(x_2)| \leq C \max_{y\in G} |K(x_1,y) - K(x_2,y)| \, \|\varphi\|_\infty \tag{12.11}$$

for all $x_1, x_2 \in G$. Now let $U \subset C(G)$ be bounded. Then from (12.10) and (12.11) we see that $\{A_n\varphi : \varphi \in U, \ n \in \mathbb{N}\}$ is bounded and equicontinuous because K is uniformly continuous on $G \times G$. Therefore, by the Arzelà–Ascoli Theorem 1.18 the sequence (A_n) is collectively compact.

Since the quadrature is convergent, for fixed $\varphi \in C(G)$ the sequence $(A_n\varphi)$ is pointwise convergent, i.e., $(A_n\varphi)(x) \to (A\varphi)(x)$, $n \to \infty$, for all $x \in G$. As a consequence of (12.11), the sequence $(A_n\varphi)$ is equicontinuous. Hence it is uniformly convergent $\|A_n\varphi - A\varphi\|_\infty \to 0$, $n \to \infty$, i.e., we have pointwise convergence $A_n\varphi \to A\varphi$, $n \to \infty$, for all $\varphi \in C(G)$ (see Problem 12.1).

For $\varepsilon > 0$ choose a function $\psi_\varepsilon \in C(G)$ with $0 \leq \psi_\varepsilon(x) \leq 1$ for all $x \in G$ such that $\psi_\varepsilon(x) = 1$ for all $x \in G$ with $\min_{j=1,\dots,n} |x - x_j| \geq \varepsilon$ and $\psi_\varepsilon(x_j) = 0$, $j = 1, \dots, n$. Then

$$\|A\varphi\psi_\varepsilon - A\varphi\|_\infty \leq \max_{x,y\in G} |K(x,y)| \int_G \{1 - \psi_\varepsilon(y)\}\, dy \to 0, \quad \varepsilon \to 0,$$

for all $\varphi \in C(G)$ with $\|\varphi\|_\infty = 1$. Using this result, we derive

$$\begin{aligned} \|A - A_n\|_\infty &= \sup_{\|\varphi\|_\infty = 1} \|(A - A_n)\varphi\|_\infty \geq \sup_{\|\varphi\|_\infty = 1} \sup_{\varepsilon > 0} \|(A - A_n)\varphi\psi_\varepsilon\|_\infty \\ &= \sup_{\|\varphi\|_\infty = 1} \sup_{\varepsilon > 0} \|A\varphi\psi_\varepsilon\|_\infty \geq \sup_{\|\varphi\|_\infty = 1} \|A\varphi\|_\infty = \|A\|_\infty, \end{aligned}$$

whence we see that the sequence (A_n) cannot be norm convergent. □

Corollary 12.9. *For a uniquely solvable integral equation of the second kind with a continuous kernel and a continuous right-hand side, the Nyström method with a convergent sequence of quadrature formulas is uniformly convergent.*

Proof. This follows from Theorems 10.12 and 12.8. □

In principle, using Theorem 10.13, it is possible to derive computable error bounds (see Problem 12.3). Because these, in general, will be too complicated to evaluate, as already mentioned in Chapter 11, in applications it will usually be sufficient to estimate the error by extrapolation from the convergence order. For the discussion of the error based on the estimate (10.11) of Corollary 10.14 we need the norm $\|(A - A_n)\varphi\|_\infty$. It can be expressed in terms of the error for the corresponding numerical quadrature by

$$\|(A - A_n)\varphi\|_\infty = \max_{x\in G} \left| \int_G K(x, y)\varphi(y)\, dy - \sum_{k=1}^{n} \alpha_k K(x, x_k)\varphi(x_k) \right|$$

and requires a uniform estimate for the error of the quadrature applied to the integration of $K(x, \cdot)\varphi$. Therefore, from the error estimate (10.11), it follows that under suitable regularity assumptions on the kernel K and the exact solution φ, the convergence order of the underlying quadrature formulas carries over to the convergence order of the approximate solutions to the integral equation. We illustrate this through the case of the trapezoidal rule. Under the assumption $\varphi \in C^2[a, b]$ and $K \in C^2([a, b] \times [a, b])$, by Theorem 12.1, we can estimate

$$\|(A - A_n)\varphi\|_\infty \le \frac{1}{12}\, h^2(b - a) \max_{x,y\in G} \left| \frac{\partial^2}{\partial y^2}\, K(x, y)\varphi(y) \right| .$$

Table 12.1 Numerical solution of (11.30) by the trapezoidal rule

n	$x = 0$	$x = 0.25$	$x = 0.5$	$x = 0.75$	$x = 1$
4	0.007146	0.008878	0.010816	0.013007	0.015479
8	0.001788	0.002224	0.002711	0.003261	0.003882
16	0.000447	0.000556	0.000678	0.000816	0.000971
32	0.000112	0.000139	0.000170	0.000204	0.000243

Example 12.10. We use the composite trapezoidal rule for approximately solving the integral equation (11.30) of Example 11.10. By the numerical results for the difference between the approximate and exact solutions given in Table 12.1 the expected convergence rate $O(h^2)$ is clearly demonstrated.

We now use the composite Simpson's rule for the integral equation (11.30). The numerical results in Table 12.2 show the convergence order $O(h^4)$, which we expect from the error estimate (10.11) and Theorem 12.2. □

After comparing the last two examples, we wish to emphasize the major advantage of Nyström's method compared with the degenerate kernel method of Chapter 11. The matrix and the right-hand side of the linear system (12.8) are obtained by just evaluating the kernel K and the given function f at the quadrature points. Therefore, without any further computational effort we can improve considerably on the approximations by choosing a more accurate numerical quadrature. For example, in the degenerate kernel method, it is much more involved to design

Table 12.2 Numerical solution of (11.30) by Simpson's rule

n	$x = 0$	$x = 0.25$	$x = 0.5$	$x = 0.75$	$x = 1$
4	0.00006652	0.00008311	0.00010905	0.00015046	0.00021416
8	0.00000422	0.00000527	0.00000692	0.00000956	0.00001366
16	0.00000026	0.00000033	0.00000043	0.00000060	0.00000086

a scheme with convergence order $O(h^4)$ rather than $O(h^2)$ from the linear spline interpolation.

Example 12.11. For the integral equation (11.35) of Example 11.12 we use the trapezoidal rule. Since we are dealing with periodic analytic functions, from Theorem 12.6 and the error estimate (12.2), we expect an exponentially decreasing error behavior that is exhibited by the numerical results in Table 12.3. Note that for periodic analytic functions the trapezoidal rule, in general, yields better approximations than Simpson's rule. □

Table 12.3 Numerical solution of (11.35) by the trapezoidal rule

	$2n$	$t = 0$	$t = \pi/2$	$t = \pi$
	4	-0.15350443	0.01354412	-0.00636277
$a = 1$	8	-0.00281745	0.00009601	-0.00004247
$b = 0.5$	16	-0.00000044	0.00000001	-0.00000001
	32	0.00000000	0.00000000	0.00000000
	4	-0.69224130	-0.06117951	-0.06216587
$a = 1$	8	-0.15017166	-0.00971695	-0.01174302
$b = 0.2$	16	-0.00602633	-0.00036043	-0.00045498
	32	-0.00000919	-0.00000055	-0.00000069

Example 12.11 is quite typical for the efficiency of the Nyström method with the composite trapezoidal rule for solving two-dimensional boundary value problems for the Laplace equation by using parameterized versions of the boundary integral equations analogous to the integral equation of Problem 6.1 for the interior Dirichlet problem. For analytic boundary curves and boundary data or for boundary curves of class C^{m+2}, in general, the trapezoidal rule yields an approximation order of the form (12.2) or (12.3), respectively. Hence, this method is extremely efficient, with the main advantage being its simplicity and its high approximation order. As a consequence of the latter, in general, the linear systems required for a sufficiently accurate approximation will be rather small, i.e., the number of quadrature points will be small. Therefore, despite the fact that they have full matrices, the linear

systems may be solved conveniently by elimination methods and do not require additional sophisticated efforts for its efficient solution.

For variants of the Nyström method for the two-dimensional double-layer potential integral equation for the Dirichlet problem in domains with corners, including an error analysis that mimics the existence approach as described in Section 6.5, we refer to Kress [138] and Kress, Sloan, and Stenger [152] (see also Atkinson [11]).

We confine ourselves to the above examples for Nyström's method for equations of the second kind with a continuous kernel. In particular, we refrain from presenting numerical examples for the use of the Gauss trapezoidal product rule (12.4) for integral equations of the second kind with continuous kernels on spheres. For a greater variety, the reader is referred to Anderssen et al. [4], Atkinson [11], Baker [14], Delves and Mohamed [42], Golberg and Chen [65], Hackbusch [76], and Prössdorf and Silbermann [200].

12.3 Weakly Singular Kernels

We will now describe the application of Nyström's method for the approximate solution of integral equations of the second kind with weakly singular kernels of the form

$$(A\varphi)(x) := \int_G w(|x-y|)K(x,y)\varphi(y)\,dy, \quad x \in G \subset \mathbb{R}^m. \tag{12.12}$$

Here, we assume the weight function $w : (0,\infty) \to \mathbb{R}$ to represent the weak singularity, i.e., w is continuous and satisfies $|w(t)| \leq Mt^{\alpha-m}$ for all $t > 0$ and some positive constants M and α. The remaining part K of the kernel is required to be continuous. We choose a sequence (Q_n) of quadrature rules

$$(Q_n g)(x) := \sum_{j=1}^n \alpha_j^{(n)}(x) g(x_j^{(n)}), \quad x \in G,$$

for the weighted integral

$$(Qg)(x) := \int_G w(|x-y|)g(y)\,dy, \quad x \in G,$$

with quadrature weights depending continuously on x. Then we approximate the weakly singular integral operator by a sequence of numerical integration operators

$$(A_n\varphi)(x) := \sum_{k=1}^n \alpha_k^{(n)}(x) K(x, x_k^{(n)})\varphi(x_k^{(n)}), \quad x \in G. \tag{12.13}$$

Appropriately modified, Theorem 12.7 for the solution of the approximating equation of the second kind, of course, remains valid. The linear system corresponding to (12.8) now assumes the form

$$\varphi_j^{(n)} - \sum_{k=1}^{n} \alpha_k^{(n)}(x_j) K(x_j, x_k) \varphi_k^{(n)} = f(x_j), \quad j = 1, \dots, n. \tag{12.14}$$

Due to the appearance of the weight function w in the quadrature Q, this form of Nyström's method for weakly singular kernels is also known as *product integration method*.

For the error analysis we will assume that the sequence (Q_n) of quadrature formulas converges, i.e., $Q_n g \to Qg$, $n \to \infty$, uniformly on G for all $g \in C(G)$. Then applying the Banach–Steinhaus theorem to the sequence of linear operators $Q_n : C(G) \to C(G)$, we observe that for convergence of the sequence (Q_n) it is necessary and sufficient that $Q_n g \to Qg$, $n \to \infty$, uniformly on G for all g in some dense subset of $C(G)$, and that there exists a constant C such that the weights satisfy

$$\sum_{k=1}^{n} |\alpha_k^{(n)}(x)| \le C \tag{12.15}$$

for all $n \in \mathbb{N}$ and all $x \in G$. Since the weights are not constant, this bound alone will not ensure collective compactness of the operator sequence (A_n). Therefore, we will also assume that the weights satisfy

$$\lim_{y \to x} \sup_{n \in \mathbb{N}} \sum_{k=1}^{n} |\alpha_k^{(n)}(y) - \alpha_k^{(n)}(x)| = 0 \tag{12.16}$$

uniformly for all $x \in G$. Then, writing

$$(A_n\varphi)(x_1) - (A_n\varphi)(x_2) = \sum_{k=1}^{n} \alpha_k^{(n)}(x_1)\{K(x_1, x_k^{(n)}) - K(x_2, x_k^{(n)})\}\varphi(x_k^{(n)})$$

$$+ \sum_{k=1}^{n} \{\alpha_k^{(n)}(x_1) - \alpha_k^{(n)}(x_2)\} K(x_2, x_k^{(n)}) \varphi(x_k^{(n)})$$

we can estimate

$$|(A_n\varphi)(x_1) - (A_n\varphi)(x_2)| \le C \max_{y \in G} |K(x_1, y) - K(x_2, y)| \, \|\varphi\|_\infty$$

$$+ \max_{x,y \in G} |K(x, y)| \sup_{n \in \mathbb{N}} \sum_{k=1}^{n} |\alpha_k^{(n)}(x_1) - \alpha_k^{(n)}(x_2)| \, \|\varphi\|_\infty.$$

Hence, analogous to Theorem 12.8, we can prove the following result.

Theorem 12.12. *Assume that the quadrature formulas (Q_n) converge and satisfy the condition (12.16). Then the sequence (A_n), given by (12.13), is collectively compact and pointwise convergent $A_n\varphi \to A\varphi$, $n \to \infty$, for all $\varphi \in C(G)$, but not norm convergent.*

For a systematic study of approximations satisfying condition (12.16) we refer to Sloan [222]. We confine ourselves to a special case by considering a weakly singular operator with a logarithmic singularity

$$(A\varphi)(t) := \frac{1}{2\pi} \int_0^{2\pi} \ln\left(4 \sin^2 \frac{t-\tau}{2}\right) K(t,\tau)\varphi(\tau)\, d\tau, \quad 0 \le t \le 2\pi, \tag{12.17}$$

in the space $C_{2\pi} \subset C(\mathbb{R})$ of 2π-periodic continuous functions. The kernel function K is assumed to be continuous and 2π-periodic with respect to both variables. The fundamental solutions to elliptic partial differential equations, for example the Laplace or the Helmholtz equation, in two space dimensions contain a logarithmic singularity. Therefore, the boundary integral equation approach to two-dimensional boundary value problems via a periodic parameterization of the boundary curve, in general, leads to such logarithmic singularities (see also Problem 7.2).

According to the general ideas outlined in Section 12.1 and following Kussmaul [154] and Martensen [161], we construct numerical quadratures for the improper integral

$$(Qg)(t) := \frac{1}{2\pi} \int_0^{2\pi} \ln\left(4 \sin^2 \frac{t-\tau}{2}\right) g(\tau)\, d\tau, \quad t \in [0, 2\pi],$$

by replacing the continuous periodic function g by its trigonometric interpolation polynomial described in Section 11.3. Using the Lagrange basis we obtain

$$(Q_n g)(t) = \sum_{j=0}^{2n-1} R_j^{(n)}(t) g(t_j), \quad t \in [0, 2\pi], \tag{12.18}$$

with the equidistant quadrature points $t_j = j\pi/n$ and the quadrature weights

$$R_j^{(n)}(t) = \frac{1}{2\pi} \int_0^{2\pi} \ln\left(4 \sin^2 \frac{t-\tau}{2}\right) L_j(\tau)\, d\tau, \quad j = 0, \ldots, 2n-1. \tag{12.19}$$

Using Lemma 8.23, from the form (11.12) of the Lagrange basis we derive

$$R_j^{(n)}(t) = -\frac{1}{n} \left\{ \sum_{m=1}^{n-1} \frac{1}{m} \cos m(t - t_j) + \frac{1}{2n} \cos n(t - t_j) \right\} \tag{12.20}$$

for $j = 0, \ldots, 2n-1$.

This quadrature is uniformly convergent for all trigonometric polynomials, since by construction Q_n integrates trigonometric polynomials of degree less than or equal to n exactly. We will now establish convergence for all 2π-periodic continuous functions and the validity of (12.16).

Using the integrals in Lemma 8.23 and Parseval's equality it can be seen that for each $t \in [0, 2\pi]$ the function

$$f_t(\tau) := \frac{1}{2\pi} \ln\left(4\sin^2\frac{t-\tau}{2}\right), \quad \tau \in \mathbb{R}, \quad \frac{t-\tau}{2\pi} \notin \mathbb{Z},$$

belongs to $L^2[0, 2\pi]$ with

$$\|f_t\|_2^2 = \frac{1}{\pi}\sum_{m=1}^{\infty}\frac{1}{m^2} = \frac{\pi}{6}.$$

For fixed $t \in [0, 2\pi]$ we choose a 2π-periodic continuous function g with $\|g\|_\infty = 1$ satisfying $g(t_j) = 1$ if $R_j^{(n)}(t) \geq 0$ and $g(t_j) = -1$ if $R_j^{(n)}(t) < 0$. Then, in view of (12.19), we can write

$$\sum_{j=0}^{2n-1} |R_j^{(n)}(t)| = \sum_{j=0}^{2n-1} R_j^{(n)}(t) g(t_j) = \frac{1}{2\pi}\int_0^{2\pi} \ln\left(4\sin^2\frac{t-\tau}{2}\right)(P_n g)(\tau)\,d\tau.$$

From this, using Lemma 11.5 and the Cauchy–Schwarz inequality, we can estimate

$$\sum_{j=0}^{2n-1} |R_j^{(n)}(t)| \leq \|f_t\|_2\,\|P_n g\|_2 \leq \sqrt{3\pi}\,\|f_t\|_2 = \frac{\pi}{\sqrt{2}},$$

and this establishes the uniform boundedness (12.15) for the weights (12.20). Similarly,

$$\sum_{j=0}^{2n-1} |R_j^{(n)}(t_1) - R_j^{(n)}(t_2)| \leq \sqrt{3\pi}\,\|f_{t_1} - f_{t_2}\|_2,$$

and again from (8.23) and Parseval's equality we see that

$$\|f_{t_1} - f_{t_2}\|_2^2 = \frac{1}{2\pi}\sum_{\substack{m=-\infty \\ m\neq 0}}^{\infty} \frac{1}{m^2}\left|e^{imt_1} - e^{imt_2}\right|^2,$$

whence the validity of condition (12.16) for the weights (12.20) follows. Hence, for the sequence

$$(A_n\varphi)(t) := \sum_{j=0}^{2n-1} R_j^{(n)}(t) K(t, t_j)\varphi(t_j) \tag{12.21}$$

generated by the quadratures (12.18) we can state the following theorem.

Theorem 12.13. *The sequence (A_n) given by (12.21) is collectively compact and pointwise convergent to the integral operator A with logarithmic singularity given by (12.17).*

Therefore our general convergence and error analysis based on Theorems 10.12 and 10.13 is applicable. By Theorem 11.7, in the case of analytic functions the quadrature error is exponentially decreasing. By the estimate (10.11), this behavior is inherited by the approximate solution to the integral equation provided that the kernel K and the exact solution φ are analytic (see Problem 12.4). The matrix entering into (12.14) is a circulant matrix, i.e.,

$$R_k^{(n)}(t_j) = R_{k-j}^{(n)}, \quad j, k = 0, \dots, 2n-1,$$

with the weights

$$R_j^{(n)} := -\frac{1}{n}\left\{\sum_{m=1}^{n-1} \frac{1}{m} \cos\frac{mj\pi}{n} + \frac{(-1)^j}{2n}\right\}, \quad j = 0, \pm 1 \dots, \pm(2n-1). \tag{12.22}$$

Example 12.14. We consider the Neumann boundary value problem for the reduced wave equation or Helmholtz equation

$$\Delta u + \kappa^2 u = 0$$

with a positive wave number κ in the open unit disk D with given normal derivative

$$\frac{\partial u}{\partial \nu} = g \quad \text{on } \partial D.$$

We assume the unit normal ν to be directed into the exterior of D. The fundamental solution to the two-dimensional Helmholtz equation is given by

$$\Phi(x, y) := \frac{i}{4}\, H_0^{(1)}(\kappa|x-y|), \quad x \neq y,$$

where $H_0^{(1)}$ denotes the Hankel function of the first kind and of order zero. We decompose

$$H_0^{(1)} = J_0 + i\, Y_0,$$

where J_0 and Y_0 are the Bessel and Neumann functions of order zero, and note the power series

$$J_0(z) = \sum_{k=0}^{\infty} \frac{(-1)^k}{(k!)^2}\left(\frac{z}{2}\right)^{2k} \tag{12.23}$$

and

$$Y_0(z) = \frac{2}{\pi}\left(\ln\frac{z}{2} + C\right) J_0(z) - \frac{2}{\pi}\sum_{k=1}^{\infty} a_k \frac{(-1)^k}{(k!)^2}\left(\frac{z}{2}\right)^{2k} \tag{12.24}$$

with $a_k = \sum_{m=1}^{k} 1/m$ and Euler's constant $C = 0.57721\dots$. From these expansions we deduce the asymptotics

$$\Phi(x, y) = \frac{1}{2\pi} \ln\frac{1}{|x-y|} + \frac{i}{4} - \frac{1}{2\pi}\left(\ln\frac{\kappa}{2} + C\right) + O\left(|x-y|^2 \ln\frac{1}{|x-y|}\right)$$

for $|x - y| \to 0$. Therefore, the fundamental solution to the Helmholtz equation has the same singular behavior as the fundamental solution of Laplace's equation. As a consequence, Green's representation Theorem 6.5 and the potential theoretic jump relations of Theorems 6.15 and 6.18 can be carried over to the Helmholtz equation. For details we refer to [31]. In particular, it can be shown that the unknown boundary values $\varphi = u$ on ∂D of the solution u satisfy the integral equation

$$\varphi(x) + 2 \int_{\partial D} \varphi(y) \frac{\partial \Phi(x, y)}{\partial \nu(y)} \, ds(y) = 2 \int_{\partial D} g(y) \Phi(x, y) \, ds(y) \tag{12.25}$$

for $x \in \partial D$. With the exception of a countable set of wave numbers κ accumulating only at infinity, for which the homogeneous Neumann problem admits nontrivial solutions, this integral equation is uniquely solvable.

We use the representation $z(t) = (\cos t, \sin t)$, $0 \le t \le 2\pi$, for the unit circle and, by straightforward calculations, transform the integral equation (12.25) into the parametric form

$$\widetilde{\varphi}(t) - \frac{1}{2\pi} \int_0^{2\pi} K(t, \tau) \widetilde{\varphi}(\tau) \, d\tau = \frac{1}{2\pi} \int_0^{2\pi} L(t, \tau) \widetilde{g}(\tau) \, d\tau \tag{12.26}$$

for $0 \le t \le 2\pi$. Here we have set $\widetilde{\varphi}(t) := \varphi(z(t))$ and $\widetilde{g}(t) := g(z(t))$, and the kernels are given by

$$K(t, \tau) := i\pi\kappa \left| \sin \frac{t - \tau}{2} \right| H_1^{(1)} \left(2\kappa \left| \sin \frac{t - \tau}{2} \right| \right)$$

and

$$L(t, \tau) := i\pi \, H_0^{(1)} \left(2\kappa \left| \sin \frac{t - \tau}{2} \right| \right)$$

for $t \neq \tau$. For computing the normal derivative of the fundamental solution we used the relation

$$H_0^{(1)\prime} = -H_1^{(1)} = -J_1 - i \, Y_1,$$

where $H_1^{(1)}$ denotes the Hankel function of the first kind and of order one represented in terms of the Bessel and Neumann functions J_1 and Y_1 of order one.

Observing the expansions (12.23) and (12.24) and their term-by-term derivatives, we can split the kernels into

$$K(t, \tau) = K_1(t, \tau) \ln \left(4 \sin^2 \frac{t - \tau}{2} \right) + K_2(t, \tau) \tag{12.27}$$

and

$$L(t, \tau) = L_1(t, \tau) \ln \left(4 \sin^2 \frac{t - \tau}{2} \right) + L_2(t, \tau), \tag{12.28}$$

where

$$K_1(t,\tau) := -\kappa \sin\frac{t-\tau}{2}\, J_1\left(2\kappa \sin\frac{t-\tau}{2}\right),$$

$$K_2(t,\tau) := K(t,\tau) - K_1(t,\tau)\ln\left(4\sin^2\frac{t-\tau}{2}\right),$$

$$L_1(t,\tau) := -J_0\left(2\kappa \sin\frac{t-\tau}{2}\right),$$

$$L_2(t,\tau) := L(t,\tau) - L_1(t,\tau)\ln\left(4\sin^2\frac{t-\tau}{2}\right).$$

The kernels K_1, K_2, L_1, and L_2 turn out to be analytic. In particular, from

$$\lim_{z\to 0} z\, H_1^{(1)}(z) = \frac{2}{\pi i}$$

and

$$\lim_{z\to 0}\left(H_0^{(1)}(z) - \frac{2i}{\pi}\ln z\, J_0(z)\right) = \frac{2i}{\pi}(C - \ln 2) + 1,$$

we deduce the diagonal terms

$$K_2(t,t) = K(t,t) = 1$$

and

$$L_2(t,t) = -2\ln\frac{\kappa}{2} - 2C + i\pi$$

for $0 \le t \le 2\pi$. Note that despite the continuity of the kernel K, for numerical accuracy it is advantageous to incorporate the quadrature (12.18) because of the logarithmic singularities of the derivatives of the kernel K. Now, in the spirit of Theorem 12.7, we approximate the integral equation (12.26) by the linear system

$$\widetilde{\varphi}_j - \sum_{k=0}^{2n-1}\left\{R_{k-j}^{(n)} K_1(t_j,t_k) + \frac{1}{2n}\, K_2(t_j,t_k)\right\}\widetilde{\varphi}_k$$

$$= \sum_{k=0}^{2n-1}\left\{R_{k-j}^{(n)} L_1(t_j,t_k) + \frac{1}{2n}\, L_2(t_j,t_k)\right\}\widetilde{g}(t_k), \quad j = 0,\dots,2n-1,$$

for approximating values $\widetilde{\varphi}_j$ for $\widetilde{\varphi}(t_j)$.

For a numerical example we consider the case in which the exact solution to the Neumann problem is given by $u(x) = Y_0(\kappa|x - x_0|)$, where $x_0 = (q, 0)$ with $q > 1$. Then the Neumann boundary data are given by

$$\widetilde{g}(t) = -\kappa\,\frac{1 - q\cos t}{v(t)}\, Y_1(\kappa v(t))$$

and the exact solution of the integral equation by $\widetilde{\varphi}(t) = Y_0(\kappa v(t))$, where

$$v^2(t) = 1 + q^2 - 2q\cos t.$$

The numerical results contained in Table 12.4 confirm the exponentially decreasing behavior of the error we expect from our general error analysis. □

Table 12.4 Numerical results for Example 12.14

	$2n$	$t = 0$	$t = \pi/2$	$t = \pi$
	4	−0.07907728	0.12761991	0.24102137
$\kappa = 1$	8	−0.01306333	0.00867450	0.01155067
$q = 2$	16	−0.00023494	0.00003924	0.00004517
	32	−0.00000019	0.00000001	0.00000000
	4	−0.13910590	0.39499045	0.68563472
$\kappa = 1$	8	−0.07111636	0.07511294	0.10540051
$q = 1.5$	16	−0.00659502	0.00277753	0.00386074
	32	−0.00005924	0.00000558	0.00000616
	8	0.35729406	0.21301358	−0.16596385
$\kappa = 5$	16	0.01138634	0.00181974	0.00377600
$q = 2$	32	−0.00000558	−0.00000006	−0.00000010
	64	0.00000003	0.00000002	0.00000000
	8	−0.54146680	−0.35298932	0.09706015
$\kappa = 5$	16	−0.05669554	−0.02916764	−0.02480085
$q = 1.5$	32	−0.00021246	−0.00000877	−0.00001332
	64	−0.00000006	0.00000000	0.00000003

As at the end of Section 12.2, we wish to point out that Example 12.14 is typical for the efficiency of the Nyström method based on trigonometric interpolatory quadratures for logarithmic singularities applied to boundary value problems for the Helmholtz equation in two dimensions. For details we refer to [32] (see also Problem 12.5). There is no immediate extension of this approach to three dimensions since as opposed to the logarithmic singularity of the fundamental solutions to the Laplace and Helmholtz equation in $\mathbb{R}^3$ the square root singularities in the three-dimensional case cannot be split off in the same elegant manner. However, it is possible to use the above technique in three dimensions for rotationally symmetric boundaries, see [137, 161].

12.4 Nyström's Method in Sobolev Spaces

In this final section we wish to illustrate that the Nyström method can also be analyzed in the Sobolev spaces $H^p[0, 2\pi]$ and in Hölder spaces. We confine our presentation to the weakly singular integral operator with logarithmic singularity defined by (12.17) and begin by investigating its mapping properties in Sobolev spaces.

Theorem 12.15. *Assume that K is infinitely differentiable and 2π-periodic with respect to both variables. Then the operator A given by (12.17) is bounded from $H^p[0, 2\pi]$ into $H^{p+1}[0, 2\pi]$ for all $p \geq 0$.*

Proof. We write

$$K(t, \tau) = \sum_{m=-\infty}^{\infty} \hat{K}_m(t) e^{im\tau}, \tag{12.29}$$

where

$$\hat{K}_m(t) := \frac{1}{2\pi} \int_0^{2\pi} K(t, \tau) e^{-im\tau} d\tau$$

denote the Fourier coefficients of $K(t, \cdot)$. The infinite differentiability of K implies that

$$\sup_{m \in \mathbb{Z}} |m|^p \|\hat{K}_m^{(k)}\|_\infty < \infty \tag{12.30}$$

for all $k \in \mathbb{N} \cup \{0\}$ and all $p \geq 0$ (see (8.6)). Therefore, the series (12.29) and its term-by-term derivatives of arbitrary order converge uniformly. Recall the trigonometric monomials $f_m(t) = e^{imt}$. Then, for trigonometric polynomials φ we have

$$(A\varphi)(t) = \sum_{m=-\infty}^{\infty} \hat{K}_m(t) \, (A_0(\varphi f_m))(t),$$

where

$$(A_0\varphi)(t) := \frac{1}{2\pi} \int_0^{2\pi} \ln\left(4 \sin^2 \frac{t - \tau}{2}\right) \varphi(\tau) \, d\tau, \quad 0 \leq t \leq 2\pi. \tag{12.31}$$

By Theorem 8.24 the operator A_0 is bounded from $H^p[0, 2\pi]$ into $H^{p+1}[0, 2\pi]$ for all $p \in \mathbb{R}$. Using Corollary 8.8 we can estimate

$$\|A\varphi\|_{p+1} \leq C_1 \sum_{m=-\infty}^{\infty} \left[\|\hat{K}_m^{(k)}\|_\infty + \|\hat{K}_m\|_\infty\right] \|A_0(\varphi f_m)\|_{p+1},$$

where k is the smallest integer larger than or equal to $p + 1$ and C_1 is some constant depending on p. The boundedness of A_0 and again Corollary 8.8 yield that

$$\|A_0(\varphi f_m)\|_{p+1} \leq C_2 \|\varphi f_m\|_p \leq C_3 |m|^k \|\varphi\|_p$$

for all $m \in \mathbb{Z}$ and some constants C_2 and C_3 depending on p. The last two inequalities, together with (12.30), imply that

$$\|A\varphi\|_{p+1} \leq C\|\varphi\|_p$$

for all trigonometric polynomials φ and some constant C depending on p. Now, in view of Theorem 8.2, the proof is complete. □

Corollary 12.16. *Under the assumptions of Theorem 12.15, the integral operator $A : H^p[0, 2\pi] \to H^p[0, 2\pi]$ is compact for all $p \geq 0$.*

Proof. This follows from the imbedding Theorem 8.3. □

Corollary 12.16 allows the application of the Riesz Theorem 3.4. In view of Theorem 8.4 and the smoothing property of A from Theorem 12.15, the inverse operator $(I - A)^{-1} : H^p[0, 2\pi] \to H^p[0, 2\pi]$ exists and is bounded if and only if the homogeneous equation $\varphi - A\varphi = 0$ has only the trivial continuous solution $\varphi = 0$.

The operator A_0 corresponds to the single-layer potential operator for the unit disk. Therefore, by Theorem 7.41 the operator A_0 is bounded from the Hölder space $C^{0,\alpha}_{2\pi}$ into $C^{1,\alpha}_{2\pi}$, and proceeding as in the proof of Theorem 12.15 we can establish the following result.

Theorem 12.17. *Under the assumptions of Theorem 12.15 the operator A is bounded from $C^{0,\alpha}_{2\pi}$ into $C^{1,\alpha}_{2\pi}$ for all $0 < \alpha < 1$.*

Theorem 12.18. *Assume that $0 \leq q \leq p$ and $p > 1/2$. Then under the assumptions of Theorem 12.15 for the operator sequence (A_n) given by (12.21) we have the estimate*

$$\|A_n\varphi - A\varphi\|_{q+1} \leq Cn^{q-p}\|\varphi\|_p \tag{12.32}$$

for all $\varphi \in H^p[0, 2\pi]$ and some constant C depending on p and q.

Proof. As in the proof of Theorem 12.15 we can write

$$(A_n\varphi)(t) - (A\varphi)(t) = \sum_{m=-\infty}^{\infty} \hat{K}_m(t)\,(A_0[P_n(\varphi f_m) - \varphi f_m])(t)$$

and estimate analogously. Using the boundedness of A_0, Theorem 11.8, and Corollary 8.8, we obtain

$$\begin{aligned}\|A_0[P_n(\varphi f_m) - \varphi f_m]\|_{q+1} &\leq c_1\|P_n(\varphi f_m) - \varphi f_m\|_q \\ &\leq c_2 n^{q-p}\|\varphi f_m\|_p \leq c_3|m|^{p+1}n^{q-p}\|\varphi\|_p\end{aligned}$$

for all $m \in \mathbb{Z}$ and some constants c_1, c_2, and c_3 depending on p and q. Now the proof can be completed as in Theorem 12.15. □

Corollary 12.19. *Under the assumptions of Theorem 12.15, for the operators $A_n : H^p[0, 2\pi] \to H^p[0, 2\pi]$, we have norm convergence $\|A_n - A\|_p \to 0$, $n \to \infty$, for all $p \geq 1$.*

Proof. Setting $q = p - 1$ in (12.32) yields

$$\|A_n\varphi - A\varphi\|_p \leq \frac{C}{n}\,\|\varphi\|_p$$

for all $\varphi \in H^p[0, 2\pi]$ and all $n \in \mathbb{N}$. This establishes the assertion. □

Now we are in the position to apply Theorem 10.1 and the error estimate (10.3) for the approximate solution of $\varphi - A\varphi = f$ by $\varphi_n - A_n\varphi_n = f_n$ in the Sobolev spaces $H^p[0, 2\pi]$ for $p \geq 1$. We note that because of the imbedding Theorem 8.4, convergence of the approximate solutions in the Sobolev space $H^p[0, 2\pi]$ for $p > 1/2$ also implies uniform convergence. As opposed to the case of the convergence and error analysis in the space of continuous functions, due to the regularity properties in the Sobolev spaces, here we do not rely on collective compactness. We leave it as an exercise to the reader to extend this analysis to the integral operator

$$(A\varphi)(t) = \int_0^{2\pi} K(t, \tau)\varphi(\tau)\,d\tau$$

with infinitely differentiable 2π-periodic kernel K and its approximations

$$(A_n\varphi)(t) = \frac{\pi}{n}\sum_{k=0}^{2n-1} K(t, t_k)\varphi(t_k)$$

by the composite trapezoidal rule. We also leave it as an exercise to carry Theorem 12.18 and its Corollary 12.19 over to Hölder spaces as in Theorem 12.17 by using the error estimate (11.20).

For an error analysis of a Nyström method for Cauchy-type singular integral equations in Sobolev spaces, which imitates the regularization of singular equations considered in Chapter 7, we refer to Kirsch and Ritter [130].

Problems

12.1. Let X and Y be normed spaces, let $U \subset X$ be compact, and let (φ_n) be an equicontinuous sequence of functions $\varphi_n : U \to Y$ converging pointwise on U to some function $\varphi : U \to Y$. Then the sequence (φ_n) converges uniformly on U.

12.2. Let $D = \mathbb{R} \times (-s, s)$ denote a strip in the complex plane. Show that the space $H(D)$ of 2π-periodic holomorphic functions defined on D with the property

$$\lim_{\sigma \to s} \int_0^{2\pi} \{|f(t + i\sigma)|^2 + |f(t - i\sigma)|^2\}\,dt < \infty$$

is a Hilbert space with the scalar product

$$(f, g) := \lim_{\sigma \to s} \int_0^{2\pi} \{f(t + i\sigma)\overline{g(t + i\sigma)} + f(t - i\sigma)\overline{g(t - i\sigma)}\}\,dt.$$

Show that the functions

$$f_n(t) = \frac{1}{(4\pi \cosh 2ns)^{1/2}}\, e^{int}, \quad n \in \mathbb{Z},$$

form a complete orthonormal system in $H(D)$. Use Cauchy's integral formula to verify that the remainder for the trapezoidal rule is a bounded linear functional on $H(D)$. Apply the Riesz representation Theorem 4.10 and Parseval's equality from Theorem 1.28 to derive the error bound

$$|R_n(g)| \le \frac{2M}{(e^{4ns} - 1)^{1/2}}$$

which is slightly better than the estimate given in Theorem 12.6 (see [47] and [134]).

12.3. Derive bounds on $\|(A_n - A)A\|_\infty$ and $\|(A_n - A)A_n\|_\infty$ for the numerical integration operators using the trapezoidal rule.

12.4. Let $D = \mathbb{R} \times (-s, s)$ denote a strip in the complex plane. Consider the Banach space $B(D)$ of 2π-periodic continuous functions defined on $\bar{D}$ that are holomorphic in D, furnished with the maximum norm

$$\|f\|_{\infty,D} := \max_{t \in \bar{D}} |f(t)|.$$

Prove that under suitable assumptions on K the integral operator given by (12.17) is compact in $B(D)$. Use this result to deduce that solutions to integral equations of the second kind with the logarithmic kernel of (12.17) are analytic provided the kernel K and the right-hand side are analytic.
Hint: By the Arzelà–Ascoli theorem, show that integral operators with analytic kernels are compact in $B(D)$. Use

$$(A\varphi)(t) := \frac{1}{2\pi} \int_0^{2\pi} \ln\left(4 \sin^2 \frac{s}{2}\right) K(t, t+s)\varphi(t+s)\, ds$$

for the definition of A on $B(D)$. Approximate $\ln \sin^2(s/2)$ by its Fourier series and use Theorem 2.22 and the Cauchy–Schwarz inequality.

12.5. Use a regular 2π-periodic parameterization $\partial D = \{x(t) : 0 \le t \le 2\pi\}$ with counterclockwise orientation for an arbitrary boundary curve ∂D to derive a parametric form of the integral equation (12.25) for the Neumann problem for the reduced wave equation corresponding to (12.26). Describe an appropriate treatment of the logarithmic singularity as in Example 12.14 (see [32, 135, 140, 154] for related integral equations).

Chapter 13
Projection Methods

The application of the quadrature method, in principle, is confined to equations of the second kind. To develop numerical methods that can also be used for equations of the first kind we will describe projection methods as a general tool for approximately solving linear operator equations. After introducing into the principal ideas of projection methods and their convergence and error analysis we shall consider collocation and Galerkin methods as special cases. We do not intend to give a complete account of the numerous implementations of collocation and Galerkin methods for integral equations that have been developed in the literature. Our presentation is meant as an introduction to these methods by studying their basic concepts and describing their numerical performance through a few typical examples.

For a more exhaustive study of projection methods in general and of collocation and Galerkin methods, we refer to Atkinson [11], Baker [14], Fenyö and Stolle [52], Hackbusch [76], Krasnoselski et al. [132], Mikhlin and Prössdorf [170], Prössdorf and Silbermann [199, 200], Rjasanow and Steinbach [207], Saranen and Vainikko [213], Sauter and Schwab [214] and Steinbach [225]. For an introduction to projection methods for boundary integral equations with an extensive bibliography, we refer to the review paper by Sloan [223].

13.1 The Projection Method

We describe the approximate solution of linear operator equations by projecting them onto subspaces, which for practical calculations we assume to be finite-dimensional.

Definition 13.1. Let X be a normed space and $U \subset X$ a nontrivial subspace. A bounded linear operator $P : X \to U$ with the property $P\varphi = \varphi$ for all $\varphi \in U$ is called a *projection operator* from X onto U.

R. Kress, *Linear Integral Equations*, Applied Mathematical Sciences 82,
DOI 10.1007/978-1-4614-9593-2_13, © Springer Science+Business Media New York 2014

Theorem 13.2. *A nontrivial bounded linear operator P mapping a normed space X into itself is a projection operator if and only if $P^2 = P$. Projection operators satisfy $\|P\| \geq 1$.*

Proof. Let $P : X \to U$ be a projection operator. Then, from $P\varphi \in U$ it follows that $P^2\varphi = P(P\varphi) = P\varphi$ for all $\varphi \in X$. Conversely, let $P^2 = P$ and set $U := P(X)$. Then, for all $\varphi \in U$ we may write $\varphi = P\psi$ for some $\psi \in X$ and obtain $P\varphi = \varphi$. Finally $P = P^2$, by Remark 2.5, implies $\|P\| \leq \|P\|^2$, whence $\|P\| \geq 1$. □

An important example for projection operators is given by the so-called orthogonal projection, i.e., by the best approximation in pre-Hilbert spaces in the sense of Theorem 1.26.

Theorem 13.3. *Let U be a nontrivial complete subspace of a pre-Hilbert space X. Then the operator P mapping each element $\varphi \in X$ into its unique best approximation with respect to U is a projection operator. It is called* orthogonal projection *onto U and satisfies $\|P\| = 1$.*

Proof. Trivially, we have $P\varphi = \varphi$ for all $\varphi \in U$. From the orthogonality condition of Theorem 1.25 for the best approximation in pre-Hilbert spaces we readily verify that P is linear and

$$\|\varphi\|^2 = \|P\varphi + (\varphi - P\varphi)\|^2 = \|P\varphi\|^2 + \|\varphi - P\varphi\|^2 \geq \|P\varphi\|^2$$

for all $\varphi \in X$. Hence $\|P\| \leq 1$, and Theorem 13.2 yields $\|P\| = 1$. □

A second important example for projection operators is given by interpolation operators.

Theorem 13.4. *The interpolation operator introduced in Theorem 11.2 is a projection operator.*

Note that for polynomial or trigonometric polynomial interpolation, because of the general non-convergence results due to Faber (see [182, 217]) combined with the uniform boundedness principle Theorem 10.6, the interpolation operators are not uniformly bounded with respect to the maximum norm (see also the estimate (11.15)).

Definition 13.5. Let X and Y be Banach spaces and let $A : X \to Y$ be an injective bounded linear operator. Let $X_n \subset X$ and $Y_n \subset Y$ be two sequences of subspaces with $\dim X_n = \dim Y_n = n$ and let $P_n : Y \to Y_n$ be projection operators. Given $f \in Y$, the *projection method*, generated by X_n and P_n, approximates the equation

$$A\varphi = f \tag{13.1}$$

for $\varphi \in X$ by the projected equation

$$P_n A\varphi_n = P_n f \tag{13.2}$$

for $\varphi_n \in X_n$. This projection method is called *convergent* for the operator A if there exists $n_0 \in \mathbb{N}$ such that for each $f \in A(X)$ the approximating equation (13.2) has a unique solution $\varphi_n \in X_n$ for all $n \geq n_0$ and these solutions converge $\varphi_n \to \varphi$, $n \to \infty$, to the unique solution φ of $A\varphi = f$.

In terms of operators, convergence of the projection method means that for all $n \geq n_0$ the finite-dimensional operators $P_nA : X_n \to Y_n$ are invertible and we have pointwise convergence $(P_nA)^{-1}P_nA\varphi \to \varphi$, $n \to \infty$, for all $\varphi \in X$. In general, we can expect convergence only if the subspaces X_n possess the denseness property

$$\inf_{\psi \in X_n} \|\psi - \varphi\| \to 0, \quad n \to \infty, \tag{13.3}$$

for all $\varphi \in X$. Therefore, in the subsequent analysis of this chapter we will always assume that this condition is fulfilled.

Since $P_nA : X_n \to Y_n$ is a linear operator between two finite-dimensional spaces, carrying out the projection method reduces to solving a finite-dimensional linear system. In the following sections we shall describe the collocation and the Galerkin method as projection methods in the sense of this definition obtained via interpolation and orthogonal projection operators, respectively. Here, we first proceed with a general convergence and error analysis.

Theorem 13.6. *A projection method with projection operators $P_n : Y \to Y_n$ converges for an injective linear operator $A : X \to Y$ from a Banach space X into a Banach space Y if and only if there exist $n_0 \in \mathbb{N}$ and a positive constant M such that for all $n \geq n_0$ the finite-dimensional operators*

$$P_nA : X_n \to Y_n$$

are invertible and the operators $(P_nA)^{-1}P_nA : X \to X_n$ are uniformly bounded, i.e.,

$$\|(P_nA)^{-1}P_nA\| \leq M \tag{13.4}$$

for all $n \geq n_0$ and some constant M. In case of convergence we have the error estimate

$$\|\varphi_n - \varphi\| \leq (1 + M) \inf_{\psi \in X_n} \|\psi - \varphi\|. \tag{13.5}$$

Proof. Provided the projection method converges for the operator A, the uniform boundedness (13.4) is a consequence of Theorem 10.6. Conversely, if the assumptions of the theorem are satisfied we can write

$$\varphi_n - \varphi = [(P_nA)^{-1}P_nA - I]\varphi.$$

Since for all $\psi \in X_n$, trivially, we have $(P_nA)^{-1}P_nA\psi = \psi$, it follows that

$$\varphi_n - \varphi = [(P_nA)^{-1}P_nA - I](\varphi - \psi).$$

Hence, we have the error estimate (13.5) and, with the aid of the denseness (13.3), the convergence follows. □

The uniform boundedness condition (13.4) is also known as the *stability condition* for the projection method. The error estimate (13.5) of Theorem 13.6 is usually referred to as *Céa's lemma* (see [26]). It indicates that the error in the projection method is determined by how well the exact solution can be approximated by elements of the subspace X_n.

We now state the main stability property of the projection method with respect to perturbations of the operator.

Theorem 13.7. *Assume that $A : X \to Y$ is a bijective bounded linear operator from a Banach space X into a Banach space Y and that the projection method generated by the projection operators $P_n : Y \to Y_n$ is convergent for A. Let $B : X \to Y$ be a bounded linear operator such that either*
(a) $\sup_{\varphi \in X_n, \|\varphi\|=1} \|P_n B\varphi\|$ *is sufficiently small for all sufficiently large n, or*
(b) *B is compact and $A + B$ is injective.*
Then the projection method with the projection operators P_n also converges for $A + B$.

Proof. The operator A satisfies the conditions of Theorem 13.6, i.e., there exists $n_0 \in \mathbb{N}$ such that for all $n \geq n_0$ the operators $P_n A : X_n \to Y_n$ are invertible and satisfy $\|(P_n A)^{-1} P_n A\| \leq M$ for some constant M. By the open mapping Theorem 10.8 the inverse operator $A^{-1} : Y \to X$ is bounded. We will show that for sufficiently large n the inverse operators of $I + (P_n A)^{-1} P_n B : X_n \to X_n$ exist and are uniformly bounded if (a) or (b) is satisfied.

In case (a), assume that $M \|A^{-1}\| \sup_{\varphi \in X_n, \|\varphi\|=1} \|P_n B\varphi\| \leq q$ for all $n \geq \tilde{n}_0$ with some $\tilde{n}_0 \geq n_0$ and some $q < 1$. Then writing

$$(P_n A)^{-1} P_n B = (P_n A)^{-1} P_n^2 B = (P_n A)^{-1} P_n A A^{-1} P_n B$$

we can estimate $\|(P_n A)^{-1} P_n B\| \leq q$ for all $n \geq n_0$. Therefore by Theorem 2.14, the inverse operators $[I + (P_n A)^{-1} P_n B]^{-1} : X_n \to X_n$ exist and are uniformly bounded for all $n \geq \tilde{n}_0$.

In case (b), by the Riesz theory $I + A^{-1}B : X \to X$ has a bounded inverse, since $A^{-1}B$ is compact. Since A^{-1} is bounded, the pointwise convergence of the sequence $(P_n A)^{-1} P_n A \to I$, $n \to \infty$, on X implies pointwise convergence $(P_n A)^{-1} P_n \to A^{-1}$, $n \to \infty$, on Y. From the pointwise convergence of $(P_n A)^{-1} P_n$ and the compactness of B, by Theorem 10.10, we derive norm convergence

$$\|[I + A^{-1}B] - [I + (P_n A)^{-1} P_n B]\| \to 0, \quad n \to \infty.$$

Therefore, by Theorem 10.1, the inverse $[I + (P_n A)^{-1} P_n B]^{-1} : X_n \to X_n$ exists and is uniformly bounded for all sufficiently large n.

We now set $S := A + B$. Then from

$$P_n S = P_n A [I + (P_n A)^{-1} P_n B]$$

it follows that $P_nS : X_n \to Y_n$ is invertible for sufficiently large n with the inverse given by

$$(P_nS)^{-1} = [I + (P_nA)^{-1}P_nB]^{-1}(P_nA)^{-1}.$$

From $(P_nS)^{-1}P_nS = [I + (P_nA)^{-1}P_nB]^{-1}(P_nA)^{-1}P_nA(I + A^{-1}B)$ we can estimate

$$\|(P_nS)^{-1}P_nS\| \leq \|[I + (P_nA)^{-1}P_nB]^{-1}\|\, M\, \|I + A^{-1}B\|$$

and observe that the condition (13.4) is satisfied for S. This completes the proof. □

Because, in general, projection methods are only semi-discrete methods, in actual numerical calculations instead of (13.2) an approximate version of the form

$$P_nA_n\widetilde{\varphi}_n = P_nf_n \tag{13.6}$$

will usually be solved, where A_n is some approximation to A and f_n approximates f. For this we can state the following result.

Theorem 13.8. *Assume that $A : X \to Y$ is a bijective bounded linear operator from a Banach space X into a Banach space Y and that for the projection operators $P_n : Y \to Y_n$ and the approximating bounded linear operators $A_n : X \to Y$ pointwise convergence*

$$P_nA_n - P_nA \to 0, \quad n \to \infty,$$

and

$$\sup_{\varphi \in X_n,\, \|\varphi\|=1} \|P_nA_n - P_nA\| \to 0, \quad n \to \infty,$$

is satisfied. Then for sufficiently large n the approximate equation (13.6) is uniquely solvable and we have an error estimate

$$\|\widetilde{\varphi}_n - \varphi\| \leq C\{\inf_{\psi \in X_n} \|\psi - \varphi\| + \|(P_nA_n - P_nA)\varphi\| + \|P_n(f_n - f)\|\} \tag{13.7}$$

for the solution φ of (13.1) and some constant C.

Proof. By setting $B = A_n - A$ in Theorem 13.7 for the case (a) it follows that for sufficiently large n the inverse operators of $P_nA_n : X_n \to Y_n$ exist and are uniformly bounded. With the aid of

$$\begin{aligned}\widetilde{\varphi}_n - \varphi_n &= [(P_nA_n)^{-1} - (P_nA)^{-1}]P_nA\varphi + (P_nA_n)^{-1}P_n(f_n - f)\\ &= (P_nA_n)^{-1}[P_nA - P_nA_n](\varphi_n - \varphi)\\ &\quad +(P_nA_n)^{-1}[P_nA - P_nA_n]\varphi + (P_nA_n)^{-1}P_n(f_n - f)\end{aligned}$$

the error estimate follows from Theorem 13.6 and the uniform boundedness principle Theorem 10.6. □

13.2 Projection Methods for Equations of the Second Kind

For an equation of the second kind

$$\varphi - A\varphi = f \tag{13.8}$$

with a bounded linear operator $A : X \to X$ we need only a sequence of subspaces $X_n \subset X$ and projection operators $P_n : X \to X_n$. Then the projection method assumes the form

$$\varphi_n - P_n A\varphi_n = P_n f. \tag{13.9}$$

Note that each solution $\varphi_n \in X$ of (13.9) automatically belongs to X_n. When A is compact, from Theorem 13.7 we have the following convergence property.

Corollary 13.9. *Let $A : X \to X$ be a compact linear operator in a Banach space X, $I - A$ be injective, and the projection operators $P_n : X \to X_n$ converge pointwise $P_n\varphi \to \varphi$, $n \to \infty$, for all $\varphi \in X$. Then the projection method converges for $I - A$.*

Proof. We apply the second case of Theorem 13.7. □

We wish to give a second proof of the latter convergence result based directly on Theorem 10.1. This will also give us the opportunity to point out that the projection method for equations of the second kind may converge without pointwise convergence of the projection operators on all of X.

Theorem 13.10. *Let $A : X \to X$ be a compact linear operator in a Banach space X and $I - A$ be injective. Assume that the projection operators $P_n : X \to X_n$ satisfy $\|P_n A - A\| \to 0$, $n \to \infty$. Then, for sufficiently large n, the approximate equation (13.9) is uniquely solvable for all $f \in X$ and we have an error estimate*

$$\|\varphi_n - \varphi\| \le M\|P_n\varphi - \varphi\| \tag{13.10}$$

for some positive constant M depending on A.

Proof. Theorems 3.4 and 10.1, applied to $I - A$ and $I - P_n A$, imply that for all sufficiently large n the inverse operators $(I-P_n A)^{-1}$ exist and are uniformly bounded. To verify the error bound, we apply the projection operator P_n to (13.8) and obtain

$$\varphi - P_n A\varphi = P_n f + \varphi - P_n\varphi.$$

Subtracting this from (13.9) we find

$$(I - P_n A)(\varphi_n - \varphi) = P_n\varphi - \varphi,$$

whence the estimate (13.10) follows. □

Note that for a compact operator A pointwise convergence $P_n\varphi \to \varphi$, $n \to \infty$, for all $\varphi \in X$, by Theorem 10.10 implies that $\|P_n A - A\| \to 0$, $n \to \infty$. But norm convergence $P_nA - A \to 0$, $n \to \infty$, may be satisfied without pointwise convergence of the projection operator sequence (P_n) as we will see in our discussion of the collocation method (see Theorems 13.15 and 13.16). Then, of course, from (13.10) we can assure convergence only if we have convergence $P_n\varphi \to \varphi$, $n \to \infty$, for the exact solution φ.

Since projection methods are only semi-discrete, in actual numerical calculations instead of (13.9) an approximate version of the form

$$\widetilde{\varphi}_n - P_nA_n\widetilde{\varphi}_n = P_nf_n \tag{13.11}$$

needs to be solved, where A_n is an approximation of A and f_n approximates f. Then we can state the following result.

Corollary 13.11. *Under the assumptions of Theorem 13.10 on the operator A and the projection operators P_n assume that pointwise convergence*

$$P_nA_n - P_nA \to 0, \quad n \to \infty,$$

and

$$\sup_{\varphi\in X_n,\, \|\varphi\|=1} \|P_nA_n - P_nA\| \to 0, \quad n \to \infty,$$

is satisfied. Then, for sufficiently large n, the approximate equation (13.11) is uniquely solvable and we have an error estimate

$$\|\widetilde{\varphi}_n - \varphi\| \leq M\{\|P_n\varphi - \varphi\| + \|P_n(A_n - A)\varphi\| + \|P_n(f_n - f)\|\} \tag{13.12}$$

for some positive constant M.

Proof. We apply Theorem 10.1 to the operators $I - P_nA : X_n \to X_n$ and $I - P_nA_n : X_n \to X_n$ to establish the existence and uniform boundedness of the inverse operators $(I - P_nA_n)^{-1} : X_n \to X_n$ for sufficiently large n. From Theorem 10.1 we also obtain the error estimate

$$\|\widetilde{\varphi}_n - \varphi_n\| \leq C\,\{\|(P_nA_n - P_nA)\varphi_n\| + \|P_n(f_n - f)\|\}$$

for some constant C. Now (13.12) follows from (13.10) and the uniform boundedness principle Theorem 10.6. □

We conclude this section with a theorem that will enable us to derive convergence results for collocation and Galerkin methods for equations of the first kind. Consider the equation

$$S\varphi - A\varphi = f, \tag{13.13}$$

where $S : X \to Y$ is assumed to be a bijective bounded linear operator mapping a Banach space X into a Banach space Y and where $A : X \to Y$ is compact and $S - A$ is injective. Then, for the projection method

$$P_n(S - A)\varphi_n = P_nf \tag{13.14}$$

with subspaces X_n and Y_n and projection operators $P_n : Y \to Y_n$ chosen as in Definition 13.5 we have the following theorem.

Theorem 13.12. *Assume that $Y_n = S(X_n)$ and $\|P_n A - A\| \to 0$, $n \to \infty$. Then, for sufficiently large n, the approximate equation (13.14) is uniquely solvable and we have an error estimate*

$$\|\varphi_n - \varphi\| \le M \|P_n S\varphi - S\varphi\| \tag{13.15}$$

for some positive constant M depending on S and A.

Proof. By Theorem 10.8 the inverse $S^{-1} : Y \to X$ is bounded. The equation (13.14) is equivalent to

$$S^{-1} P_n S(I - S^{-1}A)\varphi_n = S^{-1} P_n S S^{-1} f,$$

i.e.,

$$Q_n(I - S^{-1}A)\varphi_n = Q_n S^{-1} f,$$

where $Q_n := S^{-1} P_n S : X \to X_n$ obviously is a projection operator. Using $\|Q_n S^{-1}A - S^{-1}A\| = \|S^{-1}(P_n A - A)\|$ and $Q_n\varphi - \varphi = S^{-1}(P_n S\varphi - S\varphi)$, the assertion follows from Theorem 13.10. □

The essence of Theorem 13.12 lies in the fact that the projection method is applied directly to the given equation whereas the convergence result is based on the regularized equation of the second kind.

In actual numerical calculations, instead of (13.14) an approximate version of the form

$$P_n(S - A_n)\widetilde{\varphi}_n = P_n f_n \tag{13.16}$$

will usually be solved, where A_n and f_n are approximations of A and f, respectively. Then, analogous to Corollary 13.11, we have the following result.

Corollary 13.13. *Under the assumptions of Theorem 13.12 on the operators S and A assume that*

$$P_n A_n - P_n A \to 0, \quad n \to \infty,$$

and

$$\sup_{\varphi \in X_n, \|\varphi\|=1} \|P_n A_n - P_n A\| \to 0, \quad n \to \infty.$$

Then, for sufficiently large n, the approximate equation (13.16) is uniquely solvable and we have an error estimate

$$\|\widetilde{\varphi}_n - \varphi\| \le M\{\|P_n S\varphi - S\varphi\| + \|P_n(A_n - A)\varphi\| + \|P_n(f_n - f)\|\} \tag{13.17}$$

for some positive constant M.

13.3 The Collocation Method

The *collocation method* for approximately solving the equation

$$A\varphi = f, \tag{13.18}$$

roughly speaking, consists of seeking an approximate solution from a finite-dimensional subspace by requiring that (13.18) is satisfied only at a finite number of so-called *collocation points*. To be more precise, let $Y = C(G)$ and $A : X \to Y$ be a bounded linear operator. Let $X_n \subset X$ and $Y_n \subset Y$ denote a sequence of subspaces with $\dim X_n = \dim Y_n = n$. Choose n points $x_1^{(n)}, \dots, x_n^{(n)}$ in G (we also will write $x_1, \dots, x_n$ instead of $x_1^{(n)}, \dots, x_n^{(n)}$) such that the subspace Y_n is unisolvent (see Theorem 11.2) with respect to these points. Then the collocation method approximates the solution of (13.18) by an element $\varphi_n \in X_n$ satisfying

$$(A\varphi_n)(x_j) = f(x_j), \quad j = 1, \dots, n. \tag{13.19}$$

Let $X_n = \operatorname{span}\{u_1, \dots, u_n\}$. Then we express φ_n as a linear combination

$$\varphi_n = \sum_{k=1}^{n} \gamma_k u_k$$

and immediately see that (13.19) is equivalent to the linear system

$$\sum_{k=1}^{n} \gamma_k (Au_k)(x_j) = f(x_j), \quad j = 1, \dots, n, \tag{13.20}$$

for the coefficients $\gamma_1, \dots, \gamma_n$. The collocation method can be interpreted as a projection method with the interpolation operator $P_n : Y \to Y_n$ described in Theorem 13.4. Indeed, because the interpolating function is uniquely determined by its values at the interpolation points, equation (13.19) is equivalent to

$$P_n A\varphi_n = P_n f.$$

Hence, our general error and convergence results for projection methods apply to the collocation method.

Note that the collocation method can also be applied in function spaces other than the space $C(G)$, for example in Hölder spaces or Sobolev spaces.

For an equation of the second kind $\varphi - A\varphi = f$ in $X = C(G)$ with a bounded linear operator $A : X \to X$ we need only one sequence of subspaces $X_n \subset X$ and assume X_n to be unisolvent with respect to the collocation points. Here, the equations (13.20) assume the form

$$\sum_{k=1}^{n} \gamma_k \{u_k(x_j) - (Au_k)(x_j)\} = f(x_j), \quad j = 1, \dots, n. \tag{13.21}$$

We shall first consider the collocation method for integral equations of the second kind

$$\varphi(x) - \int_G K(x, y)\varphi(y)\, dy = f(x), \quad x \in G,$$

with a continuous or weakly singular kernel K. If we use the Lagrange basis for X_n and write

$$\varphi_n = \sum_{k=1}^{n} \gamma_k L_k$$

then $\gamma_j = \varphi_n(x_j)$, $j = 1, \ldots, n$, and the system (13.21) becomes

$$\gamma_j - \sum_{k=1}^{n} \gamma_k \int_G K(x_j, y) L_k(y)\, dy = f(x_j), \quad j = 1, \ldots, n. \tag{13.22}$$

From the last equation we observe that the collocation method for equations of the second kind is closely related to the degenerate kernel method via interpolation. Obviously, the matrix of the system (13.22) coincides with the matrix (11.25) from the degenerate kernel approach. But the right-hand sides of both systems are different and the solutions $\gamma_1, \ldots, \gamma_n$ have different meanings. However, if the right-hand side f happens to belong to X_n, then $f = P_n f$ and we may express the collocation solution φ_n also in the form

$$\varphi_n = f + \sum_{k=1}^{n} \widetilde{\gamma}_k L_k.$$

Now, for the coefficients $\widetilde{\gamma}_1, \ldots, \widetilde{\gamma}_n$ we obtain the same system as (11.25), and in this case the collocation method and the degenerate kernel method via interpolation will yield the same numerical results.

Analogous to the degenerate kernel method the collocation method is only a semi-discrete method. In principle, the simplest approach to make the collocation method fully discrete is to use a quadrature operator A_n of the form (12.6) or (12.13) from the previous chapter with the collocation points as quadrature points to approximate A in the sense of Corollary 13.11. In this case, however, the solution $\widetilde{\varphi}_n$ of

$$\widetilde{\varphi}_n - P_n A_n \widetilde{\varphi}_n = P_n f$$

will coincide at the collocation points with the solution ψ_n of the corresponding Nyström equation

$$\psi_n - A_n \psi_n = f,$$

since both equations lead to the same linear system if we use the Lagrange basis for the collocation method. In particular, we have the relation

$$\psi_n = f + A_n \widetilde{\varphi}_n.$$

This follows from the fact that $\chi_n := f + A_n\widetilde{\varphi}_n$ satisfies

$$\chi_n - A_nP_n\chi_n = f + A_n\widetilde{\varphi}_n - A_nP_n(f + A_n\widetilde{\varphi}_n) = f$$

and the observation that $A_nP_n\chi = A_n\chi$ for all χ, since the collocation and quadrature points coincide. Hence, this type of approximation can be analyzed and interpreted more conveniently as a Nyström method. Therefore, in the sequel for integral equations of the second kind we will consider only fully discrete collocation methods which are based on numerical integration of the matrix elements.

In the literature, a broad variety of collocation methods exists corresponding to various choices for the subspaces X_n, for the basis functions $u_1, \dots, u_n$, and for the collocation points $x_1, \dots, x_n$. We briefly discuss two possibilities based on linear splines and trigonometric interpolation. When we use linear splines (or more general spline interpolations) we have pointwise convergence of the corresponding interpolation operator as can be seen from the Banach–Steinhaus theorem by using Theorem 11.3 and the norm (11.9) of the linear spline interpolation operator. Therefore, in this case, Corollary 13.9 applies, and we can state the following theorem.

Theorem 13.14. *The collocation method with linear splines converges for uniquely solvable integral equations of the second kind with continuous or weakly singular kernels.*

Provided the exact solution of the integral equation is twice continuously differentiable, from Theorems 11.3 and 13.6 we derive an error estimate of the form

$$\|\varphi_n - \varphi\|_\infty \le M\|\varphi''\|_\infty h^2$$

for the linear spline collocation approximate solution φ_n. Here, M denotes some constant depending on the kernel K.

In most practical problems the evaluation of the matrix entries in (13.22) will require numerical integration. For twice continuously differentiable kernels, as in the case of the degenerate kernel method, we suggest the quadrature rule (11.27). Then, according to our remarks concerning the connection between the equations (11.25) and (13.22), the collocation method will produce the same numerical results as the degenerate kernel method. So we do not have to repeat the numerical calculations for equation (11.30) of Example 11.10 using the quadrature (11.27). However, to

Table 13.1 Collocation method for Example 11.10

n	$x = 0$	$x = 0.25$	$x = 0.5$	$x = 0.75$	$x = 1$
4	0.003984	0.004428	0.004750	0.004978	0.005135
8	0.000100	0.001112	0.001193	0.001250	0.001291
16	0.000250	0.000278	0.000298	0.000313	0.000323
32	0.000063	0.000070	0.000075	0.000078	0.000081

illustrate that it does not pay off to evaluate the matrix elements more accurately, in Table 13.1 we give the numerical results for the error between the approximate and exact solution for the spline collocation for equation (11.30) using the exact matrix entries. Comparing Tables 11.1 and 13.1, we observe that the improvement in the accuracy for the solution of the integral equation is only marginal. In general, instead of investing computing efforts in a highly accurate evaluation of the matrix elements it is more efficient to increase the number of collocation points, i.e., to increase the size of the linear system.

We proceed discussing the collocation method based on trigonometric interpolation with equidistant knots $t_j = j\pi/n$, $j = 0, .., 2n-1$.

Theorem 13.15. *The collocation method with trigonometric polynomials converges for uniquely solvable integral equations of the second kind with 2π-periodic continuously differentiable kernels and right-hand sides.*

Proof. The integral operator A with 2π-periodic continuously differentiable kernel K maps $C_{2\pi}$ into $C^1_{2\pi}$, and we have

$$\|A\varphi\|_\infty + \|(A\varphi)'\|_\infty \le 2\pi \left\{ \|K\|_\infty + \left\| \frac{\partial K}{\partial t} \right\|_\infty \right\} \|\varphi\|_\infty$$

for all $\varphi \in C_{2\pi}$. From Theorems 8.4 and 11.8 we can conclude that

$$\|P_n A\varphi - A\varphi\|_\infty \le c_1 \|P_n A\varphi - A\varphi\|_{H^q} \le c_2 n^{q-1} \|A\varphi\|_{H^1}$$

for all $\varphi \in C_{2\pi}$, all $1/2 < q < 1$, and some constants c_1 and c_2 depending on q. Combining these two inequalities and using Theorem 8.5 yields

$$\|P_n A - A\|_\infty \le C\, n^{q-1}$$

for all $1/2 < q < 1$ and some constant C depending on q. Now the assertion follows from Theorem 13.10. □

One possibility for the implementation of the collocation method is to use the monomials $f_m(t) = e^{imt}$ as basis functions. Then the integrals $\int_0^{2\pi} K(t_j, \tau) e^{ik\tau} d\tau$ have to be integrated numerically. Using fast Fourier transform techniques (see Elliott and Rao [46], Nussbaumer [186] and Rao-et-al [202]) these quadratures can be carried out very rapidly. A second even more efficient possibility is to use the Lagrange basis as already described in connection with the degenerate kernel method. This leads to the quadrature rule (11.33) for the evaluation of the matrix elements in (13.22). Because of the simple structure of (11.17), the only computational effort besides the kernel evaluation is the computation of the row sums $\sum_{m=0}^{2n-1} (-1)^m K(t_j, t_m)$ for $j = 0, \dots 2n-1$. Of course, with this quadrature the collocation method again will coincide with the degenerate kernel method via trigonometric interpolation.

We now proceed to illustrate the application of the collocation method for equations of the second kind with weakly singular kernels. As in Sections 12.3 and 12.4

we consider the operator

$$(A\varphi)(t) := \frac{1}{2\pi}\int_0^{2\pi} \ln\left(4\sin^2\frac{t-\tau}{2}\right) K(t,\tau)\varphi(\tau)\,d\tau, \quad 0 \le t \le 2\pi, \tag{13.23}$$

where the kernel function K is assumed to be 2π-periodic and infinitely differentiable.

Theorem 13.16. *The collocation method with trigonometric polynomials converges for uniquely solvable integral equations of the second kind with a logarithmic singularity of the form (13.23) provided the kernel function K is 2π-periodic and infinitely differentiable and the right-hand side is 2π-periodic and continuously differentiable.*

Proof. From Theorems 8.4, 11.8, and 12.15 it follows that

$$\|P_nA\varphi - A\varphi\|_\infty \le c_1\|P_nA\varphi - A\varphi\|_{H^q} \le c_2 n^{q-1}\|A\varphi\|_{H^1} \le c_3 n^{q-1}\|\varphi\|_{H^0} \le c_4 n^{q-1}\|\varphi\|_\infty$$

for all $\varphi \in C[0,2\pi]$, all $1/2 < q < 1$, and some constants c_1, c_2, c_3, and c_4 depending on q. Hence $\|P_nA - A\|_\infty \le c_4 n^{q-1}$ for all $1/2 < q < 1$ and the proof is completed using Theorem 13.10. □

To evaluate the matrix elements, as usual, we replace $K(t_j,\cdot)$ by its trigonometric interpolation. This yields the approximations

$$\frac{1}{2\pi}\int_0^{2\pi} \ln\left(4\sin^2\frac{t_j-\tau}{2}\right) K(t_j,\tau)L_k(\tau)\,d\tau \approx \sum_{m=0}^{2n-1} S_{k-j,m-j}K(t_j,t_m), \tag{13.24}$$

where

$$S_{km} := \frac{1}{2\pi}\int_0^{2\pi} \ln\left(4\sin^2\frac{\tau}{2}\right) L_k(\tau)L_m(\tau)\,d\tau, \quad k,m = 0,\dots,2n-1,$$

(and $S_{km} = S_{-k,m} = S_{k,-m}$). A practical numerical evaluation of the weights in (13.24) rests on the observation that the quadrature rule (12.18) with the number of grid points $2n$ doubled to $4n$ integrates trigonometric polynomials of degree less than or equal to $2n$ exactly, provided the coefficient of the term $\sin 2nt$ vanishes. Therefore we have

$$S_{km} = \sum_{j=0}^{4n-1} R_j^{(2n)} L_k(t_j^{(2n)})L_m(t_j^{(2n)}), \quad k,m = 0,\dots,2n-1.$$

With the quadrature rule (13.24) for the evaluation of the matrix elements, the system (13.22) actually corresponds to solving the equation (13.11) with

$$(A_n\varphi)(t) := \frac{1}{2\pi}\int_0^{2\pi} \ln\left(4\sin^2\frac{t-\tau}{2}\right) \sum_{m=0}^{2n-1} K(t,t_m)L_m(\tau)\varphi(\tau)\,d\tau. \quad 0 \le t \le 2\pi.$$

Since the kernel K is assumed infinitely differentiable, from Theorems 8.4 and 11.8, for the kernel function

$$K_n(t) := \sum_{m=0}^{2n-1} K(t, t_m) L_m(\tau)$$

of A_n, we have $\|K_n - K\|_\infty = O(n^{-p})$ for $p \in \mathbb{N}$, and consequently $\|A_n - A\|_\infty = O(n^{-p})$. Therefore Theorem 11.4 implies convergence $\|P_n(A_n - A)\|_\infty \to 0$, $n \to \infty$. Hence Corollary 13.11 and its error estimate (13.12) applies. In the case when the exact solution φ is infinitely differentiable, then from Theorems 8.4 and 11.8 and (13.12) we conclude that $\|\widetilde{\varphi}_n - \varphi\| = O(n^{-p})$ for each $p \in \mathbb{N}$. If both the kernel function K and the exact solution φ are analytic, then from Theorem 11.7 and (13.12) we conclude that the convergence is exponential.

Table 13.2 Numerical results for Example 13.17

	$2n$	$t = 0$	$t = \pi/2$	$t = \pi$
	4	−0.10828581	0.15071347	0.22404278
$\kappa = 1$	8	−0.01368742	0.00827850	0.01129843
$q = 2$	16	−0.00023817	0.00003702	0.00004366
	32	−0.00000019	0.00000001	0.00000000
	4	−0.19642139	0.43455120	0.66382871
$\kappa = 1$	8	−0.07271273	0.07400151	0.10462955
$q = 1.5$	16	−0.00662395	0.00275639	0.00384562
	32	−0.00005937	0.00000548	0.00000609
	8	0.18032618	0.02704260	−0.36058686
$\kappa = 5$	16	0.00617287	−0.00852966	−0.00357971
$q = 2$	32	−0.00000686	−0.00000015	−0.00000009
	64	0.00000003	0.00000002	−0.00000000
	8	−0.17339263	0.01532368	0.46622097
$\kappa = 5$	16	−0.04130869	−0.01351387	−0.01226315
$q = 1.5$	32	−0.00022670	−0.00000972	−0.00001335
	64	−0.00000006	0.00000000	0.00000003

Example 13.17. We apply the collocation method based on trigonometric polynomials for the integral equation (12.26) of Example 12.14. We use the splitting (12.27) of the kernel in a logarithmic and a smooth part, and then evaluate the corresponding matrix elements with the quadrature rules (13.24) and (11.33), respectively. The right-hand side of the system (13.22) is evaluated by numerical integration, as in Example 12.14. The numerical results for the error between the approximate and the exact solution in Table 13.2 show the exponential convergence, which we expect from our error analysis. □

In general, the implementation of the collocation method as described by our two examples can be used in all situations where the required numerical quadratures for the matrix elements can be carried out in closed form for the chosen approximating subspace and collocation points. In all these cases, of course, the quadrature formulas that are required for the related Nyström method will also be available. Because the approximation order for both methods will usually be the same, Nyström's method is preferable; it requires the least computational effort for evaluating the matrix elements.

However, the situation changes in cases where no straightforward quadrature rules for the application of Nyström's method are available. This, in particular, occurs for the boundary integral equations described in Chapter 6 in the case of three space dimensions. Here, the collocation method is the most important numerical approximation method. Usually, the boundary surface is subdivided into a finite number of segments, like curved triangles and squares. Then the approximation space is chosen to consist of some kind of low-order polynomial splines with respect to these surface elements, which possess appropriate smoothness properties across the boundary curves of the segments. Within each segment, depending on the degree of freedom in the chosen splines, a number of collocation points are selected. Then the integrals for the matrix elements in the collocation system (13.22) are evaluated using numerical integration. Since the integral equations usually have a weakly singular kernel, calculation of the improper integrals for the diagonal elements of the matrix, where the collocation points and the surface elements coincide, needs special attention. With these few remarks we have given a somewhat vague outline of a very effective method known as *boundary element method*. For a detailed description and a review of the corresponding literature we refer the reader to Atkinson [11], Brebbia, Telles, and Wrobel [20], Chen and Zhou [28], Hackbusch [76], Rjasanow and Steinbach [207], Sauter and Schwab [214] and Steinbach [225]. For global numerical approximation methods for three-dimensional boundary integral equations using spherical harmonics for the approximating subspaces and quadrature formulas for single- and double-layer potentials analogous to the Gauss trapezoidal product rule (12.4) for smooth integrands, we refer to Atkinson [11], Colton and Kress [32] and Ganesh, Graham and Sloan [61, 68]

13.4 Collocation Methods for Equations of the First Kind

We continue our study of collocation methods by indicating its applications to integral equations of the first kind, basing the convergence and error analysis on Theorem 13.12 and its Corollary 13.13.

Consider a singular integral equation of the first kind with Hilbert kernel

$$\frac{1}{2\pi}\int_0^{2\pi}\left\{\cot\frac{\tau-t}{2}+K(t,\tau)\right\}\varphi(\tau)\,d\tau=f(t),\quad 0\le t\le 2\pi, \tag{13.25}$$

as discussed in Corollary 7.27. Because of the Cauchy-type singularity, this equation, and correspondingly its error analysis, has to be treated in the Hölder space $C^{0,\alpha}_{2\pi}$ or in a Sobolev space $H^p[0, 2\pi]$. For convenience, the 2π-periodic kernel function K is assumed to be infinitely differentiable with respect to both variables. We express the leading singular part of (13.25) in the form

$$(R\varphi)(t) := \frac{1}{2\pi} \int_0^{2\pi} \left\{ \cot \frac{\tau - t}{2} + 2 \right\} \varphi(\tau) \, d\tau, \quad 0 \leq t \leq 2\pi.$$

For the constant kernel function $K = 2$, equation (13.25) corresponds to a modified version of the operator A in Example 7.14 from which we can conclude that the singular integral operator $R : C^{0,\alpha}_{2\pi} \to C^{0,\alpha}_{2\pi}$ is bounded and has a bounded inverse. Analogously, from the integrals (8.35) it can be seen that $R : H^p[0, 2\pi] \to H^p[0, 2\pi]$ is bounded and has a bounded inverse for all $p \geq 0$.

Choose X_n to be the subspace of trigonometric polynomials of the form

$$\varphi(t) = \sum_{j=0}^{n} \alpha_j \cos jt + \sum_{j=1}^{n-1} \beta_j \sin jt, \tag{13.26}$$

and note that X_n is unisolvent with respect to the equidistantly spaced grid $t_j = j\pi/n$, $j = 0, \dots, 2n - 1$ (see Section 11.3). From the integrals (8.35) it follows that $Y_n := R(X_n)$ is given by the trigonometric polynomials of the form

$$\varphi(t) = \sum_{j=0}^{n-1} \widetilde{\alpha}_j \cos jt + \sum_{j=1}^{n} \widetilde{\beta}_j \sin jt. \tag{13.27}$$

Observe that Y_n is not unisolvent with respect to the t_j, $j = 0, \dots, 2n - 1$, that, however, it is unisolvent with respect to the intersparsed set of points $\widetilde{t}_j = t_j + \pi/2n$, $j = 0, \dots, 2n - 1$. Hence, we use the latter points as collocation points.

Denote by $\widetilde{P}_n : C^{0,\alpha}_{2\pi} \to Y_n$ the trigonometric interpolation operator with respect to the points $\widetilde{t}_j$, $j = 0, \dots, 2n - 1$. Then, from the error estimate (11.20), for the integral operator A given by

$$(A\varphi)(t) := \frac{1}{2\pi} \int_0^{2\pi} K(t, \tau) \varphi(\tau) \, d\tau, \quad 0 \leq t \leq 2\pi, \tag{13.28}$$

we can conclude that

$$\|\widetilde{P}_n A - A\|_{0,\alpha} \leq c_1 \max_{0 \leq \tau \leq 2\pi} \|\widetilde{P}_n K(\cdot, \tau) - K(\cdot, \tau)\|_{0,\alpha} \leq c_2 \frac{\ln n}{n^m} \max_{0 \leq \tau \leq 2\pi} \|K(\cdot, \tau)\|_{m,\alpha}$$

for all $m \in \mathbb{N}$ and some constants c_1 and c_2 depending on m. This implies that $\|\widetilde{P}_n A - A\|_{0,\alpha} \to 0$, $n \to \infty$. Analogously, for the numerical quadrature operator A_n given by

$$(A_n\varphi)(t) := \frac{1}{2n} \sum_{k=0}^{2n-1} K(t, t_k) \varphi(t_k), \quad 0 \leq t \leq 2\pi, \tag{13.29}$$

it can be seen that

$$\|\widetilde{P}_n A_n - \widetilde{P}_n A\|_{0,\alpha} \le \|\widetilde{P}_n\|_{0,\alpha} \, \|A_n - A\|_{0,\alpha} \to 0, \quad n \to \infty.$$

Hence we can apply Theorem 13.12 and its Corollary 13.13 to obtain the following result.

Theorem 13.18. *Both the semi-discrete collocation method (with trigonometric polynomials of the form (13.26) and collocation points* $j\pi/n + \pi/2n$, $j = 0, \dots, 2n-1$*) and the corresponding fully discrete method using the numerical quadrature operator* A_n *converge in* $C^{0,\alpha}_{2\pi}$ *for uniquely solvable integral equations of the first kind of the form (13.25) with Hilbert kernel provided* K *and* f *are infinitely differentiable.*

From the error estimates (13.15) and (13.17) and Theorems 8.4 and 11.8 we can deduce convergence of order $O(n^{-m})$ for all $m \in \mathbb{N}$ if the exact solution φ is infinitely differentiable. Analogously, from Theorem 11.7 we conclude exponential convergence when the kernel function K and the exact solution φ are both analytic.

Note that here, as opposed to equations of the second kind, it makes sense to consider a fully discrete version via quadrature operators, since for equations of the first kind there is no Nyström method available. Of course, a fully discrete version of the collocation method for (13.25) can also be based on numerical evaluation of the matrix elements as described in the previous section.

We leave it as an exercise for the reader to prove a variant of Theorem 13.18 in the Sobolev spaces $H^p[0, 2\pi]$ for $p > 1/2$.

For the numerical implementation we need the integrals

$$I_{jk} := \frac{1}{2\pi} \int_0^{2\pi} \cot \frac{\tau - \widetilde{t}_j}{2} \, L_k(\tau) \, d\tau, \quad j, k = 0, \dots, 2n-1,$$

for the Lagrange basis (11.12). With the aid of the integrals (8.35) it can be seen that

$$I_{jk} = \frac{1}{2n} \{1 - \cos n(t_k - \widetilde{t}_j)\} \cot \frac{t_k - \widetilde{t}_j}{2}, \quad j, k = 0, \dots, 2n-1.$$

Then the fully discrete collocation method for the integral equation of the first kind (13.25) using the numerical quadrature operator A_n leads to the linear system

$$\sum_{k=0}^{2n-1} \left\{ I_{jk} + \frac{1}{2n} \, K(\widetilde{t}_j, t_k) \right\} \gamma_k = f(\widetilde{t}_j), \quad j = 0, \dots, 2n-1,$$

for the coefficients $\gamma_0, \dots, \gamma_{2n-1}$ in the representation $\widetilde{\varphi}_n = \sum_{k=0}^{2n-1} \gamma_k L_k$ of the approximate solution.

Example 13.19. The integral equation (13.25) with kernel

$$K(t, \tau) = 2 - \frac{(a^2 - b^2)\sin(t + \tau)}{a^2 + b^2 - (a^2 - b^2)\cos(t + \tau)}$$

corresponds to the Cauchy singular integral equation $RS_0 = 2Rf$ of Problem 7.4 for an ellipse with semi-axis $a \geq b > 0$. Therefore, in particular, we have uniqueness. Table 13.3 gives some numerical results (for the difference between the approximate and the exact solution) in the case where the exact solution is given by

$$\varphi(t) = 1 - e^{\cos t}\cos(\sin t), \quad 0 \leq t \leq 2\pi.$$

Then, setting $c = (a-b)/(a+b)$, the right-hand side becomes

$$f(t) = e^{c\cos t}\sin(c\sin t) + e^{\cos t}\sin(\sin t), \quad 0 \leq t \leq 2\pi,$$

as can be derived from (8.35) and Problem 11.4 □

Table 13.3 Numerical results for Example 13.19

	$2n$	$t = 0$	$t = \pi/2$	$t = \pi$
	4	0.16740241	−0.00528236	0.00992820
$a = 1$	8	0.00258801	−0.00022139	−0.00003859
$b = 0.5$	16	0.00000040	−0.00000004	−0.00000001
	32	0.00000000	0.00000000	0.00000000
	4	0.26639845	−0.04603478	−0.00756302
$a = 1$	8	0.06215996	−0.01726862	−0.00546249
$b = 0.2$	16	0.00267315	−0.00071327	−0.00022010
	32	0.00000409	−0.00000109	−0.00000034

Finally, we consider an integral equation of the first kind

$$S_0\varphi - A\varphi = f \tag{13.30}$$

where S_0 and A have logarithmic singularities and are given by

$$(S_0\varphi)(t) := \frac{1}{2\pi}\int_0^{2\pi}\left\{\ln\left(4\sin^2\frac{t-\tau}{2}\right) - 2\right\}\varphi(\tau)\,d\tau \tag{13.31}$$

and

$$(A\varphi)(t) := \frac{1}{2\pi}\int_0^{2\pi}\left\{K(t,\tau)\ln\left(4\sin^2\frac{t-\tau}{2}\right) + L(t,\tau)\right\}\varphi(\tau)\,d\tau \tag{13.32}$$

for $0 \leq t \leq 2\pi$. Here, we assume K and L to be infinitely differentiable and 2π-periodic with respect to both variables such that $K(t,t) = 0$ for all $0 \leq t \leq 2\pi$. For applications of integral equations with such a logarithmic singularity we recall the remarks in Section 12.3 on two-dimensional boundary value problems.

By Theorem 8.24 the operator $S_0 : H^p[0, 2\pi] \to H^{p+1}[0, 2\pi]$ is bounded and has a bounded inverse for all $p \geq 0$. From the integrals in Lemma 8.23 we conclude that S_0 maps the space of trigonometric polynomials of the form (13.26) into itself. Hence, here we can use the set $t_j = j\pi/n$, $j = 0, \ldots, 2n-1$, as collocation points.

Theorem 13.20. *The semi-discrete collocation method (with trigonometric polynomials of the form (13.26) and collocation points $j\pi/n$, $j = 0, \ldots, 2n-1$,) converges for uniquely solvable integral equations of the first kind of the form (13.30) with logarithmic kernel in the Sobolev space $H^p[0, 2\pi]$ for each $p > 1/2$ provided K, L, and f are infinitely differentiable (and $K(t, t) = 0$ for all $t \in [0, 2\pi]$).*

Proof. For the derivative of $A\varphi$ we can write

$$\frac{d}{dt}(A\varphi)(t) = \frac{1}{2\pi}\int_0^{2\pi}\left\{\widetilde{K}(t,\tau)\ln\left(4\sin^2\frac{t-\tau}{2}\right) + \widetilde{L}(t,\tau)\right\}\varphi(\tau)\,d\tau,$$

where

$$\widetilde{K}(t,\tau) = \frac{\partial K(t,\tau)}{\partial t}$$

and

$$\widetilde{L}(t,\tau) = \frac{\partial L(t,\tau)}{\partial t} + K(t,\tau)\cot\frac{t-\tau}{2}$$

are infinitely differentiable as a consequence of the assumption $K(t,t) = 0$. Therefore, for integers $p \geq 0$, from Theorem 12.15 we have that

$$\left\|\frac{d}{dt}A\varphi\right\|_{p+1} \leq c_1\|\varphi\|_p$$

for all $\varphi \in H^p[0, 2\pi]$ and some constant c_1 depending on p. This implies that A is bounded from $H^p[0, 2\pi]$ into $H^{p+2}[0, 2\pi]$ for all integers $p \geq 0$. Then, by the interpolation Theorem 8.13, it follows that $A : H^p[0, 2\pi] \to H^{p+2}[0, 2\pi]$ is bounded for arbitrary $p \geq 0$.

Now, for $p > 1/2$, from Theorem 11.8 we conclude that

$$\|P_nA\varphi - A\varphi\|_{p+1} \leq \frac{c_2}{n}\,\|A\varphi\|_{p+2} \leq \frac{c_3}{n}\,\|\varphi\|_p \qquad (13.33)$$

for all $\varphi \in H^p[0, 2\pi]$ and some constants c_2 and c_3 depending on p. This implies norm convergence $\|P_nA - A\| \to 0$, $n \to \infty$, for the operator $P_nA - A : H^p[0, 2\pi] \to H^{p+1}[0, 2\pi]$, and the assertion follows from Theorem 13.12. (Note that instead of Theorem 13.12, we could have used the stability Theorem 13.7.) □

Finally, we need to describe a fully discrete variant. For this, as in Chapter 12 (see 12.21), we use the quadrature operators defined by

$$(A_n\varphi)(t) := \sum_{k=0}^{2n-1}\left\{R_k^{(n)}(t)K(t,t_k) + \frac{1}{2n}\,L(t,t_k)\right\}\varphi(t_k), \quad 0 \leq t \leq 2\pi. \qquad (13.34)$$

We also recall their construction via interpolatory quadratures

$$(A_n\varphi)(t) = \frac{1}{2\pi}\int_0^{2\pi}\left\{[P_n(K(t,\cdot)\varphi](\tau)\ln\left(4\sin^2\frac{t-\tau}{2}\right) + [P_n(L(t,\cdot)\varphi](\tau)\right\}d\tau.$$

In order to apply Corollary 13.13 we prove the following lemma, which improves on the estimates of Theorem 12.18.

Lemma 13.21. *Assume that* $0 \le q \le p$ *and* $p > 1/2$. *Then, for the quadrature operators* A_n, *we have the estimate*

$$\|(P_nA_n - P_nA)\varphi\|_{q+1} \le C\, n^{q-p-1}\|\varphi\|_p \tag{13.35}$$

for all trigonometric polynomials φ *of degree less than or equal to n and some constant C depending on p and q.*

Proof. For simplicity we only treat the case where $L = 0$. We denote by T_n the space of all trigonometric polynomials of degree less than or equal to n. Recalling the Fourier series (12.29) from the proof of Theorem 12.15, we write

$$(P_nA_n - P_nA)\varphi = P_n\sum_{m=-\infty}^{\infty}\hat{K}_mA_0[P_n(f_m\varphi) - f_m\varphi]$$

in terms of the bounded linear operator $A_0 : H^q[0,2\pi] \to H^{q+1}[0,2\pi]$ given by (12.31) and the trigonometric monomials $f_m(t) = e^{imt}$. From this, using the property $P_n(fg) = P_n(fP_ng)$ of interpolation operators, we deduce that

$$(P_nA_n - P_nA)\varphi = P_n\sum_{m=-\infty}^{\infty}\hat{K}_mP_nA_0[P_n(f_m\varphi) - f_m\varphi].$$

By Theorem 11.8, the operators $P_n : H^{q+1}[0,2\pi] \to H^{q+1}[0,2\pi]$ are uniformly bounded. Therefore, using Corollary 8.8 and the property (12.30) of the Fourier coefficients $\hat{K}_m$ and denoting by k the smallest integer larger than or equal to q, we can estimate

$$\|(P_nA_n - P_nA)\varphi\|_{q+1} \le c_1\sum_{m=-\infty}^{\infty}\frac{1}{(1+|m|)^{3+k+p-q}}\,\|P_nA_0[P_n(f_m\varphi) - f_m\varphi]\|_{q+1}$$

for all $\varphi \in T_n$ and some constant c_1.

In the case where $|m| \ge n/2$ we can estimate

$$\|P_nA_0[P_n(f_m\varphi) - f_m\varphi]\|_{q+1} \le c_2\|f_m\varphi\|_q \le c_3|m|^k\|\varphi\|_q$$

for all $\varphi \in T_n$ and some constants c_2 and c_3. From this we conclude that

$$\sum_{|m|\ge n/2}\frac{1}{(1+|m|)^{3+k+p-q}}\,\|P_nA_0[P_n(f_m\varphi) - f_m\varphi]\|_{q+1} \le \frac{c_4}{n^{p-q+1}}\,\|\varphi\|_p \tag{13.36}$$

for some constant c_4. For $0 \leq m \leq n/2$, writing

$$\varphi = \sum_{j=0}^{2n} \hat{\varphi}_{n-j} f_{n-j}$$

in view of $P_n f_{m+n-j} = P_n f_{m-j-n}$ for $j = 0, \ldots, m$ and $P_n f_{m+n-j} = f_{m+n-j}$ for $j = m+1, \ldots, 2n$ we deduce that

$$P_n(f_m \varphi) - f_m \varphi = \sum_{j=0}^{m} \hat{\varphi}_{n-j} [P_n f_{m-j-n} - f_{m-j+n}].$$

From this, with the aid of the integrals in Lemma 8.23, we find that

$$P_n A_0 [P_n(f_m \varphi) - f_m \varphi] = \sum_{j=0}^{m} \hat{\varphi}_{n-j} \left\{ \frac{1}{n+m-j} - \frac{1}{n-m+j} \right\} P_n f_{m-j-n},$$

and, using the inequalities

$$\frac{1}{n-m+j} - \frac{1}{n+m-j} \leq \frac{4m}{n^2} \quad \text{and} \quad (n+j-m)^2 \leq n^2 \leq 4(n-j)^2$$

for $0 \leq j \leq m \leq n/2$, we can estimate

$$\|P_n A_0 [P_n(f_m \varphi) - f_m \varphi]\|_{q+1}^2 \leq \frac{c_5 m^2}{n^4} \sum_{j=0}^{m} [1 + (n-j)^2]^{q+1} |\hat{\varphi}_{n-j}|^2$$

$$\leq \frac{c_6 m^2}{n^{2p-2q+2}} \sum_{j=0}^{m} [1 + (n-j)^2]^p |\hat{\varphi}_{n-j}|^2 \leq \frac{c_6 m^2}{n^{2p-2q+2}} \|\varphi\|_p^2$$

for all $\varphi \in T_n$, all $0 \leq m \leq n/2$, and some constants c_5 and c_6. From this and the corresponding inequality for $-n/2 \leq m \leq 0$ we deduce that

$$\sum_{|m| \leq n/2} \frac{1}{(1+|m|)^{3+k+p-q}} \|P_n A_0 [P_n(f_m \varphi) - f_m \varphi]\|_{q+1} \leq \frac{c_7}{n^{p-q+1}} \|\varphi\|_p$$

for all $\varphi \in T_n$ and some constant c_7. From this, together with (13.36), the statement of the lemma follows. □

Theorem 13.22. *The fully discrete collocation method using the quadrature operators (13.34) converges for uniquely solvable integral equations of the first kind of the form (13.30) with logarithmic kernel in the Sobolev space $H^p[0, 2\pi]$ for each $p > 1/2$ provided K, L, and f are infinitely differentiable (and $K(t,t) = 0$ for all $t \in [0, 2\pi]$).*

Proof. From Theorems 11.8 and 12.18 we conclude uniform boundedness of the operators $P_n A_n - P_n A : H^p[0, 2\pi] \to H^{p+1}[0, 2\pi]$ and convergence $(P_n A_n - P_n A)\varphi \to 0$, $n \to \infty$, for all trigonometric polynomials φ. By the Banach–Steinhaus theorem this implies pointwise convergence in $H^p[0, 2\pi]$. From Lemma 13.21, setting $q = p$, we have that

$$\|(P_nA_n - P_nA)\varphi\|_{p+1} \le \frac{C}{n}\,\|\varphi\|_p, \tag{13.37}$$

for all trigonometric polynomials of the form (13.26). Now the statement follows from Corollary 13.13 in connection with Theorem 13.20. (We note that instead of Corollary 13.13 we may use Theorem 13.8.) □

From the error estimate (13.17) and Theorems 8.4 and 11.8 we have convergence of order $O(n^{-m})$ for all $m \in \mathbb{N}$ if the exact solution φ is infinitely differentiable. From Theorem 11.7 we obtain exponential convergence when the kernel functions K and L and the exact solution φ are analytic.

The above convergence and error analysis for the fully discrete trigonometric polynomial collocation for integral equations of the first kind with a logarithmic singularity follows Saranen and Vainikko [212, 213]. It differs from an earlier approach by Kress and Sloan [151] and Prössdorf and Saranen [198], because it does not require a further decomposition of the logarithmic singularity, which complicates the numerical implementation of the fully discrete scheme.

Since the quadrature operators (13.34) are exact for constant kernel functions K and L and trigonometric polynomials $\varphi \in X_n$, the fully discrete collocation method for (13.30) leads to the linear system

$$\sum_{k=0}^{2n-1} \left\{ R_{k-j}^{(n)}[1 - K(t_j, t_k)] - \frac{1}{2n}\,[L(t_j, t_k) + 2] \right\} \gamma_k = f(t_j), \quad j = 0, \ldots, 2n-1,$$

for the coefficients in $\widetilde{\varphi}_n = \sum_{k=0}^{2n-1} \gamma_k L_k$ (see (12.22)).

Example 13.23. The integral equation (13.30) with kernel $K(t, \tau) = 0$ and

$$L(t, \tau) = -\ln\{a^2 + b^2 - (a^2 - b^2)\cos(t + \tau)\} - 3 \tag{13.38}$$

essentially corresponds to the integral equation (7.59) for an ellipse with semi-axis $0 < b \le a \le 1$. Therefore, in particular, it is uniquely solvable. Table 13.4 gives some numerical results for the error between the approximate and the true solution for the case in which the exact solution is given by

$$\varphi(t) = e^{\cos t}\cos(t + \sin t), \quad 0 \le t \le 2\pi.$$

With the aid of the integrals in Lemma 8.23 and a partial integration for the integrals in Problem 11.4 it can be seen that the right-hand side becomes

$$f(t) = 2 - e^{\cos t}\cos(\sin t) - e^{c\cos t}\cos(c \sin t), \quad 0 \le t \le 2\pi,$$

with $c = (a - b)/(a + b)$. □

Table 13.4 Numerical results for Example 13.23

	$2n$	$t=0$	$t=\pi/2$	$t=\pi$
	4	-0.62927694	-0.19858702	0.20711238
$a=1$	8	-0.02649085	0.00602690	0.01355414
$b=0.5$	16	-0.00000724	0.00000110	0.00000457
	32	0.00000000	0.00000000	0.00000000
	4	-0.77913195	-0.22694775	0.20990539
$a=1$	8	-0.08896065	0.02823001	0.03008962
$b=0.2$	16	-0.00167596	0.00010607	0.00000226
	32	-0.00000114	0.00000009	-0.00000001

13.5 A Collocation Method for Hypersingular Equations

In this section we consider a collocation method for solving a certain type of hypersingular integral equations such as equation (7.62) arising from the solution of the Neumann boundary value problem for the Laplace equation via a double-layer potential approach. This method exploits the relation (7.46) between the normal derivative of the double-layer potential and the single-layer potential and uses trigonometric differentiation. To this end we begin with a short description of the latter. We denote by

$$D : g \mapsto g'$$

the differentiation operator and note that $D : H^p[0,2\pi] \to H^{p-1}[0,2\pi]$ is bounded for all p with the nullspace given by the constant functions. From Section 11.3 we recall the trigonometric interpolation operator P_n and define $D_n := DP_n$, i.e., we approximate the derivative Dg of a 2π-periodic function g by the derivative $D_n g$ of the unique trigonometric polynomial $P_n g$ of the form (13.26) that interpolates $(P_n g)(t_j) = g(t_j)$ at the interpolation points $t_j = \pi j/n$ for $j = 0,\dots,2n-1$. From the Lagrange basis (11.12), by straightforward differentiation we obtain that

$$(D_n g)(t_j) = \sum_{k=0}^{2n-1} d_{k-j}^{(n)} g(t_k), \quad j = 0,\dots,2n-1,$$

where

$$d_j^{(n)} = \begin{cases} \dfrac{(-1)^j}{2} \cot \dfrac{j\pi}{2n}, & j = \pm 1,\dots,\pm(2n-1), \\[2ex] 0, & j = 0. \end{cases} \tag{13.39}$$

From Theorem 11.8, we immediately have the error estimate

$$\|D_n g - Dg\|_{q-1} \le \frac{C}{n^{p-q}} \|g\|_p, \quad 0 \le q \le p, \ \frac{1}{2} < p, \tag{13.40}$$

for all $g \in H^p[0, 2\pi]$ and some constant C depending on p and q as consequence of the boundedness of $D : H^p[0, 2\pi] \to H^{p-1}[0, 2\pi]$.

We define

$$T_0 := DS_0D + M$$

where S_0 is given by (13.31) and M is the mean value operator given by

$$M : g \mapsto \frac{1}{2\pi} \int_0^{2\pi} g(t)\, dt.$$

In view of Theorem 8.24, the hypersingular operator $T_0 : H^p[0, 2\pi] \to H^{p-1}[0, 2\pi]$ is bounded with a bounded inverse for all $p \in \mathbb{R}$. Now we consider a hypersingular equation of the form

$$T_0\varphi - DAD\varphi - B\varphi - \lambda\varphi = f \tag{13.41}$$

where λ is a constant and A and B are of the form (13.32) with infinitely differentiable 2π-periodic kernels K_A, L_A and K_B, L_B respectively such that $K_A(t, t) = 0$ for all $0 \le t \le 2\pi$. From the integrals in Lemma 8.23 we see that T_0 maps the space of trigonometric polynomials of the form (13.26) into itself. Hence, as for equation (13.30), we can use the set $t_j = j\pi/n$, $j = 0, \dots, 2n-1$, as collocation points.

Analogous to Theorem 13.20 we now can state the following result.

Theorem 13.24. *The semi-discrete collocation method (with trigonometric polynomials of the form (13.26) and collocation points $j\pi/n$, $j = 0, \dots, 2n-1$,) converges for uniquely solvable hypersingular integral equations of the form (13.41) in the Sobolev space $H^p[0, 2\pi]$ for $p \ge 1$ provided K_A, L_A, K_B, L_B and f are infinitely differentiable (and $K_A(t, t) = 0$ for all $t \in [0, 2\pi]$).*

Proof. All constants occurring in this proof depend on p. For $p \ge 1$, by Theorem 11.8 and the boundedness of $A : H^{p-1}[0, 2\pi] \to H^{p+1}[0, 2\pi]$ as established in the proof of Theorem 13.20 we obtain that

$$\|DP_nAD\varphi - DAD\varphi\|_{p-1} \le \frac{c_1}{n}\, \|\varphi\|_p \tag{13.42}$$

for all $\varphi \in H^p[0, 2\pi]$ and some constant c_1. With the triangle inequality, for $p \ge 1$ again by Theorem 11.8 we can estimate

$$\|P_nD\psi - DP_n\psi\|_{p-1} \le \|P_nD\psi - D\psi\|_{p-1} + \|D(\psi - P_n\psi)\|_{p-1} \le \frac{c_2}{n}\, \|\psi\|_{p+1}$$

for all $\psi \in H^{p+1}[0, 2\pi]$ and some constant c_2. From this, setting $\psi = AD\varphi$, we obtain that

$$\|P_nDAD\varphi - DP_nAD\varphi\|_{p-1} \le \frac{c_3}{n}\, \|\varphi\|_p \tag{13.43}$$

for all $\varphi \in H^p[0, 2\pi]$ and some constant c_3. Now we combine (13.42) and (13.43) to obtain

$$\|P_nDAD\varphi - DAD\varphi\|_{p-1} \le \frac{c_4}{n}\, \|\varphi\|_p \tag{13.44}$$

for all $\varphi \in H^p[0, 2\pi]$ and some constant c_4.

For the second part of the operator in (13.41), using Theorem 11.8 and the boundedness of $B : H^p[0, 2\pi] \to H^{p+1}[0, 2\pi]$, for $p \geq 1$ we can estimate

$$\|P_n B\varphi - B\varphi + \lambda(P_n\varphi - \varphi)\|_{p-1} \leq \frac{c_5}{n} \|\varphi\|_p$$

for all $\varphi \in H^p[0, 2\pi]$ and some constant c_5. With this estimate and (13.44) the proof can be completed as in Theorem 13.20. □

We conclude this section by describing a fully discrete variant via trigonometric differentiation. As in the previous Section 13.4, for the approximation of A and B we use the quadrature operators A_n and B_n given analogously to (13.34) and approximate DAD by $D_nA_nD_n$. For all trigonometric polynomials φ of the form (13.26), in view of $D_n\varphi = D\varphi$, we can transform

$$P_n(D_nA_nD_n - DAD)\varphi = P_nD(P_nA_n - A)D\varphi = P_nDP_n(A_n - A)D\varphi + P_nD(P_nA - A)D\varphi.$$

Note that the interpolation operators $P_n : H^p[0, 2\pi] \to H^p[0, 2\pi]$ are bounded for $p > 1/2$ as a consequence of Theorem 11.8. Now assume that $p > 3/2$. Then with the aid of Lemma 13.21, replacing p by $p - 1$ and setting $q = p - 1$, as in the proof of Theorem 13.22, for the first term on the right-hand side we can estimate

$$\|P_nDP_n(A_n - A)D\varphi\|_{p-1} \leq \frac{c_1}{n} \|\varphi\|_p$$

for all trigonometric polynomials φ of the form (13.26) and some constant c_1. For the second term, from Theorem 11.8 and the boundedness of the operator $A : H^{p-1}[0, 2\pi] \to H^{p+1}[0, 2\pi]$, we conclude that

$$\|P_nD(P_nA - A)D\varphi\|_{p-1} \leq \frac{c_2}{n} \|\varphi\|_p$$

for all $\varphi \in H^p[0, 2\pi]$ and some constant c_2. Combining both estimates we find that

$$\|P_n(D_nA_nD_n - DAD)\varphi\|_{p-1} \leq \frac{c}{n} \|\varphi\|_p \tag{13.45}$$

for all trigonometric polynomials φ of the form (13.26) and some constant c. Using (13.45) and a corresponding estimate for the operator B that can readily be obtained from Theorem 12.18 and proceeding as in the proof of Theorem 13.22 the following result on the convergence of the fully discrete method can be proven.

Theorem 13.25. *The fully discrete collocation method using the approximate operators $D_nA_nD_n$ and B_n converges for uniquely solvable hypersingular integral equations of the form (13.41) in the Sobolev space $H^p[0, 2\pi]$ for $p > 3/2$ provided the kernels and the right-hand side are infinitely differentiable (and $K_A(t, t) = 0$ for all $t \in [0, 2\pi]$).*

As in the case of Theorem 13.22 the convergence is of order $O(n^{-m})$ for all $m \in \mathbb{N}$ if the exact solution φ is infinitely differentiable.

From the Lagrange basis (11.12) and the integrals in Lemma 8.23 we obtain that

$$(P_n T_0 \varphi)(t_j) = \sum_{k=0}^{2n-1} b_{k-j}^{(n)} \varphi(t_k), \quad j = 0, \dots, 2n-1,$$

for all trigonometric polynomials φ of degree n the form (13.26) where

$$b_j^{(n)} = \begin{cases} \dfrac{1}{2n} + \dfrac{1}{4n}\,[(-1)^j - 1]\sin^{-2}\dfrac{j\pi}{2n}, & j = \pm 1, \dots, \pm(2n-1), \\[2ex] \dfrac{n^2+1}{2n}, & j = 0. \end{cases} \tag{13.46}$$

With this the fully discrete collocation method for (13.41) leads to the linear system for the coefficients in $\widetilde{\varphi}_n = \sum_{k=0}^{2n-1} \gamma_k L_k$ of the form

$$\sum_{k=0}^{2n-1} \left(U_{jk} - V_{jk} \right) \gamma_k = f(t_j), \quad j = 0, \dots, 2n-1.$$

Here the matrix

$$U_{jk} := b_{k-j}^{(n)} - \sum_{\ell=0}^{2n-1} d_{\ell-j}^{(n)} \sum_{m=0}^{2n-1} \left\{ R_{m-\ell}^{(n)} K_A(t_\ell, t_m) + \frac{1}{2n}\, L_A(t_\ell, t_m) \right\} d_{k-m}^{(n)}$$

corresponds to the operator $T_0 - A$ and the matrix

$$V_{jk} := R_{k-j}^{(n)} K_B(t_j, t_k)] + \frac{1}{2n}\, L_B(t_j, t_k) + \lambda \delta_{jk}$$

corresponds to the operator $B + \lambda I$. Note that the two summations in the expression for U_{jk} represent multiplications by Toeplitz matrices. i.e., convolutions that can be efficiently computed via fast Fourier transform techniques (see [46, 186, 202]).

Example 13.26. The hypersingular equation (13.41) with $\lambda = 0$ and kernels $K_A = K_B = L_B = 0$, and L_A given by the right-hand side in (13.38) corresponds to the hypersingular integral equation (7.62) for the ellipse with semi-axis $0 < b \le a \le 1$. The addition of the mean value term M ensures that it is uniquely solvable. Table 13.5 gives some numerical results for the error between the approximate and exact solution for the case in which the exact solution is given by

$$\varphi(t) = e^{\cos t} \cos(\sin t), \quad 0 \le t \le 2\pi.$$

Then analogous to Example 13.23 the right-hand side becomes

$$f(t) = 1 + e^{\cos t} \cos(t + \sin t) - c e^{c \cos t} \cos(t + c \sin t), \quad 0 \le t \le 2\pi,$$

with $c = (a - b)/(a + b)$. For further numerical examples we refer to [24, 146]. □

Table 13.5 Numerical results for Example 13.26

	$2n$	$t = 0$	$t = \pi/2$	$t = \pi$
	4	0.71791866	0.17589838	-0.57725146
$a = 1$	8	0.01173558	-0.00389596	-0.00530693
$b = 0.5$	16	0.00000112	-0.00000018	-0.00000063
	32	0.00000000	0.00000000	0.00000000
	4	2.39803796	0.31205324	-2.65316812
$a = 1$	8	0.10282364	-0.04584838	-0.04338518
$b = 0.2$	16	0.00165318	-0.00038918	-0.00033041
	32	0.00000118	-0.00000028	-0.00000024

13.6 The Galerkin Method

For operator equations in Hilbert spaces the projection method via orthogonal projection into finite-dimensional subspaces leads to the *Galerkin method*, named after the Russian engineer Galerkin [60]. Let X and Y be Hilbert spaces and let $A : X \to Y$ be an injective bounded linear operator. Let $X_n \subset X$ and $Y_n \subset Y$ be subspaces with $\dim X_n = \dim Y_n = n$, and let $P_n : Y \to Y_n$ be the orthogonal projection operator described in Theorem 13.3. Then $\varphi_n \in X_n$ is a solution of the projection method, generated by X_n and P_n, for the equation $A\varphi = f$ if and only if

$$(A\varphi_n, g) = (f, g), \quad g \in Y_n. \tag{13.47}$$

This follows immediately from the fact that (13.47), by Theorem 1.25, is equivalent to $P_n(A\varphi_n - f) = 0$. The equation (13.47) is called the *Galerkin equation*.

In the literature, the Galerkin method is also known as the *Petrov–Galerkin method* and the special case where $X = Y$ and $X_n = Y_n$ is also called the *Bubnov–Galerkin method*. When the operator A is self-adjoint and positive definite, as we will briefly explain, the Bubnov–Galerkin method coincides with the *Rayleigh–Ritz method*. A bounded linear operator $A : X \to X$ is called *self-adjoint* if $A = A^*$; it is called *positive definite* if it is self-adjoint and satisfies

$$(A\varphi, \varphi) > 0 \tag{13.48}$$

for all $\varphi \in X$ with $\varphi \neq 0$. A positive definite operator clearly is injective. We can define an additional scalar product on X by

$$(\varphi, \psi)_E := (A\varphi, \psi), \tag{13.49}$$

with the corresponding norm $\|\cdot\|_E$ called the *energy norm*. Consider the so-called *energy functional*, defined on X by

$$E(\varphi) := (A\varphi, \varphi) - 2\,\mathrm{Re}(f, \varphi) \tag{13.50}$$

for all $\varphi \in X$ with $f \in X$ a given element. For $f \in A(X)$ we can transform

$$E(\varphi) - E(A^{-1}f) = \|\varphi - A^{-1}f\|_E^2.$$

Hence, solving the equation $A\varphi = f$ is equivalent to minimizing the energy functional E over X. Based on the classical work of Rayleigh [203] from 1896 and Ritz [206] from 1908, the Rayleigh–Ritz method consists of an approximation by minimizing E only over a finite-dimensional subspace X_n. This, obviously, is equivalent to finding a best approximation to the exact solution with respect to X_n in the energy norm. By Theorems 1.25 and 1.26, the best approximation exists and is uniquely determined by the orthogonality condition

$$(\varphi_n, g)_E = (A^{-1}f, g)_E$$

for all $g \in X_n$. But this coincides with the Galerkin equation (13.47), i.e., the Rayleigh–Ritz method is a special case of the Galerkin method. Note that we have convergence of the Rayleigh–Ritz method if the denseness condition (13.3) is satisfied with respect to the energy norm.

First Bubnov in 1913 and then, in more details, Galerkin [60] in 1915 approached and extended this approximation method without relying on a minimization formulation. Later, Petrov [188] first considered the general situation of the form (13.47) of the Galerkin method.

Assume that $X_n = \operatorname{span}\{u_1, \ldots, u_n\}$ and $Y_n = \operatorname{span}\{v_1, \ldots, v_n\}$. Then we express φ_n as a linear combination $\varphi_n = \sum_{k=1}^n \gamma_k u_k$ and find that solving the equation (13.47) is equivalent to the linear system

$$\sum_{k=1}^n \gamma_k (Au_k, v_j) = (f, v_j), \quad j = 1, \ldots, n, \tag{13.51}$$

for the coefficients $\gamma_1, \ldots, \gamma_n$. Usually, the scalar product will have the form of an integral. Therefore, in the case of integral equations, the system (13.51) requires a double integration for each of the matrix elements. This is a disadvantage of the Galerkin method, making it considerably inferior in numerical efficiency compared with the collocation method. Its major advantage is based on the simplicity of the orthogonal projection and the exploitation of the Hilbert space structure. In particular, for integral equations of the second kind, Corollary 13.9 can always be applied, because the denseness assumption (13.3) implies pointwise convergence of the orthogonal projection operators. Hence, we can state the following theorem.

Theorem 13.27. *Let $A : X \to X$ be a compact linear operator in a Hilbert space X and assume that $I - A$ is injective. Then the Bubnov–Galerkin method converges.*

In contrast to our previous methods, the Galerkin method applies more easily to equations of the first kind. We proceed with a few general results.

Definition 13.28. A bounded linear operator $A : X \to X$ in a Hilbert space X is called *strictly coercive* if there exists a constant $c > 0$ such that

$$\mathrm{Re}(A\varphi, \varphi) \geq c\|\varphi\|^2 \tag{13.52}$$

for all $\varphi \in X$.

Theorem 13.29 ((Lax–Milgram). *In a Hilbert space X a strictly coercive bounded linear operator $A : X \to X$ has a bounded inverse $A^{-1} : X \to X$.*

Proof. Using the Cauchy–Schwarz inequality, we can estimate

$$\|A\varphi\| \, \|\varphi\| \geq \mathrm{Re}(A\varphi, \varphi) \geq c\|\varphi\|^2.$$

Hence

$$\|A\varphi\| \geq c\|\varphi\| \tag{13.53}$$

for all $\varphi \in X$. From (13.53) we observe that $A\varphi = 0$ implies $\varphi = 0$, i.e., A is injective.

Next we show that the range $A(X)$ is closed. Let $\psi \in \overline{A(X)}$ and let (ψ_n) be a sequence from $A(X)$ with $\psi_n \to \psi$, $n \to \infty$. Then we can write $\psi_n = A\varphi_n$ with some $\varphi_n \in X$, and from (13.53) we find that

$$c\|\varphi_n - \varphi_m\| \leq \|\psi_n - \psi_m\|$$

for all $n, m \in \mathbb{N}$. Therefore, (φ_n) is a Cauchy sequence in X and consequently converges: $\varphi_n \to \varphi$, $n \to \infty$, with some $\varphi \in X$. Then $\psi = A\varphi$, since A is continuous and $A(X) = \overline{A(X)}$ is proven.

Knowing now that $A(X)$ is complete, denote by $P : X \to A(X)$ the orthogonal projection operator. Let $f \in X$ be arbitrary. Then by Theorem 1.25, we have that $Pf - f \perp A(X)$. In particular, $(Pf - f, A(Pf - f)) = 0$. Hence, from (13.52) we see that $f = Pf \in A(X)$. Therefore, A is surjective. Finally, the boundedness of the inverse $\|A^{-1}\| \leq 1/c$ is a consequence of (13.53). □

Theorem 13.30. *Let $A : X \to X$ be a strictly coercive operator. Then the Bubnov–Galerkin method converges.*

Proof. As in the previous proof we can estimate

$$\|P_nA\varphi\| \, \|\varphi\| \geq \mathrm{Re}(P_nA\varphi, \varphi) = \mathrm{Re}(A\varphi, \varphi) \geq c\|\varphi\|^2$$

for all $\varphi \in X_n$, since orthogonal projections are self-adjoint (see Problem 13.1). Hence,

$$\|P_nA\varphi\| \geq c\|\varphi\| \tag{13.54}$$

for all $\varphi \in X_n$. This implies that $\widetilde{A}_n = P_nA : X_n \to X_n$ is injective and, consequently, surjective. For all $\varphi \in X$, using (13.54) and Theorem 13.3, we can estimate

$$c\|(P_nA)^{-1}P_nA\varphi\| \leq \|P_nA(P_nA)^{-1}P_nA\varphi\| = \|P_nA\varphi\| \leq \|A\| \, \|\varphi\|.$$

Therefore, $\|\widetilde{A}_n^{-1}P_nA\| \leq \|A\|/c$ for all $n \in \mathbb{N}$, and the statement follows from Theorem 13.6. □

Theorems 13.29 and 13.30 can be extended to strictly coercive operators mapping a Hilbert space X into its dual space X^* (see Problem 13.4).

We conclude our presentation of the Galerkin method by describing one more special case.

Theorem 13.31. *Let X and Y be Hilbert spaces and let $A : X \to Y$ be an injective bounded linear operator. Let $X_n \subset X$ be a finite-dimensional subspace. Then for each $f \in Y$ there exists a unique element $\varphi_n \in X_n$ such that*

$$\|A\varphi_n - f\| = \inf_{\psi \in X_n} \|A\psi - f\|.$$

It is called the least squares solution *of $A\varphi = f$ with respect to X_n and it coincides with the Petrov–Galerkin solution for the subspaces X_n and $Y_n := A(X_n)$.*

Proof. Obviously, φ_n is a least squares solution of $A\varphi = f$ with respect to X_n if and only if $A\varphi_n$ is a best approximation of f with respect to Y_n. By Theorem 1.26, the best approximation exists and is unique. The injectivity of A then implies uniqueness for the least squares solution. By Theorem 1.25, the best approximation $A\varphi_n$ is characterized by the orthogonality condition $(A\varphi_n - f, g) = 0$ for all $g \in Y_n$. But this in turn is equivalent to the Galerkin equation (13.47) for the subspaces X_n and Y_n. □

Note that the numerical implementation of the least squares solution in the case of integral equations requires a triple integration for each matrix element and a double integration for each right-hand side of the linear system (13.51), since $g_j = Au_j$ requires an additional integration. Therefore, in general, it cannot compete in efficiency with other methods.

We now outline the use of the Galerkin method for integral equations. We start with the Bubnov–Galerkin method for equations of the second kind

$$\varphi(x) - \int_G K(x, y)\varphi(y)\, dy = f(x), \quad x \in G,$$

with continuous or weakly singular kernel K in the Hilbert space $L^2(G)$. Let $X_n = \operatorname{span}\{u_1, \dots, u_n\}$ be an n-dimensional subspace of $L^2(G)$. Then we write $\varphi = \sum_{k=1}^n \gamma_k u_k$ and the Bubnov–Galerkin equations assume the form

$$\sum_{k=1}^n \gamma_k \left\{ \int_G u_k(x)\overline{u_j(x)}\, dx - \int_G \int_G K(x, y) u_k(y)\overline{u_j(x)}\, dx dy \right\}$$
$$= \int_G f(x)\overline{u_j(x)}\, dx, \quad j = 1, \dots, n, \tag{13.55}$$

for the coefficients $\gamma_1, \dots, \gamma_n$. For this system, the first term in the expression for the matrix elements can usually be evaluated in closed form, whereas the second term needs a double numerical integration. Before we proceed, we wish to point out that the matrix of the system (13.55) coincides with the system (11.36) obtained by the degenerate kernel method via orthogonal expansion.

Without entering too deeply into the numerical implementation, we consider as an example the Galerkin method using linear splines. If we pursue the same idea as in the collocation method and use interpolatory quadratures with respect to both variables, it turns out that we solve the same approximate finite-dimensional equation on the space of spline functions as in the degenerate kernel and in the collocation method via spline interpolation. Therefore the numerical results will be the same. Only the method of solution for this equation is different: In the collocation method we equate spline functions by requiring them to coincide at the interpolation grid; in the Galerkin method we require equality of scalar products with respect to a basis. In general, we may say that most of the fully discrete implementations of Galerkin methods may be interpreted as implementations of a related collocation method.

For integral equations for periodic functions using trigonometric polynomials we note that instead of setting up the Bubnov–Galerkin method in $L^2[0, 2\pi]$ we may use the Sobolev spaces $H^p[0, 2\pi]$ with $p \geq 0$. Since the trigonometric monomials are orthogonal in each of these Sobolev spaces, the orthogonal projection onto trigonometric polynomials is always given through truncation of the Fourier series. Therefore, the Bubnov–Galerkin equations are the same for all $p \geq 0$. In particular, this means that we automatically have convergence with respect to all Sobolev norms in the case of infinitely differentiable kernels and right-hand sides.

In principle, for the Galerkin method for equations of the second kind the same remarks as for the collocation method apply. As long as numerical quadratures are available, in general, the Galerkin method cannot compete in efficiency with the Nyström method. Compared with the collocation method, it is less efficient, since its matrix elements require double integrations. Therefore, in practical problems, the collocation method is the most widely used projection method for solving integral equations of the second kind, despite the fact that its error analysis is often less satisfactory than the error analysis for the Galerkin method.

Finally, we give an indication of the Petrov–Galerkin method for equations of the form (13.25) and (13.30). From Section 13.4 we recall that for both equations the leading parts $R : H^p[0, 2\pi] \to H^p[0, 2\pi]$ and $S_0 : H^p[0, 2\pi] \to H^{p+1}[0, 2\pi]$ are isomorphisms for each $p \geq 0$.

For the first equation, as subspaces X_n and Y_n of $H^p[0, 2\pi]$, we choose trigonometric polynomials of the form (13.26) and (13.27), respectively. Because R maps X_n bijectively onto Y_n, the Petrov–Galerkin equation

$$P_n R\varphi_n = P_n f$$

has a unique solution $\varphi_n \in X_n$ given by $\varphi_n = R^{-1} P_n f$. Therefore, the Petrov–Galerkin method for R converges, and by the stability Theorem 13.7 it also converges for equation (13.25). The equation (13.30) is treated analogously, using as subspaces X_n of $H^p[0, 2\pi]$ and Y_n of $H^{p+1}[0, 2\pi]$ trigonometric polynomials of the form (13.26).

We summarize these observations in the following theorem.

Theorem 13.32. *The Petrov–Galerkin method using trigonometric polynomials converges in $H^p[0, 2\pi]$ for equations of the first kind with Hilbert or logarithmic kernels.*

13.7 The Lippmann–Schwinger Equation

We will conclude this chapter by an example for an application of a projection method to a volume integral equation and begin with presenting some basic analysis on the scattering of time harmonic waves from an inhomogeneous medium. For a more comprehensive study of this field we refer the reader to Colton and Kress [32].

The propagation of time harmonic acoustic waves in three-dimensional space is governed by the Helmholtz equation for the inhomogeneous medium

$$\Delta u + \kappa^2 n u = 0 \quad \text{in } \mathbb{R}^3 \tag{13.56}$$

where u describes the pressure p of the sound wave, depending on the space variable x and the time variable t, via $p(x, t) = \operatorname{Re}\{u(x)\, e^{-i\omega t}\}$ with a positive frequency ω. Following standard notations from scattering theory, here by n we denote the refractive index of the inhomogeneous medium given by $n(x) := c_0^2/c^2(x)$ in terms of the constant speed of sound c_0 in the background medium and the space dependent speed of sound c in the inhomogeneity. The positive wave number κ is given by $\kappa = \omega/c_0$. We assume that the contrast $a := 1-n$ has compact support and, for simplicity, is continuously differentiable in $\mathbb{R}^3$. The function u is complex-valued and describes both the amplitude and the phase of the time harmonic wave. It is decomposed into the sum $u = u^i + u^s$ of the incident wave u^i satisfying the free space Helmholtz equation $\Delta u^i + \kappa^2 u^i = 0$ in $\mathbb{R}^3$ and the scattered wave u^s. A typical example for an incident wave is a plane wave $u^i(x) = e^{i\kappa\, x\cdot d}$ with a unit vector d giving the propagation direction. In order to ensure that the scattered wave is outgoing it has to satisfy the Sommerfeld radiation condition

$$r\left(\frac{\partial u^s}{\partial r} - i\kappa u^s\right) \to 0, \quad r = |x| \to \infty, \tag{13.57}$$

uniformly for all directions.

By

$$\Phi(x, y) := \frac{1}{4\pi}\, \frac{e^{i\kappa|x-y|}}{|x-y|}\,, \quad x \neq y, \tag{13.58}$$

we now denote the fundamental solution to the Helmholtz equation $\Delta u + \kappa^2 u = 0$ in $\mathbb{R}^3$. Using Green's theorem and properties of volume potentials it can be shown that the scattering problem (13.56)–(13.57) is equivalent to solving the integral equation of the second kind

$$u(x) = u^i(x) - \kappa^2 \int_{\mathbb{R}^3} \Phi(x,y)a(y)u(y)\,dy, \quad x \in \mathbb{R}^3, \tag{13.59}$$

for u which is known as the *Lippmann–Schwinger equation*. For a proof of the following theorem we refer to [32].

Theorem 13.33. *If $u \in C^2(\mathbb{R}^3)$ is a solution of the scattering problem (13.56)–(13.57), then u solves the integral equation (13.59). Conversely, if $u \in C(\mathbb{R}^3)$ is a solution of (13.59) then $u \in C^2(\mathbb{R}^3)$ and u solves (13.56)–(13.57).*

We note that in (13.59) we can replace the region of integration by any domain D such that the support of a is contained in $\bar{D}$ and look for functions in $C(\bar{D})$ that satisfy (13.59) for $x \in \bar{D}$. Then for $x \in \mathbb{R}^3 \setminus \bar{D}$ we define $u(x)$ by the right-hand side of (13.59) and obtain a continuous solution u to the Lippmann–Schwinger equation in all of $\mathbb{R}^3$.

The integral operator in (13.59) has a weakly singular kernel and hence the Riesz theory can be applied to the integral equation in $\bar{D}$, i.e., we need to show that the homogeneous integral equation only allows the trivial solution. However, this is a nontrivial task since uniqueness of the solution to the scattering problem (13.56)–(13.57) is based on a unique continuation principle for solutions of (13.56). For a detailed analysis establishing existence and uniqueness of a solution to the Lippmann–Schwinger equation we refer to [32]. Here, we confine ourselves to prove existence of a unique solution for κ sufficiently small by the Neumann series.

Theorem 13.34. *Suppose that the support of a is contained in a ball of radius R centered at the origin and that $\kappa^2 R^2 \|a\|_\infty < 2$. Then there exists a unique solution to the Lippmann–Schwinger equation (13.59).*

Proof. It suffices to solve (13.59) for $u \in C(B)$ with the ball $B := \{x \in \mathbb{R}^3 : |x| \leq R\}$. On the Banach space $C(B)$, we define the operator $V : C(B) \to C(B)$ by

$$(Vu)(x) := \kappa^2 \int_B \Phi(x,y)a(y)u(y)\,dy, \quad x \in B. \tag{13.60}$$

From Theorem 2.13 and Problem 2.4 we conclude that

$$\|V\|_\infty \leq \frac{\kappa^2 \|a\|_\infty}{4\pi} \max_{x \in B} \int_B \frac{1}{|x-y|}\,dy.$$

Geometric considerations show that the maximum of the integral on the right-hand side is obtained for $x = 0$ leading to $\|V\|_\infty \leq \kappa^2 R^2 \|a\|_\infty / 2$. Now the proof is completed by using the Neumann series Theorem 2.14. □

Following Hohage [98, 99] and Saranen and Vainikko [213], we now present the main ideas of a fast numerical solution method for the Lippmann–Schwinger equation that was proposed by Vainikko [238]. In view of the possibility to scale the geometry, without loss of generality, we assume that the support of a is contained in a ball of radius $R \leq \pi/2$ centered at the origin. The main idea of Vainikko's approach is to periodize the integral equation in order to make it accessible to fast Fourier transform techniques. Before doing so it is convenient to transform (13.59) by multiplication with a equivalently into

$$\varphi + aV\varphi = f \tag{13.61}$$

for the unknown $\varphi := au$ and the right-hand side $f := au^i$. (Recall that it suffices to solve (13.59) on the support of a.)

Obviously the solution of the Lippmann–Schwinger equation does not change if we modify its kernel outside a ball centered at the origin with radius larger than or equal to the diameter $2R$ of the support of m. We set $B := B[0;\pi]$ and $W := [-\pi,\pi]^3$ and define a new kernel function by

$$K(x) := \begin{cases} \kappa^2 \Phi(x,0), & x \in B, \\ 0, & x \in W \setminus B. \end{cases}$$

Then we extend the functions K, f, a and φ from W to all of $\mathbb{R}^3$ as 2π-periodic functions with respect to the three variables in $x = (x_1, x_2, x_3)$ and arrive at the periodic integral equation

$$\varphi(x) + a(x) \int_W K(x-y)\varphi(y)\,dy = f(x), \quad x \in W, \tag{13.62}$$

that is equivalent to the Lippmann–Schwinger equation. We define an operator $A : L^2(W) \to L^2(W)$ by

$$(A\varphi)(x) := \int_W K(x-y)\varphi(y)\,dy, \quad x \in W, \tag{13.63}$$

and rewrite (13.62) in operator form as

$$\varphi + aA\varphi = f \tag{13.64}$$

with a interpreted as a multiplication operator.

Analogous to Chapter 8, for $0 \leq p < \infty$ we introduce the Sobolev space $H^p(W)$ as the Hilbert space of all functions $\varphi \in L^2(W)$ with

$$\sum_{m \in \mathbb{Z}^3} (1 + |m|^2)^p |\hat{\varphi}_m|^2 < \infty$$

for the Fourier coefficients

$$\hat{\varphi}_m := \frac{1}{(2\pi)^3} \int_W \varphi(x) e^{-im\cdot x} dx, \quad m \in \mathbb{Z}^3,$$

of φ. The scalar product and the norm on $H^p(W)$ are defined analogous to that on $H^p[0, 2\pi]$ in Theorem 8.2. We leave it as an exercise to formulate and prove the analogues of Theorems 8.3, 8.4 (for $p > 3/2$) and 8.5 and their corollaries for the three-dimensional case.

To establish the mapping properties of the above operator A in these Sobolev spaces we require the following lemma on the Fourier coefficients of the kernel function K.

Lemma 13.35. *The Fourier coefficients $\hat{K}_m$ of the kernel function K are given by*

$$\frac{8\pi^3}{\kappa^2}\,(\kappa^2 - |m|^2)\hat{K}_m = \left\{\cos(\pi|m|) - \frac{i\kappa}{|m|}\,\sin(\pi|m|)\right\} e^{i\kappa\pi} - 1$$

for $m \neq 0$ and

$$8\pi^3 \hat{K}_0 = (1 - i\kappa\pi)e^{i\kappa\pi} - 1.$$

Proof. With essentially the same proof as in Theorem 6.5, for a function $u \in C^2(B)$ the Green's representation formula

$$u(x) = \int_{\partial B} \left\{\frac{\partial u}{\partial \nu}(y)\,\Phi(x, y) - u(y)\,\frac{\partial \Phi(x, y)}{\partial \nu(y)}\right\} ds(y)$$

$$- \int_B \left\{\Delta u(y) + \kappa^2 u(y)\right\} \Phi(x, y)\, dy, \quad x \in B,$$

in terms of the fundamental solution (13.58) to the Helmholtz equation can be established (see also [32]). As usual ν denotes the unit normal to ∂B directed into the exterior of B. From this, inserting $u = f_{-m}$ with the trigonometric basis functions

$$f_m(z) := e^{i\, m\cdot z}, \quad z \in \mathbb{R}^3,\ m \in \mathbb{Z}^3,$$

and setting $x = 0$ we obtain that

$$\frac{1}{\kappa^2}\,(\kappa^2 - |m|^2) \int_B K f_{-m}\, dy = \frac{e^{i\kappa\pi}}{4\pi^2} \int_{\partial B} \left\{\frac{\partial f_{-m}}{\partial \nu} - \left[i\kappa - \frac{1}{\pi}\right] f_{-m}\right\} ds - 1.$$

Utilizing symmetry, elementary integrations yield

$$\int_{\partial B} f_{-m}\, ds = \frac{4\pi^2}{|m|}\,\sin(\pi|m|)$$

and

$$\int_{\partial B} \frac{\partial f_{-m}}{\partial \nu}\, ds = 4\pi^2 \cos(\pi|m|) - \frac{4\pi}{|m|}\,\sin(\pi|m|)$$

for $m \neq 0$ and

$$\int_{\partial B} f_0\, ds = 4\pi^3 \quad \text{and} \quad \int_{\partial B} \frac{\partial f_0}{\partial \nu}\, ds = 0.$$

Piecing the above together yields the assertion of the lemma. □

Theorem 13.36. *For $p \geq 0$, the operator A defined by (13.63) is a bounded linear operator from $H^p(W)$ into $H^{p+2}(W)$.*

Proof. Writing

$$\int_W K(x-y)e^{im\cdot y}\,dy = e^{im\cdot x}\int_W K(z)e^{-im\cdot z}\,dz$$

we observe that

$$Af_m = (2\pi)^3 \hat{K}_m f_m, \quad m \in \mathbb{Z}^3. \tag{13.65}$$

Then from Lemma 13.35 we conclude that

$$|\hat{K}_m| \leq \frac{c}{1+|m|^2}$$

for all $m \in \mathbb{Z}$ and some positive constant c whence the statement of the theorem follows. □

We now proceed to describe the approximate solution of (13.64) by a collocation method using trigonometric polynomials. To this end we extend the trigonometric interpolation as considered in Section 11.3 for 2π-periodic functions of one variable to the three-dimensional case. Slightly deviating from the presentation in Section 11.3, where for the case of an even number $2n$ of interpolation points we wanted the interpolation of real-valued functions to remain real-valued, for $n \in \mathbb{N}$ we define the n^3-dimensional space of trigonometric polynomials

$$T_n := \operatorname{span}\left\{f_m : m \in \mathbb{Z}^3_n\right\}$$

where

$$\mathbb{Z}^3_n := \left\{m = (m_1, m_2, m_3) \in \mathbb{Z}^3 : -\frac{n}{2} \leq m_1, m_2, m_3 < \frac{n}{2}\right\}.$$

The trigonometric interpolation operator P_n from the space of 2π-periodic continuous functions $C_{2\pi}$ into T_n is well defined by the interpolation property

$$(P_n g)(mh) = g(mh), \quad m \in \mathbb{Z}^3_n,$$

where $h := 2\pi/n$. Its Fourier coefficients $\widehat{(P_n g)}_j$ are given in terms of the nodal values of g by

$$\widehat{(P_n g)}_m = \frac{1}{n^3}\sum_{k\in\mathbb{Z}^3_n} g(kh)e^{-ihk\cdot m}, \tag{13.66}$$

i.e., by a discrete Fourier transform that can be implemented by fast Fourier transform techniques using $O(n^3 \log n)$ arithmetic operations (see [46, 186, 202]). Obviously, evaluating the nodal values $(P_n g)(mh)$ of $P_n g$ from the Fourier coefficients $\widehat{(P_n g)}_j$ again amounts to performing a discrete Fourier transform.

Analogously to Theorem 11.8 the interpolation error can be estimated by

$$\|P_n g - g\|_q \le \frac{C}{n^{p-q}}\,\|g\|_p, \quad 0 \le q \le p, \ \frac{3}{2} < p, \tag{13.67}$$

for all $g \in H^p(W)$ and some constant C depending on p and q (see [213]). As pointed out above, each function $g \in H^p(W)$ is continuous if $p > 3/2$.

Now we are in the position to apply trigonometric collocation to the periodized Lippmann–Schwinger equation (13.64) and approximate its solution φ by solving

$$\varphi_n + P_n aA\varphi_n = P_n f \tag{13.68}$$

for $\varphi_n \in T_n$. In view of Lemma 13.35 the approximating equation (13.68) is fully discrete since for $\psi \in T_n$ from (13.65) we have that

$$(A\psi)(jh) = (2\pi)^3 \sum_{m\in\mathbb{Z}_n^3} e^{ih\,j\cdot m}\hat{\psi}_m \hat{K}_m, \quad j \in \mathbb{Z}_n^3, \tag{13.69}$$

in terms of the Fourier coefficients $\hat{\psi}_m$ of ψ. Putting (13.66) and (13.69) together, we observe that solving (13.68) for the trigonometric polynomial φ_n is equivalent to solving the linear system

$$\varphi_n(jh) + a(jh)\,\frac{(2\pi)^3}{n^3} \sum_{m\in\mathbb{Z}_n^3} \sum_{k\in\mathbb{Z}_n^3} e^{ih\,(j-k)\cdot m}\hat{K}_m \varphi_n(kh) = f(jh), \quad j \in \mathbb{Z}_n^3, \tag{13.70}$$

for the nodal values of φ_n. Because of the size of the linear system (13.70) that is required for a reasonable accuracy of the approximation, iterative solution techniques need to be applied such as the two-grid iterations as described in the following Chapter 14. From the above it is obvious that in each iteration step the evaluation of $P_n aA\varphi_n$ amounts to two consecutive discrete Fourier transforms that can be performed with $O(n^3 \log n)$ arithmetic operations.

We conclude with the following convergence result.

Theorem 13.37. *Assume that the contrast a is twice continuously differentiable. Then the collocation method (13.68) with trigonometric polynomials for the Lippmann–Schwinger equation (13.64) converges uniformly.*

Proof. From the error estimate (13.67), the three-dimensional versions of Theorem 8.4 and Corollary 8.8, and Theorem 13.36 it follows that

$$\|P_n(aA\varphi) - aA\varphi\|_\infty \le c_1\|P_n(aA\varphi) - aA\varphi\|_{H^q} \le c_2 n^{q-2}\|aA\varphi\|_{H^2}$$

$$\le c_3 n^{q-2}\|A\varphi\|_{H^2} \le c_4 n^{q-2}\|\varphi\|_{H^0} \le c_5 n^{q-2}\|\varphi\|_\infty$$

for all $\varphi \in C_{2\pi}$, all $3/2 < q < 2$, and some constants c_1, c_2, c_3, c_4, and c_5 depending on q. Hence $\|P_n A - A\|_\infty \le c_5\, n^{q-2}$ for all $3/2 < q < 2$ and the proof is completed using Theorem 13.10. □

Problems

13.1. Show that orthogonal projection operators are self-adjoint.

13.2. Let $A : X \to X$ be a bounded positive self-adjoint operator in a Hilbert space X. Choose $u_0 \in X$ and define $u_j = Au_{j-1}$ for $j = 1, \dots, n-1$. Show that the Bubnov–Galerkin equations for $A\varphi = f$ with respect to the so-called *Krylov subspaces* $X_n = \text{span}\{u_0, \dots, u_{n-1}\}$ are uniquely solvable for each $n \in \mathbb{N}$. Moreover, if f is in the closure of $\text{span}\{A^j u_0 : j = 0, 1, \dots\}$, then the Bubnov–Galerkin approximation φ_n converges to the solution of $A\varphi = f$.

Show that in the special case $u_0 = f$ the approximations φ_n can be computed iteratively by the formulas $\varphi_0 = 0$, $p_0 = f$ and

$$\begin{aligned}
\varphi_{n+1} &= \varphi_n - \alpha_n p_n, \\
p_n &= r_n + \beta_{n-1} p_{n-1}, \\
r_n &= r_{n-1} - \alpha_{n-1} A p_{n-1}, \\
\alpha_{n-1} &= (r_{n-1}, p_{n-1})/(Ap_{n-1}, p_{n-1}), \\
\beta_{n-1} &= -(r_n, Ap_{n-1})/(Ap_{n-1}, p_{n-1}).
\end{aligned}$$

Here r_n is the residual $r_n = A\varphi_n - f$. This is the *conjugate gradient method* of Hestenes and Stiefel [92].

13.3. Under the assumptions of Theorem 13.27 let φ_n be the Bubnov–Galerkin solution. Consider the iterated Bubnov–Galerkin solution $\widetilde{\varphi}_n$, defined by

$$\widetilde{\varphi}_n := A\varphi_n + f$$

and show that

$$\|\widetilde{\varphi}_n - \varphi\| \le c_n \|\varphi_n - \varphi\|$$

where $c_n \to 0$, $n \to \infty$. This more rapid convergence of the iterated Bubnov–Galerkin solution is called *superconvergence* (see Sloan [221]).
Hint: Show first that $\widetilde{\varphi}_n - AP_n\widetilde{\varphi}_n = f$, then that $\|AP_n - A\| \to 0$, $n \to \infty$, and finally that $\widetilde{\varphi}_n - \varphi = (I - AP_n)^{-1}(AP_n - A)(\varphi - P_n\varphi)$.

13.4. A bounded linear operator $A : X \to X^*$ mapping a Hilbert space X into its dual space X^* is called *strictly coercive* if there exists a constant $c > 0$ such that

$$\text{Re}(A\varphi)(\varphi) \ge c\,\|\varphi\|^2$$

for all $\varphi \in X$. Formulate and prove extensions of Theorems 13.29 and 13.30.

13.5. Apply the results of Problem 13.4 to the integral equation (13.30) in the Sobolev space $H^{-1/2}[0, 2\pi]$.

Chapter 14
Iterative Solution and Stability

The approximation methods for integral equations described in Chapters 11–13 lead to full linear systems. Only if the number of unknowns is reasonably small may these equations be solved by direct methods like Gaussian elimination. But, in general, a satisfying accuracy of the approximate solution to the integral equation will require a comparatively large number of unknowns, in particular for integral equations in more than one dimension. Therefore iterative methods for the resulting linear systems will be preferable. For this, in principle, in the case of positive definite symmetric matrices the classical conjugate gradient method (see Problem 13.2) can be used. In the general case, when the matrix is not symmetric more general Krylov subspace iterations may be used among which a method called *generalized minimum residual method (GMRES)* due to Saad and Schultz [211] is widely used. Since there is a large literature on these and other general iteration methods for large linear systems (see Freud, Golub, and Nachtigal [56], Golub and van Loan [67], Greenbaum [69], Saad [210], and Trefethen and Bau [235], among others), we do not intend to present them in this book. At the end of this chapter we will only briefly describe the main idea of the *panel clustering methods* and the *fast multipole methods* based on iterative methods and on a speed-up of matrix-vector multiplications for the matrices arising from the discretization of integral equations.

Instead of describing general iterative methods for linear systems, however, we will discuss two-grid and multigrid iterations that are especially suitable for solving the linear systems arising from the numerical solution of integral equations of the second kind.

14.1 Stability of Linear Systems

We will start this chapter by briefly considering the question of stability of the linear systems arising in the discretization of integral equations. Usually, in the course of any numerical procedure for solving linear systems, errors will be introduced. For example, in the collocation and Galerkin method, errors occur through the numerical

R. Kress, *Linear Integral Equations*, Applied Mathematical Sciences 82,
DOI 10.1007/978-1-4614-9593-2_14, © Springer Science+Business Media New York 2014

approximations of the matrix elements and the right-hand side of the linear systems. A linear system is called *stable* if small changes in the data of the system cause only small changes in the solution. We will make this notion more precise by introducing the concept of a condition number.

Definition 14.1. Let X and Y be normed spaces and let $A : X \to Y$ be a bounded linear operator with a bounded inverse $A^{-1} : Y \to X$. Then

$$\operatorname{cond}(A) := \|A\|\,\|A^{-1}\|$$

is called the *condition number* of A.

Since $1 = \|I\| = \|AA^{-1}\| \le \|A\|\,\|A^{-1}\|$, we always have $\operatorname{cond}(A) \ge 1$. The following theorem shows that the condition number may serve as a measure for stability.

Theorem 14.2. *Let X and Y be Banach spaces, let $A : X \to Y$ be a bijective bounded linear operator, and let $A^\delta : X \to Y$ be a bounded linear operator such that $\|A^{-1}\|\,\|A^\delta - A\| < 1$. Assume that φ and φ^δ are solutions of the equations*

$$A\varphi = f \tag{14.1}$$

and

$$A^\delta \varphi^\delta = f^\delta, \tag{14.2}$$

respectively. Then

$$\frac{\|\varphi^\delta - \varphi\|}{\|\varphi\|} \le \frac{\operatorname{cond}(A)}{1 - \operatorname{cond}(A)\dfrac{\|A^\delta - A\|}{\|A\|}} \left\{ \frac{\|f^\delta - f\|}{\|f\|} + \frac{\|A^\delta - A\|}{\|A\|} \right\}. \tag{14.3}$$

Proof. Note that the inverse operator A^{-1} is bounded by the open mapping Theorem 10.8. Writing $A^\delta = A[I + A^{-1}(A^\delta - A)]$, by Theorem 10.1 we observe that the inverse operator $(A^\delta)^{-1} = [I + A^{-1}(A^\delta - A)]^{-1}A^{-1}$ exists and is bounded by

$$\|(A^\delta)^{-1}\| \le \frac{\|A^{-1}\|}{1 - \|A^{-1}\|\,\|A^\delta - A\|}. \tag{14.4}$$

From (14.1) and (14.2) we find that

$$A^\delta(\varphi^\delta - \varphi) = f^\delta - f - (A^\delta - A)\varphi,$$

whence

$$\varphi^\delta - \varphi = (A^\delta)^{-1}\{f^\delta - f - (A^\delta - A)\varphi\}$$

follows. Now we can estimate

$$\|\varphi^\delta - \varphi\| \le \|(A^\delta)^{-1}\| \left\{ \|f^\delta - f\| + \|A^\delta - A\|\,\|\varphi\| \right\}$$

and insert (14.4) to obtain

$$\frac{\|\varphi^\delta - \varphi\|}{\|\varphi\|} \le \frac{\operatorname{cond}(A)}{1 - \|A^{-1}\|\,\|A^\delta - A\|}\left\{\frac{\|f^\delta - f\|}{\|A\|\,\|\varphi\|} + \frac{\|A^\delta - A\|}{\|A\|}\right\}.$$

From this the assertion follows with the aid of $\|A\|\,\|\varphi\| \ge \|f\|$. □

For finite-dimensional approximations of a given operator equation we have to distinguish three condition numbers: namely, the condition numbers of the original operator, of the approximating operator as mappings from the Banach space X into the Banach space Y, and of the finite-dimensional system that has to be set up for the actual numerical solution. This latter system we can influence, for example, in the collocation and Galerkin methods by the choice of the basis for the approximating subspaces.

Consider an equation of the second kind $\varphi - A\varphi = f$ in a Banach space X and approximating equations $\varphi_n - A_n\varphi_n = f_n$ under the assumptions of Theorem 10.1, i.e., norm convergence, or Theorem 10.12, i.e., collective compactness and pointwise convergence. Then from Theorems 10.1, 10.6, and 10.12 it follows that the condition numbers $\operatorname{cond}(I - A_n)$ are uniformly bounded. Hence, for the condition of the approximating scheme, we mainly have to be concerned with the condition of the linear system for the actual computation of the solution of $\varphi_n - A_n\varphi_n = f_n$.

For the discussion of the condition number for the Nyström method we recall the notations of Section 12.2. We have to relate the condition number for the numerical quadrature operator

$$(A_n\varphi)(x) := \sum_{k=1}^{n} \alpha_k K(x, x_k)\varphi(x_k), \quad x \in G,$$

as operator $A_n : C(G) \to C(G)$ in the Banach space $C(G)$ to the condition number for the matrix operator $\widetilde{A}_n : \mathbb{C}^n \to \mathbb{C}^n$ given by

$$(\widetilde{A}_n\Phi)_j := \sum_{k=1}^{n} \alpha_k K(x_j, x_k)\Phi_k, \quad j = 1, \dots, n,$$

for $\Phi = (\Phi_1, \dots, \Phi_n) \in \mathbb{C}^n$. We choose functions $L_j \in C(G)$ with the properties $\|L_j\|_\infty = 1$ and $L_j(x_j) = 1$ for $j = 1, \dots, n$, such that $L_j(x_k) = 0$ for $j \ne k$. Then we introduce linear operators $R_n : C(G) \to \mathbb{C}^n$ and $M_n : \mathbb{C}^n \to C(G)$ by

$$R_n f := (f(x_1), \dots, f(x_n)), \quad f \in C(G),$$

and

$$M_n\Phi := \sum_{j=1}^{n} \Phi_j L_j, \quad \Phi = (\Phi_1, \dots, \Phi_n) \in \mathbb{C}^n,$$

and have $\|R_n\|_\infty = \|M_n\|_\infty = 1$ (see Problem 14.1). From Theorem 12.7 we conclude that

$$(I - \widetilde{A}_n) = R_n(I - A_n)M_n$$

and

$$(I - \widetilde{A}_n)^{-1} = R_n(I - A_n)^{-1}M_n.$$

From these relations we immediately obtain the following theorem.

Theorem 14.3. *For the Nyström method the condition numbers for the linear system are uniformly bounded.*

This theorem states that the Nyström method essentially preserves the stability of the original integral equation.

For the collocation method we recall Section 13.3 and introduce operators $E_n, \widetilde{A}_n : \mathbb{C}^n \to \mathbb{C}^n$ by

$$(E_n\Phi)_j := \sum_{k=1}^{n} u_k(x_j)\Phi_k, \quad j = 1, \dots, n,$$

and

$$(\widetilde{A}_n\Phi)_j := \sum_{k=1}^{n} (Au_k)(x_j)\Phi_k, \quad j = 1, \dots, n,$$

for $\Phi = (\Phi_1, \dots, \Phi_n) \in \mathbb{C}^n$. Since $X_n = \operatorname{span}\{u_1, \dots, u_n\}$ is assumed to be unisolvent, the operator E_n is invertible. In addition, let the operator $W_n : \mathbb{C}^n \to C(G)$ be defined by

$$W_n\gamma := \sum_{k=1}^{n} \gamma_k u_k$$

for $\gamma = (\gamma_1, \dots, \gamma_n)$ and recall the operators R_n and M_n from above. Then we have $W_n = P_n M_n E_n$. From (13.19) and (13.21) we can conclude that

$$(E_n - \widetilde{A}_n) = R_n P_n(I - A)W_n$$

and

$$(E_n - \widetilde{A}_n)^{-1} = E_n^{-1}R_n(I - P_nA)^{-1}P_nM_n.$$

From these three relations and the fact that by Theorems 10.6 and 13.10 the sequence of operators $(I - P_nA)^{-1}P_n$ is uniformly bounded, we obtain the following theorem.

Theorem 14.4. *Under the assumptions of Theorem 13.10, for the collocation method the condition number of the linear system satisfies*

$$\operatorname{cond}(E_n - \widetilde{A}_n) \le C\|P_n\|_\infty^2 \operatorname{cond} E_n$$

for all sufficiently large n and some constant C.

This theorem suggests that the basis functions must be chosen with caution. For a poor choice, like monomials, the condition number of E_n can grow quite rapidly. However, for the Lagrange basis, E_n is the identity matrix with condition number one. In addition, $\|P_n\|$ enters in the estimate on the condition number of the linear system, and, for example, for polynomial or trigonometric polynomial interpolation we have $\|P_n\| \to \infty$, $n \to \infty$ (see the estimate (11.15)). A discussion of the condition number for the Galerkin method can be carried out quite similarly (see Problem 14.2).

Summarizing, we can state that for equations of the second kind the Nyström method is generically stable whereas the collocation and Galerkin methods may suffer from instabilities due to a poor choice of the basis for the approximating subspaces.

14.2 Two-Grid Methods

Let X be a Banach space and let $A : X \to X$ be a bounded linear operator such that $I-A$ is bijective. As described in previous chapters we replace the operator equation of the second kind

$$\varphi - A\varphi = f \tag{14.5}$$

by a sequence of approximating equations

$$\varphi_n - A_n\varphi_n = f_n \tag{14.6}$$

leading to finite-dimensional linear systems. We assume the sequence (A_n) of bounded linear operators $A_n : X \to X$ to be either norm convergent or collectively compact and pointwise convergent such that either Theorem 10.1 or Theorem 10.12 may be applied to yield existence and uniqueness of a solution to the approximating equation (14.6) for all sufficiently large n. Our analysis includes projection methods in the setting of Theorem 13.10.

We note that the index n indicates different *levels* of discretization, i.e., different numbers of quadrature points in the Nyström method, or different dimensions of the approximating subspaces in the degenerate kernel or projection methods. Referring to the quadrature and collocation methods, we will use the term *grid* instead of level. Deviating from the notation in preceding chapters, here the number of unknowns in the linear system on the level n will be denoted by z_n and will be different from n. Frequently, in particular for one-dimensional integral equations, the number of unknowns z_n on the level n will double the number of unknowns z_{n-1} on the preceding level $n-1$ as in all our numerical tables in Chapters 11–13.

Assume that we already have an approximate solution $\varphi_{n,0}$ of (14.6) with a *residual*

$$r_n := f_n - (I - A_n)\varphi_{n,0}.$$

Then we try to improve on the accuracy by writing

$$\varphi_{n,1} = \varphi_{n,0} + \delta_n. \tag{14.7}$$

For $\varphi_{n,1}$ to solve equation (14.6) the correction term δ_n has to satisfy the *residual correction* equation

$$\delta_n - A_n \delta_n = r_n.$$

We observe that the correction term δ_n, in general, will be small compared with $\varphi_{n,0}$, and therefore it is unnecessary to solve the residual correction equation exactly. Hence we write

$$\delta_n = B_n r_n,$$

where the bounded linear operator B_n is some approximation for the inverse operator $(I - A_n)^{-1}$ of $I - A_n$. Plugging this into (14.7) we obtain

$$\varphi_{n,1} = [I - B_n(I - A_n)]\varphi_{n,0} + B_n f_n$$

as our new approximate solution to (14.6). Repeating this procedure yields the *defect correction iteration* defined by

$$\varphi_{n,i+1} := [I - B_n(I - A_n)]\varphi_{n,i} + B_n f_n, \quad i = 0, 1, 2, \ldots, \tag{14.8}$$

for the solution of (14.6). By Theorem 2.15, the iteration (14.8) converges to the unique solution ψ_n of $B_n(I - A_n)\psi_n = B_n f_n$ provided

$$\|I - B_n(I - A_n)\| < 1,$$

or by Theorem 10.16, if the spectral radius of $I - B_n(I - A_n)$ is less than one. Since the unique solution φ_n of $\varphi_n - A_n\varphi_n = f_n$ trivially satisfies $B_n(I - A_n)\varphi_n = B_n f_n$, if the iteration scheme (14.8) converges it converges to the unique solution of (14.6). For a rapid convergence it is desirable that the norm or spectral radius be close to zero. For a more complete introduction into this residual correction principle, we refer to Stetter [227].

Keeping in mind that the operators A_n approximate the operator A, an obvious choice for an approximate inverse B_n is given by the correct inverse $B_n = (I - A_m)^{-1}$ of $I - A_m$ for some level $m < n$ corresponding to a coarser discretization. For this so-called *two-grid method*, we can simplify (14.8) using

$$I - (I - A_m)^{-1}(I - A_n) = (I - A_m)^{-1}(A_n - A_m).$$

Following Brakhage [19] and Atkinson [10], we will consider the two special cases where we use either the preceding level $m = n - 1$ or the coarsest level $m = 0$.

Theorem 14.5. *Assume that the sequence of operators $A_n : X \to X$ is either norm convergent $\|A_n - A\| \to 0$, $n \to \infty$, or collectively compact and pointwise convergent*

$A_n\varphi \to A\varphi$, $n \to \infty$, for all $\varphi \in X$. Then the two-grid iteration

$$\varphi_{n,i+1} := (I - A_{n-1})^{-1}\{(A_n - A_{n-1})\varphi_{n,i} + f_n\}, \quad i = 0, 1, 2, \ldots, \tag{14.9}$$

using two consecutive grids converges, provided n is sufficiently large. The two-grid iteration

$$\varphi_{n,i+1} := (I - A_0)^{-1}\{(A_n - A_0)\varphi_{n,i} + f_n\}, \quad i = 0, 1, 2, \ldots, \tag{14.10}$$

using a fine and a coarse grid converges, provided the approximation A_0 is already sufficiently close to A.

Proof. Let (A_n) be norm convergent. Then from the estimate (10.1) we observe that $\|(I - A_n)^{-1}\| \le C$ for all sufficiently large n with some constant C. Then the statement on the scheme (14.9) follows from

$$\left\|(I - A_{n-1})^{-1}(A_n - A_{n-1})\right\| \le C\|A_n - A_{n-1}\| \to 0, \quad n \to \infty. \tag{14.11}$$

For the scheme (14.10) we assume that the coarsest grid is chosen such that

$$\|A_n - A_0\| < 1/2C$$

for all $n \ge 0$. Then $\|(I - A_0)^{-1}(A_n - A_0)\| < 1$ for all $n \in \mathbb{N}$. Note that

$$\left\|(I - A_0)^{-1}(A_n - A_0)\right\| \to \left\|(I - A_0)^{-1}(A - A_0)\right\| > 0, \quad n \to \infty. \tag{14.12}$$

Now assume that the sequence (A_n) is collectively compact and pointwise convergent. Then the limit operator A is also compact and, by Theorem 10.12, the inverse operators $(I - A_n)^{-1}$ exist and are uniformly bounded for sufficiently large n. Hence the sequence $(\widetilde{A}_n)$ defined by

$$\widetilde{A}_n := (I - A_{n-1})^{-1}(A_n - A_{n-1})$$

is collectively compact. From the pointwise convergence $A_n\varphi - A_{n-1}\varphi \to 0$, $n \to \infty$, for all $\varphi \in X$, by Theorem 10.10, we conclude that

$$\left\|\{(I - A_{n-1})^{-1}(A_n - A_{n-1})\}^2\right\| \to 0, \quad n \to \infty. \tag{14.13}$$

Again by Theorem 10.10, we may choose the coarsest grid such that

$$\left\|(I - A_0)^{-1}(A_m - A)(I - A_0)^{-1}(A_n - A_0)\right\| < \frac{1}{2}$$

for all $m, n \ge 0$. This implies

$$\left\|\left\{(I - A_0)^{-1}(A_n - A_0)\right\}^2\right\| < 1 \tag{14.14}$$

for all $n \in \mathbb{N}$. Now the statement follows from (14.13) and (14.14) and the variant of Theorem 2.15 discussed in Problem 2.2. □

Note that, in view of Theorem 12.8, for the Nyström method we cannot expect that

$$\left\|(I - A_{n-1})^{-1}(A_n - A_{n-1})\right\| \to 0, \quad n \to \infty,$$

instead of (14.13) or that

$$\left\|(I - A_0)^{-1}(A_n - A_0)\right\| < 1$$

instead of (14.14). Based on this observation, Brakhage [19] suggested a further variant of the defect correction iteration (see Problem 14.3).

Comparing the two variants described in Theorem 14.5, we observe from (14.11) to (14.14) that, in general, we have a more rapid convergence in the first case. Actually the convergence rate for (14.9) tends to zero when $n \to \infty$. But for a value judgment we also have to take into account the computational effort for each iteration step. For the actual computation we write the defect correction equation in the form

$$\varphi_{n,i+1} - A_m \varphi_{n,i+1} = (A_n - A_m)\varphi_{n,i} + f_n, \tag{14.15}$$

indicating that, given $\varphi_{n,i}$, we have to solve a linear system for the unknown $\varphi_{n,i+1}$. From this we observe that for the evaluation of the right-hand side in (14.15) we need to multiply a $z_n \times z_n$ matrix with a z_n vector requiring, in general, $O(z_n^2)$ operations. Then solving (14.15) for $\varphi_{n,i+1}$ means directly solving a linear system on the mth level with z_m unknowns and, in general, requires $O(z_m^3)$ operations. Therefore, for the second variant, working with the coarsest grid we may neglect the computational effort for the solution of the linear system and remain with $O(z_n^2)$ operations for each iteration step. However, for the first variant, the effort for the solution of the linear system dominates: Each iteration step needs $O(z_{n-1}^3)$ operations. In particular, when the discretization is set up such that the number of unknowns is doubled when we proceed from the level $n-1$ to the next level n, then the computational effort for each iteration step in (14.9) is roughly 1/8th of the effort for solving the linear system on the level n directly. Hence, we have rapid convergence but each iteration step is still very costly.

We indicate the numerical implementation of the defect correction iteration for the Nyström method using the notations introduced in Chapter 12. Here the approximate operators are given by

$$(A_n\varphi)(x) = \sum_{k=1}^{z_n} \alpha_k^{(n)} K(x, x_k^{(n)})\varphi(x_k^{(n)}), \quad x \in G.$$

Each iteration step first requires the evaluation of the right-hand side

$$g_{n,i} := f_n + (A_n - A_m)\varphi_{n,i}$$

of (14.15) at the z_m quadrature points $x_j^{(m)}$, $j = 1, \dots, z_m$, on the level m and at the z_n quadrature points $x_j^{(n)}$, $j = 1, \dots, z_n$, on the level n. The corresponding computations

$$g_{n,i}(x_j^{(m)}) = f_n(x_j^{(m)}) + \sum_{k=1}^{z_n} \alpha_k^{(n)} K(x_j^{(m)}, x_k^{(n)}) \varphi_{n,i}(x_k^{(n)})$$
$$- \sum_{k=1}^{z_m} \alpha_k^{(m)} K(x_j^{(m)}, x_k^{(m)}) \varphi_{n,i}(x_k^{(m)}), \quad j = 1, \dots, z_m,$$

and

$$g_{n,i}(x_j^{(n)}) = f_n(x_j^{(n)}) + \sum_{k=1}^{z_n} \alpha_k^{(n)} K(x_j^{(n)}, x_k^{(n)}) \varphi_{n,i}(x_k^{(n)})$$
$$- \sum_{k=1}^{z_m} \alpha_k^{(m)} K(x_j^{(n)}, x_k^{(m)}) \varphi_{n,i}(x_k^{(m)}), \quad j = 1, \dots, z_n,$$

require $O(z_n^2)$ operations. Then, according to Theorem 12.7, we have to solve the linear system

$$\varphi_{n,i+1}(x_j^{(m)}) - \sum_{k=1}^{z_m} \alpha_k^{(m)} K(x_j^{(m)}, x_k^{(m)}) \varphi_{n,i+1}(x_k^{(m)}) = g_{n,i}(x_j^{(m)})$$

for the values $\varphi_{n,i+1}(x_j^{(m)})$ at the z_m quadrature points $x_j^{(m)}$, $j = 1, \dots, z_m$. The direct solution of this system needs $O(z_m^3)$ operations. Finally, the values at the z_n quadrature points $x_j^{(n)}$, $j = 1, \dots, z_n$, are obtained from the Nyström interpolation

$$\varphi_{n,i+1}(x_j^{(n)}) = \sum_{k=1}^{z_m} \alpha_k^{(m)} K(x_j^{(n)}, x_k^{(m)}) \varphi_{n,i+1}(x_k^{(m)}) + g_{n,i}(x_j^{(n)})$$

for $j = 1, \dots, z_n$, requiring $O(z_n z_m)$ operations. All together, in agreement with our previous operation count, we need $O(z_n^2) + O(z_m^3)$ operations. It is left as an exercise to set up the corresponding equations for the degenerate kernel and the collocation method (see Problem 14.4).

14.3 Multigrid Methods

Multigrid methods have been developed as a very efficient iteration scheme for solving the sparse linear systems that arise from finite difference or finite element discretization of elliptic boundary value problems. Following Hackbusch [75] and Schippers [216] we now briefly sketch how multigrid methods can also be applied for integral equations of the second kind. The two-grid methods described in Theorem 14.5 use only two levels; the multigrid methods use $n + 1$ levels.

Definition 14.6. The *multigrid iteration* is a defect correction iteration of the form (14.8) with the approximate inverses defined recursively by

$$\begin{aligned} B_0 &:= (I - A_0)^{-1}, \\ B_n &:= \sum_{m=0}^{p-1} [I - B_{n-1}(I - A_{n-1})]^m B_{n-1}, \quad n = 1, 2, \dots, \end{aligned} \tag{14.16}$$

with some $p \in \mathbb{N}$.

Apparently, B_0 stands for the exact solution on the coarsest grid. The approximate inverse B_n for $n+1$ levels represents the application of p steps of the multigrid iteration B_{n-1} on n levels for approximately solving the residual correction equation starting with initial element zero, since it is given by the pth partial sum of the Neumann series. This observation, simultaneously, is the motivation for the definition of the multigrid method and the basis for its actual recursive numerical implementation. In practical calculations $p = 1, 2,$ or 3 are appropriate values for the iteration number.

It is our aim to illustrate that the multigrid iteration combines the advantages of the two-grid methods of Theorem 14.5. It has the fast convergence rate of the scheme (14.9) and the computational effort for one step is essentially of the same order as for one step of the scheme (14.10). First we note that the approximate inverses satisfy

$$I - B_n(I - A_{n-1}) = [I - B_{n-1}(I - A_{n-1})]^p \tag{14.17}$$

for all $n \in \mathbb{N}$. For the iteration operator M_n corresponding to one step of the multigrid iteration using $n + 1$ levels, from Definition 14.6 and from (14.17), we deduce that

$$\begin{aligned} M_n &= I - B_n(I - A_n) = I - B_n(I - A_{n-1})(I - A_{n-1})^{-1}(I - A_n) \\ &= I - (I - A_{n-1})^{-1}(I - A_n) + \{I - B_n(I - A_{n-1})\}(I - A_{n-1})^{-1}(I - A_n) \\ &= (I - A_{n-1})^{-1}(A_n - A_{n-1}) + M_{n-1}^p(I - A_{n-1})^{-1}(I - A_n). \end{aligned}$$

After introducing the two-grid iteration operator

$$T_n := (I - A_{n-1})^{-1}(A_n - A_{n-1}), \tag{14.18}$$

we can write the recursion for the iteration operator M_n in the form

$$\begin{aligned} M_1 &= T_1, \\ M_{n+1} &= T_{n+1} + M_n^p(I - T_{n+1}), \quad n = 1, 2, \dots. \end{aligned} \tag{14.19}$$

Theorem 14.7. *Assume that*

$$\|T_n\| \le q^{n-1} C \tag{14.20}$$

for all $n \in \mathbb{N}$ and some constants $q \in (0, 1]$ and $C > 0$ satisfying

$$C \le \frac{1}{2q}\left(\sqrt{1+q^2} - 1\right). \tag{14.21}$$

Then, if $p \ge 2$, we have

$$\|M_n\| \le 2q^{n-1} C < 1, \quad n \in \mathbb{N}.$$

Proof. Note that $\sqrt{1+q^2} - 1 < q$ for all $q > 0$. Therefore the statement is correct for $n = 1$. Assume that it is proven for some $n \in \mathbb{N}$. Then, using the recurrence relation (14.19), we can conclude that

$$\begin{aligned}
\|M_{n+1}\| &\le \|T_{n+1}\| + \|M_n\|^p (1 + \|T_{n+1}\|)\\
&\le q^n C + (2q^{n-1} C)^2 (1 + q^n C)\\
&\le q^n C \left\{1 + \frac{4}{q} C(1 + qC)\right\} \le 2q^n C < 1,
\end{aligned}$$

since $4t(1 + qt) \le q$ for all $0 \le t \le \left(\sqrt{1+q^2} - 1\right)/2q$. □

In particular, setting $q = 1$, we obtain that

$$\sup_{n\in\mathbb{N}} \|T_n\| \le \frac{1}{2}\left(\sqrt{2} - 1\right) = 0.207\ldots$$

implies that $\|M_n\| \le \sqrt{2} - 1 < 1$ for all $n \in \mathbb{N}$ and all $p \ge 2$. Roughly speaking, this ensures convergence of the multigrid method, provided the approximation on the coarsest level is sufficiently accurate.

More generally, Theorem 14.7 implies that the convergence rate of the multigrid method has the same behavior as the two-grid method based on the two finest grids. Consider as a typical case an approximation of order s in the number z_n of grid points on the nth level

$$\|A_n - A\| = O\left(\frac{1}{z_n^s}\right)$$

and assume a relation

$$z_{n-1} = \zeta z_n \tag{14.22}$$

between the number of grid points on two consecutive levels with a constant $\zeta < 1$. Then

$$\|A_n - A_{n-1}\| = O(\zeta^{ns}).$$

Hence, in this case, (14.20) is satisfied with $q = \zeta^s$. In one dimension, typically the number of grid points on each level will be doubled, i.e., $\zeta = 1/2$. In this case for an approximation of second order $s = 2$ we have $q = 1/4$ and (14.21) requires $C \leq \left(\sqrt{17} - 4\right)/2 = 0.061\ldots$.

For a discussion of the computational complexity, let a_n denote the number of operations necessary for the matrix-vector multiplication involved in the application of A_n. In most cases we will have $a_n = az_n^2$, where a is some constant. Denote by b_n the number of operations for one step of the multigrid iteration with $n + 1$ levels. Then, from the recursive definition, we have

$$b_n = a_n + pb_{n-1}, \tag{14.23}$$

since each step first requires the evaluation of the residual on the level n and then performs p steps of the multigrid iteration with n levels. Neglecting the effort for the direct solutions on the coarsest grid, from (14.22) and (14.23) we obtain that

$$b_n \leq \frac{a}{1 - p\zeta^2}\, z_n^2 \tag{14.24}$$

provided $p\zeta^2 < 1$. In particular, for the canonical case where $p = 2$ and $\zeta = 1/2$ we have $b_n \leq 2az_n^2$. In general, provided the number of grid points grows fast enough in relation to the iteration number p, the computational complexity for the multigrid method is of the same order as for the two-grid method using the finest and coarsest grids.

Using iterative methods for the approximate solution of $\varphi_n - A_n\varphi_n = f_n$, in general, it is useless to try and make the iteration error $\|\varphi_{n,i} - \varphi_n\|$ smaller than the discretization error $\|\varphi_n - \varphi\|$ to the exact solution of $\varphi - A\varphi = f$. Therefore, the number of iterations should be chosen such that both errors are of the same magnitude. Unfortunately, the quantitative size of the discretization error, in most practical situations, is not known beforehand, only its order of convergence is known. However, the *nested iteration*, or full multigrid scheme, which we will describe briefly, constructs approximations with iteration errors roughly of the same size as the discretization error. The basic idea is to provide a good initial element $\varphi_{n,0}$ for the multigrid iteration on $n + 1$ levels by multigrid iterations on coarser levels as described in the following definition.

Definition 14.8. Starting with $\widetilde{\varphi}_0 := (I - A_0)^{-1} f_0$, the *full multigrid scheme* constructs a sequence $(\widetilde{\varphi}_n)$ of approximations by performing k steps of the multigrid iteration on $n + 1$ levels using the preceding $\widetilde{\varphi}_{n-1}$ as initial element.

First, we note that the computational complexity of the full multigrid method is still of order $O(z_n^2)$. Using (14.23) and (14.24), the total number c_n of operations up to $n + 1$ levels can be estimated by

$$c_n = k \sum_{m=1}^{n} b_m \leq k\, \frac{a}{1 - p\zeta^2}\, \frac{1}{1 - \zeta^2}\, z_n^2.$$

Theorem 14.9. *Assume that the discretization error satisfies*

$$\|\varphi_n - \varphi\| \le Cq^n \tag{14.25}$$

for some constants $0 < q < 1$ *and* $C > 0$, *and that* $t := \sup_{n\in\mathbb{N}} \|M_n\|$ *satisfies*

$$t^k < q. \tag{14.26}$$

Then for the approximation $\widetilde{\varphi}_n$ *obtained by the full multigrid method we have that*

$$\|\widetilde{\varphi}_n - \varphi_n\| \le C\,\frac{(q+1)t^k}{q - t^k}\,q^n, \quad n \in \mathbb{N}. \tag{14.27}$$

Proof. We set

$$\alpha := \frac{(q+1)t^k}{q - t^k}$$

and note that

$$(\alpha + 1 + q)t^k = q\alpha.$$

Trivially, (14.27) is true for $n = 0$. Assume that it has been proven for some $n \ge 0$. Since $\widetilde{\varphi}_{n+1}$ is obtained through k steps of the defect correction iteration (14.8) on the level $n + 1$ with $\widetilde{\varphi}_n$ as initial element, we can write

$$\widetilde{\varphi}_{n+1} = M_{n+1}^k \widetilde{\varphi}_n + \sum_{m=0}^{k-1} M_{n+1}^m B_{n+1} f_{n+1}.$$

From $\varphi_{n+1} - A_{n+1}\varphi_{n+1} = f_{n+1}$ we deduce that

$$B_{n+1} f_{n+1} = B_{n+1}(I - A_{n+1})\varphi_{n+1} = \varphi_{n+1} - M_{n+1}\varphi_{n+1}$$

and can insert this into the previous equation to arrive at

$$\widetilde{\varphi}_{n+1} - \varphi_{n+1} = M_{n+1}^k(\widetilde{\varphi}_n - \varphi_{n+1}).$$

Then we can estimate

$$\begin{aligned}
\|\widetilde{\varphi}_{n+1} - \varphi_{n+1}\| &= \|M_{n+1}^k(\widetilde{\varphi}_n - \varphi_{n+1})\| \le t^k \|\widetilde{\varphi}_n - \varphi_{n+1}\| \\
&\le t^k\left(\|\widetilde{\varphi}_n - \varphi_n\| + \|\varphi_n - \varphi\| + \|\varphi - \varphi_{n+1}\|\right) \\
&\le t^k C(\alpha + 1 + q)q^n = \alpha C q^{n+1},
\end{aligned}$$

and the proof is complete. □

Now, indeed, Theorem 14.9 implies that the iteration error, even for $k = 1$, is of the same size as the discretization error, provided the approximation on the coarsest level is sufficiently accurate to ensure (14.26) by Theorem 14.7. In the typical situation where $s = 2$ and $\zeta = 1/2$, an estimate of the form (14.25) is satisfied with $q = 1/4$.

In connection with the two-grid iterations of Theorem 14.5 we observed already that for the Nyström method we cannot expect $\|T_n\| \to 0$, $n \to \infty$. Therefore, in view of Theorem 14.7, modifications of the multigrid scheme are necessary for the quadrature method. The multigrid iteration proposed by Hemker and Schippers [90] replaces (14.16) by

$$\widetilde{B}_0 := (I - A_0)^{-1}$$

$$\widetilde{B}_n := \sum_{m=0}^{p-1} [I - \widetilde{B}_{n-1}(I - A_{n-1})]^m \widetilde{B}_{n-1}(I - A_{n-1} + A_n), \quad n = 1, 2, \ldots.$$

This means that at each level a smoothing operation $I - A_{n-1} + A_n$ is included before the application of p steps of the multigrid iteration at the preceding level. We write

$$\widetilde{B}_n = Q_n(I - A_{n-1} + A_n)$$

and, as above, we have the property

$$I - Q_n(I - A_{n-1}) = [I - \widetilde{B}_{n-1}(I - A_{n-1})]^p.$$

Proceeding as in the derivation of (14.19), it can be shown that the iteration operators satisfy the recurrence relation $\widetilde{M}_1 = \widetilde{T}_1$ and

$$\widetilde{M}_{n+1} = \widetilde{T}_{n+1} + \widetilde{M}_n^p (I - \widetilde{T}_{n+1}), \quad n = 1, 2, \ldots,$$

where the two-grid iteration operator $\widetilde{T}_n$ is defined by

$$\widetilde{T}_n := (I - A_{n-1})^{-1}(A_n - A_{n-1})A_n, \quad n = 1, 2, \ldots.$$

Now Theorem 14.7 holds with T_n and M_n replaced by $\widetilde{T}_n$ and $\widetilde{M}_n$. From Corollary 10.11 we have that $\|\widetilde{T}_n\| \to 0, n \to \infty$.

For further variants and a more detailed study of multigrid methods for equations of the second kind including numerical examples, we refer the reader to Hackbusch [75].

14.4 Fast Matrix-Vector Multiplication

If A is a compact operator then the eigenvalues of $I - A$ are clustered at one (see Theorem 3.9). Furthermore, from Theorems 14.3 and 14.4, we observe that the condition number of the linear systems arising from the Nyström method (or the collocation method with piecewise linear interpolation) is independent of the size n of the matrix, i.e., of the degree of discretization. Therefore, using conjugate gradient-type methods like GMRES (or other Krylov subspace methods) for the iterative solution of these systems, we can expect that the number of iterations required for a certain accuracy is independent of n (see [70]). Hence, the computational complexity of the iterative solution of integral equations of the second kind by these iteration methods is determined by the amount of work required for the computation of matrix-vector products. Note that these and subsequent considerations also apply to the matrix-vector multiplications occurring in the two-grid and multigrid methods of the two previous sections.

Since the discretization of integral equations leads to dense matrices, each matrix-vector multiplication, if it is performed in the naive way, requires $O(n^2)$ operations. Therefore, even solving the linear systems by iteration methods such as the conjugate gradient method or GMRES might become too expensive for realistic applied problems. A remedy is to try and perform the matrix-vector multiplication only approximately within some tolerance. Indeed, in many cases this concept allows a reduction of the computational complexity to almost linear growth of order $O(n)$. Examples of this type are *the panel clustering methods* suggested by Hackbusch and Nowak [77] and the closely related *fast multipole methods* of Rokhlin [208] and Greengard and Rokhlin [71]. Another approach has been initiated through Beylkin, Coifman, and Rokhlin [18] by using a wavelet basis for spline spaces in the Galerkin method for integral equations of the second kind. In the latter method, the main idea is to replace the exact Galerkin matrix through a matrix that is nearly sparse by setting all entries below a certain threshold equal to zero. For a survey on these wavelet methods we refer to Dahmen [38]. Since developing the main ideas of *wavelets* is beyond the aim of this introduction, we will confine ourselves to explaining the basic principle of the panel clustering and fast multipole methods by considering a simple model problem of a one-dimensional integral equation.

For one-dimensional integral equations, in general, the discrete linear system can be solved efficiently by either elimination methods or the two-grid and multigrid methods with traditional matrix-vector multiplication. Nevertheless, the following example is suitable for explaining the basic ideas of panel clustering and fast multipole methods. This is due to the fact that in higher dimensions there are more technical details to consider, which distract from the basic principles. However, these basic principles do not depend on the dimension of the integral equation.

Denote by Γ the unit circle in $\mathbb{R}^2$ and consider an integral equation of the second kind

$$\varphi(x) - \int_\Gamma K(x, y)\varphi(y)\,ds(y) = f(x), \quad x \in \Gamma, \tag{14.28}$$

with sufficiently smooth kernel K and right-hand side f. Using the composite trapezoidal rule in the Nyström method the integral equation (14.28) is approximated by the linear system

$$\varphi_j - \frac{2\pi}{n}\sum_{k=1}^{n} K(x_j, x_k)\varphi_k = f(x_j), \quad j = 1, \dots, n, \tag{14.29}$$

with the equidistant discretization points $x_j = (\cos(2j\pi/n), \sin(2j\pi/n))$, $j = 1, \dots, n$. Assume that for each $z \in \Gamma$ we have degenerate kernels

$$K_M(x, y; z) = \sum_{m=1}^{M} a_m(x; z) b_m(y; z) \tag{14.30}$$

available such that

$$|K(x, y) - K_M(x, y; z)| \le \frac{c}{a^M} \tag{14.31}$$

for $|y-z| \ge 2|x-z|$ and some constants $c > 0$ and $a > 1$. Such degenerate kernels can be obtained, for example, by Taylor series expansions. In the fast multipole method for the kernel of the logarithmic double-layer potential operator the approximating kernels are obtained by taking the real part of Taylor expansions of the complex logarithm in $\mathbb{C}$. Now the main idea of both the panel clustering and the fast multipole methods consists in approximating the exact kernel in (14.29) by the degenerate kernels (14.30) for x_k away from x_j. As a result, large submatrices of the $n \times n$ matrix $K(x_j, x_k)$ are well approximated by matrices of low rank. Multiplying an $n \times n$ matrix of rank M with a vector requires only $O(Mn)$ operations as opposed to $O(n^2)$ for a matrix of full rank. This observation leads to a scheme for the approximate evaluation of the matrix-vector multiplication in (14.29) with the computational cost reduced to order $O(Mn \log n)$.

To explain this scheme we assume that $n = 2^p$ with an integer $p \ge 3$. For $q = 1, \dots, p-1$ and $\ell = 1, 2, \dots, 2^{q+1}$ we introduce subsets $A_\ell^q, B_\ell^q \subset \Gamma$ by setting

$$A_\ell^q := \left\{ (\cos\varphi, \sin\varphi) : \left| \varphi - \frac{\ell\pi}{2^q} \right| < \frac{\pi}{2^q} \right\} \quad \text{and} \quad B_\ell^q := \Gamma \setminus A_\ell^q.$$

Then we have that $A_{2\ell}^q \subset A_\ell^{q-1}$ and $A_{2\ell-1}^q \subset A_\ell^{q-1}$ and also that $B_{2\ell}^q \supset B_\ell^{q-1}$ and $B_{2\ell-1}^q \supset B_\ell^{q-1}$ for $q = 2, \dots, p-1$ and $\ell = 1, 2, \dots, 2^q$. Each set A_ℓ^q contains a total of $2^{p-q} - 1$ of the discretization points centered around

$$z_\ell^q := x_{2^{p-q-1}\ell}.$$

Correspondingly, the complement set B_ℓ^q contains $2^p - 2^{p-q} + 1$ discretization points. For notational convenience we also define $A_\ell^p := \{x_\ell\}$ for $\ell = 1, \dots, n$. Finally, we set $C_\ell^1 := B_\ell^1$ for $\ell = 1, \dots, 4$, and recalling that $B_{2\ell}^q \supset B_\ell^{q-1}$ and $B_{2\ell-1}^q \supset B_\ell^{q-1}$, we further define

$$C_{2\ell}^q := B_{2\ell}^q \setminus B_\ell^{q-1}, \quad C_{2\ell-1}^q := B_{2\ell-1}^q \setminus B_\ell^{q-1}$$

for $q = 2, \dots, p-1$ and $\ell = 1, 2, \dots, 2^q$. Then, for $q \ge 2$, the set C_ℓ^q contains 2^{p-q} discretization points.

Now, we set $S_0(x_j) := 0$, $j = 1, \dots, n$, and compute recursively

$$S_q(x_j) := S_{q-1}(x_j) + \sum_{m=1}^{M} a_m(x_j; z_\ell^q) \sum_{x_k \in C_\ell^q} b_m(x_k; z_\ell^q)\varphi_k, \quad x_j \in A_\ell^q, \tag{14.32}$$

for $q = 1, \dots, p-1$ and $\ell = 1, 2, \dots, 2^{q+1}$. Since each of the sets C_ℓ^q contains 2^{p-q} discretization points and since we have 2^{q+1} sets C_ℓ^q, the computations of the $S_q(x_j)$ from the preceding $S_{q-1}(x_j)$ requires a total of $O(Mn)$ operations. Therefore, the total cost to compute $S_{p-1}(x_j)$ for $j = 1, \dots, n$ is of order $O(Mpn)$, i.e., of order $O(Mn \log n)$.

Since $2|x_j - z_\ell^q| \le |x_k - z_\ell^q|$ for $x_j \in A_{2\ell}^{q+1}$ and $x_k \in C_\ell^q$, from (14.31) we can conclude that

$$\left| \sum_{x_k \in C_\ell^q} \left\{ \sum_{m=1}^{M} a_m(x_j; z_\ell^q) b_m(x_k; z_\ell^q) - K(x_j, x_k) \right\} \varphi_k \right| \le \frac{cn}{2^q a^M} \|\varphi\|_\infty$$

for $x_j \in A_{2\ell}^{q+1}$ and $\ell = 1, \dots, 2^{q+1}$. From this, by induction, we obtain that

$$\left| S_q(x_j) - \sum_{x_k \in B_\ell^q} K(x_j, x_k)\varphi_k \right| \le \frac{cn}{a^M} \|\varphi\|_\infty \sum_{r=1}^{q} \frac{1}{2^r}$$

for $x_j \in A_{2\ell}^{q+1}$ and $\ell = 1, \dots, 2^{q+1}$. In particular, for $q = p-1$, we have that

$$\left| \frac{1}{n} S_{p-1}(x_j) - \frac{1}{n} \sum_{\substack{k=1 \\ k \ne j}}^{n} K(x_j, x_k)\varphi_k \right| \le \frac{c}{a^M} \|\varphi\|_\infty, \quad j = 1, \dots, n, \tag{14.33}$$

since $A_\ell^{p-1} = \{x_\ell\}$ for $\ell = 1, \dots, n$. Therefore, the above scheme indeed computes the required matrix-vector multiplication with a total computational cost of order $O(Mn \log n)$.

From (14.33) we also observe that the computations correspond to replacing the matrix with the entries $K(x_j, x_k)/n$ by an approximating matrix such that the difference has maximum norm less than or equal to ca^{-M}. This, by Theorem 14.2, implies that the resulting solution of the perturbed linear systems differs from the solution of (14.29) by an error of order $O(a^{-M})$. We want to ensure that the additional error induced by the approximation of the matrix elements is of the same magnitude as the error for the approximation of the integral equation (14.28) by the Nyström method (14.29). If for the latter we assume a convergence order $O(n^{-m})$ for some $m \in \mathbb{N}$, then we need to make sure that $a^{-M} = O(n^{-m})$, i.e., we have to choose $M = O(m \log n)$. Therefore, we have achieved a computational scheme preserving the accuracy of the numerical solution of the integral equation with the computational cost of order $O(m\,n(\log n)^2)$.

For analytic kernels and right-hand sides, for the Nyström method, we have exponential convergence $O(e^{-sn})$ for some $s > 0$. Therefore, in this case, the above argument leads to the requirement that $M = O(n)$, i.e., we do not obtain a reduction of the computational cost compared to the naive matrix-vector multiplications. Hence, as a rule of thumb, panel clustering and fast multipole methods are only efficient for low-order approximations.

For further studies of panel clustering and fast multipole methods, we refer to Greengard and Rokhlin [72], Hackbusch [76], Hackbusch and Sauter [78], and Liu [159], and the references therein.

Problems

14.1. For the operator M_n, from the proof of Theorem 14.3, show that

$$\|M_n\|_\infty = \sup_{\|\Phi\|_\infty=1} \|M_n\Phi\|_\infty = 1$$

and verify the existence of the functions L_j used in the definition of M_n.

14.2. For the condition number of the Bubnov–Galerkin method for an equation of the second kind, define matrix operators $E_n, \widetilde{A}_n : \mathbb{C}^n \to \mathbb{C}^n$ by

$$(E_n\gamma)_j := \sum_{k=1}^n (u_j, u_k)\gamma_k$$

and

$$(\widetilde{A}_n\gamma)_j := \sum_{k=1}^n (u_j, Au_k)\gamma_k$$

for $j = 1, \dots, n$. Proceding as in Theorem 14.4, show that in the Euclidean norm the condition number of the linear system satisfies

$$\operatorname{cond}(E_n - \widetilde{A}_n) \le C \operatorname{cond} E_n$$

for all sufficiently large n and some constant C.

14.3. Brakhage [19] suggested using $B_n = I + (I - A_m)^{-1}A_n$ as an approximate inverse for the Nyström method in the defect correction method (compare also the proof of Theorem 10.12). Prove a variant of Theorem 14.5 for this case.

14.4. Set up the equations for one step of the two-grid iteration for the collocation method.

14.5. Formulate and prove an analog of Theorem 14.7 for the case of exponential convergence $\|T_n\| = O(\exp(-sz_{n-1}))$ with some $s > 0$.

Chapter 15
Equations of the First Kind

Compact operators cannot have a bounded inverse. Therefore, equations of the first kind with a compact operator provide a typical example for so-called *ill-posed problems*. This chapter is intended as an introduction into the basic ideas on ill-posed problems and regularization methods for their stable solution. We mainly confine ourselves to linear equations of the first kind with compact operators in a Hilbert space setting and will base our presentation on the singular value decomposition. For a more comprehensive study of ill-posed problems, we refer to Baumeister [15], Engl, Hanke, and Neubauer [49], Groetsch [74], Kabanikhin [113], Kaltenbacher, Neubauer and Scherzer [115], Kirsch [126], Louis [160], Morozov [176], Rieder [204], Tikhonov and Arsenin [234] and Wang, Yagola and Yang [243].

15.1 Ill-Posed Problems

For problems in mathematical physics, in particular for initial and boundary value problems for partial differential equations, Hadamard [79] postulated three properties:

(1) Existence of a solution.

(2) Uniqueness of the solution.

(3) Continuous dependence of the solution on the data.

The third postulate is motivated by the fact that in all applications the data will be measured quantities. Therefore, one wants to make sure that small errors in the data will cause only small errors in the solution. A problem satisfying all three requirements is called *well-posed*. The potential theoretic boundary value problems of Chapter 6 and the initial boundary value problem for the heat equation of Chapter 9 are examples of well-posed problems. We will make Hadamard's concept of well-posedness more precise through the following definition.

R. Kress, *Linear Integral Equations*, Applied Mathematical Sciences 82,
DOI 10.1007/978-1-4614-9593-2_15, © Springer Science+Business Media New York 2014

Definition 15.1. Let $A : U \to V$ be an operator from a subset U of a normed space X into a subset V of a normed space Y. The equation

$$A\varphi = f \tag{15.1}$$

is called *well-posed* or *properly posed* if $A : U \to V$ is bijective and the inverse operator $A^{-1} : V \to U$ is continuous. Otherwise, the equation is called *ill-posed* or *improperly posed.*

According to this definition we may distinguish three types of ill-posedness. If A is not surjective, then equation (15.1) is not solvable for all $f \in V$ (*nonexistence*). If A is not injective, then equation (15.1) may have more than one solution (*nonuniqueness*). Finally, if $A^{-1} : V \to U$ exists but is not continuous, then the solution φ of (15.1) does not depend continuously on the data f (*instability*). The latter case of instability is the one of primary interest in the study of ill-posed problems. We note that the three properties, in general, are not independent. For example, if $A : X \to Y$ is a bounded linear operator mapping a Banach space X bijectively onto a Banach space Y, then by the Banach open mapping Theorem 10.8 the inverse operator $A^{-1} : Y \to X$ is bounded and therefore continuous. For a long time the research on improperly posed problems was neglected, since they were not considered relevant to the proper treatment of applied problems.

Note that the well-posedness of a problem is a property of the operator A together with the solution space X and the data space Y including the norms on X and Y. Therefore, if an equation is ill-posed one could try to restore stability by changing the spaces X and Y and their norms. But, in general, this approach is inadequate, since the spaces X and Y, including their norms are determined by practical needs. In particular, the space Y and its norm must be suitable to describe the measured data and, especially, the measurement errors.

Example 15.2. A classic example of an ill-posed problem, given by Hadamard, is the following Cauchy problem (or initial value problem) for the Laplace equation: Find a harmonic function u in $\mathbb{R} \times [0, \infty)$ satisfying the initial conditions

$$u(\cdot\,, 0) = 0, \quad \frac{\partial}{\partial y}\, u(\cdot\,, 0) = f,$$

where f is a given continuous function. By Holmgren's Theorem 6.7 we have uniqueness of the solution to this Cauchy problem. If we choose as data

$$f_n(x) = \frac{1}{n} \sin nx, \quad x \in \mathbb{R},$$

for $n \in \mathbb{N}$, then we obtain the solution

$$u_n(x, y) = \frac{1}{n^2} \sin nx \sinh ny, \quad (x, y) \in \mathbb{R} \times [0, \infty).$$

Obviously, the data sequence (f_n) converges uniformly to zero, whereas the solution sequence (u_n) does not converge in any reasonable norm. Therefore, the solution to the Cauchy problem for harmonic functions is ill-posed. But a number of applications lead to such a Cauchy problem, for example, the extrapolation of measured data for the gravity potential from parts of the surface of the earth to the space above or below this surface. □

Example 15.3. A second typical example of an ill-posed problem is the backward heat conduction. Consider the heat equation

$$\frac{\partial u}{\partial t} = \frac{\partial^2 u}{\partial x^2}$$

in a rectangle $[0, \pi] \times [0, T]$ subject to the boundary conditions

$$u(0, t) = u(\pi, t) = 0, \quad 0 \leq t \leq T,$$

and the initial condition

$$u(x, 0) = \varphi(x), \quad 0 \leq x \leq \pi,$$

where φ is a given function. For existence and uniqueness, we refer back to Chapter 9. Here, the solution can be obtained by separation of variables in the form

$$u(x, t) = \sum_{n=1}^{\infty} \hat{\varphi}_n e^{-n^2 t} \sin nx \tag{15.2}$$

with the Fourier coefficients

$$\hat{\varphi}_n = \frac{2}{\pi} \int_0^{\pi} \varphi(y) \sin ny \, dy \tag{15.3}$$

of the given initial values. This initial value problem is well-posed: The final temperature $f := u(\cdot\, , T)$ depends continuously on the initial temperature. For example, in an L^2 setting, using Parseval's equality, we have

$$\|f\|_2^2 = \frac{\pi}{2} \sum_{n=1}^{\infty} \hat{\varphi}_n^2 e^{-2n^2 T} \leq \frac{\pi}{2} e^{-2T} \sum_{n=1}^{\infty} \hat{\varphi}_n^2 = e^{-2T} \|\varphi\|_2^2.$$

However, the inverse problem, i.e., the determination of the initial temperature φ from the knowledge of the final temperature f, is ill-posed. Here, we can write

$$u(x, t) = \sum_{n=1}^{\infty} \hat{f}_n e^{n^2 (T-t)} \sin nx$$

with the Fourier coefficients

$$\hat{f}_n = \frac{2}{\pi} \int_0^{\pi} f(y) \sin ny \, dy$$

of the given final temperature. Then we have

$$\|\varphi\|_2^2 = \frac{\pi}{2}\sum_{n=1}^{\infty} \hat{f}_n^2 e^{2n^2T}$$

and there exists no positive constant C such that

$$\|\varphi\|_2^2 \leq C\|f\|_2^2 = C\,\frac{\pi}{2}\sum_{n=1}^{\infty} \hat{f}_n^2.$$

We may interpret this inverse problem as an initial value problem for the backward heat equation, where t is replaced by $-t$. The ill-posedness reflects the fact that heat conduction is an irreversible physical process. □

Inserting (15.3) into (15.2), we see that the last example can be put into the form of an integral equation of the first kind

$$\int_0^{\pi} K(x,y)\varphi(y)\,dy = f(x), \quad 0 \leq x \leq \pi,$$

where the kernel is given by

$$K(x,y) = \frac{2}{\pi}\sum_{n=1}^{\infty} e^{-n^2T}\sin nx \sin ny, \quad 0 \leq x, y \leq \pi.$$

In general, linear integral equations of the first kind with continuous or weakly singular kernels provide typical examples for ill-posed problems with respect to both the maximum and the mean square norm. They are special cases of the following theorem which is a reformulation of Theorem 2.26.

Theorem 15.4. *Let X and Y be normed spaces and let $A : X \to Y$ be a compact linear operator. Then the equation of the first kind $A\varphi = f$ is improperly posed if X is not of finite dimension.*

The ill-posed nature of an equation, of course, has consequences for its numerical treatment. We may view a numerical solution of a given equation as the solution to perturbed data. Therefore, straightforward application of the methods developed in previous chapters to ill-posed equations of the first kind will usually generate numerical nonsense. In terms of the concepts introduced in Section 14.1, the fact that an operator does not have a bounded inverse means that the condition numbers of its finite-dimensional approximations grow with the quality of the approximation. Hence, a careless discretization of ill-posed problems leads to a numerical behavior that only at first glance seems to be paradoxical. Namely, increasing the degree of discretization, i.e., increasing the accuracy of the approximation for the operator A, will cause the approximate solution to the equation $A\varphi = f$ to become less reliable.

15.2 Regularization of Ill-Posed Problems

Methods for constructing a stable approximate solution of ill-posed problems are called *regularization methods*. Note that in the context of improperly posed problems the term *regularization* has a different meaning than in the theory of singular equations described in Chapter 5. It is our aim to introduce a few of the classical regularization concepts for linear equations of the first kind.

In the sequel, we will mostly assume that the linear operator A is injective. This is not a principal loss of generality, since uniqueness for a linear equation can always be achieved by a suitable modification of the solution space X. We wish to approximate the solution φ to the equation $A\varphi = f$ from the knowledge of a perturbed right-hand side f^δ with a known error level

$$\|f^\delta - f\| \leq \delta. \tag{15.4}$$

When f belongs to the range $A(X)$, there exists a unique solution φ of $A\varphi = f$. For the perturbed right-hand side, in general, we cannot expect $f^\delta \in A(X)$. Using the erroneous data f^δ we want to construct a reasonable approximation φ^δ to the exact solution φ of the unperturbed equation $A\varphi = f$. Of course, we want this approximation to be stable, i.e., we want φ^δ to depend continuously on the actual data f^δ. Therefore our task requires finding an approximation of the unbounded inverse operator $A^{-1} : A(X) \to X$ by a bounded linear operator $R : Y \to X$.

Definition 15.5. Let X and Y be normed spaces and let $A : X \to Y$ be an injective bounded linear operator. Then a family of bounded linear operators $R_\alpha : Y \to X$, $\alpha > 0$, with the property of pointwise convergence

$$\lim_{\alpha \to 0} R_\alpha A\varphi = \varphi, \quad \varphi \in X, \tag{15.5}$$

is called a *regularization scheme* for the operator A. The parameter α is called the *regularization parameter.*

Of course, (15.5) is equivalent to $R_\alpha f \to A^{-1}f$, $\alpha \to 0$, for all $f \in A(X)$. Occasionally, we will use regularization parameter sets other than the positive real numbers.

The following theorem illustrates properties that cannot be fulfilled by regularization schemes for compact operators.

Theorem 15.6. *Let X and Y be normed spaces, $A : X \to Y$ be a compact linear operator, and* $\dim X = \infty$. *Then for a regularization scheme the operators R_α cannot be uniformly bounded with respect to α, and the operators $R_\alpha A$ cannot be norm convergent as $\alpha \to 0$.*

Proof. For the first statement, assume that $\|R_\alpha\| \leq C$ for all $\alpha > 0$ with some constant C. Then from $R_\alpha f \to A^{-1}f$, $\alpha \to 0$, for all $f \in A(X)$ we deduce $\|A^{-1}f\| \leq C\|f\|$, i.e., $A^{-1} : A(X) \to X$ is bounded. By Theorem 15.4 this is a contradiction to $\dim X = \infty$.

For the second statement, assume that we have norm convergence. Then there exists $\alpha > 0$ such that $\|R_\alpha A - I\| < 1/2$. Now for all $f \in A(X)$ we can estimate

$$\|A^{-1}f\| \le \|A^{-1}f - R_\alpha AA^{-1}f\| + \|R_\alpha f\| \le \frac{1}{2}\,\|A^{-1}f\| + \|R_\alpha\|\,\|f\|,$$

whence $\|A^{-1}f\| \le 2\|R_\alpha\|\,\|f\|$ follows. Therefore $A^{-1} : A(X) \to X$ is bounded, and we have the same contradiction as above. □

The regularization scheme approximates the solution φ of $A\varphi = f$ by the regularized solution

$$\varphi_\alpha^\delta := R_\alpha f^\delta. \tag{15.6}$$

Then, for the total approximation error, writing

$$\varphi_\alpha^\delta - \varphi = R_\alpha f^\delta - R_\alpha f + R_\alpha A\varphi - \varphi,$$

by the triangle inequality we have the estimate

$$\|\varphi_\alpha^\delta - \varphi\| \le \delta\|R_\alpha\| + \|R_\alpha A\varphi - \varphi\|. \tag{15.7}$$

This decomposition shows that the error consists of two parts: the first term reflects the influence of the incorrect data and the second term is due to the approximation error between R_α and A^{-1}. Under the assumptions of Theorem 15.6, the first term cannot be estimated uniformly with respect to α and the second term cannot be estimated uniformly with respect to φ. Typically, the first term will be increasing as $\alpha \to 0$ due to the ill-posed nature of the problem whereas the second term will be decreasing as $\alpha \to 0$ according to (15.5). Every regularization scheme requires a strategy for choosing the parameter α in dependence on the error level δ and on the given data f^δ in order to achieve an acceptable total error for the regularized solution. On one hand, the accuracy of the approximation asks for a small error $\|R_\alpha A\varphi - \varphi\|$, i.e., for a small parameter α. On the other hand, the stability requires a small $\|R_\alpha\|$, i.e., a large parameter α. An optimal choice would try and make the right-hand side of (15.7) minimal. The corresponding parameter effects a compromise between accuracy and stability. For a reasonable regularization strategy we expect the regularized solution to converge to the exact solution when the error level tends to zero. We express this requirement through the following definition.

Definition 15.7. A *strategy* for a regularization scheme R_α, $\alpha > 0$, i.e., the choice of the regularization parameter $\alpha = \alpha(\delta)$ depending on the error level δ and on f^δ, is called *regular* if for all $f \in A(X)$ and all $f^\delta \in Y$ with $\|f^\delta - f\| \le \delta$ we have

$$R_{\alpha(\delta)} f^\delta \to A^{-1}f, \quad \delta \to 0.$$

In the discussion of regularization schemes, one usually has to distinguish between an *a priori* and an *a posteriori* choice of the regularization parameter α. An a priori choice would be based on some information on smoothness properties of the exact solution that, in practical problems, will not generally be available.

Therefore, a posteriori strategies based on some considerations of the data error level δ are more practical.

A natural a posteriori strategy is given by the *discrepancy* or *residue principle* introduced by Morozov [174, 175]. Its motivation is based on the consideration that, in general, for erroneous data the residual $\|A\varphi_\alpha^\delta - f^\delta\|$ should not be smaller than the accuracy of the measurements of f, i.e., the regularization parameter α should be chosen such that

$$\|AR_\alpha f^\delta - f^\delta\| = \gamma\delta$$

with some fixed parameter $\gamma \geq 1$ multiplying the error level δ. In the case of a regularization scheme R_m with a regularization parameter $m = 1, 2, 3, \ldots$ taking only discrete values, m should be chosen as the smallest integer satisfying

$$\|AR_m f^\delta - f^\delta\| \leq \gamma\delta.$$

Finally, we also need to note that quite often the only choice for selecting the regularization parameter will be *trial and error*, i.e., one uses a few different parameters α and then picks the most reasonable result based on appropriate information on the expected solution.

In the sequel we will describe some regularization schemes including regular strategies in a Hilbert space setting. Our approach will be based on the singular value decomposition.

15.3 Compact Self-Adjoint Operators

Throughout the remainder of this chapter, X and Y will always denote Hilbert spaces. From Theorem 4.11 recall that every bounded linear operator $A : X \to Y$ possesses an adjoint operator $A^* : Y \to X$, i.e.,

$$(A\varphi, \psi) = (\varphi, A^*\psi)$$

for all $\varphi \in X$ and $\psi \in Y$. For the norms we have $\|A\| = \|A^*\|$.

Theorem 15.8. *For a bounded linear operator we have*

$$A(X)^\perp = N(A^*) \quad \textit{and} \quad N(A^*)^\perp = \overline{A(X)}.$$

Proof. $g \in A(X)^\perp$ means $(A\varphi, g) = 0$ for all $\varphi \in X$. This is equivalent to $(\varphi, A^*g) = 0$ for all $\varphi \in X$, which is equivalent to $A^*g = 0$, i.e., $g \in N(A^*)$. Hence, $A(X)^\perp = N(A^*)$. We abbreviate $U = A(X)$ and, trivially, have $\bar{U} \subset (U^\perp)^\perp$. Denote by $P : Y \to \bar{U}$ the orthogonal projection operator. For arbitrary $\varphi \in (U^\perp)^\perp$, by Theorem 1.25, we have orthogonality $P\varphi - \varphi \perp U$. But we also have $P\varphi - \varphi \perp U^\perp$, because we already know that $\bar{U} \subset (U^\perp)^\perp$. Therefore it follows that $\varphi = P\varphi \in \bar{U}$, whence $\bar{U} = (U^\perp)^\perp$, i.e., $\overline{A(X)} = N(A^*)^\perp$. □

An operator $A : X \to X$ mapping a Hilbert space X into itself is called *self-adjoint* if $A = A^*$, i.e., if

$$(A\varphi, \psi) = (\varphi, A\psi)$$

for all $\varphi, \psi \in X$. Note that for a self-adjoint operator the scalar product $(A\varphi, \varphi)$ is real-valued for all $\varphi \in X$, since $\overline{(A\varphi, \varphi)} = (\varphi, A\varphi) = (A\varphi, \varphi)$.

Theorem 15.9. *For a bounded self-adjoint operator we have*

$$\|A\| = \sup_{\|\varphi\|=1} |(A\varphi, \varphi)|. \tag{15.8}$$

Proof. We abbreviate the right-hand side of (15.8) by q. By the Cauchy–Schwarz inequality we find

$$|(A\varphi, \varphi)| \le \|A\varphi\| \, \|\varphi\| \le \|A\|$$

for all $\|\varphi\| = 1$, whence $q \le \|A\|$ follows. On the other hand, for all $\varphi, \psi \in X$ we have

$$(A(\varphi + \psi), \varphi + \psi) - (A(\varphi - \psi), \varphi - \psi) = 2\{(A\varphi, \psi) + (A\psi, \varphi)\} = 4\,\mathrm{Re}(A\varphi, \psi).$$

Therefore we can estimate

$$4\,\mathrm{Re}(A\varphi, \psi) \le q\left\{\|\varphi + \psi\|^2 + \|\varphi - \psi\|^2\right\} = 2q\left\{\|\varphi\|^2 + \|\psi\|^2\right\}.$$

Now let $\|\varphi\| = 1$ and $A\varphi \ne 0$. Then choose $\psi = \|A\varphi\|^{-1}A\varphi$ to obtain

$$\|A\varphi\| = \mathrm{Re}(A\varphi, \psi) \le q.$$

This implies $\|A\| \le q$ and completes the proof. □

Recall the spectral theoretic notions of Definition 3.8.

Theorem 15.10. *All eigenvalues of a self-adjoint operator are real and eigenelements to different eigenvalues are orthogonal.*

Proof. $A\varphi = \lambda\varphi$ and $\varphi \ne 0$ imply $\lambda(\varphi, \varphi) = (A\varphi, \varphi) \in \mathbb{R}$, whence $\lambda \in \mathbb{R}$. Let $A\varphi = \lambda\varphi$ and $A\psi = \mu\psi$ with $\lambda \ne \mu$. Then, from

$$(\lambda - \mu)\,(\varphi, \psi) = (A\varphi, \psi) - (\varphi, A\psi) = 0$$

it follows that $(\varphi, \psi) = 0$. □

Theorem 15.11. *The spectral radius of a bounded self-adjoint operator A satisfies*

$$r(A) = \|A\|. \tag{15.9}$$

If A is compact then there exists at least one eigenvalue with $|\lambda| = \|A\|$.

Proof. For each bounded linear operator we have $r(A) \le \|A\|$, since all $\lambda \in \mathbb{C}$ with $|\lambda| > \|A\|$ are regular by the Neumann series Theorem 2.14. By Theorem 15.9 there exists a sequence (φ_n) in X with $\|\varphi_n\| = 1$ such that

$$|(A\varphi_n, \varphi_n)| \to \|A\|, \quad n \to \infty.$$

We may assume that $(A\varphi_n, \varphi_n) \to \lambda$, $n \to \infty$, where λ is real and $|\lambda| = \|A\|$. Then

$$\begin{aligned}
0 &\le \|A\varphi_n - \lambda\varphi_n\|^2 \\
&= \|A\varphi_n\|^2 - 2\lambda(A\varphi_n, \varphi_n) + \lambda^2\|\varphi_n\|^2 \\
&\le \|A\|^2 - 2\lambda(A\varphi_n, \varphi_n) + \lambda^2 \\
&= 2\lambda\{\lambda - (A\varphi_n, \varphi_n)\} \to 0, \quad n \to \infty.
\end{aligned}$$

Therefore

$$A\varphi_n - \lambda\varphi_n \to 0, \quad n \to \infty. \tag{15.10}$$

This implies that λ is not a regular value because if it was we would have the contradiction

$$1 = \|\varphi_n\| = \|(\lambda I - A)^{-1}(\lambda\varphi_n - A\varphi_n)\| \to 0, \quad n \to \infty.$$

Hence, $r(A) \ge |\lambda| = \|A\|$.

If A is compact, then the bounded sequence (φ_n) contains a subsequence $(\varphi_{n(k)})$ such that $A\varphi_{n(k)} \to \psi$, $k \to \infty$, for some $\psi \in X$. We may assume that $A \neq 0$, since for $A = 0$ the statement of the theorem is trivial. Then, from (15.10) it follows that $\varphi_{n(k)} \to \varphi$, $k \to \infty$, for some $\varphi \in X$, and $\|\varphi_{n(k)}\| = 1$ for all k implies that $\|\varphi\| = 1$. Finally, again from (15.10), by the continuity of A, we obtain $A\varphi = \lambda\varphi$, and the proof is finished. □

Now we are ready to summarize our results into the following *spectral theorem* for self-adjoint compact operators.

Theorem 15.12. *Let X be a Hilbert space and let $A : X \to X$ be a self-adjoint compact operator (with $A \neq 0$). Then all eigenvalues of A are real. A has at least one eigenvalue different from zero and at most a countable set of eigenvalues accumulating only at zero. All eigenspaces $N(\lambda I - A)$ for nonzero eigenvalues λ have finite dimension and eigenspaces to different eigenvalues are orthogonal. Assume the sequence (λ_n) of the nonzero eigenvalues to be ordered such that*

$$|\lambda_1| \ge |\lambda_2| \ge |\lambda_3| \ge \cdots$$

and denote by $P_n : X \to N(\lambda_n I - A)$ the orthogonal projection operator onto the eigenspace for the eigenvalue λ_n. Then

$$A = \sum_{n=1}^{\infty} \lambda_n P_n \tag{15.11}$$

in the sense of norm convergence. Let $Q : X \to N(A)$ denote the orthogonal projection operator onto the nullspace $N(A)$. Then

$$\varphi = \sum_{n=1}^{\infty} P_n\varphi + Q\varphi \tag{15.12}$$

for all $\varphi \in X$. (When there are only finitely many eigenvalues, the series (15.11) and (15.12) degenerate into finite sums.)

Proof. The first part of the statement is a consequence of Theorems 3.1, 3.9, 15.10, and 15.11. The orthogonal projection operators P_n are self-adjoint by Problem 13.1 and bounded with $\|P_n\| = 1$ by Theorem 13.3. Therefore, by Theorem 2.23, they are compact, since the eigenspaces $N(\lambda_n I - A)$ have finite dimension. Hence the operators $A_m := A - \sum_{n=1}^{m} \lambda_n P_n$ are self-adjoint and compact.

Let $\lambda \neq 0$ be an eigenvalue of A_m with eigenelement φ, i.e., $A_m\varphi = \lambda\varphi$. Then for $1 \leq n \leq m$ we have

$$\lambda P_n\varphi = P_n A_m\varphi = P_n(A\varphi - \lambda_n\varphi),$$

since $P_n P_k = 0$ for $n \neq k$. From this it follows that

$$\lambda^2\|P_n\varphi\|^2 = (A\varphi - \lambda_n\varphi, P_n(A\varphi - \lambda_n\varphi)) = (\varphi, (A - \lambda_n I)P_n(A\varphi - \lambda_n\varphi)) = 0,$$

since $P_n(A\varphi - \lambda_n\varphi) \in N(\lambda_n I - A)$. Therefore $P_n\varphi = 0$, whence $A\varphi = \lambda\varphi$. Conversely, let $\varphi \in N(\lambda_n I - A)$. Then $A_m\varphi = \lambda_n\varphi$ if $n > m$ and $A_m\varphi = 0$ if $n \leq m$. Therefore, the eigenvalues of A_m different from zero are given by $\lambda_{m+1}, \lambda_{m+2}, \ldots$. Now Theorem 15.11 yields $\|A_m\| = |\lambda_{m+1}|$, whence (15.11) follows.

From

$$\left\| \varphi - \sum_{n=1}^{m} P_n\varphi \right\|^2 = \|\varphi\|^2 - \sum_{n=1}^{m} \|P_n\varphi\|^2$$

we observe that $\sum_{n=1}^{\infty} \|P_n\varphi\|^2$ converges in $\mathbb{R}$. As in the proof of Theorem 8.2 this implies convergence of the series $\sum_{n=1}^{\infty} P_n\varphi$ in the Hilbert space X. Then, by the continuity of A and (15.11), we obtain

$$A\left(\varphi - \sum_{n=1}^{\infty} P_n\varphi\right) = A\varphi - \sum_{n=1}^{\infty} \lambda_n P_n\varphi = 0.$$

Observing that $QP_n = 0$ for all n completes the proof of (15.12). □

Let U be a finite-dimensional subspace of X and let $\varphi_1, \ldots, \varphi_m$ be an *orthonormal basis*, i.e., $(\varphi_n, \varphi_k) = 0$, $n, k = 1, \ldots, m$, $n \neq k$ and $\|\varphi_n\| = 1$, $n = 1, \ldots, m$. Then the orthogonal projection operator $P : X \to U$ has the representation

$$P\varphi = \sum_{n=1}^{m} (\varphi, \varphi_n)\varphi_n.$$

Hence, we can rewrite the series (15.11) and (15.12) into a more explicit form. For this, deviating from the numbering in Theorem 15.12, we repeat each eigenvalue in the sequence (λ_n) according to its *multiplicity*, i.e., according to the dimension of the eigenspace $N(\lambda_n I - A)$. Assume (φ_n) to be a sequence of corresponding orthonormal eigenelements. Then for each $\varphi \in X$ we can expand

$$A\varphi = \sum_{n=1}^{\infty} \lambda_n(\varphi, \varphi_n)\varphi_n \tag{15.13}$$

and

$$\varphi = \sum_{n=1}^{\infty} (\varphi, \varphi_n)\varphi_n + Q\varphi. \tag{15.14}$$

By Theorem 15.12, the orthonormal eigenelements of a compact self-adjoint operator, including those for the possible eigenvalue zero, are complete in the sense of Theorem 1.28.

Example 15.13. Consider the integral operator $A : L^2[0, \pi] \to L^2[0, \pi]$ with continuous kernel

$$K(x, y) := \begin{cases} \dfrac{1}{\pi}\,(\pi - x)\,y, & 0 \le y \le x \le \pi, \\[2ex] \dfrac{1}{\pi}\,(\pi - y)\,x, & 0 \le x \le y \le \pi. \end{cases}$$

This kernel is the so-called *Green's function* to the boundary value problem for the simple ordinary differential equation

$$\varphi''(x) = -f(x), \quad 0 \le x \le \pi,$$

with homogeneous boundary condition $\varphi(0) = \varphi(\pi) = 0$. To each $f \in C[0, \pi]$ there exists a unique solution $\varphi \in C^2[0, \pi]$ of the boundary value problem, which is given by

$$\varphi(x) = \int_0^{\pi} K(x, y) f(y)\, dy, \quad 0 \le x \le \pi. \tag{15.15}$$

Uniqueness of the solution is obvious, and straightforward differentiation shows that (15.15) indeed represents the solution. The compact integral operator A with this so-called triangular kernel is self-adjoint, since its real-valued kernel is symmetric $K(x, y) = K(y, x)$. The eigenvalue equation $A\varphi = \lambda\varphi$ is equivalent to the differential equation

$$\lambda\varphi''(x) + \varphi(x) = 0, \quad 0 \le x \le \pi,$$

with homogeneous boundary condition $\varphi(0) = \varphi(\pi) = 0$. In particular, taking $\lambda = 0$ yields a trivial nullspace $N(A) = \{0\}$. The only nontrivial solutions to the boundary value problem are given through

$$\lambda_n = \frac{1}{n^2} \quad \text{and} \quad \varphi_n(x) = \sqrt{\frac{2}{\pi}}\, \sin nx, \quad 0 \le x \le \pi,$$

for $n \in \mathbb{N}$. Therefore, in this example, (15.13) becomes

$$\int_0^\pi K(x,y)\varphi(y)\,dy = \frac{2}{\pi}\sum_{n=1}^\infty \frac{1}{n^2}\sin nx \int_0^\pi \sin ny\,\varphi(y)\,dy$$

corresponding to the Fourier expansion of the kernel K. □

We now wish to characterize the eigenvalues of compact self-adjoint operators by a minimum-maximum principle. We confine our presentation to the case of a nonnegative operator. A self-adjoint operator $A : X \to X$ is called *nonnegative* if

$$(A\varphi, \varphi) \geq 0, \quad \varphi \in X.$$

Note that nonnegative operators only have nonnegative eigenvalues.

Theorem 15.14 (Courant). *Let X be a Hilbert space, $A : X \to X$ be a nonnegative self-adjoint compact operator, and (λ_n) denote the nonincreasing sequence of the nonzero eigenvalues repeated according to their multiplicity. Then*

$$\lambda_1 = \|A\| = \sup_{\|\varphi\|=1} (A\varphi, \varphi) \tag{15.16}$$

and

$$\lambda_{n+1} = \inf_{\psi_1,\ldots,\psi_n \in X} \sup_{\substack{\varphi \perp \psi_1,\ldots,\psi_n \\ \|\varphi\|=1}} (A\varphi, \varphi), \quad n = 1, 2, \ldots. \tag{15.17}$$

Proof. We need only to prove (15.17). To this end, we abbreviate the right-hand side of (15.17) by μ_{n+1}. If we choose $\psi_1 = \varphi_1, \ldots, \psi_n = \varphi_n$, then for all $\varphi \perp \psi_1, \ldots, \psi_n$ from (15.13) and (15.14) we obtain that

$$(A\varphi, \varphi) = \sum_{k=n+1}^\infty \lambda_k |(\varphi, \varphi_k)|^2 \leq \lambda_{n+1} \sum_{k=n+1}^\infty |(\varphi, \varphi_k)|^2 \leq \lambda_{n+1}\,\|\varphi\|^2,$$

with equality holding for $\varphi = \varphi_{n+1}$. Therefore $\mu_{n+1} \leq \lambda_{n+1}$.

On the other hand, to each choice $\psi_1, \ldots, \psi_n \in X$ there exists an element

$$\varphi \in \operatorname{span}\{\varphi_1, \ldots, \varphi_{n+1}\}$$

with $\|\varphi\| = 1$ and $\varphi \perp \psi_1, \ldots, \psi_n$. To see this we write $\varphi = \sum_{k=1}^{n+1} \gamma_k \varphi_k$ and have to solve the linear system of n equations

$$\sum_{k=1}^{n+1} \gamma_k (\varphi_k, \psi_j) = 0, \quad j = 1, \ldots, n,$$

for the $n + 1$ unknowns $\gamma_1, \ldots, \gamma_{n+1}$. This system always allows a solution that can be normalized such that

$$\|\varphi\|^2 = \sum_{k=1}^{n+1} |\gamma_k|^2 = 1.$$

Then, again from (15.13) and (15.14), it follows that

$$(A\varphi, \varphi) = \sum_{k=1}^{n+1} \lambda_k |(\varphi, \varphi_k)|^2 \geq \lambda_{n+1} \sum_{k=1}^{n+1} |(\varphi, \varphi_k)|^2 = \lambda_{n+1} \|\varphi\|^2 = \lambda_{n+1}.$$

Hence $\mu_{n+1} \geq \lambda_{n+1}$, and the proof is complete. □

15.4 Singular Value Decomposition

We now will describe modified forms of the expansions (15.11) and (15.12) for arbitrary compact operators in Hilbert spaces. From Theorem 4.12 we recall that the adjoint operator of a compact linear operator is compact.

Definition 15.15. Let X and Y be Hilbert spaces, $A : X \to Y$ be a compact linear operator, and $A^* : Y \to X$ be its adjoint. The nonnegative square roots of the eigenvalues of the nonnegative self-adjoint compact operator $A^*A : X \to X$ are called *singular values* of A.

Theorem 15.16. *Let (μ_n) denote the sequence of the nonzero singular values of the compact linear operator A (with $A \neq 0$) repeated according to their multiplicity, i.e., according to the dimension of the nullspaces $N(\mu_n^2 I - A^*A)$. Then there exist orthonormal sequences (φ_n) in X and (g_n) in Y such that*

$$A\varphi_n = \mu_n g_n, \quad A^* g_n = \mu_n \varphi_n \tag{15.18}$$

for all $n \in \mathbb{N}$. For each $\varphi \in X$ we have the singular value decomposition

$$\varphi = \sum_{n=1}^{\infty} (\varphi, \varphi_n)\varphi_n + Q\varphi \tag{15.19}$$

with the orthogonal projection operator $Q : X \to N(A)$ and

$$A\varphi = \sum_{n=1}^{\infty} \mu_n (\varphi, \varphi_n) g_n. \tag{15.20}$$

Each system (μ_n, φ_n, g_n), $n \in \mathbb{N}$, with these properties is called a singular system *of A. When there are only finitely many singular values, the series (15.19) and (15.20) degenerate into finite sums. (Note that for an injective operator A the orthonormal system $\{\varphi_n : n \in \mathbb{N}\}$ provided by the singular system is complete in X.)*

Proof. Let (φ_n) denote an orthonormal sequence of the eigenelements of A^*A, i.e.,

$$A^*A\varphi_n = \mu_n^2 \varphi_n$$

and define a second orthonormal sequence by

$$g_n := \frac{1}{\mu_n} A\varphi_n.$$

Straightforward computations show that the system (μ_n, φ_n, g_n), $n \in \mathbb{N}$, satisfies (15.18). Application of the expansion (15.14) to the self-adjoint compact operator A^*A yields

$$\varphi = \sum_{n=1}^{\infty} (\varphi, \varphi_n)\varphi_n + Q\varphi, \quad \varphi \in X,$$

where Q denotes the orthogonal projection operator from X onto $N(A^*A)$. Let $\psi \in N(A^*A)$. Then

$$(A\psi, A\psi) = (\psi, A^*A\psi) = 0,$$

and this implies that $N(A^*A) = N(A)$. Therefore, (15.19) is proven and (15.20) follows by applying A to (15.19). □

Note that the singular value decomposition implies that for all $\varphi \in X$ we have

$$\|\varphi\|^2 = \sum_{n=1}^{\infty} |(\varphi, \varphi_n)|^2 + \|Q\varphi\|^2 \tag{15.21}$$

and

$$\|A\varphi\|^2 = \sum_{n=1}^{\infty} \mu_n^2 |(\varphi, \varphi_n)|^2. \tag{15.22}$$

Theorem 15.17. *Let $A, B : X \to Y$ be compact linear operators. Then for the nonincreasing sequence of singular values we have*

$$\mu_1(A) = \|A\| = \sup_{\|\varphi\|=1} \|A\varphi\| \tag{15.23}$$

and

$$\mu_{n+1}(A) = \inf_{\psi_1,\dots,\psi_n \in X} \sup_{\substack{\varphi \perp \psi_1,\dots,\psi_n \\ \|\varphi\|=1}} \|A\varphi\|, \quad n = 1, 2, \dots. \tag{15.24}$$

Furthermore

$$\mu_{n+m+1}(A + B) \le \mu_{n+1}(A) + \mu_{m+1}(B), \quad n, m = 0, 1, 2, \dots. \tag{15.25}$$

Proof. (15.23) and (15.24) follow immediately from Theorem 15.14, since the squares μ_n^2 of the singular values of A are given by the eigenvalues of the nonnegative self-adjoint operator A^*A, and since $(A^*A\varphi, \varphi) = \|A\varphi\|^2$ for all $\varphi \in X$. The inequality (15.25) is a consequence of

$$\begin{aligned}
&\inf_{\psi_1,\dots,\psi_{n+m}\in X} \sup_{\substack{\varphi\perp\psi_1,\dots,\psi_{n+m}\\ \|\varphi\|=1}} \|(A+B)\varphi\| \\
&\le \inf_{\psi_1,\dots,\psi_{n+m}\in X} \sup_{\substack{\varphi\perp\psi_1,\dots,\psi_n\\ \|\varphi\|=1}} \|A\varphi\| + \inf_{\psi_1,\dots,\psi_{n+m}\in X} \sup_{\substack{\varphi\perp\psi_{n+1},\dots,\psi_{n+m}\\ \|\varphi\|=1}} \|B\varphi\| \\
&\le \inf_{\psi_1,\dots,\psi_n\in X} \sup_{\substack{\varphi\perp\psi_1,\dots,\psi_n\\ \|\varphi\|=1}} \|A\varphi\| + \inf_{\psi_{n+1},\dots,\psi_{n+m}\in X} \sup_{\substack{\varphi\perp\psi_{n+1},\dots,\psi_{n+m}\\ \|\varphi\|=1}} \|B\varphi\|
\end{aligned}$$

and (15.24). □

In the following theorem, we express the solution to an equation of the first kind with a compact operator in terms of a singular system.

Theorem 15.18 (Picard). *Let $A : X \to Y$ be a compact linear operator with singular system (μ_n, φ_n, g_n). The equation of the first kind*

$$A\varphi = f \tag{15.26}$$

is solvable if and only if f belongs to the orthogonal complement $N(A^*)^\perp$ and satisfies

$$\sum_{n=1}^{\infty} \frac{1}{\mu_n^2} |(f, g_n)|^2 < \infty. \tag{15.27}$$

In this case a solution is given by

$$\varphi = \sum_{n=1}^{\infty} \frac{1}{\mu_n} (f, g_n)\varphi_n. \tag{15.28}$$

Proof. The necessity of $f \in N(A^*)^\perp$ follows from Theorem 15.8. If φ is a solution of (15.26), then

$$\mu_n(\varphi, \varphi_n) = (\varphi, A^*g_n) = (A\varphi, g_n) = (f, g_n),$$

and (15.21) implies

$$\sum_{n=1}^{\infty} \frac{1}{\mu_n^2} |(f, g_n)|^2 = \sum_{n=1}^{\infty} |(\varphi, \varphi_n)|^2 \le \|\varphi\|^2,$$

whence the necessity of (15.27) follows.

Conversely, assume that $f \perp N(A^*)$ and (15.27) is fulfilled. Then, as in the proof of Theorems 8.2 and 15.12, the convergence of the series (15.27) in $\mathbb{R}$ implies convergence of the series (15.28) in the Hilbert space X. We apply A to (15.28), use (15.19) with the singular system (μ_n, g_n, φ_n) of the operator A^*, and observe $f \in N(A^*)^\perp$ to obtain

$$A\varphi = \sum_{n=1}^{\infty} (f, g_n) g_n = f.$$

This ends the proof. □

Picard's theorem demonstrates the ill-posed nature of the equation $A\varphi = f$. If we perturb the right-hand side f by $f^\delta = f + \delta g_n$ we obtain the solution $\varphi^\delta = \varphi + \delta\mu_n^{-1}\varphi_n$. Hence, the ratio $\|\varphi^\delta - \varphi\|/\|f^\delta - f\| = 1/\mu_n$ can be made arbitrarily large due to the fact that the singular values tend to zero. The influence of errors in the data f is obviously controlled by the rate of this convergence. In this sense we may say that the equation is *mildly ill-posed* if the singular values decay slowly to zero and that it is *severely ill-posed* if they decay very rapidly.

Example 15.19. Consider the integral operator $A : L^2[0,1] \to L^2[0,1]$ defined by

$$(A\varphi)(x) := \int_0^x \varphi(y)\, dy, \quad 0 \le x \le 1.$$

Then the inverse operator A^{-1} corresponds to differentiation. The adjoint operator is given by

$$(A^*\psi)(x) = \int_x^1 \psi(y)\, dy, \quad 0 \le x \le 1.$$

Hence

$$(A^*A\varphi)(x) = \int_x^1 \int_0^y \varphi(z)\, dz dy, \quad 0 \le x \le 1,$$

and the eigenvalue equation $A^*A\varphi = \mu^2\varphi$ is equivalent to the boundary value problem for the ordinary differential equation

$$\mu^2\varphi'' + \varphi = 0$$

with homogeneous boundary conditions $\varphi(1) = \varphi'(0) = 0$. The nontrivial solutions are given by

$$\mu_n = \frac{2}{(2n-1)\pi}, \quad \varphi_n(x) = \sqrt{2}\cos\frac{(2n-1)\pi x}{2}, \quad n \in \mathbb{N}.$$

The singular system is completed by $g_n = \mu_n^{-1} A\varphi_n$ through

$$g_n(x) = \sqrt{2}\sin\frac{(2n-1)\pi x}{2}.$$

In this example we have a decay $\mu_n = O(1/n)$ for an integral operator with a discontinuity of the kernel along the diagonal $x = y$. The kernel of the integral operator in Example 15.13 has a discontinuity in the first derivatives at the diagonal and the singular values decay $\mu_n = O(1/n^2)$. □

In general, for compact integral operators the smoothness of the kernel controls the degree of ill-posedness. Roughly speaking, the smoothness of the kernel of the operator A determines the smoothness of the range of A, and this effects the regularity condition on f, which is required for the solvability of $A\varphi = f$. We illustrate this statement with the following result due to Little and Reade [158].

Theorem 15.20. *Let $A : L^2[-1,1] \to L^2[-1,1]$ be an integral operator with analytic kernel on $[-1,1]\times[-1,1]$. Then the singular values of A decay at least exponentially $\mu_n = O(R^{-n})$ for some constant $R > 1$.*

Proof. Let K denote the kernel of A and

$$K_n(x,y) := \frac{1}{2}\, T_0(x)a_0(y) + \sum_{m=1}^{n} T_m(x)a_m(y)$$

its approximation by the orthonormal Chebyshev expansion as considered in Section 11.5. For the integral operator A_n with the degenerate kernel K_n, by (11.40) we have $\|A_n - A\|_\infty = O(R^{-n})$ with some $R > 1$. Since $\dim A_n(X) \le n+1$, from the singular value decomposition (15.20) we observe that A_n has at most $n+1$ nonzero singular values. Therefore we can apply Theorem 15.17 to obtain

$$\mu_{n+2}(A) \le \mu_1(A - A_n) + \mu_{n+2}(A_n) = \|A - A_n\|_2 = O(R^{-n}),$$

and the proof is complete. □

15.5 Regularization Schemes

As already pointed out, Picard's Theorem 15.18 illustrates the fact that the ill-posedness of an equation of the first kind with a compact operator stems from the behavior of the singular values $\mu_n \to 0$, $n \to \infty$. This suggests trying to regularize the equation by damping or filtering out the influence of the factor $1/\mu_n$ in the solution formula (15.28).

Theorem 15.21. *Let $A : X \to Y$ be an injective compact linear operator with singular system (μ_n, φ_n, g_n), $n \in \mathbb{N}$, and let $q : (0,\infty)\times(0,\|A\|] \to \mathbb{R}$ be a bounded function such that for each $\alpha > 0$ there exists a positive constant $c(\alpha)$ with*

$$|q(\alpha,\mu)| \le c(\alpha)\mu, \quad 0 < \mu \le \|A\|, \tag{15.29}$$

and

$$\lim_{\alpha\to 0} q(\alpha,\mu) = 1, \quad 0 < \mu \le \|A\|. \tag{15.30}$$

Then the bounded linear operators $R_\alpha : Y \to X$, $\alpha > 0$, defined by

$$R_\alpha f := \sum_{n=1}^{\infty} \frac{1}{\mu_n}\, q(\alpha,\mu_n)\,(f,g_n)\varphi_n, \quad f \in Y, \tag{15.31}$$

describe a regularization scheme with

$$\|R_\alpha\| \le c(\alpha). \tag{15.32}$$

Proof. From (15.21) and (15.29) we have

$$\|R_\alpha f\|^2 = \sum_{n=1}^{\infty} \frac{1}{\mu_n^2}\,[q(\alpha,\mu_n)]^2\,|(f,g_n)|^2 \le [c(\alpha)]^2 \sum_{n=1}^{\infty} |(f,g_n)|^2 \le [c(\alpha)]^2 \|f\|^2$$

for all $f \in Y$. Therefore, the series (15.31) converges, i.e., the operator R_α is well defined, and R_α is bounded by (15.32). With the aid of

$$(R_\alpha A\varphi, \varphi_n) = \frac{1}{\mu_n}\,q(\alpha,\mu_n)\,(A\varphi, g_n) = q(\alpha,\mu_n)\,(\varphi,\varphi_n)$$

and the singular value decomposition for $R_\alpha A\varphi - \varphi$ we obtain

$$\|R_\alpha A\varphi - \varphi\|^2 = \sum_{n=1}^{\infty} |(R_\alpha A\varphi - \varphi, \varphi_n)|^2 = \sum_{n=1}^{\infty} [q(\alpha,\mu_n) - 1]^2\,|(\varphi,\varphi_n)|^2.$$

Here we have used the fact that A is injective. Let $\varphi \in X$ with $\varphi \neq 0$ and $\varepsilon > 0$ be given and let M denote a bound for q. Then there exists $N(\varepsilon) \in \mathbb{N}$ such that

$$\sum_{n=N+1}^{\infty} |(\varphi,\varphi_n)|^2 < \frac{\varepsilon}{2(M+1)^2}\,.$$

By the convergence condition (15.30), there exists $\alpha_0(\varepsilon) > 0$ such that

$$[q(\alpha,\mu_n) - 1]^2 < \frac{\varepsilon}{2\|\varphi\|^2}$$

for all $n = 1,\dots,N$ and all $0 < \alpha \le \alpha_0$. Splitting the series in two parts and using (15.21), it follows that

$$\|R_\alpha A\varphi - \varphi\|^2 < \frac{\varepsilon}{2\|\varphi\|^2} \sum_{n=1}^{N} |(\varphi,\varphi_n)|^2 + \frac{\varepsilon}{2} \le \varepsilon$$

for all $0 < \alpha \le \alpha_0$. Thus we have established that $R_\alpha A\varphi \to \varphi$, $\alpha \to 0$, for all $\varphi \in X$, and the proof is complete. □

Remark 15.22. *If we replace condition (15.29) by the stronger condition*

$$q(\alpha,\mu) \le \tilde{c}(\alpha)\mu^2, \quad 0 < \mu \le \|A\|, \tag{15.33}$$

and denote by M a bound on q, then instead of (15.32) we have

$$\|R_\alpha\| \le \sqrt{M\tilde{c}(\alpha)}. \tag{15.34}$$

Proof. From

$$AR_\alpha f = \sum_{n=1}^{\infty} q(\alpha, \mu_n)\,(f, g_n)g_n$$

we find

$$\|AR_\alpha f\|^2 = \sum_{n=1}^{\infty} [q(\alpha, \mu_n)]^2\, |(f, g_n)|^2 \le M^2 \sum_{n=1}^{\infty} |(f, g_n)|^2 \le M^2\|f\|^2,$$

whence $\|AR_\alpha\| \le M$ follows. For the operators $\widetilde{R}_\alpha : Y \to Y$, $\alpha > 0$, defined by

$$\widetilde{R}_\alpha f := \sum_{n=1}^{\infty} \frac{1}{\mu_n^2}\, q(\alpha, \mu_n)\,(f, g_n)g_n, \quad f \in Y,$$

as in the proof of the previous Theorem 15.21, the condition (15.33) implies the bound $\|\widetilde{R}_\alpha\| \le \widetilde{c}(\alpha)$. Now, using $A^*\widetilde{R}_\alpha = R_\alpha$, by the Cauchy–Schwarz inequality we obtain

$$\|R_\alpha f\|^2 = (R_\alpha f, A^*\widetilde{R}_\alpha f) = (AR_\alpha f, \widetilde{R}_\alpha f) \le \|AR_\alpha f\|\, \|\widetilde{R}_\alpha f\| \le M\widetilde{c}(\alpha)\|f\|^2$$

for all $f \in Y$, whence (15.34) follows. □

We now describe some classical regularization schemes by choosing the damping or filter function q appropriately.

Theorem 15.23. *Let $A : X \to Y$ be a compact linear operator. Then for each $\alpha > 0$ the operator $\alpha I + A^*A : X \to X$ has a bounded inverse. Furthermore, if A is injective then*

$$R_\alpha := (\alpha I + A^*A)^{-1}A^*$$

describes a regularization scheme with $\|R_\alpha\| \le 1/2\sqrt{\alpha}$.

Proof. From

$$\alpha\|\varphi\|^2 \le (\alpha\varphi + A^*A\varphi, \varphi), \quad \varphi \in X,$$

we conclude that for $\alpha > 0$ the operator $\alpha I + A^*A$ is injective. Hence, by the Riesz Theorem 3.4 we have a bounded inverse $(\alpha I + A^*A)^{-1}$, because A^*A is compact.

Now assume that A is injective and let (μ_n, φ_n, g_n), $n \in \mathbb{N}$, be a singular system for A. Then the unique solution φ_α of

$$\alpha\varphi_\alpha + A^*A\varphi_\alpha = A^*f$$

can be written in the form

$$\varphi_\alpha = \sum_{n=1}^{\infty} \frac{\mu_n}{\alpha + \mu_n^2}\,(f, g_n)\varphi_n.$$

Indeed, using $A^*A\varphi_n = \mu_n^2\varphi_n$ and the singular value decomposition (15.20) applied to A^*f, we find

$$(\alpha I + A^*A)\varphi_\alpha = \sum_{n=1}^{\infty} \mu_n(f, g_n)\varphi_n = A^*f.$$

Hence, R_α can be brought into the form (15.31) with

$$q(\alpha, \mu) = \frac{\mu^2}{\alpha + \mu^2}\,.$$

This function q is bounded by $0 < q(\alpha, \mu) < 1$ and satisfies the conditions (15.29) and (15.30) with

$$c(\alpha) = \frac{1}{2\sqrt{\alpha}}$$

because of the arithmetic-geometric mean inequality

$$\sqrt{\alpha}\mu \leq \frac{\alpha + \mu^2}{2}\,.$$

The statement of the theorem now follows from Theorem 15.21. □

The regularization described in Theorem 15.23 is called *Tikhonov regularization*, since it was introduced by Tikhonov [232]. We will analyze it in more detail in Chapter 16.

Theorem 15.24. *Let $A : X \to Y$ be an injective compact linear operator with singular system (μ_n, φ_n, g_n), $n \in \mathbb{N}$. Then the* spectral cut-off

$$R_m f := \sum_{\mu_n \geq \mu_m} \frac{1}{\mu_n}\,(f, g_n)\varphi_n \tag{15.35}$$

describes a regularization scheme with regularization parameter $m \to \infty$ and

$$\|R_m\| = \frac{1}{\mu_m}\,.$$

Proof. The function q with $q(m, \mu) = 1$ for $\mu \geq \mu_m$ and $q(m, \mu) = 0$ otherwise satisfies the conditions (15.29) and (15.30) with $q(m, \mu) \leq \mu/\mu_m$. For the norm we can estimate

$$\|R_m f\|^2 = \sum_{\mu_n \geq \mu_m} \frac{1}{\mu_n^2}\,|(f, g_n)|^2 \leq \frac{1}{\mu_m^2} \sum_{\mu_n \geq \mu_m} |(f, g_n)|^2 \leq \frac{1}{\mu_m^2}\,\|f\|^2,$$

whence $\|R_m\| \leq 1/\mu_m$. Equality follows from $R_m(g_m) = \varphi_m/\mu_m$. □

Here, the regularization parameter m determines the number of terms in the sum (15.35). Accuracy of the approximation requires this number to be large, and stability requires it to be small. In particular, the following discrepancy principle turns out to be a regular a posteriori strategy for determining the stopping point for the spectral cut-off.

Theorem 15.25. *Let $A : X \to Y$ be an injective compact linear operator with dense range, $f \in Y$, and let $\delta > 0$. Then, for the spectral cut-off, there exists a smallest integer m, depending on f and δ, such that*

$$\|AR_m f - f\| \leq \delta.$$

Proof. By Theorem 15.8, the dense range $\overline{A(X)} = Y$ implies that A^* is injective. Hence, the singular value decomposition (15.19) with the singular system (μ_n, g_n, φ_n) for the adjoint operator A^* yields

$$f = \sum_{n=1}^{\infty} (f, g_n) g_n, \quad f \in Y, \tag{15.36}$$

and consequently

$$\|(AR_m - I)f\|^2 = \sum_{\mu_n < \mu_m} |(f, g_n)|^2 \to 0, \quad m \to \infty. \tag{15.37}$$

From this we now can conclude that there exists a smallest integer m such that $\|AR_m f - f\| \leq \delta$. □

From (15.36) and (15.37), we see that

$$\|AR_m f - f\|^2 = \|f\|^2 - \sum_{\mu_n \geq \mu_m} |(f, g_n)|^2.$$

This allows a stable determination of the stopping parameter $m(\delta)$ by terminating the sum when the right-hand side becomes smaller than or equal to δ^2 for the first time.

The regularity of the discrepancy principle for the spectral cut-off described through Theorem 15.25 is established in the following theorem.

Theorem 15.26. *Let $A : X \to Y$ be an injective compact linear operator with dense range. Let $f \in A(X)$ and $f^\delta \in Y$ satisfy $\|f^\delta - f\| \leq \delta$ with $\delta > 0$ and let $\gamma > 1$. Then, for the spectral cut-off, there exists a smallest integer $m = m(\delta)$, depending on f^δ and δ, such that*

$$\|AR_{m(\delta)} f^\delta - f^\delta\| \leq \gamma\delta \tag{15.38}$$

is satisfied and

$$R_{m(\delta)} f^\delta \to A^{-1} f, \quad \delta \to 0. \tag{15.39}$$

Proof. In view of Theorem 15.25, we only need to establish the convergence (15.39). We first note that (15.37) implies $\|I - AR_m\| = 1$ for all $m \in \mathbb{N}$. Therefore, writing

$$(AR_m f^\delta - f^\delta) - (AR_m f - f) = (AR_m - I)(f^\delta - f)$$

we have the triangle inequalities

$$\|AR_m f - f\| \le \delta + \|AR_m f^\delta - f^\delta\| \tag{15.40}$$

and

$$\|AR_m f^\delta - f^\delta\| \le \delta + \|AR_m f - f\|. \tag{15.41}$$

From (15.38) and (15.40) we obtain

$$\|AR_{m(\delta)} f - f\| \le \delta + \|AR_{m(\delta)} f^\delta - f^\delta\| \le (1+\gamma)\delta \to 0, \quad \delta \to 0. \tag{15.42}$$

Therefore, from the expansion (15.37), we conclude that either the cut-off number $m(\delta) \to \infty$, $\delta \to 0$, or the expansion for f degenerates into a finite sum

$$f = \sum_{\mu_n \ge \mu_{m_0}} (f, g_n) g_n$$

and $m(\delta) \ge m_0$. In the first case, from

$$\|AR_{m(\delta)-1} f^\delta - f^\delta\| > \gamma\delta$$

and (15.41), we conclude that

$$\gamma\delta < \delta + \|A(R_{m(\delta)-1} f - A^{-1} f)\|,$$

whence

$$\delta < \frac{1}{\gamma - 1} \|A(R_{m(\delta)-1} f - A^{-1} f)\| \tag{15.43}$$

follows. In order to establish the convergence (15.39), in this case, in view of (15.7) and (15.43), it suffices to show that

$$\|R_m\| \, \|A(R_{m-1} A\varphi - \varphi)\| \to 0, \quad m \to \infty,$$

for all $\varphi \in X$. But the latter property is obvious from

$$\|R_m\|^2 \, \|A(R_{m-1} A\varphi - \varphi)\|^2 = \frac{1}{\mu_m^2} \sum_{\mu_n < \mu_{m-1}} \mu_n^2 |(\varphi, \varphi_n)|^2 \le \sum_{\mu_n \le \mu_m} |(\varphi, \varphi_n)|^2.$$

In the case where f has a finite expansion then clearly

$$A^{-1} f = \sum_{\mu_n \ge \mu_{m_0}} \frac{1}{\mu_n} (f, g_n) \varphi_n = R_m f$$

for all $m \ge m_0$. Hence

$$\|AR_m f^\delta - f^\delta\| = \|(AR_m - I)(f^\delta - f)\| \le \|f^\delta - f\| \le \delta < \gamma\delta$$

for all $m \geq m_0$, and therefore $m(\delta) \leq m_0$. This implies $m(\delta) = m_0$ since $m(\delta) \geq m_0$ as noted above. Now observing

$$\|R_{m(\delta)} f^\delta - A^{-1} f\| = \|R_{m_0}(f^\delta - f)\| \leq \frac{\delta}{\mu_{m_0}} \to 0, \quad \delta \to 0,$$

the proof is finished. □

Theorem 15.27. *Let $A : X \to Y$ be an injective compact linear operator and let $0 < a < 1/\|A\|^2$. Then the bounded linear operators*

$$R_m := a \sum_{k=0}^{m} (I - aA^*A)^k A^*$$

describe a regularization scheme with regularization parameter $m \to \infty$ and

$$\|R_m\| \leq \sqrt{a(m+1)}.$$

Proof. Using a singular system for A and the singular value decomposition (15.20) applied to A^*f, with the aid of

$$a\mu^2 \sum_{k=0}^{m} (1 - a\mu^2)^k = 1 - (1 - a\mu^2)^{m+1},$$

we can write

$$R_m f = \sum_{n=1}^{\infty} \frac{1}{\mu_n} \left\{1 - (1 - a\mu_n^2)^{m+1}\right\} (f, g_n)\varphi_n. \tag{15.44}$$

The corresponding function q is given by

$$q(m, \mu) = 1 - (1 - a\mu^2)^{m+1}.$$

It is bounded by $0 \leq q(m, \mu) \leq 1$ for $0 < \mu \leq \|A\|$, and it satisfies the conditions (15.33) and (15.30) with $q(m, \mu) \leq a(m+1)\mu^2$. Hence, the proof is completed by Remark 15.22. □

The evaluation of the approximation $\psi_m = R_m f$ corresponds to m steps of the iteration scheme

$$\psi_k := (I - a\, A^*A)\psi_{k-1} + a\, A^* f, \quad k = 1, 2, \ldots, m, \tag{15.45}$$

starting with $\psi_0 := a\, A^* f$. This scheme goes back to Landweber [155] and Fridman [57] and consequently is known as the *Landweber–Fridman iteration*. The regularization parameter is given by the number m of iteration steps. Accuracy of the approximation requires m to be large, and stability requires m to be small. Again, we consider the discrepancy principle as a strategy for terminating the iterations.

Theorem 15.28. *Let $A : X \to Y$ be an injective compact linear operator with dense range, $f \in Y$, and $\delta > 0$. Then, for the Landweber–Fridman iterations, there exists a smallest integer m, depending on f and δ, such that*

$$\|AR_m f - f\| \le \delta.$$

Proof. (Compare the proof of Theorem 15.25.) From (15.44) it follows that

$$\|AR_m f - f\|^2 = \sum_{n=1}^{\infty} (1 - a\mu_n^2)^{2m+2} |(f, g_n)|^2. \tag{15.46}$$

Now, because of $a\mu_n^2 \le a\|A\|^2 < 1$ for $n \in \mathbb{N}$, we can choose $N \in \mathbb{N}$ such that

$$\sum_{n=N+1}^{\infty} (1 - a\mu_n^2)^{2m+2} |(f, g_n)|^2 \le \sum_{n=N+1}^{\infty} |(f, g_n)|^2 \le \frac{\delta^2}{2}.$$

Since $(1 - a\mu_n^2)^{2m+2} \to 0$, $m \to \infty$, for $n = 1, \dots, N$, there exists $m_0 \in \mathbb{N}$ such that

$$\sum_{n=1}^{N} (1 - a\mu_n^2)^{2m+2} |(f, g_n)|^2 \le \frac{\delta^2}{2}$$

for all $m \ge m_0$. In view of (15.46), the last two inequalities imply that $\|AR_m f - f\| \le \delta$ for all $m \ge m_0$, and the theorem is proven. □

In the next theorem we establish the regularity of this discrepancy principle for the Landweber–Fridman iteration.

Theorem 15.29. *Let $A : X \to Y$ be an injective compact linear operator with dense range. Let $f \in A(X)$ and $f^\delta \in Y$ satisfy $\|f^\delta - f\| \le \delta$ with $\delta > 0$ and let $\gamma > 1$. Then, for the Landweber–Fridman iterations, there exists a smallest integer $m = m(\delta)$, depending on f^δ and δ, such that*

$$\|AR_{m(\delta)} f^\delta - f^\delta\| \le \gamma\delta \tag{15.47}$$

is satisfied and

$$R_{m(\delta)} f^\delta \to A^{-1} f, \quad \delta \to 0. \tag{15.48}$$

Proof. We only need to establish the convergence (15.48). From (15.46) we conclude that $\|AR_m - I\| \le 1$ for all $m \in \mathbb{N}$. Therefore, proceeding as in the proof of Theorem 15.26 (see 15.42) we obtain that $\|AR_{m(\delta)} f - f\| \to 0$ as $\delta \to 0$. Hence, from (15.46) we conclude that either $m(\delta) \to \infty$, $\delta \to 0$, or $f = 0$. Since in the latter case the statement of the theorem is obvious, we only need to be concerned with the case where $m(\delta) \to \infty$, $\delta \to 0$. Analogously to (15.43) we obtain

$$(\gamma - 1)\delta < \|AR_{m(\delta)-1} f - f\|,$$

and hence

$$(\gamma - 1) \|R_{m(\delta)}\|\delta \le \sqrt{a(m(\delta) + 1)} \, \|AR_{m(\delta)-1}f - f\|.$$

Therefore, in view of (15.7) the proof is completed by showing that

$$\sqrt{m+1} \, \|AR_{m-1}A\varphi - A\varphi\| \to 0, \quad m \to \infty, \tag{15.49}$$

for all $\varphi \in X$. From (15.46) we conclude that

$$(m+1) \, \|AR_{m-1}A\varphi - A\varphi\|^2 = \sum_{n=1}^{\infty} (m+1)(1 - a\mu_n^2)^{2m} \mu_n^2 |(\varphi, \varphi_n)|^2.$$

By induction, it can be seen that

$$(m+1)(1 - a\mu^2)^{2m} \mu^2 \le \frac{1}{2a}, \quad 0 < a\mu^2 \le \frac{1}{2},$$

for all $m \in \mathbb{N}$. Proceeding as in the proof of Theorem 15.21, with the help of

$$(m+1)(1 - a\mu^2)^{2m} \to 0, \quad m \to \infty,$$

for all $0 < a\mu^2 < 1$, we can establish (15.49). □

Problems

15.1. Prove the Riemann–Lebesgue lemma

$$\int_0^{\pi} K(\cdot, y) \sin ny \, dy \to 0, \quad n \to \infty,$$

in the mean square norm for a kernel $K \in L^2([0, \pi] \times [0, \pi])$. How can this result be used to illustrate the ill-posedness of integral equations of the first kind?

15.2. Let X be a Hilbert space and $A : X \to X$ be a compact self-adjoint operator. With the aid of the spectral Theorem 15.12, solve the equation of the second kind

$$\lambda\varphi - A\varphi = f, \quad \lambda \ne 0.$$

15.3. The Poisson integral

$$u(\rho \cos t, \rho \sin t) = \frac{1}{2\pi} \int_0^{2\pi} \frac{1 - \rho^2}{1 + \rho^2 - 2\rho \cos(t - \tau)} \, \varphi(\tau) \, d\tau, \quad 0 \le \rho < 1, \, 0 \le t \le 2\pi,$$

gives the solution to the Dirichlet problem for the Laplace equation $\Delta u = 0$ in the unit disk

$$D = \{x \in \mathbb{R}^2 : |x| < 1\}$$

with boundary values

$$u(\cos t, \sin t) = \varphi(t), \quad 0 \le t \le 2\pi,$$

on ∂D in polar coordinates. Therefore, the continuation of a harmonic function given on a concentric disk of radius $r < 1$ to a harmonic function on the unit disk is equivalent to the solution of the integral equation of the first kind

$$\frac{1}{2\pi}\int_0^{2\pi} \frac{1-r^2}{1+r^2-2r\cos(t-\tau)}\,\varphi(\tau)\,d\tau = f(t), \quad 0 \le t \le 2\pi,$$

with $f(t) = u(r\cos t, r\sin t)$, $0 \le t \le 2\pi$. Determine the singular values.

15.4. Determine the singular values of the operator $A : L^2[0,1] \to L^2[0,1]$ given by

$$(A\varphi)(x) := \int_0^x (x-y)\varphi(y)\,dy, \quad 0 \le x \le 1.$$

What is the inverse of A?

15.5. Consider the central difference quotient

$$(R_h f)(x) := \frac{1}{h}\left\{f\left(x+\frac{h}{2}\right) - f\left(x-\frac{h}{2}\right)\right\}, \quad 0 \le x \le 1,$$

for functions f that are odd with respect to $x = 0$ and even with respect to $x = 1$. Show that R_h, $h > 0$, can be interpreted as a regularization scheme in the sense of Theorem 15.21 for the integral operator of Example 15.19.

Chapter 16
Tikhonov Regularization

This chapter will continue the study of Tikhonov regularization and will be based on its classical interpretation as a penalized residual minimization. For this we will consider the more general case of merely bounded linear operators. In particular, we shall explain the concepts of quasi-solutions and minimum norm solutions as strategies for the selection of the regularization parameter. We then proceed with a discussion of the classical regularization of integral equations of the first kind as introduced by Tikhonov [232] and Phillips [189]. The final section of this chapter is devoted to an application of Tikhonov regularization to severely ill-posed integral equations arising in a solution method for boundary value problems for the Laplace equation due to Kupradze [153].

16.1 Weak Convergence

As a prerequisite of the subsequent analysis we need to introduce the notion of weak convergence.

Definition 16.1. A sequence (φ_n) of elements from a Hilbert space X is called *weakly convergent* to an element $\varphi \in X$ if

$$\lim_{n\to\infty} (\psi, \varphi_n) = (\psi, \varphi)$$

for all $\psi \in X$.

For a weakly convergent sequence we will write $\varphi_n \rightharpoonup \varphi$, $n \to \infty$. Note that norm convergence $\varphi_n \to \varphi$, $n \to \infty$, always implies weak convergence $\varphi_n \rightharpoonup \varphi$, $n \to \infty$, whereas simple examples show that the converse of this statement is generally false. In particular, the elements u_n of a complete orthonormal system (see Theorem 1.28) converge weakly to the zero element as a consequence of Parseval's equality. But

R. Kress, *Linear Integral Equations*, Applied Mathematical Sciences 82,
DOI 10.1007/978-1-4614-9593-2_16,

they are not norm convergent to the zero element since they all have norm one. We leave it to the reader to verify that a sequence in a Hilbert space cannot weakly converge to two different elements.

Theorem 16.2. *A weakly convergent sequence in a Hilbert space is bounded.*

Proof. Let (φ_n) be a weakly convergent sequence. Then the sequence of bounded linear functionals $F_n : X \to \mathbb{C}$, defined by $F_n(\psi) := (\psi, \varphi_n)$ for $\psi \in X$, is pointwise convergent. Therefore, by the uniform boundedness principle Theorem 10.6 it is uniformly bounded, i.e., $\|F_n\| \le C$ for all $n \in \mathbb{N}$ and some constant C. Hence from $\|\varphi_n\|^2 = F_n(\varphi_n) \le \|F_n\|\|\varphi_n\| \le C\|\varphi_n\|$ we have $\|\varphi_n\| \le C$ for all $n \in \mathbb{N}$. □

Theorem 16.3. *Every bounded sequence in a Hilbert space contains a weakly convergent subsequence.*

Proof. Let (φ_n) be a bounded sequence, i.e., $\|\varphi_n\| \le C$ for all $n \in \mathbb{N}$ and some constant C. Then, for each $i \in \mathbb{N}$ the sequence (φ_i, φ_n) is bounded in $\mathbb{C}$. Therefore, by the standard diagonalization procedure (see the proof of Theorem 1.18) we can select a subsequence $(\varphi_{n(k)})$ such that $(\varphi_i, \varphi_{n(k)})$ converges in $\mathbb{C}$ as $k \to \infty$ for each $i \in \mathbb{N}$. Hence, the linear functional F given by

$$F(\psi) := \lim_{k\to\infty} (\psi, \varphi_{n(k)})$$

is well defined on $U := \operatorname{span}\{\varphi_i : i \in \mathbb{N}\}$ and by continuity it is also well defined on the closure $\bar{U}$. By decomposing an arbitrary $\psi \in X$ into the form

$$\psi = P\psi + \psi - P\psi$$

where $P : X \to \bar{U}$ denotes the orthogonal projection operator, we finally find that F is well defined on all of X. Furthermore, F is bounded by $\|F\| \le C$. Therefore, by the Riesz representation Theorem 4.10, there exists a unique element $\varphi \in X$ such that $F(\psi) = (\psi, \varphi)$ for all $\psi \in X$. Hence,

$$\lim_{k\to\infty} (\psi, \varphi_{n(k)}) = (\psi, \varphi)$$

for all $\psi \in X$, that is, $(\varphi_{n(k)})$ converges weakly to φ as $k \to \infty$. □

In an obvious meaning we may reformulate Theorem 16.3 by saying that in a Hilbert space each bounded set is *relatively weakly sequentially compact* (see Section 1.4).

16.2 The Tikhonov Functional

The following theorem presents another aspect of the Tikhonov regularization complementing its introduction in Theorem 15.23. Throughout this chapter X and Y will always denote Hilbert spaces.

Theorem 16.4. *Let $A : X \to Y$ be a bounded linear operator and let $\alpha > 0$. Then for each $f \in Y$ there exists a unique $\varphi_\alpha \in X$ such that*

$$\|A\varphi_\alpha - f\|^2 + \alpha\|\varphi_\alpha\|^2 = \inf_{\varphi\in X}\left\{\|A\varphi - f\|^2 + \alpha\|\varphi\|^2\right\}. \tag{16.1}$$

The minimizer φ_α is given by the unique solution of the equation

$$\alpha\varphi_\alpha + A^*A\varphi_\alpha = A^*f \tag{16.2}$$

and depends continuously on f.

Proof. From the equation

$$\begin{aligned}\|A\varphi - f\|^2 + \alpha\|\varphi\|^2 = {} & \|A\varphi_\alpha - f\|^2 + \alpha\|\varphi_\alpha\|^2\\ & +2\,\mathrm{Re}(\varphi - \varphi_\alpha, \alpha\varphi_\alpha + A^*(A\varphi_\alpha - f))\\ & +\|A(\varphi - \varphi_\alpha)\|^2 + \alpha\|\varphi - \varphi_\alpha\|^2,\end{aligned}$$

which is valid for all $\varphi \in X$, we observe that the condition (16.2) is necessary and sufficient for φ_α to minimize the *Tikhonov functional* defined by (16.1).

Consider the operator $T_\alpha : X \to X$, given by $T_\alpha := \alpha I + A^*A$. Since

$$\alpha\|\varphi\|^2 \le \alpha\|\varphi\|^2 + \|A\varphi\|^2 = \mathrm{Re}(T_\alpha\varphi, \varphi), \quad \varphi \in X, \tag{16.3}$$

the operator T_α is strictly coercive and therefore, by the Lax–Milgram Theorem 13.29, has a bounded inverse $T_\alpha^{-1} : X \to X$. □

The equation (16.2), of course, coincides with the Tikhonov regularization introduced in Theorem 15.23. The interpretation of the Tikhonov regularization as a minimizer of the Tikhonov functional allows us to extend the regularization property from Theorem 15.23 to the case of merely bounded operators.

Theorem 16.5. *Let $A : X \to Y$ be an injective bounded linear operator. Then*

$$R_\alpha := (\alpha I + A^*A)^{-1}A^*$$

describes a regularization scheme with

$$\|R_\alpha\| \le \frac{\|A\|}{\alpha}.$$

Proof. The bound on R_α follows from (16.3) (see the proof of the Lax–Milgram Theorem 13.29) and the fact that $\|A^*\| = \|A\|$. Considering the case $A = I$ it can be seen that the order of this estimate is optimal.

Let $f \in A(X)$ and set

$$\varphi_\alpha := R_\alpha f.$$

Since φ_α minimizes the Tikhonov functional we can estimate

$$\alpha\|\varphi_\alpha\|^2 \le \alpha\|\varphi_\alpha\|^2 + \|A\varphi_\alpha - f\|^2 \le \alpha\|A^{-1}f\|^2.$$

From this we conclude that

$$\|\varphi_\alpha\| \le \|A^{-1}f\| \tag{16.4}$$

for all $\alpha > 0$ and

$$\|A\varphi_\alpha - f\| \to 0, \quad \alpha \to 0.$$

Now let $g \in Y$ be arbitrary. Then we have

$$|(\varphi_\alpha - A^{-1}f, A^*g)| = |(A\varphi_\alpha - f, g)| \le \|A\varphi_\alpha - f\|\,\|g\| \to 0, \quad \alpha \to 0.$$

This implies weak convergence $\varphi_\alpha \rightharpoonup A^{-1}f$, $\alpha \to 0$, since for the injective operator A the range $A^*(Y)$ is dense in X by Theorem 15.8 and since φ_α is bounded according to (16.4). Finally we can estimate

$$\|\varphi_\alpha - A^{-1}f\|^2 = \|\varphi_\alpha\|^2 - 2\,\mathrm{Re}(\varphi_\alpha, A^{-1}f) + \|A^{-1}f\|^2$$

$$\le 2\,\mathrm{Re}(A^{-1}f - \varphi_\alpha, A^{-1}f) \to 0, \quad \alpha \to 0,$$

and the proof is complete. □

In general, convergence $\varphi_\alpha \to A^{-1}f$ for $\alpha \to 0$ will be slow. The optimal convergence rate is described in the following theorem.

Theorem 16.6. *Let $A : X \to Y$ be an injective compact linear operator. Then for $f \in A(X)$ the condition $f \in AA^*A(X)$ is necessary and sufficient for*

$$\|\varphi_\alpha - A^{-1}f\| = O(\alpha), \quad \alpha \to 0.$$

Proof. Let (μ_n, φ_n, g_n), $n = 1, 2, \ldots$, be a singular system for the operator A. Then by Picard's Theorem 15.18 we have

$$A^{-1}f = \sum_{n=1}^{\infty} \frac{1}{\mu_n}\,(f, g_n)\varphi_n.$$

From the proof of Theorem 15.23 we recall that

$$\varphi_\alpha = \sum_{n=1}^{\infty} \frac{\mu_n}{\alpha + \mu_n^2}\,(f, g_n)\varphi_n. \tag{16.5}$$

The expansions for $A^{-1}f$ and φ_α together imply that

$$\|\varphi_\alpha - A^{-1}f\|^2 = \sum_{n=1}^{\infty} \frac{\alpha^2}{\mu_n^2(\alpha + \mu_n^2)^2}\,|(f, g_n)|^2. \tag{16.6}$$

Now let $f = AA^*Ag$ for some $g \in X$. Then

$$(f, g_n) = (AA^*Ag, g_n) = (g, A^*AA^*g_n) = \mu_n^3(g, \varphi_n),$$

and from (16.6) it follows that

$$\|\varphi_\alpha - A^{-1}f\|^2 = \sum_{n=1}^{\infty} \frac{\alpha^2\mu_n^4}{(\alpha + \mu_n^2)^2} |(g, \varphi_n)|^2 \le \alpha^2 \sum_{n=1}^{\infty} |(g, \varphi_n)|^2 = \alpha^2\|g\|^2,$$

that is, $\|\varphi_\alpha - A^{-1}f\| = O(\alpha)$, $\alpha \to 0$.

Conversely, assume that $\|\varphi_\alpha - A^{-1}f\| = O(\alpha)$, $\alpha \to 0$. Then from (16.6) we observe that there exists a constant $M > 0$ such that

$$\sum_{n=1}^{\infty} \frac{|(f, g_n)|^2}{\mu_n^2(\alpha + \mu_n^2)^2} \le M$$

for all $\alpha > 0$. Passing to the limit $\alpha \to 0$, this implies that

$$\sum_{n=1}^{\infty} \frac{1}{\mu_n^6} |(f, g_n)|^2 \le M.$$

Hence,

$$g := \sum_{n=1}^{\infty} \frac{1}{\mu_n^3} (f, g_n)\varphi_n \in X$$

is well defined and $AA^*Ag = f$. □

The condition $f \in AA^*A(X)$ of the preceding theorem can be interpreted as a regularity assumption on f because it controls how fast the Fourier coefficients (f, g_n) tend to zero as $n \to \infty$.

Remark 16.7. *For an injective bounded linear operator $A : X \to Y$ with dense range we have convergence $R_\alpha f \to \varphi \in X$ as $\alpha \to 0$ if and only if $f \in A(X)$.*

Proof. On one hand, $f \in A(X)$ implies convergence since R_α is a regularization scheme by Theorem 16.5. On the other hand, convergence implies that the limit φ satisfies $A^*A\varphi = A^*f$. From this if follows that $A\varphi = f$ since A^* is injective as a consequence of the dense range of A. □

However, in the image space we have the following convergence results.

Theorem 16.8. *Let $A : X \to Y$ be an injective bounded linear operator with dense range. Then*

$$\|A\varphi_\alpha - f\| \to 0, \quad \alpha \to 0,$$

for all $f \in Y$.

Proof. Since $A(X)$ is dense in Y, for every $\varepsilon > 0$ there exists an element $\varphi_\varepsilon \in X$ such that $\|A\varphi_\varepsilon - f\|^2 < \varepsilon/2$. Choose δ such that $\delta\|\varphi_\varepsilon\|^2 \leq \varepsilon/2$. Then, using Theorem 16.4, for all $\alpha < \delta$ we have

$$\|A\varphi_\alpha - f\|^2 \leq \|A\varphi_\alpha - f\|^2 + \alpha\|\varphi_\alpha\|^2 \leq \|A\varphi_\varepsilon - f\|^2 + \alpha\|\varphi_\varepsilon\|^2 < \varepsilon.$$

This implies convergence $A\varphi_\alpha \to f$, $\alpha \to 0$. □

Theorem 16.9. *Under the assumptions of Theorem 16.8 assume that $f \in A(X)$. Then*

$$\|A\varphi_\alpha - f\| = o(\sqrt{\alpha}), \quad \alpha \to 0.$$

Proof. We write $f = A\varphi$ and, by Theorem 16.5, we have $\varphi_\alpha \to \varphi$, $\alpha \to 0$. Then

$$\|A\varphi_\alpha - f\|^2 = (\varphi_\alpha - \varphi, A^*[A\varphi_\alpha - f]) = \alpha(\varphi - \varphi_\alpha, \varphi_\alpha), \tag{16.7}$$

which implies $\|A\varphi_\alpha - f\|^2 = o(\alpha)$, and the proof is complete. □

The optimal rate of convergence is described in the following theorem.

Theorem 16.10. *Let $A : X \to Y$ be an injective compact linear operator with dense range. Then the condition $f \in AA^*(Y)$ is necessary and sufficient for*

$$\|A\varphi_\alpha - f\| = O(\alpha), \quad \alpha \to 0.$$

Proof. Let (μ_n, φ_n, g_n), $n = 1, 2, \ldots$, be a singular system for the operator A. Since A^* is injective, by the singular value decomposition Theorem 15.16, the $\{g_n : n \in \mathbb{N}\}$ form a complete orthonormal system in Y, i.e., we can expand

$$f = \sum_{n=1}^{\infty} (f, g_n) g_n.$$

This expansion of f and the expansion (16.5) for φ_α together imply that

$$\|A\varphi_\alpha - f\|^2 = \sum_{n=1}^{\infty} \frac{\alpha^2}{(\alpha + \mu_n^2)^2} |(f, g_n)|^2. \tag{16.8}$$

Now let $f = AA^*g$ for some $g \in X$. Then

$$(f, g_n) = (AA^*g, g_n) = (g, AA^*g_n) = \mu_n^2(g, g_n), \tag{16.9}$$

and from (16.8), as in the proof of Theorem 16.6, it follows that $\|A\varphi_\alpha - f\|^2 \leq \alpha^2\|g\|^2$.

Conversely, assume that $\|A\varphi_\alpha - f\| = O(\alpha)$, $\alpha \to 0$. Then from (16.8), again as in the proof of Theorem 16.6, we obtain that there exists a constant $M > 0$ such that

$$\sum_{n=1}^{\infty} \frac{1}{\mu_n^4} |(f, g_n)|^2 \leq M.$$

Hence,

$$g := \sum_{n=1}^{\infty} \frac{1}{\mu_n^2} (f, g_n) g_n \in Y$$

is well defined and $AA^*g = f$. □

We note that under the assumptions of Theorem 16.10 as a consequence of (16.6), (16.9) and the arithmetic-geometric mean inequality we have that

$$\|\varphi_\alpha - A^{-1}f\| = O(\sqrt{\alpha}), \quad \alpha \to 0,$$

for all $f \in AA^*(Y)$ as an intermediate convergence order between Theorems 16.5 and 16.6.

By the interpretation of the Tikhonov regularization as minimizer of the Tikhonov functional, its solution keeps the residual $\|A\varphi_\alpha - f\|^2$ small and is stabilized through the penalty term $\alpha\|\varphi_\alpha\|^2$. Although Tikhonov regularization itself is not a penalty method, such a view nevertheless suggests the following *constraint* optimization problems:

(a) For given $\rho > 0$, minimize the defect $\|A\varphi - f\|$ subject to the constraint that the norm is bounded by $\|\varphi\| \leq \rho$.

(b) For given $\delta > 0$, minimize the norm $\|\varphi\|$ subject to the constraint that the defect is bounded by $\|A\varphi - f\| \leq \delta$.

The first interpretation leads to the concept of *quasi-solutions* and the second to the concept of *minimum norm solutions* and the *discrepancy principle.*

16.3 Quasi-Solutions

The principal idea underlying the concept of quasi-solutions as introduced by Ivanov [106] is to stabilize an ill-posed problem by restricting the solution set to some subset $U \subset X$ exploiting suitable a priori informations on the solution of $A\varphi = f$. For perturbed right-hand sides, in general, we cannot expect a solution in U. Therefore, instead of trying to solve the equation exactly, we minimize the residual. For simplicity we restrict our presentation to the case where $U = B[0;\rho]$ is a closed ball of radius ρ with some $\rho > 0$. This choice requires some a priori knowledge on the norm of the solution.

Definition 16.11. Let $A : X \to Y$ be a bounded injective linear operator and let $\rho > 0$. For a given $f \in Y$ an element $\varphi_0 \in X$ is called a *quasi-solution* of $A\varphi = f$ with constraint ρ if $\|\varphi_0\| \leq \rho$ and

$$\|A\varphi_0 - f\| = \inf_{\|\varphi\| \leq \rho} \|A\varphi - f\|.$$

Note that φ_0 is a quasi-solution to $A\varphi = f$ with constraint ρ if and only if $A\varphi_0$ is a best approximation to f with respect to the set $V := A(B[0;\rho])$. It is obvious how the definition of a quasi-solution can be extended to more general constraint sets. The injectivity of the operator A is essential for uniqueness properties of the quasi-solution. For the sake of clarity, in the following analysis on quasi-solutions, we confine ourselves to the case where the operator A has dense range in Y. By Theorem 15.8 this is equivalent to assuming that the adjoint operator $A^* : Y \to X$ is injective.

Theorem 16.12. *Let $A : X \to Y$ be a bounded injective linear operator with dense range and let $\rho > 0$. Then for each $f \in Y$ there exists a unique quasi-solution of $A\varphi = f$ with constraint ρ.*

Proof. Since A is linear, the set $V = A(B[0;\rho])$ clearly is convex. By Theorem 1.27 there exists at most one best approximation to f with respect to V. Since A is injective, this implies uniqueness of the quasi-solution.

If $f \in A(B[0;\rho])$, then there exists φ_0 with $\|\varphi_0\| \le \rho$ and $A\varphi_0 = f$. Clearly, φ_0 is a quasi-solution. Therefore, we only need to be concerned with the case where $f \notin A(B[0;\rho])$. We will establish existence of the quasi-solution by constructing an element φ_0 that satisfies the sufficient condition for the best approximation given in Theorem 1.27. In the case of approximation with respect to V, this condition reads $\mathrm{Re}(f - A\varphi_0, A\varphi - A\varphi_0) \le 0$, that is,

$$\mathrm{Re}(A^*(f - A\varphi_0), \varphi - \varphi_0) \le 0 \tag{16.10}$$

for all φ with $\|\varphi\| \le \rho$. Obviously, any element φ_0 with $\|\varphi_0\| = \rho$ satisfying

$$\alpha\varphi_0 + A^*A\varphi_0 = A^*f, \tag{16.11}$$

for some $\alpha > 0$, fulfills the condition (16.10) and therefore provides a quasi-solution. For $f \notin A(B[0;\rho])$ we will show that α can be chosen such that the unique solution φ_0 of (16.11) (see Theorem 16.4) satisfies $\|\varphi_0\| = \rho$.

Define a function $F : (0, \infty) \to \mathbb{R}$ by

$$F(\alpha) := \|\varphi_\alpha\|^2 - \rho^2,$$

where φ_α denotes the unique solution of (16.2). We have to show that F has a zero. By a Neumann series argument, using Theorem 10.1, the function F can be seen to be continuous. From Theorem 16.5 we have that $\|R_\alpha\| \le \|A\|/\alpha$ whence

$$\varphi_\alpha = R_\alpha f \to 0, \quad \alpha \to \infty, \tag{16.12}$$

follows. This implies $F(\alpha) \to -\rho^2 < 0$, $\alpha \to \infty$.

Now assume that $F(\alpha) \le 0$, i.e., $\|\varphi_\alpha\| \le \rho$ for all $\alpha > 0$. Then, by Theorem 16.3, we can choose a sequence (α_n) with $\alpha_n \to 0$, $n \to \infty$, such that we have weak convergence $\varphi_n := \varphi_{\alpha_n} \rightharpoonup \varphi$, $n \to \infty$, with some $\varphi \in X$. From

$$\|\varphi\|^2 = \lim_{n\to\infty} (\varphi_n, \varphi) \le \rho\|\varphi\|$$

we obtain $\|\varphi\| \le \rho$. Writing $(A\varphi_n, \psi) = (\varphi_n, A^*\psi)$ we conclude weak convergence $A\varphi_n \rightharpoonup A\varphi$, $n \to \infty$. Finally, we can use Theorem 16.8 to find

$$\|A\varphi - f\|^2 = \lim_{n\to\infty} (A\varphi_n - f, A\varphi - f) \le \lim_{n\to\infty} \|A\varphi_n - f\|\,\|A\varphi - f\| = 0,$$

which is a contradiction to $f \notin A(B[0;\rho])$. Therefore there exists α such that $F(\alpha) = \|\varphi_\alpha\|^2 - \rho^2 > 0$. Now the continuity of F implies the existence of a zero of F, and the proof is completed. □

We note that in the case when A is compact a simpler proof of Theorem 16.12 based on the singular value decomposition can be given (see [32, Theorem 4.18]).

The quasi-solution can be shown to restore stability in the sense that it depends weakly continuously on the right-hand side f (see Problem 16.1).

In applications, errors in the data f generally ensure that $f \notin A(B[0;\rho])$. Then the quasi-solution with constraint ρ can be obtained numerically by Newton's method for solving $F(\alpha) = 0$. By writing

$$\frac{1}{h}(\varphi_{\alpha+h} - \varphi_\alpha) = -(\alpha I + A^*A)^{-1}\varphi_{\alpha+h}$$

it can be seen that the solution φ_α to (16.2) is differentiable with respect to α and the derivative $d\varphi_\alpha/d\alpha$ satisfies the equation

$$\alpha \frac{d\varphi_\alpha}{d\alpha} + A^*A \frac{d\varphi_\alpha}{d\alpha} = -\varphi_\alpha. \tag{16.13}$$

Then the derivative of F is given by

$$F'(\alpha) = 2\,\mathrm{Re}\left(\frac{d\varphi_\alpha}{d\alpha}, \varphi_\alpha\right).$$

We may view the quasi-solution as described in Theorem 16.12 as an a posteriori strategy for the choice of the parameter α in the Tikhonov regularization: Given a perturbed right-hand side f^δ of $f \in A(X)$ with $\|f^\delta - f\| \le \delta$, we choose α such that

$$\varphi_\alpha^\delta = (\alpha I + A^*A)^{-1}A^*f^\delta$$

satisfies $\|\varphi_\alpha^\delta\| = \rho$ with some a priori known bound ρ on the norm of the exact solution. Then we can deduce that

$$\begin{aligned}\alpha\rho^2 = \alpha(\varphi_\alpha^\delta, \varphi_\alpha^\delta) &= (A\varphi_\alpha^\delta, f^\delta - A\varphi_\alpha^\delta)\\ &\le \rho\,\|A\|\,\|f^\delta - A\varphi_\alpha^\delta\|\\ &\le \rho\|A\|\,\|f^\delta - AA^{-1}f\|\\ &\le \rho\,\|A\|\,\delta\end{aligned}$$

provided $\|A^{-1}f\| \le \rho$. Therefore, in this case, we have the estimate

$$\alpha\rho \le \|A\|\,\delta, \tag{16.14}$$

which may serve as a starting value for the Newton iteration to find the zero of F. The following theorem answers the question for regularity in the sense of Definition 15.7 for this strategy.

Theorem 16.13. *Let $A : X \to Y$ be a bounded injective linear operator with dense range and let $f \in A(X)$ and $\rho \ge \|A^{-1}f\|$. For $f^\delta \in Y$ with $\|f^\delta - f\| \le \delta$, let φ^δ denote the quasi-solution to $A\varphi = f^\delta$ with constraint ρ. Then we have weak convergence*

$$\varphi^\delta \rightharpoonup A^{-1}f, \quad \delta \to 0.$$

If $\rho = \|A^{-1}f\|$, then we have norm convergence

$$\varphi^\delta \to A^{-1}f, \quad \delta \to 0.$$

Proof. Let $g \in Y$ be arbitrary. Then, since $\|A^{-1}f\| \le \rho$, we can estimate

$$\begin{aligned} |(A\varphi^\delta - f, g)| &\le \left\{\|A\varphi^\delta - f^\delta\| + \|f^\delta - f\|\right\}\|g\| \\ &\le \left\{\|AA^{-1}f - f^\delta\| + \|f^\delta - f\|\right\}\|g\| \\ &\le 2\delta\,\|g\|. \end{aligned} \tag{16.15}$$

Hence, $(\varphi^\delta - A^{-1}f, A^*g) \to 0$, $\delta \to 0$, for all $g \in Y$. This implies weak convergence $\varphi^\delta \rightharpoonup A^{-1}f$, $\delta \to 0$, because for the injective operator A the range $A^*(Y)$ is dense in X by Theorem 15.8 and because φ^δ is bounded by $\|\varphi^\delta\| \le \rho$.

When $\rho = \|A^{-1}f\|$, we have

$$\begin{aligned} \|\varphi^\delta - A^{-1}f\|^2 &= \|\varphi^\delta\|^2 - 2\,\mathrm{Re}(\varphi^\delta, A^{-1}f) + \|A^{-1}f\|^2 \\ &\le 2\,\mathrm{Re}(A^{-1}f - \varphi^\delta, A^{-1}f) \to 0, \quad \delta \to 0, \end{aligned}$$

and the proof is complete. □

Note that we cannot expect weak convergence if $\rho < \|A^{-1}f\|$, since then we would have the contradiction

$$\|A^{-1}f\|^2 = \lim_{\delta\to 0}(\varphi^\delta, A^{-1}f) \le \rho\|A^{-1}f\| < \|A^{-1}f\|^2.$$

In general, we also cannot expect norm convergence if $\rho > \|A^{-1}f\|$ because generically we will have $\|\varphi^\delta\| = \rho$ for all δ. Thus, for regularity we need an exact a priori information on the norm of the exact solution.

Under additional conditions on f, which may be interpreted as regularity conditions, we can obtain results on the order of convergence.

Theorem 16.14. *Under the assumptions of Theorem 16.13, let $f \in AA^*(Y)$ and $\rho = \|A^{-1}f\|$. Then*

$$\|\varphi^\delta - A^{-1}f\| = O(\delta^{1/2}), \quad \delta \to 0.$$

Proof. We can write $A^{-1}f = A^*g$ with some $g \in Y$. Therefore, the last inequality in the proof of Theorem 16.13, together with (16.15), yields

$$\|\varphi^\delta - A^{-1}f\|^2 \le 2\,\mathrm{Re}(f - A\varphi^\delta, g) \le 4\delta\|g\|,$$

and this is the desired result. □

Example 16.15. The following counterexample shows that the result of Theorem 16.14 is optimal. Let A be a compact injective operator with $\dim A(X) = \infty$ and singular system (μ_n, φ_n, g_n). Consider $f = \mu_1 g_1$ and $f^{\delta_n} = \mu_1 g_1 + \delta_n g_n$ with $\delta_n = \mu_n^2$. Then $A^{-1}f = \varphi_1$ and

$$\begin{aligned}\varphi^{\delta_n} &= (\alpha_n I + A^*A)^{-1}A^*(\mu_1 g_1 + \delta_n g_n)\\ &= \frac{\mu_1^2}{\alpha_n + \mu_1^2}\,\varphi_1 + \frac{\delta_n\mu_n}{\alpha_n + \mu_n^2}\,\varphi_n,\end{aligned} \tag{16.16}$$

where α_n must satisfy

$$\frac{\mu_1^4}{(\alpha_n + \mu_1^2)^2} + \frac{\delta_n^2\mu_n^2}{(\alpha_n + \mu_n^2)^2} = 1$$

so that φ^{δ_n} is the quasi-solution with constraint $\rho = 1$. Assume now that we have convergence order

$$\|\varphi^{\delta_n} - A^{-1}f\| = o\,(\delta_n^{1/2}), \quad n \to \infty.$$

Then, using $\delta_n = \mu_n^2$, from (16.16) we find

$$\frac{\delta_n}{\alpha_n + \delta_n} = \frac{\delta_n^{1/2}\mu_n}{\alpha_n + \mu_n^2} \to 0, \quad n \to \infty,$$

whence $\alpha_n/\delta_n \to \infty$, $n \to \infty$ follows. But this is a contradiction to the inequality (16.14). □

For a compact integral operator $A : L^2[a, b] \to L^2[a, b]$, the Tikhonov residual functional corresponds to

$$\int_a^b |(A\varphi)(x) - f(x)|^2 dx + \alpha \int_a^b |\varphi(x)|^2 dx. \tag{16.17}$$

The regularized equation (16.2) is an integral equation of the second kind. For its numerical solution the methods of Chapters 12 and 13 are available.

Example 16.16. Consider the integral equation

$$\int_0^1 e^{xy}\varphi(y)\,dy = \frac{1}{x}\,(e^x - 1), \quad 0 < x \le 1.$$

Its unique solution (see Problem 16.4) is given by $\varphi(x) = 1$. For the corresponding integral operator $A : L^2[0,1] \to L^2[0,1]$ elementary calculations yield

$$(A^*A\varphi)(x) = \int_0^1 H(x,y)\varphi(y)\,dy, \quad 0 \le x \le 1,$$

where

$$H(x,y) = \int_0^1 e^{(x+y)z}\,dz = \frac{1}{x+y}\,(e^{x+y} - 1)$$

for $x + y > 0$ and $H(0,0) = 1$. Our numerical results are obtained by discretizing the integral equation of the second kind with $\alpha > 0$ by Nyström's method using Simpson's rule with eight equidistant intervals. The integral for the right-hand side is also evaluated numerically by Simpson's rule. We have assumed a regular error distribution $f_i^\delta = f_i + (-1)^i\delta$ at the grid points and used the norm $\rho = 1$ of the exact solution. Table 16.1 gives the values of the regularization parameter α (obtained by Newton's method) and the mean square error $E := \|\varphi_\alpha^\delta - \varphi\|_{L^2}$ between the regularized solution φ_α^δ and the exact solution φ depending on the error level δ. In addition, the quotient $q := E/\delta^{1/2}$ is listed. □

Table 16.1 Numerical results for Example 16.16

δ	α	E	q
0.02	0.000055	0.112	0.795
0.04	0.000076	0.161	0.805
0.06	0.000093	0.198	0.809
0.08	0.000109	0.229	0.812
0.10	0.000123	0.257	0.814

16.4 Minimum Norm Solutions

As already observed at the end of Section 15.2, the principal motivation for the discrepancy principle as introduced by Morozov [174, 175] is based on the observation that, in general, for erroneous data it does not make too much sense to try and make the residual $\|A\varphi - f\|$ smaller than the error in f. Assume that we have some a priori bound δ on the error in f. Then we look for elements φ satisfying $\|A\varphi - f\| \le \delta$ and stabilize by making the norm $\|\varphi\|$ small.

Definition 16.17. Let $A : X \to Y$ be a bounded linear operator and let $\delta > 0$. For a given $f \in Y$ an element $\varphi_0 \in X$ is called a *minimum norm solution* of $A\varphi = f$ with *discrepancy* δ if $\|A\varphi_0 - f\| \le \delta$ and

$$\|\varphi_0\| = \inf_{\|A\varphi - f\| \le \delta} \|\varphi\|.$$

Note that φ_0 is a minimum norm solution to $A\varphi = f$ with discrepancy δ if and only if φ_0 is a best approximation to the zero element of X with respect to $U_f := \{\varphi \in X : \|A\varphi - f\| \le \delta\}$. As in the discussion of quasi-solutions, for the sake of simplicity, in dealing with existence of minimum norm solutions we confine ourselves to operators with dense range and note again that by Theorem 15.8 this is equivalent to assuming that the adjoint operator $A^* : Y \to X$ is injective.

Theorem 16.18. *Let $A : X \to Y$ be a bounded linear operator with dense range and let $\delta > 0$. Then for each $f \in Y$ there exists a unique minimum norm solution of $A\varphi = f$ with discrepancy δ.*

Proof. From

$$\|A(\lambda\varphi_1 + (1-\lambda)\varphi_2) - f\| \le \lambda\|A\varphi_1 - f\| + (1-\lambda)\|A\varphi_2 - f\|$$

for all $\varphi_1, \varphi_2 \in X$ and all $\lambda \in (0,1)$, we observe that U_f is convex. Then, by Theorem 1.27, there exists at most one best approximation of the zero element with respect to U_f.

If $\|f\| \le \delta$, then clearly $\varphi_0 = 0$ is the minimum norm solution with discrepancy δ. Therefore, we only need to consider the case where $\|f\| > \delta$. Since A has dense range in Y, the set U_f is not empty and we will again establish existence of the minimum norm solution by constructing an element φ_0 that satisfies the sufficient condition for the best approximation given in Theorem 1.27. For approximation with respect to U_f, this condition reads

$$\mathrm{Re}(\varphi_0, \varphi_0 - \varphi) \le 0 \tag{16.18}$$

for all $\varphi \in X$ with $\|A\varphi - f\| \le \delta$. Assume that φ_0 satisfies

$$\alpha\varphi_0 + A^*A\varphi_0 = A^*f \tag{16.19}$$

for some $\alpha > 0$ and $\|A\varphi_0 - f\| = \delta$. Then

$$\begin{aligned}\alpha\,\mathrm{Re}(\varphi_0, \varphi_0 - \varphi) &= \mathrm{Re}(A^*(f - A\varphi_0), \varphi_0 - \varphi)\\ &= \mathrm{Re}(A\varphi_0 - f, A\varphi - f) - \|A\varphi_0 - f\|^2\\ &\le \delta\,(\|A\varphi - f\| - \delta) \le 0.\end{aligned}$$

Hence φ_0 satisfies the condition (16.18) and therefore is a minimum norm solution. We will show that α can be chosen such that the unique solution φ_0 of (16.19) (see Theorem 16.4) satisfies $\|A\varphi_0 - f\| = \delta$.

Define a function $G : (0, \infty) \to \mathbb{R}$ by

$$G(\alpha) := \|A\varphi_\alpha - f\|^2 - \delta^2,$$

where φ_α denotes the unique solution of equation (16.2). We have to show that the continuous function G has a zero. From the convergence (16.12), we observe that

$G(\alpha) \to \|f\|^2 - \delta^2 > 0$, $\alpha \to \infty$. On the other hand, from Theorem 16.8 we obtain that $G(\alpha) \to -\delta^2 < 0$, $\alpha \to 0$. This completes the proof. □

We note that in the case when A is compact a simpler proof of Theorem 16.18 based on the singular value decomposition can be given (see [32, Theorem 4.15]).

The minimum norm solution can be proven to depend weakly continuously on the right-hand side f (see Problem 16.2).

In general, we will have data satisfying $\|f\| > \delta$, i.e., data exceeding the error level. Then the minimum norm solution with discrepancy δ can be obtained numerically by Newton's method for solving $G(\alpha) = 0$. After rewriting

$$\|f - A\varphi_\alpha\|^2 = (f - A\varphi_\alpha, f) - (A^*(f - A\varphi_\alpha), \varphi_\alpha) = \|f\|^2 - (\varphi_\alpha, A^*f) - \alpha\|\varphi_\alpha\|^2,$$

we get

$$G(\alpha) = \|f\|^2 - (\varphi_\alpha, A^*f) - \alpha\|\varphi_\alpha\|^2 - \delta^2$$

and

$$G'(\alpha) = -\left(\frac{d\varphi_\alpha}{d\alpha}, A^*f\right) - \|\varphi_\alpha\|^2 - 2\alpha \operatorname{Re}\left(\frac{d\varphi_\alpha}{d\alpha}, \varphi_\alpha\right),$$

where the derivative $d\varphi_\alpha/d\alpha$ is given by (16.13).

We may look at the minimum norm solution as described in Theorem 16.18 as an a posteriori strategy for the choice of the parameter α in the Tikhonov regularization: Given a perturbed right-hand side f^δ of an element $f \in A(X)$ with a known error level $\|f^\delta - f\| \le \delta < \|f^\delta\|$, we choose α such that

$$\varphi_\alpha^\delta = (\alpha I + A^*A)^{-1}A^*f^\delta$$

satisfies $\|A\varphi_\alpha^\delta - f^\delta\| = \delta$. Then, using $\alpha\varphi_\alpha^\delta + A^*A\varphi_\alpha^\delta = A^*f^\delta$, we find

$$\|f^\delta\| - \delta = \|f^\delta\| - \|A\varphi_\alpha^\delta - f^\delta\| \le \|A\varphi_\alpha^\delta\| = \frac{1}{\alpha}\|AA^*(f^\delta - A\varphi_\alpha^\delta)\| \le \frac{\|A\|^2\delta}{\alpha}$$

provided $\|f^\delta\| > \delta$. Hence, we have established the estimate

$$\alpha\,(\|f^\delta\| - \delta) \le \|A\|^2\delta, \tag{16.20}$$

which we may use as a starting value for the Newton iteration to find the zero of G.

The following theorem answers the question of regularity for this *discrepancy principle* for the Tikhonov regularization.

Theorem 16.19. *Let $A : X \to Y$ be a bounded injective linear operator with dense range and let $\delta > 0$ and $f \in A(X)$. For $f^\delta \in Y$ with $\|f^\delta - f\| \le \delta$ and $\delta < \|f^\delta\|$, let φ^δ denote the minimum norm solution with discrepancy δ. Then*

$$\varphi^\delta \to A^{-1}f, \quad \delta \to 0.$$

Proof. Since $\|f^\delta\| > \delta$, from the proof of Theorem 16.18 we know that φ^δ minimizes the Tikhonov functional. Therefore,

$$\begin{aligned}\delta^2 + \alpha\|\varphi^\delta\|^2 &= \|A\varphi^\delta - f^\delta\|^2 + \alpha\|\varphi^\delta\|^2 \\ &\le \|AA^{-1}f - f^\delta\|^2 + \alpha\|A^{-1}f\|^2 \\ &\le \delta^2 + \alpha\|A^{-1}f\|^2,\end{aligned}$$

whence

$$\|\varphi^\delta\| \le \|A^{-1}f\| \tag{16.21}$$

follows. This inequality is also trivially satisfied when $\|f^\delta\| \le \delta$, since in this case $\varphi^\delta = 0$.

Now let $g \in Y$ be arbitrary. Then we can estimate

$$|(A\varphi^\delta - f, g)| \le \left\{\|A\varphi^\delta - f^\delta\| + \|f^\delta - f\|\right\}\|g\| \le 2\delta\,\|g\|.$$

As in the proof of Theorem 16.13, this implies $\varphi^\delta \rightharpoonup A^{-1}f$, $\delta \to 0$. Then using (16.21), we obtain

$$\begin{aligned}\|\varphi^\delta - A^{-1}f\|^2 &= \|\varphi^\delta\|^2 - 2\,\mathrm{Re}(\varphi^\delta, A^{-1}f) + \|A^{-1}f\|^2 \\ &\le 2\{\|A^{-1}f\|^2 - \mathrm{Re}(\varphi^\delta, A^{-1}f)\} \to 0, \quad \delta \to 0,\end{aligned} \tag{16.22}$$

which finishes the proof. □

Theorem 16.20. *Under the assumptions of Theorem 16.19, let $f \in AA^*(Y)$. Then*

$$\|\varphi^\delta - A^{-1}f\| = O(\delta^{1/2}), \quad \delta \to 0.$$

Proof. Writing $A^{-1}f = A^*g$ with some $g \in Y$ from (16.22) we deduce

$$\begin{aligned}\|\varphi^\delta - A^{-1}f\|^2 &\le 2\,\mathrm{Re}(A^{-1}f - \varphi^\delta, A^{-1}f) \\ &= 2\,\mathrm{Re}\,(f - A\varphi^\delta, g) \\ &\le 2\{\|f - f^\delta\| + \|f^\delta - A\varphi^\delta\|\}\|g\| \le 4\delta\|g\|,\end{aligned}$$

and the proof is finished. □

Using the same example as in connection with Theorem 16.14, it can be shown that the result of Theorem 16.20 is optimal (see Problem 16.3).

Example 16.21. We apply the discrepancy principle to the integral equation of Example 16.16. Table 16.2 gives the values of the regularization parameter α, the mean square error $E := \|\varphi_\alpha^\delta - \varphi\|_{L^2}$, and the quotient $q := E/\delta^{1/2}$ depending on the error level δ. □

Table 16.2 Numerical results for Example 16.21

δ	α	E	q
0.02	0.0059	0.067	0.479
0.04	0.0148	0.108	0.542
0.06	0.0252	0.131	0.536
0.08	0.0359	0.145	0.514
0.10	0.0466	0.155	0.492

16.5 Classical Tikhonov Regularization

In our examples of integral equations of the first kind, so far, we have interpreted the integral operator as a mapping $A : L^2[a,b] \to L^2[a,b]$ corresponding to the Tikhonov residual functional in the form (16.17). Here, the choice of the data space $Y = L^2[a,b]$, in general, is determined by the need to adequately measure the error in the data. But we have more flexibility concerning the solution space X, in particular, when additional regularity properties of the exact solution are a priori known.

In his pioneering papers on integral equations of the first kind in 1963, Tikhonov [232, 233] suggested damping out highly oscillating parts in the approximate solution by incorporating the derivative into the penalty term, i.e., to replace (16.17) by

$$\int_a^b |(A\varphi)(x) - f(x)|^2 dx + \alpha \int_a^b \left\{ |\varphi(x)|^2 + |\varphi'(x)|^2 \right\} dx. \tag{16.23}$$

To include this approach into our general theory we need a Hilbert space with norm corresponding to the penalty term in (16.23). Since in Chapter 8 we introduced the Sobolev spaces only via Fourier expansion, we briefly discuss the definition for the space $H^1[a,b]$ based on the concept of weak derivatives. Compare also to the notion of weak solutions to the Laplace equation from Section 8.3.

Definition 16.22. A function $\varphi \in L^2[a,b]$ is said to possess a *weak derivative* $\varphi' \in L^2[a,b]$ if

$$\int_a^b \varphi\psi' \, dx = -\int_a^b \varphi'\psi \, dx \tag{16.24}$$

for all $\psi \in C^1[a,b]$ with $\psi(a) = \psi(b) = 0$.

By partial integration, it follows that (16.24) is satisfied for $\varphi \in C^1[a,b]$. Hence, weak differentiability generalizes classical differentiability.

From the denseness of $\left\{\psi \in C^1[a,b] : \psi(a) = \psi(b) = 0\right\}$ in $L^2[a,b]$, or from the Fourier series for the odd extension for φ, it can be seen that the weak derivative, if it exists, is unique. From the denseness of $C[a,b]$ in $L^2[a,b]$, or from the Fourier series for the even extension of φ, it follows that each function with vanishing weak derivative must be constant (almost everywhere). The latter, in particular, implies

$$\varphi(x) = \int_a^x \varphi'(\xi)d\xi + c \tag{16.25}$$

for almost all $x \in [a, b]$ and some constant c, since by Fubini's theorem

$$\int_a^b \left(\int_a^x \varphi'(\xi)\,d\xi \right) \psi'(x)\,dx = \int_a^b \varphi'(\xi) \left(\int_\xi^b \psi'(x)dx \right) d\xi = -\int_a^b \varphi'(\xi)\psi(\xi)\,d\xi$$

for all $\psi \in C^1[a, b]$ with $\psi(a) = \psi(b)$. Hence both sides of (16.25) have the same weak derivative.

Theorem 16.23. *The linear space*

$$H^1[a, b] := \left\{ \varphi \in L^2[a, b] : \varphi' \in L^2[a, b] \right\}$$

endowed with the scalar product

$$(\varphi, \psi)_{H^1} := \int_a^b (\varphi\bar{\psi} + \varphi'\bar{\psi}')\,dx \tag{16.26}$$

is a Hilbert space.

Proof. It is readily checked that $H^1[a, b]$ is a linear space and that (16.26) defines a scalar product. Let (φ_n) denote a H^1 Cauchy sequence. Then (φ_n) and (φ_n') are both L^2 Cauchy sequences. From the completeness of $L^2[a, b]$ we obtain the existence of $\varphi \in L^2[a, b]$ and $\chi \in L^2[a, b]$ such that $\|\varphi_n - \varphi\|_{L^2} \to 0$ and $\|\varphi_n' - \chi\|_{L^2} \to 0$ as $n \to \infty$. Then for all $\psi \in C^1[a, b]$ with $\psi(a) = \psi(b) = 0$, we can estimate

$$\int_a^b (\varphi\psi' + \chi\psi)\,dx = \int_a^b \{(\varphi - \varphi_n)\psi' + (\chi - \varphi_n')\psi\}\,dx$$

$$\leq \|\varphi - \varphi_n\|_{L^2}\|\psi'\|_{L^2} + \|\chi - \varphi_n'\|_{L^2}\|\psi\|_{L^2} \to 0, \quad n \to \infty.$$

Therefore, $\varphi \in H^1[a, b]$ with $\varphi' = \chi$ and $\|\varphi - \varphi_n\|_{H^1} \to 0$, $n \to \infty$, which completes the proof. □

Theorem 16.24. *$C^1[a, b]$ is dense in $H^1[a, b]$.*

Proof. Since $C[a, b]$ is dense in $L^2[a, b]$, for each $\varphi \in H^1[a, b]$ and $\varepsilon > 0$ there exists $\chi \in C[a, b]$ such that $\|\varphi' - \chi\|_{L^2} < \varepsilon$. Then we define $\psi \in C^1[a, b]$ by

$$\psi(x) := \varphi(a) + \int_a^x \chi(\xi)\,d\xi, \quad x \in [a, b],$$

and using (16.25), we have

$$\varphi(x) - \psi(x) = \int_a^x \{\varphi'(\xi) - \chi(\xi)\}\, d\xi, \quad x \in [a,b].$$

By the Cauchy–Schwarz inequality this implies $\|\varphi - \psi\|_{L^2} < (b-a)\varepsilon$, and the proof is complete. □

Theorem 16.25. *$H^1[a,b]$ is contained in $C[a,b]$ with compact imbedding.*

Proof. From (16.25) we have

$$\varphi(x) - \varphi(y) = \int_y^x \varphi'(\xi)\, d\xi, \tag{16.27}$$

whence by the Cauchy–Schwarz inequality,

$$|\varphi(x) - \varphi(y)| \le |x-y|^{1/2} \|\varphi'\|_{L^2} \tag{16.28}$$

follows for all $x, y \in [a,b]$. Therefore, every function $\varphi \in H^1[a,b]$ belongs to $C[a,b]$, or more precisely, it coincides almost everywhere with a continuous function.

Choose $y \in [a,b]$ such that $|\varphi(y)| = \min_{a \le x \le b} |\varphi(x)|$. Then from

$$(b-a) \min_{a \le x \le b} |\varphi(x)|^2 \le \int_a^b |\varphi(x)|^2 dx$$

and (16.27), again by the Cauchy–Schwarz inequality, we find that

$$\|\varphi\|_\infty \le C\|\varphi\|_{H^1} \tag{16.29}$$

for all $\varphi \in H^1[a,b]$ and some constant C. The latter inequality means that the H^1 norm is stronger than the maximum norm (in one space dimension!). The inequalities (16.28) and (16.29), in particular, imply that each bounded set in $H^1[a,b]$ is a bounded and equicontinuous subset of $C[a,b]$. Hence, by the Arzelà–Ascoli Theorem 1.18, the imbedding operator from $H^1[a,b]$ into $C[a,b]$ is compact. (See also Theorem 8.4.) □

Now we return to integral equations of the first kind and interpret the integral operator with smooth kernel as a mapping from $H^1[a,b]$ into $L^2[a,b]$. Then our complete theory on regularization including convergence and regularity remains applicable in this setting. Since the imbedding from $H^1[a,b]$ into $L^2[a,b]$ clearly is bounded (by Theorem 16.25 it is even compact) the integral operator $A : H^1[a,b] \to L^2[a,b]$ is compact by Theorem 2.21. Only the adjoint operator $\widetilde{A}^* : L^2[a,b] \to H^1[a,b]$ looks different. We denote it by $\widetilde{A}^*$ to distinguish it from the adjoint $A^* : L^2[a,b] \to L^2[a,b]$ of $A : L^2[a,b] \to L^2[a,b]$. We avoid its explicit calculation through the observation that the regularized equation

$$\alpha\varphi_\alpha + \widetilde{A}^* A\varphi_\alpha = \widetilde{A}^* f \tag{16.30}$$

is equivalent to

$$\alpha(\varphi_\alpha, \psi)_{H^1} + (A^*A\varphi_\alpha, \psi)_{L^2} = (A^*f, \psi)_{L^2} \tag{16.31}$$

for all $\psi \in H^1[a, b]$. This follows from the fact that

$$(\widetilde{A}^*\chi, \psi)_{H^1} = (\chi, A\psi)_{L^2} = (A^*\chi, \psi)_{L^2}$$

for all $\chi \in L^2[a, b]$ and $\psi \in H^1[a, b]$. By Theorem 16.4, there exists a unique solution φ_α to (16.30). In the following theorem we will show that φ_α is the solution to a boundary value problem for an integro-differential equation.

Theorem 16.26. *Assume that the integral operator A has continuous kernel. Then the unique solution φ_α of the regularized equation (16.30) belongs to $C^2[a, b]$ and satisfies the* integro-differential equation

$$\alpha\,(\varphi_\alpha - \varphi_\alpha'') + A^*A\varphi_\alpha = A^*f \tag{16.32}$$

and the boundary condition

$$\varphi_\alpha'(a) = \varphi_\alpha'(b) = 0. \tag{16.33}$$

Proof. First we define $g \in C^1[a, b]$ by

$$g(x) := -\int_x^b \{(A^*A\varphi_\alpha)(\xi) + \alpha\varphi_\alpha(\xi) - (A^*f)(\xi)\}\, d\xi, \quad x \in [a, b],$$

and then $\psi \in H^1[a, b]$ by

$$\psi(x) := \int_a^x \{\alpha\varphi_\alpha'(\xi) - g(\xi)\}\, d\xi, \quad x \in [a, b].$$

Then, by partial integration, since $g(b) = \psi(a) = 0$, we get

$$\begin{aligned} \|\alpha\varphi_\alpha' - g\|_{L^2}^2 &= \int_a^b (\alpha\varphi_\alpha' - g)\,\bar{\psi}'\, dx = \int_a^b (\alpha\varphi_\alpha'\bar{\psi}' + g'\bar{\psi})\, dx \\ &= \alpha(\varphi_\alpha, \psi)_{H^1} + (A^*A\varphi_\alpha, \psi)_{L^2} - (A^*f, \psi)_{L^2} = 0. \end{aligned}$$

Note that by the denseness result of Theorem 16.24 partial integration can be carried over from $C^1[a, b]$ into $H^1[a, b]$. Our last equation implies $\alpha\varphi_\alpha' = g$. From this, it follows that $\varphi_\alpha \in C^2[a, b]$ and that the integro-differential equation (16.32) is satisfied. Inserting (16.32) into (16.31) and performing a partial integration yields

$$\varphi_\alpha'(b)\overline{\psi(b)} - \varphi_\alpha'(a)\overline{\psi(a)} = 0$$

for all $\psi \in H^1[a, b]$. This implies that the boundary conditions (16.33) are fulfilled. □

Note that by partial integration a solution of (16.32) and (16.33) also solves (16.31). Hence, the boundary value problem for the integro-differential equation and the regularized equation (16.30) are equivalent.

Example 16.27. We compare the Tikhonov regularization based on the L^2 norm and the H^1 norm penalty term for the integral equation of Example 16.16. For the discretization of the regularized equation we use the Petrov–Galerkin method with linear splines. Let $x_j = jh,\ j = 0, \dots, n$, be an equidistant grid with step size $h = 1/n$ and let $L_j,\ j = 0, \dots, n$, denote the corresponding Lagrange basis for linear spline interpolation as introduced by (11.8). Then straightforward calculations yield the tridiagonal matrices

$$V = \frac{1}{h}\begin{pmatrix} 1 & -1 & & & & \\ -1 & 2 & -1 & & & \\ & -1 & 2 & -1 & & \\ & \cdot & \cdot & \cdot & \cdot & \cdot \\ & & & -1 & 2 & -1 \\ & & & & -1 & 1 \end{pmatrix}$$

for the weights $v_{mk} = \int_0^1 L_m'(y)L_k'(y)\,dy$ and

$$W = \frac{h}{6}\begin{pmatrix} 2 & 1 & & & & \\ 1 & 4 & 1 & & & \\ & 1 & 4 & 1 & & \\ & \cdot & \cdot & \cdot & \cdot & \cdot \\ & & & 1 & 4 & 1 \\ & & & & 1 & 2 \end{pmatrix}$$

for the weights $w_{mk} = \int_0^1 L_m(y)L_k(y)\,dy$. Writing the approximate solution in the form

$$\varphi_n = \sum_{k=0}^{n} \gamma_k L_k,$$

we have to solve the linear system

$$\sum_{k=0}^{n} \gamma_k \left\{ \alpha\,(w_{jk} + v_{jk}) + \int_0^1\!\!\int_0^1 H(x,y)L_j(x)L_k(y)\,dxdy \right\}$$

$$= \int_0^1\!\!\int_0^1 K(y,x)L_j(x)f(y)\,dxdy, \quad j = 0, \dots, n,$$

in the case of the H^1 norm penalty term. Here, K and H denote the kernels of the integral operators A and A^*A, respectively (see Example 16.16). In the case of the L^2 norm the matrix V has to be omitted. Note that the weight matrix V indicates how oscillations in the approximate solution are penalized by the regularization using the derivative. For numerical evaluation of the matrix elements we apply interpolatory quadratures as described in detail in Chapters 11 and 13.

Table 16.3 compares the mean square error between the exact and approximate solutions for both regularizations for an error distribution as in Example 16.16 with $\delta = 0.01$.

Table 16.3 Numerical results for Example 16.27

$\log\alpha$	−7	−6	−5	−4	−3	−2	−1
L^2 penalty	41.83	6.541	3.268	1.567	0.250	0.040	0.107
H^1 penalty	3.503	2.353	0.550	0.063	0.015	0.015	0.019

In closing this section, we wish to point out that in a paper predating that of Tikhonov, Phillips [189] proposed penalizing only by the L^2 norm of the derivative, i.e., to replace (16.23) by

$$\int_a^b |(A\varphi)(x) - f(x)|^2 dx + \alpha \int_a^b |\varphi'(x)|^2 dx. \tag{16.34}$$

Note that this differs from the Tikhonov regularization because the penalty term is only a semi-norm rather than a norm. For an analysis of this approach we refer to Groetsch [74]. In view of Theorem 16.26 it is no surprise that minimizing (16.34) is equivalent to the boundary value problem for the integro-differential equation

$$-\alpha\varphi_\alpha'' + A^*A\varphi_\alpha = A^*f$$

with boundary condition $\varphi_\alpha'(a) = \varphi_\alpha'(b) = 0$.

16.6 Ill-Posed Integral Equations in Potential Theory

In our treatment of the potential theoretic boundary value problems in Chapters 6–8 we have derived integral equations by seeking the solutions in the form of single- or double-layer potentials over the boundary of the domain in which the boundary value problem is posed. In the sequel we shall refer to the integral equations of the first and second kind obtained by this approach as *boundary integral equations*. Due to the singularity of the fundamental solution, these boundary integral equations, in general, are weakly singular. Since this causes some difficulties with their numerical solution, it is certainly tempting to try and solve the boundary value problems by using potentials with densities on curves or surfaces different from the actual boundary of the underlying domain. We shall illustrate this idea by looking at the interior two-dimensional Dirichlet problem.

Let $D \subset \mathbb{R}^2$ be a bounded domain with a connected boundary ∂D of class C^2. Consider the Dirichlet problem for a solution $u \in C^2(D) \cap C(\bar{D})$ of Laplace's equation

$$\triangle u = 0 \quad \text{in } D \tag{16.35}$$

with boundary condition

$$u = f \quad \text{on } \partial D, \tag{16.36}$$

where f is a given continuous function. We choose a closed curve Γ of class C^2 such that the closure $\bar{D}$ is contained in the interior of Γ and denote its outward unit normal by ν. Then we seek the solution of the Dirichlet problem (16.35), (16.36) in the form of a double-layer potential

$$u(x) = \int_\Gamma \varphi(y) \frac{\partial \Phi(x, y)}{\partial \nu(y)}\, ds(y), \quad x \in \bar{D}, \tag{16.37}$$

where we recall the fundamental solution to Laplace's equation

$$\Phi(x, y) := \frac{1}{2\pi} \ln \frac{1}{|x - y|}, \quad x \neq y,$$

in $\mathbb{R}^2$ and $\varphi \in L^2(\Gamma)$ is an unknown density. The harmonic function u solves the Dirichlet problem provided the density φ is a solution to the integral equation of the first kind

$$\int_\Gamma \varphi(y) \frac{\partial \Phi(x, y)}{\partial \nu(y)}\, ds(y) = f(x), \quad x \in \partial D. \tag{16.38}$$

We introduce the double-layer operator $A : L^2(\Gamma) \to L^2(\partial D)$ by

$$(A\varphi)(x) := \int_\Gamma \varphi(y) \frac{\partial \Phi(x, y)}{\partial \nu(y)}\, ds(y), \quad x \in \partial D, \tag{16.39}$$

and rewrite the integral equation (16.38) in the short form $A\varphi = f$. Since the integral operator A has a continuous kernel, it is compact, and therefore the integral equation (16.38) is ill-posed. Furthermore, the kernel is analytic with respect to x, and therefore (16.38) is severely ill-posed (compare Theorem 15.20). This ill-posedness can also be seen from the observation that the double-layer potential (16.37) defines a harmonic function in all of the interior of Γ, and not only in D. Hence, for a given function f, the integral equation (16.38) can have a solution only if the solution to the Dirichlet problem in D with boundary values f can be continued across the boundary ∂D as a harmonic function into the interior of Γ. A closer examination shows that this condition, together with the property that the boundary data of the continued function belong to $L^2(\Gamma)$, is also sufficient for the solvability. In general, however, there are no means available to decide for given boundary values whether such a continuation is possible.

In order to apply the regularization results of this chapter we establish the following theorem.

Theorem 16.28. *The compact integral operator $A : L^2(\Gamma) \to L^2(\partial D)$, defined by (16.39), is injective and has dense range.*

Proof. Let φ satisfy $A\varphi = 0$ and define u in $\mathbb{R}^2 \setminus \Gamma$ by (16.37). Then $u = 0$ on ∂D and, by the uniqueness Theorem 6.12, it follows that $u = 0$ in D. The analyticity of harmonic functions from Theorem 6.6 then yields $u = 0$ in the interior of Γ. From the jump relation (6.52) for the double-layer potential with L^2 densities we derive that $\varphi - K\varphi = 0$, where K denotes the double-layer boundary operator on the curve Γ. Since the kernel of K is continuous (see Problem 6.1), from $\varphi - K\varphi = 0$ we conclude that $\varphi \in C(\Gamma)$. Now Theorem 6.21 implies that $\varphi = 0$, hence A is injective.

The adjoint operator $A^* : L^2(\partial D) \to L^2(\Gamma)$ of A is given by

$$(A^*\psi)(x) = \int_{\partial D} \psi(y) \frac{\partial \Phi(x, y)}{\partial \nu(x)} \, ds(y), \quad x \in \Gamma.$$

When $A^*\psi = 0$, then, with the aid of Example 6.17, we have

$$(1, \psi)_{L^2(\partial D)} = -(A1, \psi)_{L^2(\partial D)} = -(1, A^*\psi)_{L^2(\Gamma)} = 0,$$

i.e., $\int_{\partial D} \psi \, ds = 0$. Hence, the single-layer potential

$$v(x) = \int_{\partial D} \psi(y) \Phi(x, y) \, ds(y), \quad x \in \mathbb{R}^2,$$

vanishes at infinity. Because $A^*\psi = 0$ implies $\partial v / \partial \nu = 0$ on Γ, by the uniqueness Theorem 6.13, we have $v = 0$ in the exterior of Γ, and by analyticity it follows that $v = 0$ in the exterior of ∂D. Since by Theorem 7.34 logarithmic single-layer potentials with L^2 densities are continuous in all of $\mathbb{R}^2$, the uniqueness for the interior Dirichlet problem tells us that $v = 0$ everywhere. Then the L^2 jump relation (6.53) for the normal derivative of single-layer potentials implies $\psi = 0$. Therefore A^* is injective, whence, by Theorem 15.8, the range of A is dense in $L^2(\partial D)$. □

We now apply Tikhonov regularization, i.e., we approximate (16.38) by the regularized equation

$$\alpha \varphi_\alpha + A^* A \varphi_\alpha = A^* f \tag{16.40}$$

with some regularization parameter $\alpha > 0$. Note that for the solution of (16.40) we have $\varphi_\alpha \in C(\Gamma)$, since A^* maps $L^2(\partial D)$ into $C(\Gamma)$. As is always the case with Tikhonov regularization, we can expect convergence of the solution φ_α to (16.40) as $\alpha \to 0$ only when the original equation (16.38) has a solution (see Remark 16.7). Fortunately, in our current situation, in order to obtain approximate solutions to the boundary value problem by

$$u_\alpha(x) := \int_\Gamma \varphi_\alpha(y) \frac{\partial \Phi(x, y)}{\partial \nu(y)} \, ds(y), \quad x \in \bar{D},$$

we do not need convergence of the densities φ_α. In order to satisfy the boundary condition approximately, what we actually want is convergence $A\varphi_\alpha \to f$ as $\alpha \to 0$

on ∂D which is ensured by Theorem 16.8. Then, since the Dirichlet problem itself is well-posed, small deviations $u_\alpha - f$ in the boundary values ensure small deviations $u_\alpha - u$ in the solution in all of D. The following remark makes this comment more precise.

Remark 16.29. *For each compact set $W \subset D$ there exists a constant C (depending on D and W) such that*

$$\|u\|_{C(W)} \leq C\|u\|_{L^2(\partial D)}$$

for all harmonic functions $u \in C(\bar{D})$.

Proof. We define the double-layer potential operator $V : C(\partial D) \to C(D)$ by

$$(V\psi)(x) := \int_{\partial D} \psi(y) \frac{\partial \Phi(x, y)}{\partial \nu(y)} \, ds(y), \quad x \in D.$$

Then, by Theorems 6.22 and 6.23, we can write

$$u = -2V(I - K)^{-1} u|_{\partial D}$$

with the double-layer integral operator K for the boundary ∂D. By the Cauchy–Schwarz inequality, there exists a constant c depending on D and W such that

$$\|V\psi\|_{C(W)} \leq c\|\psi\|_{L^2(\partial D)}$$

for all $\psi \in C(\partial D)$. Since K has a continuous kernel, from Theorems 2.28 and 6.21, and the Riesz Theorem 3.4 it follows that $(I - K)^{-1}$ is bounded with respect to the L^2 norm. □

For the convergence on the boundary we can make use of Theorems 16.9 and 16.10. As already mentioned, the conditions $f \in A(L^2(\Gamma))$ and $f \in AA^*(L^2(\partial D))$ of these two theorems can be interpreted as regularity assumptions on f because they control how fast the Fourier coefficients (f, g_n) tend to zero as $n \to \infty$. They may also be considered as conditions on how far the solution to the Dirichlet problem can be continued across the boundary ∂D. The following example will shed some light on these remarks.

Example 16.30. Consider the simple situation where D is the unit disk and Γ a concentric circle with radius $R > 1$. We parameterize $x \in \partial D$ by

$$x(t) = (\cos t, \sin t), \quad 0 \leq t \leq 2\pi,$$

and $x \in \Gamma$ by

$$x(t) = (R\cos t, R\sin t), \quad 0 \leq t \leq 2\pi,$$

and write $\psi(t) := \varphi(x(t))$ and $g(t) := f(x(t))$. Then we transform the integral equation (16.38) into

$$\frac{R}{2\pi} \int_0^{2\pi} \frac{\cos(t - \tau) - R}{1 - 2R\cos(t - \tau) + R^2} \psi(\tau) \, d\tau = g(t), \quad 0 \leq t \leq 2\pi, \tag{16.41}$$

which we rewrite in operator notation $\widetilde{A}\psi = g$ with an obvious meaning for the self-adjoint operator $\widetilde{A} : L^2[0, 2\pi] \to L^2[0, 2\pi]$. By decomposing

$$\frac{2(\cos t - R)e^{int}}{1 - 2R\cos t + R^2} = \frac{e^{int}}{e^{it} - R} + \frac{e^{int}}{e^{-it} - R},$$

we derive the integrals

$$\frac{R}{2\pi}\int_0^{2\pi} \frac{(\cos t - R)e^{int}}{1 - 2R\cos t + R^2}\, dt = \begin{cases} \dfrac{-1}{2R^{|n|}}, & n = \pm 1, \pm 2, \dots, \\ -1, & n = 0. \end{cases}$$

Hence, for $\psi_n(t) = e^{int}$ we have

$$\widetilde{A}\psi_n = \frac{-1}{2R^{|n|}}\,(1 + \delta_{n0})\psi_n, \quad n \in \mathbb{Z}.$$

Therefore, in polar coordinates $x = (r\cos\theta, r\sin\theta)$, a singular system (μ_n, φ_n, g_n) of A is given by

$$\mu_n = \frac{1}{2R^{|n|+1/2}}\,(1 + \delta_{n0}), \quad \varphi_n(x) = \frac{-1}{\sqrt{2\pi R}}\, e^{in\theta}, \quad g_n(x) = \frac{1}{\sqrt{2\pi}}\, e^{in\theta}$$

for $n = 0, \pm 1, \pm 2, \dots$. The approximate solution to the Dirichlet problem obtained by (16.40) can now be written in the form

$$u_\alpha(r, \theta) = \frac{1}{\sqrt{2\pi}} \sum_{n=-\infty}^{\infty} \frac{\mu_n^2}{\alpha + \mu_n^2}\,(f, g_n)\, r^{|n|} e^{in\theta}$$

compared with the exact separation of variables solution

$$u(r, \theta) = \frac{1}{\sqrt{2\pi}} \sum_{n=-\infty}^{\infty} (f, g_n)\, r^{|n|} e^{in\theta}. \tag{16.42}$$

The solvability condition $f \in A(L^2(\Gamma))$ now corresponds to

$$\sum_{n=-\infty}^{\infty} R^{2|n|} |(f, g_n)|^2 < \infty,$$

and therefore, by the Cauchy–Schwarz inequality, the series (16.42) can be seen to converge uniformly in each disk with radius $\rho < R$ centered at the origin. Hence, in this case the solution to the Dirichlet problem in the unit disk with boundary values f can be harmonically extended into the interior of the concentric circle of radius R. The stronger condition $f \in AA^*(L^2(\partial D))$ corresponds to

$$\sum_{n=-\infty}^{\infty} R^{4|n|} |(f, g_n)|^2 < \infty$$

and, similarly, implies a harmonic continuation into the interior of the concentric circle of radius R^2. □

To illustrate our results numerically we consider the boundary condition given by $f(x) = \ln|x - x_0|$, where $x_0 = (q, 0)$ with $q > 1$. Here, the solution to the boundary value problem clearly allows a harmonic continuation into the open disk of radius q centered at the origin. Table 16.4 gives the error $\|A\varphi_\alpha - f\|$ between the exact solution and the Tikhonov regularization for (16.41), approximated by Nyström's method using the trapezoidal rule with 32 knots. The radius of Γ is chosen to be $R = 2$. The dependence of the error on the parameters α and q confirms the convergence behavior predicted by our theoretical results.

Table 16.4 Numerical results for Example 16.30

$\log\alpha$	$q = 10$	$q = 4$	$q = 3$	$q = 2$	$q = 1.5$
−5	0.00006	0.00009	0.0001	0.001	0.01
−4	0.00060	0.00091	0.0014	0.006	0.02
−3	0.00642	0.00873	0.0125	0.030	0.08
−2	0.06255	0.07370	0.0955	0.162	0.27

This approach, of course, can be extended to cover the interior and exterior Dirichlet and Neumann problem in two and three space dimensions. Also, instead of the double-layer potential, a single-layer potential can be used. For an application of these ideas to the Helmholtz equation, see Kress and Mohsen [148].

In a discretized version of the approach for the solution of the interior Dirichlet problem in D with boundary values $u = f$ via a single-layer potential on a curve Γ located in $\mathbb{R}^2 \setminus \bar{D}$, in principle, one tries to approximate the solution u by a finite sum of fundamental solutions

$$u_n(x) = \sum_{k=1}^{n} \varphi_k \Phi(x, y_k)$$

with source points $y_k \in \mathbb{R}^2 \setminus \bar{D}$ and real coefficients φ_k, $k = 1, \dots, n$. For a given choice of the source points, the collocation $u_n(x_j) = f(x_j)$ of the boundary condition at collocation points $x_j \in \partial D$ for $j = 1, \dots, n$ provides a linear system for determining the coefficients φ_k, $k = 1, \dots, n$. For obvious reasons this approach to the approximate solution of boundary value problems is known as the *method of fundamental solutions*. Because of its close relation to the integral equations discussed above it suffers from severe ill-posedness which compensates for the advantage of its computational simplicity (see [50, 66]).

By Green's Theorem 6.3, for each harmonic function u in $C^2(D) \cap C^1(\bar{D})$ we have that

$$\int_{\partial D} \left\{ \frac{\partial u}{\partial \nu}(y)\, \Phi(x, y) - u(y)\, \frac{\partial \Phi(x, y)}{\partial \nu(y)} \right\} ds(y) = 0$$

for all $x \in \mathbb{R}^2 \setminus D$. From this, we derive the ill-posed integral equation of the first kind

$$\int_{\partial D} \psi(y)\Phi(x,y)\,ds(y) = \int_{\partial D} f(y)\,\frac{\partial\Phi(x,y)}{\partial\nu(y)}\,ds(y), \quad x \in \Gamma,$$

for the unknown normal derivative $\psi = \partial u/\partial\nu$ on ∂D. This method to solve the potential theoretic boundary value problems is due to Kupradze [153]. A fairly complete description and extensive references are given by Christiansen [29].

The integral equations described in this section are simple, since as opposed to the boundary integral equations they possess smooth kernels. Hence, they allow the application of simple quadrature rules without the need to treat singular integrals. On the other hand, in general, we can expect only poor convergence as the regularization parameter tends to zero. Therefore, this approach may serve as a very instructive example for the use of Tikhonov regularization but, though it has its supporters in the literature, it cannot be considered as a serious competitor for the numerically well-established boundary integral equations. However, in Chapter 18 we will illustrate that integral equations of this nature have some merits in solving inverse problems where the boundary is not known.

Problems

16.1. Show that the quasi-solution given by Theorem 16.12 depends weakly continuously on the right-hand side, i.e., norm convergence $f_n \to f$, $n \to \infty$, implies weak convergence for the corresponding quasi-solutions $\varphi_n \rightharpoonup \varphi$, $n \to \infty$. Show that in the complement of $A(B[0;\rho])$ the quasi-solution depends continuously on the right-hand side. How are these results related to the regularity Theorem 16.13?
Hint: By Theorem 16.3 the sequence (φ_n) contains a subsequence that converges weakly to some element ψ in X. Establish that $\|\psi\| \le \rho$ and $\|A\psi - f\| \le \|A\varphi - f\|$, i.e., ψ is a quasi-solution, and by uniqueness it must coincide with φ. Use this result to show that the sequence (φ_n) itself converges weakly to φ. Finally, conclude norm convergence from $\|\varphi_n\| = \rho$.

16.2. Show that the minimum norm solution given by Theorem 16.18 depends weakly continuously on the right-hand side. Show that in the complement of $B[0;\delta] \subset Y$ the minimum norm solution depends continuously on the right-hand side. How are these results related to the regularity Theorem 16.19?
Hint: Show that there exists an element $\chi \in X$ with $\chi \in U_{f_n}$ for all sufficiently large n. Therefore the sequence (φ_n) is bounded and contains a weakly convergent subsequence with limit $\psi \in X$. Show that $\|A\psi - f\| \le \delta$ and $\|\psi\| \le \|\varphi\|$, i.e., ψ is a minimum norm solution, and by uniqueness it must coincide with φ. Use this result to show that the sequence (φ_n) itself converges weakly to φ. Finally, conclude norm convergence from $\|A\varphi_n - f_n\| = \delta$.

16.3. Show that the convergence order given in Theorem 16.20 is optimal.

16.4. Show that the solution of the integral equation in Example 16.16 is unique.
Hint: Show that $\int_0^1 y^n\varphi(y)\,dy = 0$, $n = 0, 1, 2, \ldots$, for each solution φ of the homogeneous equation.

16.5. Show that a linear operator $A : X \to Y$ from a Hilbert space X into a Hilbert space Y is compact if and only if weak convergence $\varphi_n \rightharpoonup \varphi$ for $n \to \infty$ implies norm convergence $A\varphi_n \to A\varphi$ for $n \to \infty$.

Chapter 17
Regularization by Discretization

We briefly return to the study of projection methods and will consider their application to ill-posed equations of the first kind. In particular we will present an exposition of the moment discretization method. For further studies of regularization through discretization, we refer to Baumeister [15], Kirsch [126], Louis [160], Natterer [183] and Rieder [204].

17.1 Projection Methods for Ill-Posed Equations

Solving an equation of the first kind

$$A\varphi = f \tag{17.1}$$

with an injective compact operator $A : X \to Y$ from a Banach space X into a Banach space Y numerically without regularization usually means approximately solving it by a projection method. The ill-posedness of the equation (17.1) then will cause the condition number of the discrete linear system to grow with the dimension n of the subspace used for the projection method. Increasing n will make the error due to the discretization smaller, but due to ill-conditioning the computation will be contaminated by errors in the given data f. Note that in actual numerical computations such errors will automatically occur because of round-off effects for the data. On the other hand, if n is small, then the approximation is robust against errors in f but will be inaccurate due to a large discretization error. This dilemma calls for a compromise in the choice of the discretization parameter n.

Recalling the projection methods from Chapter 13, let $X_n \subset X$ and $Y_n \subset Y$ be two sequences of subspaces with $\dim X_n = \dim Y_n = n$ and let $P_n : Y \to Y_n$ be projection operators. For given $f \in A(X)$ the projection method approximates the solution $\varphi \in X$ of $A\varphi = f$ by the solution $\varphi_n \in X_n$ of the projected equation

$$P_n A\varphi_n = P_n f. \tag{17.2}$$

R. Kress, *Linear Integral Equations*, Applied Mathematical Sciences 82,
DOI 10.1007/978-1-4614-9593-2_17,

For the remainder of this chapter we assume that the projection method is convergent for exact right-hand sides $f \in A(X)$. This, by Theorem 13.6, implies that the finite-dimensional operators $P_nA : X_n \to Y_n$ are invertible and that the sequence of operators $(P_nA)^{-1}P_nA : X \to X_n$ is uniformly bounded. Then the operators $R_n := (P_nA)^{-1}P_n : Y \to X_n$ provide a regularization scheme in the sense of Definition 15.5 with regularization parameter n. The approximation

$$\varphi_n^\delta := R_n f^\delta \tag{17.3}$$

corresponding to (15.6) is obtained by numerically performing the projection method for the inexact right-hand side f^δ.

Theorem 17.1. *Let $A : X \to Y$ be an injective compact linear operator, $f \in A(X)$, and $f^\delta \in Y$ satisfy $\|f^\delta - f\| \leq \delta$. Then the approximation $\varphi_n^\delta := R_n f^\delta$ by the projection method satisfies*

$$\|\varphi_n^\delta - A^{-1}f\| \leq \|R_n\|\,\delta + C \inf_{\psi \in X_n} \|\psi - A^{-1}f\| \tag{17.4}$$

for some constant C.

Proof. This follows from Theorem 13.6 by decomposing

$$\begin{aligned}\|\varphi_n^\delta - A^{-1}f\| &\leq \|R_n\|\,\|f^\delta - f\| + \|R_n f - A^{-1}f\|\\ &\leq \|R_n\|\,\delta + (1 + M) \inf_{\psi \in X_n} \|\psi - A^{-1}f\|,\end{aligned}$$

where M is a bound on $\|R_nA\|$. □

The error estimate (17.4) corresponds to (15.7) and illustrates that the total error consists of two parts: the influence of the incorrect data and the discretization error of the projection method. Recall that for the compact operator A, by Theorem 15.6, the operators R_n cannot be uniformly bounded.

At this point we wish to emphasize the fact that for the analysis of the discretization error we may use any Banach space structure for the range space. In particular, we may choose an image space such that the equation becomes well-posed. Hence, we can apply the convergence results on projection methods given in Chapter 13. But for the influence of the error in the right-hand side we have to stick with an image space as dictated by the needs of the particular problem. We want to make these remarks more precise through the following theorem.

Theorem 17.2. *Let $A : X \to Y$ be an injective compact operator from a Banach space X into a Banach space Y and assume an additional Banach space $Z \subset Y$ with continuous imbedding such that $A : X \to Z$ is an isomorphism. Let $X_n \subset X$ and $Y_n \subset Z \subset Y$ be two sequences of subspaces with $\dim X_n = \dim Y_n = n$ and let $P_n : Y \to Y_n$ be projection operators (bounded with respect to the norm on Y) and assume that the projection method is convergent for all $f \in A(X) = Z$. Then the* *operators* $R_n : Y \to X$ *satisfy*

$$\|R_n\| \le C\,\|P_n\| \sup_{f\in Y_n,\,\|f\|_Y=1} \|f\|_Z$$

for some constant C.

Proof. For abbreviation we set

$$\gamma_n := \sup_{f\in Y_n,\,\|f\|_Y=1} \|f\|_Z.$$

Since by Theorem 1.6 all norms on the finite-dimensional subspace Y_n are equivalent, we have that γ_n is finite. Then, from $\|f\|_Z \le \gamma_n \|f\|_Y$ for all $f \in Y_n$, it follows that

$$\|P_n f\|_Z \le \gamma_n \|P_n f\|_Y \le \gamma_n \|P_n\|\,\|f\|_Y \le M\gamma_n \|P_n\|\,\|f\|_Z \tag{17.5}$$

for all $f \in Z$ and some constant M, since the imbedding from Z into Y is assumed to be bounded. Hence, the operators P_n are also bounded with respect to the norm on Z and we can interpret the projection scheme for $A : X \to Y$ also as a projection scheme in the structure $A : X \to Z$. Since we assume convergence for all $f \in A(X)$, by Theorem 13.6, we know that the operators $(P_nA)^{-1}P_nA : X \to X$ are uniformly bounded. Therefore, for all $f \in Y$ with $\|f\|_Y = 1$ we can estimate

$$\|R_n f\| = \|(P_nA)^{-1}P_n f\| = \|(P_nA)^{-1}P_nAA^{-1}P_n f\|$$

$$\le \|(P_nA)^{-1}P_nA\|\,\|A^{-1}\|_{Z\to X}\,\|P_n f\|_Z \le C\gamma_n\|P_n\|$$

for some constant C. □

We illustrate the use of Theorem 17.2 by considering the following example.

Example 17.3. Let $A : L^2[0, 2\pi] \to L^2[0, 2\pi]$ be given by

$$(A\varphi)(t) = \frac{1}{2\pi}\int_0^{2\pi} \left\{ \ln 4 \sin^2 \frac{t-\tau}{2} + K(t,\tau) \right\} \varphi(\tau)\,d\tau, \quad 0 \le t \le 2\pi, \tag{17.6}$$

where we assume that K is infinitely differentiable and 2π-periodic with respect to both variables. As a consequence of Theorem 8.24, it can be seen that $A : L^2[0, 2\pi] \to H^1[0, 2\pi]$ is an isomorphism, provided A is injective. In this case, from Theorem 13.32 we can conclude that the Petrov–Galerkin method with the orthogonal projection operator P_n from $L^2[0, 2\pi]$ onto the space Y_n of trigonometric polynomials of degree less than or equal to n converges for $A : L^2[0, 2\pi] \to H^1[0, 2\pi]$. (Here we need to use the fact that by Theorem 1.25 it can be observed that P_n coincides with the orthogonal projection operator from $H^1[0, 2\pi]$ onto Y_n.) In particular, if the right-hand side also is infinitely differentiable, we have a convergence order $O(n^{-m})$ for all $m \in \mathbb{N}$. On the other hand, the orthogonal projection operator has norm $\|P_n\| = 1$ and for the subspace Y_n we readily see that $\gamma_n = \sqrt{1+n^2}$. Hence, in this case we have $\|R_n\| = O(n)$ and the equation of the first kind with the operator

(17.6) is only mildly ill-posed. In particular, the decay of the approximation error $O(n^{-m})$ dominates the increase $O(n)$ of the data error. Therefore, provided the error of the right-hand side is reasonably small, we conclude that it is numerically safe to apply the Petrov–Galerkin method in this case because we need only a comparatively small n to achieve an acceptable approximation error (see Example 13.23 and Problem 17.1). For a more detailed discussion of the regularization of integral equations of the first kind with a logarithmic singularity and in particular Symm's integral equation we refer to [126]. □

To make the remarks in the preceding Example 17.3 more precise, following ideas developed by Kaltenbacher [114], we briefly consider the selection of the discretization level via the discrepancy principle for a special type of mildly ill-posed equations. We note that in view of Theorem 11.8 the assumptions of the following theorem are satisfied for the integral equation (17.6) for a projection method with X_n the space of trigonometric polynomials of the form (13.26) both for the collocation and the Petrov–Galerkin method.

Theorem 17.4. *As in Theorem 17.2 let X, Y and Z be Banach spaces such that $Z \subset Y$ with continuous imbedding and let $A : X \to Y$ be an injective compact linear operator of the form $A = A_0 + B$ such that $A_0 : X \to Z$ is an isomorphism and $B : X \to Z$ is compact. Further let $X_n \subset X$ be a sequence of finite-dimensional subspaces, define subspaces $Y_n \subset Y$ by $Y_n := A_0(X_n)$ and let $P_n : Y \to Y_n$ be projection operators that are uniformly bounded. Assume that the corresponding projection method converges for all $f \in A(X) = Z$ and that*

$$\beta_n := \sup_{g \in Z,\ \|g\|_Z = 1} \|P_n g - g\|_Y \quad and \quad \gamma_n := \sup_{f \in Y_n,\ \|f\|_Y = 1} \|f\|_Z$$

satisfy $1/M \le \beta_n \gamma_n \le M$ for all $n \in \mathbb{N}$ and some constant $M > 1$. Then we can choose a positive constant τ such that for each $f \in A(X)$ and $f^\delta \in Y$ with $\|f^\delta - f\| \le \delta$ and $\delta > 0$ there exists a smallest integer $n = n(\delta)$, depending on f^δ and δ, such that

$$\|AR_{n(\delta)} f^\delta - f^\delta\| \le \tau\delta. \tag{17.7}$$

Proof. From the assumptions on A, A_0 and B, by the Riesz theory we have that $A : X \to Z$ is an isomorphism. By the definition of Y_n, clearly, $P_n A_0 : X_n \to Y_n$ coincides with $A_0 : X_n \to Y_n$ and therefore $(P_n A_0)^{-1} = A_0^{-1}$. From the proof of Theorem 13.7 we know that the inverse operators $[I + (P_n A_0)^{-1} P_n B]^{-1} : X_n \to X_n$ exist and are uniformly bounded for all sufficiently large n and therefore $(P_n A)^{-1} : Y_n \to X_n$ is given by

$$(P_n A)^{-1} = [I + (P_n A_0)^{-1} P_n B]^{-1} A_0^{-1}.$$

From this, recalling that $R_n = (P_n A)^{-1} P_n$, as in the proof of Theorem 17.2 we infer that

$$\|R_n g\| \le c_1 \gamma_n \|g\| \tag{17.8}$$

for all $g \in Y$, all sufficiently large $n \in \mathbb{N}$ and some positive constant c_1.

Now, in view of $P_nAR_nf^\delta = P_nf^\delta$, we compute

$$AR_nf^\delta - f^\delta = (I - P_n)(AR_nf^\delta - f^\delta) = (I - P_n)[(AR_n - I)(f^\delta - f) + (AR_n - I)f].$$

From this, using the estimates

$$\|(I - P_n)AR_n(f^\delta - f)\| \le \beta_n\|A\|_{X\to Z}\|R_n(f^\delta - f)\| \le c_1\beta_n\gamma_n\|A\|_{X\to Z}\delta$$

and

$$\|(I - P_n)(f^\delta - f)\| \le \|I - P_n\|\,\delta$$

and the assumptions on the uniform boundedness of the P_n and the products $\beta_n\gamma_n$, we obtain that

$$\|AR_nf^\delta - f^\delta\| \le C\delta + \|(I - P_n)A(R_nf - A^{-1}f)\| \tag{17.9}$$

for all sufficiently large n and some positive constant C. Choosing $\tau > C$, now the assertion of the theorem follows from the convergence $R_nf \to A^{-1}f$ as $n \to \infty$ for $f \in A(X)$. □

Theorem 17.5. *Under the assumptions of Theorem 17.4 the discrepancy principle is a regular regularization strategy, i.e, we have that*

$$R_{n(\delta)}f^\delta \to A^{-1}f, \quad \delta \to 0. \tag{17.10}$$

Proof. We choose $m(\delta)$ as the smallest $m \in \mathbb{N}$ with the property

$$\beta_m\|A\|_{X\to Z}\|R_mf - A^{-1}f\| \le (\tau - C)\delta.$$

From (17.9) we observe that

$$(\tau - C)\delta < \|(I - P_n)A(R_nf - A^{-1}f)\| \le \beta_n\|A\|_{X\to Z}\|R_nf - A^{-1}f\|$$

for all $n < n(\delta)$ and consequently $n(\delta) \le m(\delta)$. After transforming

$$R_{n(\delta)}f^\delta - A^{-1}f = R_{m(\delta)}P_{m(\delta)}(AR_{n(\delta)}f^\delta - f^\delta) + R_{m(\delta)}(f^\delta - f) + R_{m(\delta)}f - A^{-1}f$$

and using (17.7) and (17.8), we can estimate

$$\|R_{n(\delta)}f^\delta - A^{-1}f\| \le c_1(\tau + 1)\gamma_{m(\delta)}\delta + \|R_{m(\delta)}f - A^{-1}f\|.$$

From this, in view of the definition of $m(\delta)$ and the boundedness of the products $\beta_n\gamma_n$ from below, we obtain that

$$\|R_{n(\delta)}f^\delta - A^{-1}f\| \le c_2\gamma_{m(\delta)}\delta \tag{17.11}$$

with a positive constant c_2. Using again the definition of $m(\delta)$ and the boundedness of the products $\beta_n\gamma_n$, from (17.11) we further find that

$$\|R_{n(\delta)}f^\delta - A^{-1}f\| \le c_3\|R_{m(\delta)-1}f - A^{-1}f)\| \tag{17.12}$$

with a positive constants c_3. Now have to distinguish two cases. When $m(\delta) \to \infty$ for $\delta \to 0$ then from (17.12) it immediately follows that $R_{n(\delta)}f^\delta \to A^{-1}f$ as $\delta \to 0$. If $m(\delta)$ does not tend to infinity as $\delta \to 0$ assume that there exists a null sequence δ_k such that $R_{n(\delta_k)}f^{\delta_k}$ does not converge to $A^{-1}f$ as $k \to \infty$. Then by selecting a subsequence we may assume that there exists $N \in \mathbb{N}$ such that $m(\delta_k) \le N$ for all $k \in \mathbb{N}$ and from (17.11) we obtain the contradiction

$$\|R_{n(\delta_k)}f^{\delta_k} - A^{-1}f\| \le c_2\gamma_N\delta_k \to 0, \quad k \to \infty,$$

and this ends the proof. □

For further analysis on the discrepancy principle for regularization via discretization we refer to [15, 85, 190, 204, 239].

For projection methods the norm on the image space Y might not always be the most appropriate measure to describe the inaccuracy of the data. Consider the Petrov–Galerkin method for an operator $A : X \to Y$ between two Hilbert spaces X and Y. With subspaces $X_n = \operatorname{span}\{u_1, \dots, u_n\}$ and $Y_n = \operatorname{span}\{v_1, \dots, v_n\}$ we write $\varphi_n = \sum_{k=1}^n \gamma_k u_k$ and then have to solve the Petrov–Galerkin equations

$$\sum_{k=1}^n \gamma_k(Au_k, v_j) = (f, v_j), \quad j = 1, \dots, n, \tag{17.13}$$

for the coefficients $\gamma_1, \dots, \gamma_n$. Here, it makes more sense to measure the error of the discrete data vector $F = ((f, v_1), \dots, (f, v_n))$ in the Euclidean norm on $\mathbb{C}^n$ (compare the stability analysis of Section 14.1 for equations of the second kind). Let $F^d \in \mathbb{C}^n$ be a perturbed right-hand side for (17.13) with

$$\|F^d - F\|_2 \le d$$

and denote the Petrov–Galerkin solution to these inexact data by φ_n^d. After introducing the matrix operators $E_n, \widetilde{A}_n : \mathbb{C}^n \to \mathbb{C}^n$ by

$$(E_n\gamma)_j := \sum_{k=1}^n (u_k, u_j)\gamma_k, \quad j = 1, \dots, n,$$

and

$$(\widetilde{A}_n\gamma)_j := \sum_{k=1}^n (Au_k, v_j)\gamma_k, \quad j = 1, \dots, n,$$

for $\gamma = (\gamma_1, \dots, \gamma_n) \in \mathbb{C}^n$ we can formulate the following error estimate which replaces Theorem 17.1.

Theorem 17.6. *The error in the Petrov–Galerkin method can be estimated by*

$$\|\varphi_n^d - A^{-1}f\| \le \frac{d}{\sqrt{\lambda_n \nu_n}} + C \inf_{\psi \in X_n} \|\psi - A^{-1}f\|, \tag{17.14}$$

where λ_n and ν_n denote the smallest singular values of $\widetilde{A}_n$ and $\widetilde{A}_n E_n^{-1}$, respectively, and C is some positive constant.

Proof. Using the singular value decomposition Theorem 15.16, we can estimate

$$\begin{aligned}\|\varphi_n\|^2 &= \sum_{j,k=1}^{n} \gamma_j \bar{\gamma}_k (u_j, u_k) = (E_n\gamma, \gamma) = (E_n \widetilde{A}_n^{-1} F, \widetilde{A}_n^{-1} F)\\ &\le \|E_n \widetilde{A}_n^{-1}\|_2 \, \|\widetilde{A}_n^{-1}\|_2 \, \|F\|_2^2 \le \frac{1}{\lambda_n \nu_n} \, \|F\|_2^2.\end{aligned}$$

Now we apply this inequality to the difference $\varphi_n^d - \varphi_n$ and obtain (17.14) by decomposing the error as in the proof of Theorem 17.1. □

For the least squares method as described in Theorem 13.31, i.e., for $Y_n = A(X_n)$, the matrix operator $\widetilde{A}_n$ assumes the form

$$(\widetilde{A}_n \gamma)_j := \sum_{k=1}^{n} (Au_k, Au_j)\gamma_k, \quad j = 1, \dots, n,$$

and is self-adjoint and positive definite. In particular, for an orthonormal basis $u_1, \dots, u_n$ of X_n, we have $E_n = I$. Hence, in this case we only need to be concerned with the eigenvalues of $\widetilde{A}_n$. Let (μ_m, φ_m, g_m), $m = 1, 2, \dots$, be a singular system of the compact operator A. Then we can choose an element $\varphi = \sum_{j=1}^{n} \gamma_j u_j$ in X_n such that $(\varphi, \varphi_m) = 0$, $m = 1, \dots, n-1$, and $\|\varphi\| = 1$. Now, by the singular value decomposition Theorem 15.16, we obtain

$$\mu_n^2 \ge (A^*A\varphi, \varphi) = (\widetilde{A}_n \gamma, \gamma) \ge \lambda_n,$$

since $\|\gamma\|_2 = \|\varphi\| = 1$. Therefore,

$$\frac{1}{\lambda_n} \ge \frac{1}{\mu_n^2} \, .$$

This again exhibits the influence of the singular values on the degree of ill-posedness.

For the integral equation of Example 16.27, by Theorem 15.20, the singular values decay at least exponentially. Hence, projection methods cannot be recommended for this equation. More generally, for severely ill-posed equations projection methods should only be applied in connection with further regularization, for example, by Tikhonov regularization.

17.2 The Moment Method

We proceed by describing the moment method to approximately solve an equation of the first kind. Let $A : X \to Y$ be an injective bounded linear operator from a Hilbert space X into another Hilbert space Y. Choose a finite-dimensional subspace $Y_n \subset Y$ with $\dim Y_n = n$ and, given an element $f \in Y$, define the affine linear subspace $U_n \subset X$ by

$$U_n := \{\varphi \in X : (A\varphi, g) = (f, g),\ g \in Y_n\}.$$

We assume U_n is nonempty. By the Riesz representation Theorem 4.10, bounded linear functionals in a Hilbert space can be equivalently expressed by scalar products. Therefore, we may interpret U_n as the set of all elements $\varphi \in X$ for which n linearly independent functionals vanish when applied to the residual $A\varphi - f$.

Definition 17.7. An element $\varphi_n \in U_n$ is called a *moment solution* to $A\varphi = f$ with respect to Y_n if

$$\|\varphi_n\| = \inf_{\varphi \in U_n} \|\varphi\|.$$

Note that φ_n is a moment solution with respect to Y_n if and only if it is a best approximation to the zero element in X with respect to the closed affine linear subspace U_n. Finding a best approximation with respect to an affine linear subspace can be equivalently reduced to finding a best approximation with respect to a linear subspace. Therefore, by Theorem 1.26 there exists a unique moment solution, and by Theorem 1.25 it can be characterized as the unique element $\varphi_n \in U_n$ satisfying

$$(\varphi_n - \varphi, \varphi_n) = 0 \tag{17.15}$$

for all $\varphi \in U_n$. The following theorem gives an interpretation of the moment method as a projection method.

Theorem 17.8. *Let $A : X \to Y$ be an injective bounded linear operator with dense range. Then the moment approximate solution of $A\varphi = f$ with respect to a subspace $Y_n \subset Y$ coincides with the Petrov–Galerkin approximation corresponding to the subspaces $X_n := A^*(Y_n) \subset X$ and $Y_n \subset Y$.*

Proof. By assumption we have $\overline{A(X)} = Y$ and therefore, by Theorem 15.8, the adjoint operator $A^* : Y \to X$ is injective. Hence, $\dim X_n = \dim Y_n = n$. Let $P_n : Y \to Y_n$ be the orthogonal projection operator. We will show that the Petrov–Galerkin operator $P_nA : X_n \to Y_n$ is injective. Indeed, assume that $\varphi \in X_n$ satisfies $P_nA\varphi = 0$. Then we can write $\varphi = A^*g$ for some $g \in Y_n$ and have

$$\|\varphi\|^2 = (A^*g, A^*g) = (AA^*g, g) = (P_nA\varphi, g) = 0.$$

The injectivity of P_nA ensures that the Petrov–Galerkin approximation φ_n exists and is unique. It can be written in the form $\varphi_n = A^*g_n$ with $g_n \in Y_n$, and it satisfies

$$(AA^*g_n, g) = (f, g) \tag{17.16}$$

for all $g \in Y_n$. This implies $\varphi_n \in U_n$ and

$$(\varphi_n - \varphi, \varphi_n) = (\varphi_n - \varphi, A^* g_n) = (A\varphi_n - A\varphi, g_n) = 0$$

for all $\varphi \in U_n$. Therefore, by (17.15), the Petrov–Galerkin approximation φ_n is the moment solution with respect to U_n. □

When $f \in A(X)$ we write $A\varphi = f$ and have (17.16) to be equivalent to

$$(A^* g_n, A^* g) = (\varphi, A^* g)$$

for all $g \in Y_n$. This implies that the moment method corresponds to the least squares method for the adjoint (or dual) equation $A^* h = \varphi$ with respect to Y_n. Therefore it is also known as the *dual least squares method* (for the least squares method, recall Theorem 13.31).

We continue our analysis with a convergence result on the moment method. Note that the least squares method, in general, will not converge (see the counterexample in Problem 17.2).

Theorem 17.9. *Let A be as in Theorem 17.8 and let $f \in A(X)$. Assume that the sequence of subspaces $Y_n \subset Y$ has the denseness property*

$$\inf_{h \in Y_n} \|h - g\| \to 0, \quad n \to \infty, \tag{17.17}$$

for all $g \in Y$. Then the moment method converges, i.e., $\varphi_n \to A^{-1} f$, $n \to \infty$.

Proof. Let $P_n : Y \to Y_n$ and $Q_n : X \to X_n = A^*(Y_n)$ be the orthogonal projection operators. We will show that (17.17) implies convergence

$$Q_n \varphi \to \varphi, \quad n \to \infty,$$

for all $\varphi \in X$. Since A is injective, by Theorem 15.8, the range $A^*(Y)$ is dense in X. Therefore, given $\varphi \in X$ and $\varepsilon > 0$, there exists $g \in Y$ such that $\|A^* g - \varphi\| < \varepsilon/2$. As a consequence of (17.17) we can choose $N \in \mathbb{N}$ such that $\|P_n g - g\| < \varepsilon \|A\|/2$ for all $n \geq N$. Since $A^* P_n g \in X_n$, for all $n \geq N$ we can estimate

$$\|Q_n \varphi - \varphi\| \leq \|A^* P_n g - \varphi\| \leq \|A^*(P_n g - g)\| + \|A^* g - \varphi\| < \varepsilon.$$

Now the assertion of the theorem follows from the fact that for $f \in A(X)$ the dual least squares solution φ_n is given by $\varphi_n = Q_n A^{-1} f$. Indeed, the orthogonal projection satisfies

$$(Q_n A^{-1} f, A^* g) = (A^{-1} f, A^* g)$$

for all $g \in Y_n$. This is equivalent to $(A Q_n A^{-1} f, g) = (f, g)$ for all $g \in Y_n$. Hence $Q_n A^{-1} f$ solves the Petrov–Galerkin equation. □

For the dual least squares method, from $X_n = A^*(Y_n)$ we observe that the matrix operators occurring in Theorem 17.6 are related by $E_n = \widetilde{A}_n$. Therefore, in this case the error estimate (17.14) can be simplified to

$$\|\varphi_n^d - A^{-1}f\| \leq \frac{d}{\sqrt{\lambda_n}} + C \inf_{\psi \in A^*(Y_n)} \|\psi - A^{-1}f\|. \tag{17.18}$$

17.3 Hilbert Spaces with Reproducing Kernel

For an application of the above results to collocation methods for equations of the first kind, we obviously need a Hilbert space in which the point evaluation functionals are bounded. Therefore, we briefly introduce the concept of Hilbert spaces with reproducing kernel.

Definition 17.10. Let H be a Hilbert space of real- or complex-valued functions f defined on an interval $[a, b]$. A function M on $[a, b] \times [a, b]$ is called a reproducing kernel if $h_x := M(x, \cdot)$ belongs to H for all $x \in [a, b]$ and

$$(f, h_x) = f(x) \tag{17.19}$$

for all $x \in [a, b]$ and all $f \in H$.

Theorem 17.11. *A Hilbert space H has a reproducing kernel if and only if the evaluation functionals*

$$f \mapsto f(x)$$

are bounded for all $x \in [a, b]$.

Proof. If H has a reproducing kernel, then (17.19) implies continuity for the evaluation functionals. Conversely, if the evaluation functionals are bounded, then by the Riesz Theorem 4.10 for each $x \in [a, b]$ there exists an element $h_x \in H$ such that (17.19) holds for all $f \in H$. Hence $M(x, \cdot) := h_x$ is a reproducing kernel. □

Obviously, $L^2[a, b]$ is not a Hilbert space with reproducing kernel, whereas $H^1[a, b]$ is, since by (16.29) the maximum norm is weaker than the H^1 norm.

For each injective linear operator $A : X \to Y$ acting between two Hilbert spaces X and Y we can make the range $H := A(X)$ a Hilbert space with the scalar product given by

$$(f, g)_H = (A^{-1}f, A^{-1}g)_X, \quad f, g \in A(X). \tag{17.20}$$

Indeed, since A is injective, (17.20) defines a scalar product on $A(X)$. Assume (f_n) is a Cauchy sequence in H. Then we can write $f_n = A\varphi_n$ with $\varphi_n \in X$ and find that (φ_n) is a Cauchy sequence in X because

$$\|\varphi_n - \varphi_m\|_X = \|f_n - f_m\|_H.$$

Since X is a Hilbert space, there exists $\varphi \in X$ with $\varphi_n \to \varphi$, $n \to \infty$. Then we have $f = A\varphi \in H$ and

$$\|f_n - f\|_H = \|\varphi_n - \varphi\|_X \to 0, \quad n \to \infty.$$

Hence, with the scalar product given by (17.20) the range $A(X)$ is a Hilbert space. By (17.20), the linear operator $A : X \to H$ can be seen to be bounded with the adjoint operator $A^* : H \to X$ given by $A^* = A^{-1}$. In the case of an integral operator this Hilbert space turns out to have a reproducing kernel.

Theorem 17.12. *Let $A : L^2[a,b] \to L^2[a,b]$ denote an injective integral operator*

$$(A\varphi)(x) = \int_a^b K(x,y)\varphi(y)\,dy, \quad a \le x \le b,$$

with continuous kernel K. Then the range $H = A(L^2[a,b])$ furnished with the scalar product (17.20) is a Hilbert space with reproducing kernel

$$M(x,y) = \int_a^b \overline{K(x,z)}K(y,z)\,dz, \quad x,y \in [a,b]. \tag{17.21}$$

Proof. Define M by (17.21) and set $h_x := M(x,\cdot)$ and $k_x := \overline{K(x,\cdot)}$. Then

$$(Ak_x)(y) = \int_a^b K(y,z)\overline{K(x,z)}\,dz = M(x,y),$$

i.e., $Ak_x = h_x$ for all $x \in [a,b]$. Therefore, writing $f = A\varphi$, we find

$$(f, h_x)_H = (A^{-1}f, A^{-1}h_x)_{L^2} = (\varphi, k_x)_{L^2}$$

$$= \int_a^b K(x,y)\varphi(y)\,dy = (A\varphi)(x) = f(x).$$

Hence H is a Hilbert space with reproducing kernel M. □

The statement of Theorem 17.12 remains valid for a weakly singular kernel K with the property that M, defined by (17.21), is continuous.

For a detailed study of Hilbert spaces with reproducing kernels, we refer to Aronszajn [8] and Meschkowski [168].

17.4 Moment Collocation

For the numerical solution of the integral equation of the first kind

$$\int_a^b K(x,y)\varphi(y)\,dy = f(x), \quad a \le x \le b, \tag{17.22}$$

with continuous kernel K by moment collocation we choose n collocation points $x_j \in [a,b]$, $j = 1,\dots,n$. We approximate the solution φ of (17.22) by a function φ_n

with minimal L^2 norm satisfying the integral equation at the collocation points

$$\int_a^b K(x_j, y)\varphi_n(y)\,dy = f(x_j), \quad j = 1, \ldots, n. \tag{17.23}$$

With the kernel function given in Theorem 17.12 we can rewrite (17.23) in the form

$$(A\varphi_n, h_{x_j})_H = (f, h_{x_j})_H, \quad j = 1, \ldots, n. \tag{17.24}$$

Hence, the moment collocation coincides with the moment method applied to the integral operator $A : L^2[a, b] \to H$ with the subspaces $H_n \subset H$ given by $H_n := \operatorname{span}\{h_{x_1}, \ldots, h_{x_n}\}$. In particular, from the analysis in Sections 17.2 and 17.3, we conclude that a unique solution to (17.23) with minimal norm exists and is given by the orthogonal projection of the exact solution φ onto the subspace

$$X_n = A^*(H_n) = A^{-1}(H_n) = \operatorname{span}\{\overline{K(x_1, \cdot)}, \ldots, \overline{K(x_n, \cdot)}\}.$$

We may assume that the kernel functions $\overline{K(x_1, \cdot)}, \ldots, \overline{K(x_n, \cdot)}$ are linearly independent, since otherwise the equations (17.23) would be linearly dependent. Then, writing

$$\varphi_n = \sum_{k=1}^{n} \gamma_k \overline{K(x_k, \cdot)} \tag{17.25}$$

we have to solve the linear system

$$\sum_{k=1}^{n} \gamma_k \int_a^b \overline{K(x_k, y)} K(x_j, y)\,dy = f(x_j), \quad j = 1, \ldots, n, \tag{17.26}$$

with a positive definite matrix for the coefficients $\gamma_1, \ldots, \gamma_n$.

Theorem 17.13. *Assume that the sequence (x_n) of collocation points is dense in the interval $[a, b]$. Then the corresponding moment collocation solutions converge, i.e., $\|\varphi_n - A^{-1}f\|_{L^2} \to 0, \quad n \to \infty$.*

Proof. By Theorem 17.9, applied to $A : L^2[a, b] \to H$, it suffices to show that $U := \operatorname{span}\{h_{x_n} : n \in \mathbb{N}\}$ is dense in H. Note that each $g \in H$ is continuous, since A has continuous kernel K. Denote by $P : H \to \bar{U}$ the orthogonal projection operator. For $g \in H$, by Theorem 1.25, we have $Pg - g \perp U$. This implies $(Pg - g, h_{x_n})_H = 0$ for all $n \in \mathbb{N}$, i.e., $(Pg)(x_n) = g(x_n)$ for all $n \in \mathbb{N}$. From this, by the denseness of the collocation points and the continuity of Pg and g, it follows that $g = Pg \in \bar{U}$. □

For the error we can apply the special case (17.18) of Theorem 17.6. Let $F^d \in \mathbb{C}^n$ be a perturbation of the data vector $F = (f(x_1), \ldots, f(x_n))$ with $\|F^d - F\|_2 \leq d$ and denote the moment collocation solution to these inexact data via (17.25) and (17.26) by φ_n^d. Then we have the following error estimate.

Theorem 17.14. *The error in the moment collocation can be estimated by*

$$\|\varphi_n^d - A^{-1}f\|_{L^2} \le \frac{d}{\sqrt{\lambda_n}} + C \inf_{\psi \in X_n} \|\psi - A^{-1}f\|_{L^2},$$

where λ_n denotes the smallest eigenvalue of the positive definite matrix of the linear system (17.26) and C is some positive constant.

The quality of the discretization depends on how well the exact solution can be approximated by the subspace X_n, i.e., (besides the smoothness of the exact solution) it depends on the smoothness of the kernel: the smoother the kernel, the better the approximation. On the other hand, the smoothness of the kernel also controls the asymptotics of the singular values of the integral operator and, consequently, the behavior of the eigenvalues λ_n (see Theorem 15.20). In general, the decay of the singular values increases with the smoothness of the kernel. Therefore, we can expect the moment collocation to deliver reliable approximations only for kernels with poor smoothness. For an example see Problem 17.5.

For more details on the moment collocation, we refer to Engl [48], Groetsch [74], and Nashed and Wahba [181], and the literature therein.

Problems

17.1. Formulate and prove a variant of Theorem 17.6 for the collocation method of Section 13.4. Apply the result to equation (13.30).

17.2. Let $A : X \to X$ be a nonnegative self-adjoint compact operator with one-dimensional nullspace $N(A) = \text{span}\{u_0\}$ and spectral decomposition

$$A\varphi = \sum_{j=1}^{\infty} \lambda_j(\varphi, u_j)u_j,$$

where $\lambda_j \neq 0$ for all $j \in \mathbb{N}$. Consider the compact operator $S : X \to X$ with

$$S\varphi := A\varphi + (\varphi, u_0)g,$$

where $g \in X$ with $(g, u_0) = 1$ and $(g, u_j) \neq 0$ for all $j \in \mathbb{N}$. For the equation $S\varphi = f$ with right-hand side $f = Ag + g$ and exact solution $\varphi = g$ show that the coefficients γ_j, $j = 0, \dots, n$, for the least squares approximation (see Theorem 13.31)

$$\varphi_n = \sum_{j=0}^{n} \gamma_j u_j$$

with respect to the subspace $X_n = \text{span}\{u_0, \dots, u_n\}$ satisfy

$$(\gamma_0 - 1)\{1 + \|g - P_n g\|^2\} = (Ag, g - P_n g)$$

and

$$\lambda_j\{\gamma_j - (g, u_j)\} = (1 - \gamma_0)(g, u_j), \quad j = 1, \dots, n.$$

Here $P_n : X \to X_n$ denotes the orthogonal projection. Use these results to establish the inequalities

$$\|\varphi_n - g\| \geq |\gamma_n - (g, u_n)| \geq \frac{|(Ag, g - P_n g)|\,|(g, u_n)|}{\{1 + \|g\|^2\}|\lambda_n|} \geq \frac{|\lambda_{n+1}|}{|\lambda_n|} |(g, u_{n+1})|^2 |(g, u_n)|.$$

Use this to show that A and g can be chosen such that the least squares solutions φ_n do not converge to the exact solution (see Seidman [218]).

17.3. Show that in a Hilbert space with a reproducing kernel a weakly convergent sequence is pointwise convergent. Show that a norm convergent sequence converges uniformly provided the reproducing kernel is continuous.

17.4. Determine the reproducing kernel of the Sobolev space $H^1[a, b]$.

17.5. Apply the moment collocation to the Volterra integral operator of Example 15.19 with an equidistant mesh $x_j = jh$, $j = 1, \ldots, n$, where $h = 1/n$. Show that the inverse matrix for the linear system (17.26) in this case is given by the tridiagonal matrix

$$n \begin{pmatrix} 2 & -1 & & & & \\ -1 & 2 & -1 & & & \\ & -1 & 2 & -1 & & \\ & \cdot & \cdot & \cdot & \cdot & \cdot \\ & & & -1 & 2 & -1 \\ & & & & -1 & 1 \end{pmatrix}$$

and that the total error in the spirit of Theorem 17.14 can be estimated by

$$\|\varphi_n^d - A^{-1} f\|_{L^2} \leq c_1 n^{1/2} d + \frac{c_2}{n}$$

for some constants c_1 and c_2 provided the exact solution is twice continuously differentiable.

Chapter 18
Inverse Boundary Value Problems

In this book, so far, we have considered only so-called *direct* boundary value problems where, given a differential equation, its domain, and a boundary condition, we want to determine its solution. From Chapters 6–9 we know that the classical boundary value problems for the Laplace equation and the heat equation are well-posed. However, if for a given differential equation the problem consists of determining its domain, i.e., the boundary of the domain from a knowledge of the type of boundary condition and information on one (or several) of its solutions, then this *inverse* boundary value problem will often be ill-posed in any reasonable setting. Hence, the study of inverse boundary value problems is closely interlinked with the investigation of ill-posed problems.

Since integral equation methods are of fundamental importance in the solution of direct boundary value problems it is no surprise that they also play a prominent role in the study of inverse problems. Thus an inclusion of a chapter on inverse boundary value problems is imperative for a book on integral equations.

In the fifteen years since the second edition of this book was written the field of inverse boundary value problems has experienced a number of new developments. To end this book, in order to acquaint the reader with current research directions in inverse boundary value problems, we shall study an inverse Dirichlet problem for the Laplace equation as a model problem. To create more variety in the presentation, occasionally we also will consider the corresponding Neumann problem.

Of course, in a single chapter it is impossible to give a complete account of inverse boundary value problems. Hence we shall content ourselves with developing some of the main principles. For a detailed study of inverse boundary value problems, we refer to Colton and Kress [32], Isakov [105] and Potthast [196].

18.1 An Inverse Problem for the Laplace Equation

To introduce the reader to inverse boundary value problems, for the sake of simplicity, we will confine ourselves to the presentation of an inverse boundary value problem for the Laplace equation. We recall the fundamental solution

R. Kress, *Linear Integral Equations*, Applied Mathematical Sciences 82,
DOI 10.1007/978-1-4614-9593-2_18, © Springer Science+Business Media New York 2014

$$\Phi(x,y) := \frac{1}{2\pi} \ln \frac{1}{|x-y|}, \quad x \neq y,$$

to Laplace's equation in $\mathbb{R}^2$. Let $D \subset \mathbb{R}^2$ be a bounded domain with a connected boundary ∂D of class C^2. Then we consider the two-dimensional exterior Dirichlet problem for a bounded function $u \in C^2(\mathbb{R}^2 \setminus \bar{D}) \cap C(\mathbb{R}^2 \setminus D)$ satisfying Laplace's equation

$$\Delta u = 0 \quad \text{in } \mathbb{R}^2 \setminus \bar{D} \tag{18.1}$$

and the Dirichlet boundary condition

$$u = -\Phi(\cdot\,,\zeta) \quad \text{on } \partial D \tag{18.2}$$

for some $\zeta \in \mathbb{R}^2 \setminus \bar{D}$. From Chapter 6 we know that this exterior Dirichlet problem has a unique solution and that this solution depends continuously on the boundary data, i.e., on the location of the source point ζ. We note that analogous to Section 8.3, this boundary value problem can also be treated in a Sobolev space setting. On occasion we also will replace the Dirichlet condition (18.2) by the Neumann condition

$$\frac{\partial u}{\partial \nu} = -\frac{\partial \Phi(\cdot\,,\zeta)}{\partial \nu} \quad \text{on } \partial D \tag{18.3}$$

with the normal vector ν directed into the exterior of D together with the condition $u(x) \to 0$ as $|x| \to \infty$. Again from Chapter 6 we recall existence and uniqueness for this exterior Neumann problem. Note that the sovability condition (6.49) is satisfied.

Now, let $B \subset \mathbb{R}^2$ be an additional bounded domain with connected boundary ∂B of class C^2 containing the closure $\bar{D}$. Then the inverse problem we want to consider is, given $u|_{\partial B}$ on ∂B for a solution u of (18.1)–(18.2), determine the shape of D, i.e., we want to construct the shape of a perfectly conducting infinite cylinder with cross section D from the response to an incident static electric field of a line source parallel to the cylinder axis passing through a point ζ in the exterior of D. We may view this inverse problem as a model problem for inverse boundary value problems of similar structure occurring in various applications such as remote sensing, nondestructive testing, ultrasound medicine and seismic imaging. As we shall illustrate in the sequel, this inverse boundary value problem associated with (18.1)–(18.2) is nonlinear and improperly posed. It is nonlinear, since the solution of (18.1)–(18.2) depends nonlinearly on the boundary ∂D. Roughly speaking, the inverse problem is ill-posed, since perturbations of $u|_{\partial B}$ in the maximum norm (or L^2 norm) may lead to functions that are no longer analytic and therefore fail to be the restriction of a harmonic function on ∂B (in case of an analytic ∂B). Hence, for small perturbations of $u|_{\partial B}$, in general, the inverse problem is not solvable, and if it is solvable the solution cannot depend continuously on the data.

An important question to ask about the inverse problem is uniqueness, i.e., is the boundary curve ∂D uniquely determined by the values of the harmonic function u, solving (18.1)–(18.2), on the curve ∂B.

Theorem 18.1. *Let D_1 and D_2 be two bounded domains with connected boundaries ∂D_1 and ∂D_2 of class C^2, respectively, such that $\bar{D}_1 \subset B$ and $\bar{D}_2 \subset B$. Denote by u_1 and u_2 the solutions to the exterior Dirichlet problem (18.1)–(18.2) for the domains D_1 and D_2, respectively, and assume that $u_1 = u_2$ on ∂B. Then $D_1 = D_2$.*

Proof. Assume that $D_1 \neq D_2$. From $u_1 = u_2$ on ∂B, by the uniqueness of the solution to the two-dimensional exterior Dirichlet problem (Theorem 6.12), it follows that $u_1 = u_2$ in $\mathbb{R}^2 \setminus B$. From this, since harmonic functions are analytic (Theorem 6.6), we can conclude that $u_1 = u_2$ in the unbounded component W of the complement of $\bar{D}_1 \cup \bar{D}_2$. Without loss of generality, we can assume that $W^* := (\mathbb{R}^2 \setminus \overline{W}) \setminus \bar{D}_2$ is nonempty. Then u_2 is defined in W^*, since it describes the solution of (18.1)–(18.2) for D_2. The function $v := \Phi(\cdot\,,\zeta) + u_2$ solves the Laplace equation in W^*, is continuous in $\overline{W^*}$, and satisfies the homogeneous boundary condition $v = 0$ on ∂W^*. This boundary condition follows from the observation that each boundary point of W^* either belongs to ∂D_2 or to $\partial W \cap \partial D_1$. For $x \in \partial D_2$ we have $\Phi(x,\zeta) + u_2(x) = 0$ as a consequence of the boundary condition for u_2, and for $x \in \partial W \cap \partial D_1$ we have $u_2(x) = u_1(x)$ and therefore

$$\Phi(x,\zeta) + u_2(x) = \Phi(x,\zeta) + u_1(x) = 0$$

as a consequence of the boundary condition for u_1. Now, by the maximum-minimum principle for harmonic functions we can conclude that $\Phi(\cdot\,,\zeta) + u_2 = 0$ in W^* and consequently, by analyticity, $\Phi(\cdot\,,\zeta)+u_2 = 0$ in $\mathbb{R}^2 \setminus (D_2 \cup \{\zeta\})$. Since u_2 is harmonic in all of $\mathbb{R}^2 \setminus \bar{D}_2$, this contradicts the singularity of the fundamental solution at the point ζ, and the proof is complete. □

The idea of the proof of Theorem 18.1 actually goes back to a private communication of Schiffer referenced in [157]. Due to regularity issues, the proof does not carry over to the Neumann boundary condition (18.3). However, in two dimensions uniqueness can be settled by reducing the Neumann boundary condition for u to a Dirichlet boundary condition for the conjugate harmonic v defined via the Cauchy–Riemann equations for the real and imaginary part of holomorphic functions by $\partial v/\partial x_1 := -\partial u/\partial x_2$ and $\partial v/\partial x_2 := \partial u/\partial x_1$ (see [80]).

Roughly speaking, most approaches for approximately solving an inverse boundary value problem like our model problem (18.1)–(18.2) belong to one group of the following trilogy. In a first group of methods, denoted as *direct methods* or *iterative methods*, the inverse boundary value problem is considered as an ill-posed nonlinear operator equation and iterative techniques are employed for its solution. For fixed location ζ of the source point, the solution to the Dirichlet problem (18.1)–(18.2) defines an operator

$$F : \partial D \mapsto u|_{\partial B} \tag{18.4}$$

that maps the boundary curve ∂D onto the trace of the solution u on ∂B. In terms of this operator F, which we also will denote as the *boundary to data operator*, given $u|_{\partial B}$, the inverse boundary value problem consists in solving the operator equation

$$F(\partial D) = u|_{\partial B} \tag{18.5}$$

for the unknown boundary ∂D. For the approximate solution of (18.5), it seems natural to apply iterative methods based on linearization, such as Newton's method. However, since the linearized version of (18.5) inherits the ill-posedness from the inverse problem, the Newton iterations need to be regularized. This approach has the advantage that, in principle, it is conceptually simple and that it leads to highly accurate reconstructions. However, as disadvantage we note that the numerical implementation requires the forward solution of the Dirichlet problem (18.1)–(18.2) in each step of the Newton iteration and good a priori information is needed for the initial approximation.

In the second group of methods, known as *decomposition methods*, the inverse problem is decomposed or separated into a linear ill-posed part for the construction of the harmonic function u from the given data $u|_{\partial B}$ and a nonlinear and at most mildly ill-posed part for finding the boundary as the location where the boundary condition

$$u + \Phi(\cdot\,, \zeta) = 0$$

is satisfied. The decomposition methods have the advantage that their numerical implementation does not require the solution for the boundary value problem (18.1)–(18.2) in each iteration step.

A third group of methods, denoted as *sampling methods*, is based on designing an appropriate indicator function f on B such that its value $f(z)$ decides whether $z \in B$ lies inside or outside the unknown domain D. Usually such an indicator function is defined in terms of the solvability of a certain ill-posed linear integral equation of the first kind depending on the location parameter z. As a main advantage, sampling methods do not need any a priori informations on the geometry of the unknown domain D and on the boundary condition. In particular, sampling methods also can be applied to the case where D has a finite number of connected components. However, in general, they need more data such as the trace $u|_{\partial B}$ for a large number of source points ζ whereas the iterative and decomposition methods, in general, work with one source point only. In the remainder of this chapter we will consider typical examples for each of the three groups as applied to our model problem.

18.2 Decomposition Methods

We begin with a method belonging to the second group. In its first part, as in Theorem 6.24 we try to represent the solution u of (18.1)–(18.2) by a modified double-layer potential with a density $\varphi \in L^2(\Gamma)$ on some auxiliary closed curve Γ contained in the unknown domain D. The knowledge of such an internal curve Γ requires weak a priori information about D. Slightly deviating from the analysis of Theorem 6.24, we introduce an additional positive weight function $\omega \in C(\Gamma)$. In terms of a regular 2π-periodic parameterization $\Gamma = \{z(t) : t \in [0, 2\pi]\}$ we choose ω via

$$\omega \circ z = \frac{1}{|z'|}\,, \tag{18.6}$$

i.e., $\omega(z(t)) = 1/|z'(t)|$ for $t \in [0, 2\pi]$, and seek the solution of exterior Dirichlet problems in the form

$$u(x) = \int_\Gamma \varphi(y) \left\{ \frac{\partial \Phi(x,y)}{\partial \nu(y)} + \omega(y) \right\} ds(y), \quad x \in \mathbb{R}^2 \setminus \Gamma, \tag{18.7}$$

where ν denotes the outward unit normal to Γ. The existence proof of Theorem 6.25 carries over to this variant with the weight function which will lead to slightly simpler parametrized versions of the potential and the integral equation. If we define an integral operator $A : L^2(\Gamma) \to L^2(\partial B)$ by the restriction of the double-layer potential (18.7) on ∂B, i.e., by

$$(A\varphi)(x) := \int_\Gamma \varphi(y) \left\{ \frac{\partial \Phi(x,y)}{\partial \nu(y)} + \omega(y) \right\} ds(y), \quad x \in \partial B, \tag{18.8}$$

then, given the values $u|_{\partial B}$ on ∂B, we are looking for a density φ satisfying the equation

$$A\varphi = u|_{\partial B}. \tag{18.9}$$

The integral operator A has a smooth kernel, and therefore equation (18.9) is a severely ill-posed equation of the first kind. As for the integral equation (16.38), its solvability can again be related to the analytic continuation of the potential u across the boundary ∂D. The integral equation (18.9) is solvable if and only if u can be extended as a harmonic function into the exterior of Γ with boundary data in $L^2(\Gamma)$. As in Example 16.30 it can be shown that the singular values of A for the special case where Γ is the unit circle and B a concentric disk of radius R decay exponentially (see Problem 18.3).

Theorem 18.2. *The integral operator A, defined by (18.8), is injective and has dense range.*

Proof. The proof is analogous to that of Theorem 16.28. For the injectivity of A, the uniqueness part of the proof of Theorem 6.24 has to be used. For the dense range of A we need to show injectivity of the adjoint operator A^* given by

$$(A^*\psi)(x) = \int_{\partial B} \psi(y) \left\{ \frac{\partial \Phi(x,y)}{\partial \nu(x)} + \omega(x) \right\} ds(y), \quad x \in \Gamma.$$

For this we observe that from $A^*\psi = 0$, with the aid of Example 6.17, it follows that

$$(1, \psi)_{L^2(\partial B)} \int_\Gamma \omega \, ds = (A1, \psi)_{L^2(\partial B)} = (1, A^*\psi)_{L^2(\Gamma)} = 0$$

and consequently $\int_{\partial B} \psi \, ds = 0$. Hence, the single-layer potential v with density ψ on ∂B satisfies $\partial v / \partial \nu = 0$ on Γ. Now the interior uniqueness Theorem 6.13 and analyticity imply that v is constant in B. From the jump relation (6.53) for the normal derivative of single-layer potentials with L^2 densities we derive $\psi + K'\psi = 0$, where K' denotes the normal derivative boundary operator on the curve ∂B. Now the continuity of the kernel of K' (see Problem 6.1) and Theorem 6.21 imply that $\psi = 0$, since we already know that $\int_{\partial B} \psi \, ds = 0$. Hence A^* is injective. □

The Tikhonov regularization technique may be applied for the numerical solution of the ill-posed equation (18.9). Recall that solving the regularized equation

$$\alpha\varphi_\alpha + A^*A\varphi_\alpha = A^*u|_{\partial B} \tag{18.10}$$

with regularization parameter $\alpha > 0$ is equivalent to minimizing the penalized residual

$$\mu_1(\varphi;\alpha) := \|A\varphi - u|_{\partial B}\|^2_{L^2(\partial B)} + \alpha\|\varphi\|^2_{L^2(\Gamma)}$$

over all $\varphi \in L^2(\Gamma)$.

After we have determined φ_α as the solution of (18.10), we use the corresponding double-layer potential $u_\alpha = V\varphi_\alpha$ as an approximation for the solution u of (18.1)–(18.2). Here, for convenience, we use the abbreviation

$$(V\varphi)(x) := \int_\Gamma \varphi(y)\left\{\frac{\partial\Phi(x,y)}{\partial\nu(y)} + \omega(y)\right\} ds(y), \quad x \in \mathbb{R}^2 \setminus \Gamma,$$

for the double-layer potential with density $\varphi \in L^2(\Gamma)$ as defined through (18.7). Then, in the second part of our method, we seek the boundary of D as the location of the zeros of $\Phi(\cdot\,,\zeta) + u_\alpha$ in a minimum norm sense, i.e., we approximate ∂D by minimizing the defect

$$\|\Phi(\cdot\,,\zeta) + u_\alpha\|_{L^2(\Lambda)}$$

over some suitable class U of admissible curves Λ. In the following, we will choose U to be a compact subset (with respect to the $C^2_{2\pi}$ norm) of the set of all starlike closed C^2 curves, described by

$$z(t) = r(t)\,(\cos t, \sin t), \quad t \in [0, 2\pi], \tag{18.11}$$

where $r \in C^2_{2\pi}$ satisfies the a priori assumption

$$0 < r_i(t) \le r(t) \le r_e(t), \quad t \in [0, 2\pi].$$

Here, r_i and r_e are given functions representing curves Λ_i and Λ_e such that the internal auxiliary curve Γ is contained in the interior of Λ_i, the boundary ∂D of the unknown scatterer D is contained in the annulus between Λ_i and Λ_e, and Λ_e is contained in B. For a sequence of curves we understand convergence $\Lambda_n \to \Lambda$, $n \to \infty$, in the sense that $\|r_n - r\|_{C^2_{2\pi}} \to 0$, $n \to \infty$, for the functions r_n and r representing Λ_n and Λ via (18.11). We say that a sequence of functions f_n from $C(\Lambda_n)$ is L^2 convergent to a function f in $C(\Lambda)$ if

$$\int_0^{2\pi} |f_n(r_n(t)\,(\cos t, \sin t)) - f(r(t)\,(\cos t, \sin t))|^2\, dt \to 0, \quad n \to \infty.$$

Now we introduce the functional

$$\mu_2(\Lambda, \varphi) := \|\Phi(\cdot\,, \zeta) + V\varphi\|^2_{L^2(\Lambda)}.$$

Then, given φ_α, we seek an approximation to ∂D by minimizing the defect $\mu_2(\Lambda, \varphi_\alpha)$ over all $\Lambda \in U$. Due to the continuity of $\mu_2(\cdot\,, \varphi)$ and the compactness of U, this minimization problem has a solution (see Problem 1.3).

Since, in general, we do not have convergence of the densities φ_α as $\alpha \to 0$, for a theoretically satisfactory reformulation of the inverse boundary value problem as an optimization problem we combine the two steps and minimize the sum

$$\mu(\varphi, \Lambda; \alpha) := \mu_1(\varphi; \alpha) + \gamma\mu_2(\Lambda, \varphi) \tag{18.12}$$

simultaneously over all $\varphi \in L^2(\Gamma)$ and $\Lambda \in U$, i.e., we call $\Lambda_0 \in U$ *optimal* if there exists $\varphi_0 \in L^2(\Gamma)$ such that $\mu(\varphi_0, \Lambda_0; \alpha) = M(\alpha)$, where

$$M(\alpha) := \inf_{\varphi \in L^2(\Gamma),\, \Lambda \in U} \mu(\varphi, \Lambda; \alpha).$$

Here, $\gamma > 0$ denotes a coupling parameter, which has to be chosen appropriately for the numerical implementation to make the two terms in (18.12) of the same magnitude. In the sequel, for theoretical purposes we may assume that $\gamma = 1$. For our reformulation of the inverse boundary value problem into a nonlinear optimization problem we can state the following results. Note that in the existence Theorem 18.3 we need not assume that $u|_{\partial B}$ is the exact restriction of a harmonic function.

Theorem 18.3. *For each $\alpha > 0$, the optimization formulation of the inverse boundary value problem given through (18.12) has a solution.*

Proof. Let (φ_n, Λ_n) be a minimizing sequence in $L^2(\Gamma) \times U$, i.e.,

$$\lim_{n\to\infty} \mu(\varphi_n, \Lambda_n; \alpha) = M(\alpha).$$

Since U is compact, we can assume that $\Lambda_n \to \Lambda \in U$, $n \to \infty$. From $\alpha > 0$ and

$$\alpha\|\varphi_n\|^2_{L^2(\Gamma)} \leq \mu(\varphi_n, \Lambda_n; \alpha) \to M(\alpha), \quad n \to \infty,$$

we conclude that the sequence (φ_n) is bounded, i.e., $\|\varphi_n\|_{L^2(\Gamma)} \leq c$ for all n and some constant c. Hence, by Theorem 16.3, we can assume weak convergence $\varphi_n \rightharpoonup \varphi$ with some $\varphi \in L^2(\Gamma)$ as $n \to \infty$. Since $A : L^2(\Gamma) \to L^2(\partial B)$ and $V : L^2(\Gamma) \to L^2(\Lambda)$ represent compact operators, by Problem 16.5 it follows that both $A\varphi_n \to A\varphi$ and $V\varphi_n \to V\varphi$ as $n \to \infty$. With functions r_n and r representing Λ_n and Λ via (18.11), by Taylor's formula we can estimate

$$\left| \frac{\partial\Phi(r_n(t)\,(\cos t, \sin t), y)}{\partial\nu(y)} - \frac{\partial\Phi(r(t)\,(\cos t, \sin t), y)}{\partial\nu(y)} \right| \leq L\,|r_n(t) - r(t)|$$

for $t \in [0, 2\pi]$ and $y \in \Gamma$. Here, L denotes a bound on $\operatorname{grad}_x\{\partial\Phi(x, y)/\partial\nu(y)\}$ on $W \times \Gamma$, where W is the closed annular domain between the two curves Λ_i and Λ_e.

Then, using the Cauchy–Schwarz inequality, we find that

$$|(V\varphi_n)(r_n(t)\,(\cos t, \sin t)) - (V\varphi_n)(r(t)\,(\cos t, \sin t))| \le cL\sqrt{|\Gamma|}\,|r_n(t) - r(t)|$$

for $t \in [0, 2\pi]$. Therefore, from $\|V\varphi_n - V\varphi\|^2_{L^2(\Lambda)} \to 0$, $n \to \infty$, we can deduce that

$$\|\Phi(\cdot\,,\zeta) + V\varphi_n\|^2_{L^2(\Lambda_n)} \to \|\Phi(\cdot\,,\zeta) + V\varphi\|^2_{L^2(\Lambda)}, \quad n \to \infty.$$

This now implies

$$\alpha\|\varphi_n\|^2_{L^2(\Gamma)} \to M(\alpha) - \|A\varphi - u|_{\partial B}\|^2_{L^2(\partial B)} - \|\Phi(\cdot\,,\zeta) + V\varphi\|^2_{L^2(\Lambda)} \le \alpha\|\varphi\|^2_{L^2(\Gamma)}$$

for $n \to \infty$. Since we already know weak convergence $\varphi_n \rightharpoonup \varphi$, $n \to \infty$, it follows that

$$\lim_{n\to\infty} \|\varphi_n - \varphi\|^2_{L^2(\Gamma)} = \lim_{n\to\infty} \|\varphi_n\|^2_{L^2(\Gamma)} - \|\varphi\|^2_{L^2(\Gamma)} \le 0,$$

i.e., we also have norm convergence $\varphi_n \to \varphi$, $n \to \infty$. Finally, by continuity we obtain that

$$\mu(\varphi, \Lambda; \alpha) = \lim_{n\to\infty} \mu(\varphi_n, \Lambda_n; \alpha) = M(\alpha),$$

and this completes the proof. □

In order to prove regularity (in the sense of Definition 15.7) of the regularization included in the above optimization reformulation of the inverse boundary value problem, we need the following two lemmas.

Lemma 18.4. *Let $u|_{\partial B}$ be the restriction to ∂B of the solution u to (18.1)–(18.2) for a domain D such that ∂D is contained in U. Then*

$$\lim_{\alpha\to 0} M(\alpha) = 0. \tag{18.13}$$

Proof. By Theorem 18.2, applied to the double-layer operator from $L^2(\Gamma)$ into $L^2(\partial D)$, given $\varepsilon > 0$ there exists $\varphi \in L^2(\Gamma)$ such that

$$\|\Phi(\cdot\,,\zeta) + V\varphi\|_{L^2(\partial D)} < \varepsilon.$$

Analogous to Remark 16.29 we can estimate

$$\|A\varphi - u\|_{L^2(\partial B)} \le c\|V\varphi - u\|_{L^2(\partial D)}$$

with some constant c. From $\Phi(\cdot\,,\zeta) + u = 0$ on ∂D we then deduce that

$$\mu(\varphi, \partial D; \alpha) \le (1 + c^2)\varepsilon^2 + \alpha\|\varphi\|^2_{L^2(\Gamma)} \to (1 + c^2)\varepsilon^2, \quad \alpha \to 0.$$

Since ε is arbitrary, (18.13) follows. □

Lemma 18.5. *Let (Λ_n) be a sequence of starlike curves of class C^2 that converges to a starlike curve Λ of class C^2 and let u_n and u be bounded harmonic functions in* *the exterior of Λ_n and Λ, respectively. Assume that u_n and u are continuous in the closure of the exterior of Λ_n and Λ, respectively, and that the boundary values of u_n on Λ_n are L^2 convergent to the boundary values of u on Λ. Then $\|u_n - u\|_{C(\partial B)} \to 0$, $n \to \infty$.*

Proof. We use the modified double-layer potential (18.7), i.e., we represent u in the form

$$u(x) = \int_\Lambda \varphi(y) \left\{ \frac{\partial \Phi(x,y)}{\partial \nu(y)} + \omega(y) \right\} ds(y), \quad x \in \partial B,$$

with a density $\varphi \in C(\Lambda)$ and, analogously, we write u_n as a double-layer potential with density $\varphi_n \in C(\Lambda_n)$. With the aid of the parametric representation (18.11) of Λ we transform the integral equation corresponding to (6.42) into the parameterized form $\psi + L\psi = f$, where $\psi = \varphi \circ z$ and $f = 2u \circ z$. Analogously, we have an integral equation $\psi_n + L_n\psi_n = f_n$ corresponding to u_n. By assumption, $\|f_n - f\|_{L^2[0,2\pi]} \to 0$ as $n \to \infty$.

We denote the kernel functions of L and L_n by k and k_n, respectively. By Problem 6.1, the kernel k is given by

$$k(t,\tau) = \frac{1}{\pi} \frac{[z'(\tau)]^\perp \cdot [z(t) - z(\tau)]}{|z(t) - z(\tau)|^2} + 2, \quad t \neq \tau,$$

and k_n is given by the analogous expression in terms of the parametric representation $z_n(t) = r_n(t)(\cos t, \sin t)$. As before, for a vector $a = (a_1, a_2)$ in $\mathbb{R}^2$ we denote $a^\perp = (a_2, -a_1)$. From Taylor's formula

$$z(t) - z(\tau) = (t - \tau) \int_0^1 z'(\tau + \lambda(t - \tau))\, d\lambda \tag{18.14}$$

we observe that the function

$$g(t,\tau) := \begin{cases} \dfrac{|z(t) - z(\tau)|}{|t - \tau|}, & t \neq \tau, \\ |z'(t)|, & t = \tau, \end{cases}$$

is continuous in $\mathbb{R}^2$. Therefore, in view of $z(t) \neq z(\tau)$ for $0 < |t - \tau| \leq \pi$ and $|z'(t)|^2 = [r(t)]^2 + [r'(t)]^2 > 0$ for all t, there exists a constant c_1, depending on r, such that

$$|z(t) - z(\tau)| \geq c_1 |t - \tau|, \quad |t - \tau| \leq \pi. \tag{18.15}$$

From this, using (18.14) applied to $z_n - z$, we deduce that

$$|[z_n(t) - z_n(\tau)] - [z(t) - z(\tau)]| \leq c_2 \|r_n - r\|_{C^1_{2\pi}} |z(t) - z(\tau)| \tag{18.16}$$

for $|t - \tau| \leq \pi$ and some constant c_2 depending on r. By Taylor's formula we have

$$\left| \frac{1}{|x_n|^2} - \frac{1}{|x|^2} \right| \leq \frac{16}{|x|^3} |x_n - x|, \quad x \neq 0, \ 2|x_n - x| \leq |x|.$$

From this, setting $x = z(t) - z(\tau)$, $x_n = z_n(t) - z_n(\tau)$, and using (18.15) and (18.16) we find the estimate

$$\left| \frac{1}{|z_n(t) - z_n(\tau)|^2} - \frac{1}{|z(t) - z(\tau)|^2} \right| \le c_3 \, \frac{\|r_n - r\|_{C^1_{2\pi}}}{(t-\tau)^2}, \tag{18.17}$$

which is valid for sufficiently small $\|r_n - r\|_{C^1_{2\pi}}$, for $|t - \tau| \le \pi$, and for some constant c_3 depending on r. Using Taylor's formula

$$z(t) - z(\tau) = (t-\tau)z'(\tau) + (t-\tau)^2 \int_0^1 (1-\lambda) z''(\tau + \lambda(t-\tau))\, d\lambda,$$

we can write

$$[z'(\tau)]^\perp \cdot [z(t) - z(\tau)] = (t-\tau)^2 \, [z'(\tau)]^\perp \cdot \int_0^1 (1-\lambda) z''(\tau + \lambda(t-\tau))\, d\lambda.$$

Hence, we can estimate

$$|[z'(\tau)]^\perp \cdot [z(t) - z(\tau)]| \le c_4 (t-\tau)^2$$

and

$$|[z_n'(\tau)]^\perp \cdot [z_n(t) - z_n(\tau)] - [z'(\tau)]^\perp \cdot [z(t) - z(\tau)]| \le c_5 (t-\tau)^2 \|r_n - r\|_{C^2_{2\pi}}$$

for all $|t - \tau| \le \pi$ and some constants c_4 and c_5 depending on r. Combining these two estimates with (18.15)–(18.17), by decomposing

$$k_n(t,\tau) - k(t,\tau) = \frac{1}{\pi} \, [z_n'(\tau)]^\perp \cdot [z_n(t) - z_n(\tau)] \left\{ \frac{1}{|z_n(t) - z_n(\tau)|^2} - \frac{1}{|z(t) - z(\tau)|^2} \right\}$$

$$+ \frac{1}{\pi} \, \frac{[z_n'(\tau)]^\perp \cdot [z_n(t) - z_n(\tau)] - [z'(\tau)]^\perp \cdot [z(t) - z(\tau)]}{|z(t) - z(\tau)|^2},$$

we finally obtain that

$$|k_n(t,\tau) - k(t,\tau)| \le C \|r_n - r\|_{C^2_{2\pi}}, \quad |t - \tau| \le \pi, \tag{18.18}$$

for sufficiently small $\|r_n - r\|_{C^2_{2\pi}}$ and some constant C depending on r.

The estimate (18.18) now implies norm convergence

$$\|L_n - L\|_{L^2[0,2\pi]} \le 2\pi C \|r_n - r\|_{C^2_{2\pi}} \to 0, \quad n \to \infty.$$

Therefore, using Theorem 10.1 we can conclude that $\|\psi_n - \psi\|_{L^2[0,2\pi]} \to 0$, $n \to \infty$. From this, the statement of the lemma follows by parameterizing the double-layer potential representation for $u_n - u$ and then using the Cauchy–Schwarz inequality. □

Now we are ready to establish regularity of the combined optimization scheme through the following convergence result.

Theorem 18.6. *Let (α_n) be a null sequence and let (Λ_n) be a corresponding sequence of solutions to the optimization problem with regularization parameter α_n. Assume that $u|_{\partial B}$ is the restriction to ∂B of the solution u to (18.1)–(18.2) for a domain D such that ∂D belongs to U. Then $\Lambda_n \to \partial D$, $n \to \infty$.*

Proof. Since by assumption U is compact, there exists a convergent subsequence $\Lambda_{n(k)} \to \Lambda^* \in U$, $k \to \infty$. Let u^* denote the unique solution to the boundary value problem (18.1)–(18.2) for the exterior domain with boundary Λ^*, i.e., the boundary condition reads

$$\Phi(\cdot\,,\zeta) + u^* = 0 \quad \text{on } \Lambda^*. \tag{18.19}$$

Since $\Lambda_{n(k)}$ is optimal for the parameter $\alpha_{n(k)}$, there exists $\varphi_{n(k)} \in L^2(\Gamma)$ such that $\mu(\varphi_{n(k)}, \Lambda_{n(k)}; \alpha_{n(k)}) = M(\alpha_{n(k)})$ for $k = 1, 2, \ldots$. By Lemma 18.4, we have

$$\|\Phi(\cdot\,,\zeta) + V\varphi_{n(k)}\|^2_{L^2(\Lambda_{n(k)})} \leq M(\alpha_{n(k)}) \to 0, \quad k \to \infty. \tag{18.20}$$

Now, by Lemma 18.5, from (18.19) and (18.20), we deduce that

$$\|A\varphi_{n(k)} - u^*\|_{L^2(\partial B)} \to 0, \quad k \to \infty.$$

By Lemma 18.4 we also have

$$\|A\varphi_{n(k)} - u\|^2_{L^2(\partial B)} \leq M(\alpha_{n(k)}) \to 0, \quad k \to \infty.$$

Therefore, we conclude $u = u^*$ on ∂B, and by Theorem 18.1 we have $\partial D = \Lambda^*$.

Now, by a standard argument, assuming that the sequence (Λ_n) does not converge to ∂D leads to a contradiction, since by the above argument any subsequence Λ_n contains a subsequence that converges to ∂D. □

The extension of the above method to inverse obstacle scattering problems for time-harmonic waves, i.e., for an inverse boundary value problems for the Helmholtz equation is due to Kirsch and Kress [128] and actually predated its application in potential theory. For a detailed discussion of the method as applied to inverse scattering theory and for numerical examples, we refer to [32, 129].

In a hybrid method combining ideas of the above decomposition method and Newton iterations as discussed in the following two sections the auxiliary internal curve Γ is viewed as an approximation for the unknown boundary ∂D. Then, keeping the double-layer potential u_α resulting via (18.7) from a regularized solution of (18.9) fixed, Γ is updated via linearization of the boundary condition $\Phi(\cdot\,,\zeta)+u_\alpha = 0$ around Γ. If we assume again that Γ is starlike with radial function r and look for an update $\widetilde{\Gamma} = \{\tilde{z}(t) : t \in [2\pi]\}$ that is starlike with radial function $r + q$, the update is found by solving the linear equation

$$\{\Phi(\cdot\,,\zeta) + u_\alpha\}\big|_\Gamma + \operatorname{grad}\{\Phi(\cdot\,,\zeta) + u_\alpha\}\big|_\Gamma \cdot \left((\tilde{z} - z) \circ z^{-1}\right) = 0 \tag{18.21}$$

for the radial function q determining the update $\tilde{z}$. In an obvious way, the two steps of alternatingly solving (18.9) by Tikhonov regularization and solving (18.21) in the least squares sense are iterated. For the numerical implementation, the terms $u_\alpha|_\Gamma$ and $\operatorname{grad} u_\alpha\big|_\Gamma$ in (18.21) are evaluated with the aid of the jump relations. From numerical examples (see [27]) it can be concluded that the quality of the reconstructions is similar to that of Newton iterations as discussed in the next sections. Again the application of this approach in potential theory was predated by that in inverse obstacle scattering (see Kress [144]). For numerical examples in inverse obstacle scattering we refer to [144, 150, 219] in two dimensions and to [220] in three dimensions.

We conclude this section on decomposition methods by describing a complex analysis approach that was developed in a series of papers by Akduman, Haddar and Kress [3, 80, 81, 145] for a slight modification of the inverse problem related to the Dirichlet problem (18.1)–(18.2). As we will see this approach can be viewed as a decomposition method with the sequential order of the nonlinear part and the ill-posed linear part reversed.

For this we consider the inverse problem to determine the boundary curve ∂D from the Cauchy data

$$f = u\,|_{\partial B} \quad \text{and} \quad g = \frac{\partial u}{\partial \nu}\bigg|_{\partial B}$$

of a harmonic function u in $B \setminus \bar{D}$ satisfying the homogeneous Dirichlet boundary condition $u = 0$ on ∂D. Here, as usual, ν denotes the outward unit normal to ∂B. Clearly, after renaming the unknowns, the inverse problem related to (18.1)–(18.2) can be considered as a special case of this modification after determining the normal derivative on ∂B by additional measurements or by solving a Dirichlet problem in the exterior of B. We note that the corresponding direct Dirichlet problem in $B \setminus \bar{D}$ with Dirichlet condition on both ∂D and ∂B is uniquely solvable (see Problem 18.1). Also the uniqueness Theorem 18.1 carries over to this case (see Problem 18.2).

We will identify $\mathbb{R}^2$ and $\mathbb{C}$ and introduce the annulus Ω bounded by two concentric circles Γ_0 with radius $0 < \rho < 1$ and Γ_1 with radius one centered at the origin. By the Riemann conformal mapping theorem for doubly connected domains (see [244]) there exists a uniquely determined radius ρ and a holomorphic function Ψ that maps Ω bijectively onto $B \setminus \bar{D}$ such that the boundaries Γ_0 and Γ_1 are mapped onto ∂D and ∂B, respectively, with all boundary curves in counterclockwise orientation. The function Ψ is unique up to a rotation of the annulus Ω. We represent ∂B by a regular parameterization

$$\partial B = \{\eta(t) : t \in [0, 2\pi]\}$$

and fix the freedom in rotating Ω by prescribing $\Psi(1) = \eta(0)$. Then we define a boundary correspondence function $\varphi : [0, 2\pi] \to [0, 2\pi]$ by setting

$$\varphi(t) := \eta^{-1}(\Psi(e^{it})), \quad t \in [0, 2\pi]. \tag{18.22}$$

Clearly, the boundary values φ uniquely determine Ψ as the solution to the Cauchy problem for a holomorphic function with $\Psi|_{\Gamma_1}$ given by $\Psi(e^{it}) = \eta(\varphi(t))$. The function $\chi : [0, 2\pi] \to \mathbb{C}$ defined by

$$\chi(t) := \Psi(\rho e^{it}), \quad t \in [0, 2\pi], \tag{18.23}$$

describes ∂D and therefore determining χ solves the inverse problem.

We now derive a non-local ordinary differential equation for the boundary correspondence function φ. To this end, we denote by $A_\rho : H^{1/2}[0, 2\pi] \to H^{-1/2}[0, 2\pi]$ the Dirichlet-to-Neumann operator for the annulus Ω given by

$$A_\rho : F \mapsto \frac{\partial w}{\partial \nu} \circ \eta,$$

where $w \in H^1(\Omega)$ is the unique harmonic function with boundary values in the trace sense given by

$$w \circ \eta = F \quad \text{on } \Gamma_1$$

and $w = 0$ on Γ_0 (see Section 8.3). Via $v := u \circ \Psi$ we associate the harmonic function u in D with a harmonic function v in B. Locally, the conjugate harmonics $\tilde{u}$ and $\tilde{v}$ are well defined and also related via $\tilde{v} := \tilde{u} \circ \Psi$. Then

$$\frac{d}{dt}\,\tilde{v}(e^{it}) = \varphi'(t)\,|\eta'(\varphi(t))|\,\frac{\partial \tilde{u}}{\partial s}\,(\eta(\varphi(t)))$$

with the tangential derivative $\partial\tilde{u}/\partial s$ and from the Cauchy–Riemann equations for u and v and their harmonic conjugates $\tilde{u}$ and $\tilde{v}$, respectively, we conclude that

$$\frac{\partial v}{\partial \nu}\,(e^{it}) = \varphi'(t)\,|\eta'(\varphi(t))|\,\frac{\partial u}{\partial \nu}\,(\eta(\varphi(t))), \quad t \in [0, 2\pi].$$

Therefore we have the non-local differential equation

$$\varphi' = \frac{A_\rho(f \circ \eta \circ \varphi)}{|\eta' \circ \varphi|\; g \circ \eta \circ \varphi} \tag{18.24}$$

for the boundary correspondence function φ in terms of the given Cauchy data f and g. The differential equation has to be complemented by the boundary conditions $\varphi(0) = 0$ and $\varphi(2\pi) = 2\pi$. It is non-local since the evaluation of its right-hand side at a point t requires the knowledge of φ everywhere in $[0, 2\pi]$.

Applying Green's second integral theorem in Ω to v and $x \mapsto \ln|x|$ we obtain the equation

$$\rho = \exp\left(-\frac{\int_0^{2\pi} f \circ \eta \circ \varphi\, dt}{\int_{\partial D} g\, ds}\right) \tag{18.25}$$

for the radius ρ of the annulus Ω. In [3] it is shown that under appropriate conditions the two equations (18.24) and (18.25) can be solved simultaneously for the boundary correspondence function φ and the radius ρ by successive approximations.

The equations (18.24) and (18.25) suffer from two drawbacks. Zeros in the Neumann data g will cause difficulties through the appearance of g in the denominator of (18.24), and (18.25) fails for g with mean value zero. Remedies for these deficiencies have been proposed in [80] by using a pair of Cauchy data and additional formulas for the radius. Here we refrain from describing the details of this modification.

Once the radius ρ and the boundary correspondence function φ are known we expand $\eta \circ \varphi$ in a Fourier series to obtain

$$\Psi(e^{it}) = \sum_{k=-\infty}^{\infty} \hat{\varphi}_k e^{ikt}, \quad t \in [0, 2\pi],$$

whence the Laurent series

$$\Psi(z) = \sum_{k=-\infty}^{\infty} \hat{\varphi}_k z^k, \quad z \in \Omega, \tag{18.26}$$

follows. From this we find

$$\chi(t) = \sum_{k=-\infty}^{\infty} \hat{\varphi}_k \rho^k e^{ikt}, \quad t \in [0, 2\pi], \tag{18.27}$$

as a parameterization of the unknown boundary ∂D. The series (18.27) exhibits the ill-posedness of the inverse problem, since small errors in the Fourier coefficients $\hat{\varphi}_k$ for $k < 0$ will be amplified by the exponentially increasing factors ρ^k. Therefore we incorporate a Tikhonov type regularization replacing χ by

$$\chi_\alpha(t) := \sum_{k=0}^{\infty} \hat{\varphi}_k \rho^k e^{ikt} + \sum_{k=1}^{\infty} \hat{\varphi}_{-k} \frac{\rho^k}{\alpha + \rho^{2k}} e^{-ikt} \tag{18.28}$$

where $\alpha > 0$ serves as a regularization parameter.

We conclude this short outline on the conformal mapping method by noting that it also has been used for the solution of an inverse scattering problem [82]. It also can be applied for the solution of the direct conformal mapping problem, namely the construction of a conformal mapping $\Psi : \Omega \to B \setminus \bar{D}$ for given B and D (see [145]).

18.3 Differentiability with Respect to the Boundary

We now turn to approximation methods of the first group mentioned at the end of Section 18.1. The foundation of iterative methods for the solution of the nonlinear operator equation (18.5) such as Newton methods requires the investigation of the differentiability of the boundary to data operator F with respect to the boundary and to characterize its derivative.

A mapping $M : U \subset X \to Y$ from an open subset of a normed space X into a normed space Y is said to be *Fréchet differentiable* at the point $\varphi \in U$ if there exists a bounded linear operator $M'(\varphi; \cdot) : X \to Y$ such that

$$\lim_{h \to 0} \frac{1}{\|h\|} \|M(\varphi + h) - M(\varphi) - M'(\varphi; h)\| = 0.$$

The bounded linear operator $M'(\varphi; \cdot)$ is called the *Fréchet derivative* of M at the point φ. The mapping M is called Fréchet differentiable on U if it is Fréchet differentiable for all $\varphi \in U$. Instead of $M'(\varphi; \cdot)$ we also will write M'_φ for the derivative at $\varphi \in U$. For an introduction to the concept of the Fréchet derivative and its basic properties we refer to Berger [17].

So far we have not specified a domain of definition for the operator F. For this, a parameterization of the boundary curve is required. For the sake of simplicity, as in Section 18.2, we will restrict our analysis to the case of starlike domains of the form (18.11) and consider F as a mapping from the open set $U := \left\{r \in C^2_{2\pi} : r > 0\right\}$ of positive functions in $C^2_{2\pi}$ into $C(\partial B)$. We will also write $F(r)$ synonymously for $F(\partial D)$. For notational convenience we associate with each 2π-periodic scalar function q a vector function h_q by setting

$$h_q(t) := q(t) \,(\cos t, \sin t), \quad t \in [0, 2\pi]. \tag{18.29}$$

We note that the function h_r corresponds to the parameterization (18.11) and maps $[0, 2\pi)$ bijectively onto ∂D, that is, we can use its inverse h_r^{-1}.

We now establish differentiability of F with respect to the boundary. For this we first need to examine the corresponding differentiability of the boundary integral operators used in the existence proof for the solution of the Dirichlet problem. We recall the double-layer approach for the exterior Dirichlet problem from Theorem 6.25 in the modified form (18.7) with the weight function ω and, as in the proof of Theorem 6.25, denote the corresponding integral operator by $\widetilde{K} : C(\partial D) \to C(\partial D)$. As in the proof of Lemma 18.5, via (18.11) and (18.29) we introduce its parameterized form

$$(L_r\psi)(t) = \int_0^{2\pi} k(t, \tau; r)\psi(\tau)\, d\tau, \quad t \in [0, 2\pi],$$

with kernel

$$k(t, \tau; r) = 2\,[h_r'(\tau)]^\perp \cdot \operatorname{grad}_y \Phi(h_r(t), h_r(\tau)) + 2, \quad t \neq \tau,$$

(see Problem 6.1). The subscript indicates the dependence of L_r on the boundary curve.

Theorem 18.7. *The mapping $r \mapsto L_r$ is Fréchet differentiable from $U \subset C^2_{2\pi}$ into the space $L(C_{2\pi}, C_{2\pi})$ of bounded linear operators from $C_{2\pi}$ into $C_{2\pi}$.*

Proof. For $t \neq \tau$, the Fréchet derivative of $r \mapsto k(\cdot\,, \cdot\,; r)$ is given by

$$k'(t, \tau; r, q) = 2[h_r'(\tau)]^\perp \cdot \Phi_y''(h_r(t), h_r(\tau))[h_q(\tau) - h_q(t)] + 2[h_q'(\tau)]^\perp \cdot \operatorname{grad}_y \Phi(h_r(t), h_r(\tau))$$

where $\Phi''_y(x, y)$ denotes the Hessian of $\Phi(x, y)$ with respect to y and the function $q \in C^2_{2\pi}$ need to be such that $r + q > 0$. To establish this, proceeding as in the proof of the estimate (18.18) in Lemma 18.5, straightforward calculations using Taylor's formula show that $k'(\cdot, \cdot; r, q)$ is continuous on $\mathbb{R}^2$ and 2π-periodic with respect to both variables and satisfies

$$\|k(\cdot, \cdot; r + q) - k(\cdot, \cdot; r) - k'(\cdot, \cdot; r, q)\|_\infty \le c\, \|q\|^2_{C^2_{2\pi}} \tag{18.30}$$

for all sufficiently small $\|q\|^2_{C^2_{2\pi}}$ and some constant c depending on r (see Problem 18.4). From this, it follows that the integral operator $L'_{r,q}$ with kernel $k'(\cdot, \cdot; r, q)$ satisfies

$$\|L_{r+q} - L_r - L'_{r,q}\|_\infty \le 2\pi c\, \|q\|^2_{C^2_{2\pi}},$$

and therefore the linear mapping $q \mapsto L'_{r,q}$ is the Fréchet derivative of $r \mapsto L_r$. □

The parameterized form of the double-layer operator is given by the integral operator

$$(V_r\psi)(x) = \int_0^{2\pi} v(x, \tau; r)\psi(\tau)\, d\tau, \quad x \in \mathbb{R}^2 \setminus \bar{D},$$

with kernel

$$v(x, \tau; r) := [h'_r(\tau)]^\perp \cdot \operatorname{grad}_y \Phi(x, h_r(\tau)) + 1, \quad x \in \mathbb{R}^2 \setminus \bar{D}, \tau \in [0, 2\pi].$$

We introduce the integral operator $V'_{r,q}$ with kernel

$$v'(x, \tau; r, q) = [h'_r(\tau)]^\perp \cdot \Phi''_y(x, h_r(\tau))\, h_q(\tau) + [h'_q(\tau)]^\perp \cdot \operatorname{grad}_y \Phi(x, h_r(\tau)).$$

Then, since the kernel of V_r has no singularities, straightforward differentiation shows that for closed subsets W of $\mathbb{R}^2 \setminus \bar{D}$ the Fréchet derivative of the mapping $r \mapsto V_r$ from $U \subset C^2_{2\pi}$ into the space $L(C_{2\pi}, C(W))$ of bounded linear operators from $C_{2\pi}$ into $C(W)$ is given by $q \mapsto V'_{r,q}$.

Theorem 18.8. *Assume that $\psi \in C^{1,\alpha}_{2\pi}$. Then $V'_{r,q}\psi$ is a bounded harmonic function in $\mathbb{R}^2 \setminus \bar{D}$ that is continuous in $\mathbb{R}^2 \setminus D$ with boundary values*

$$V'_{r,q}\psi + \left(h_q \circ h_r^{-1}\right) \cdot \operatorname{grad}(V_r\psi) = \frac{1}{2}\,(L'_{r,q}\psi) \circ h_r^{-1} \quad \text{on } \partial D. \tag{18.31}$$

Proof. From the form of the kernel $v'(\cdot, \cdot; r, q)$, it is obvious that $V'_{r,q}\psi$ is bounded at infinity and harmonic. To show that it is continuous up to the boundary ∂D and to compute its boundary values we use the vector identity

$$\operatorname{grad} \operatorname{div} w = \Delta w + [\operatorname{grad} \operatorname{div} w^\perp]^\perp,$$

Laplace's equation for Φ, and $\operatorname{grad}_y \Phi = -\operatorname{grad}_x \Phi$ to transform

$$[h_r'(\tau)]^\perp \cdot \Phi_y''(x, h_r(\tau))\, h_q(\tau) = [h_r'(\tau)]^\perp \cdot \operatorname{grad}_x \operatorname{div}_x\{\Phi(x, h_r(\tau))\, h_q(\tau)\}$$

$$= -\operatorname{div}_x \left\{ [h_q(\tau)]^\perp \, \frac{d}{d\tau}\, \Phi(x, h_r(\tau)) \right\}$$

(compare the proof of Theorem 7.32). From this, by partial integration, we find that

$$(V_{r,q}'\psi)(x) = \operatorname{div}_x \int_0^{2\pi} \Phi(x, h_r(\tau))\, [h_q(\tau)]^\perp \psi'(\tau)\, d\tau, \quad x \in \mathbb{R}^2 \setminus \bar{D}, \tag{18.32}$$

i.e., we can interpret $V_{r,q}'\psi$ as a derivative of a single-layer potential. Therefore, from Theorem 7.30 it follows that $V_{r,q}'\psi$ is continuous up to the boundary with boundary values

$$(V_{r,q}'\psi)(h_r(t)) = -\int_0^{2\pi} \operatorname{grad}_y \Phi(h_r(t), h_r(\tau)) \cdot [h_q(\tau)]^\perp \psi'(\tau)\, d\tau$$

$$-\frac{\psi'(t)}{2\,|h_r'(t)|}\, h_r'(t) \cdot h_q(t), \quad t \in [0, 2\pi].$$

Proceeding analogously for the gradient of the double-layer potential, i.e., from Theorem 7.32, we find that

$$h_q(t) \cdot \operatorname{grad}(V_r\psi)(h_r(t)) = \int_0^{2\pi} \operatorname{grad}_y \Phi(h_r(t), h_r(\tau)) \cdot [h_q(t)]^\perp \psi'(\tau)\, d\tau$$

$$+\frac{\psi'(t)}{2\,|h_r'(t)|}\, h_r'(t) \cdot h_q(t), \quad t \in [0, 2\pi].$$

Adding the last two equations and performing an analogous partial integration in the integral representing $L_{r,q}'\psi$ now yields (18.31). For the latter partial integration one has to take proper care of the singularity at $t = \tau$. □

Note that (18.31) can be interpreted as formal differentiation of the jump relation $2V_r\psi = (I + L_r)\psi$ with respect to r. Analogously, the boundary condition (18.33) in the following theorem can be obtained by formally differentiating the boundary condition (18.2) with respect to r.

Theorem 18.9. *The boundary to data operator $F : r \mapsto u|_{\partial B}$ is Fréchet differentiable from $U \subset C_{2\pi}^2$ into $C(\partial B)$. The Fréchet derivative is given by*

$$F'(r; q) = w|_{\partial B},$$

where $w \in C^2(\mathbb{R}^2 \setminus \bar{D}) \cap C(\mathbb{R}^2 \setminus D)$ denotes the uniquely determined bounded harmonic function satisfying the Dirichlet boundary condition

$$w = -\nu \cdot \left(h_q \circ h_r^{-1}\right) \frac{\partial}{\partial \nu} \{\Phi(\cdot\,,\zeta) + u\} \quad \text{on } \partial D. \tag{18.33}$$

Proof. In view of Theorem 6.25, we write the solution to (18.1)–(18.2) in the form

$$u = -2V_r(I + L_r)^{-1} f_r \quad \text{in } \mathbb{R}^2 \setminus \bar{D},$$

where $f_r(t) := \Phi(h_r(t), \zeta)$. We note that differentiability of $r \mapsto L_r$ implies differentiability of $r \mapsto (I + L_r)^{-1}$ with the derivative given by $-(I + L_r)^{-1} L'_{r,q}(I + L_r)^{-1}$. Then, using the chain rule for the Fréchet derivative, we obtain

$$u' = -2V'_{r,q}(I + L_r)^{-1} f_r + 2V_r(I + L_r)^{-1} L'_{r,q}(I + L_r)^{-1} f_r - 2V_r(I + L_r)^{-1} b_{r,q}$$

in $\mathbb{R}^2 \setminus \bar{D}$, where $b_{r,q}(t) = h_q(t) \cdot \operatorname{grad} \Phi(h_r(t), \zeta)$ represents the Fréchet derivative of the boundary values with respect to the boundary, i.e., of $r \mapsto f_r$. From this, it is obvious that u' is bounded and harmonic. To derive the boundary condition (18.33) we use the jump relation $2V_r(I + L_r)^{-1} = I$ on ∂D for the double-layer potential. Now (18.33) follows, since (18.31) implies that

$$-2V'_{r,q}(I + L_r)^{-1} f_r = -L'_{r,q}(I + L_r)^{-1} f_r - h_q \cdot (\operatorname{grad} u \circ h_r).$$

This completes the proof, provided we justify the application of (18.31) by showing that $(I + L_r)^{-1} f_r \in C^{1,\alpha}_{2\pi}$.

From the proofs of Theorems 7.5 and 7.6 we know that the double-layer integral operator K maps $C(\partial D)$ boundedly into $C^{0,\alpha}(\partial D)$ and that it maps $C^{0,\alpha}(\partial D)$ boundedly into $C^{1,\alpha}(\partial D)$. Then, from $\Phi(\cdot\,,\zeta)|_{\partial D} \in C^{1,\alpha}(\partial D)$, it follows that the solution ψ of the integral equation $\psi + \widetilde{K}\psi = 2\Phi(\cdot\,,\zeta)$ is in $C^{1,\alpha}(\partial D)$. □

The double-layer potential operator $\widetilde{K}$ from the proof of Theorem 6.25 in the modified form (18.7) has a continuous kernel (in two dimensions). Therefore it is compact from $L^2(\partial D)$ into itself and $(I + \widetilde{K})^{-1} : L^2(\partial D) \to L^2(\partial D)$ exists and is bounded. Denote by $V : L^2(\partial D) \to L^2(\partial B)$ the operator that takes densities on ∂D onto the trace on ∂B of their double-layer potential (and has the parameterized form V_r as used above). Then we can express

$$F'(r; q) = -2V(I + \widetilde{K})^{-1} \left(\nu \cdot \left(h_q \circ h_r^{-1}\right) \frac{\partial}{\partial \nu} \{\Phi(\cdot\,,\zeta) + u\} \right). \tag{18.34}$$

From this representation we observe that $F'(r; \cdot)$ can be extended as a bounded linear operator from $L^2[0, 2\pi]$ into $L^2(\partial B)$. For this extension we have the following theorem.

Theorem 18.10. *The linear operator $F'(r; \cdot) : L^2[0, 2\pi] \to L^2(\partial B)$ is injective and has dense range.*

Proof. Relabeling the curves in Theorem 18.2, we observe that V is injective and has dense range. Assume that $F'(r;q) = 0$. Then from (18.34) and the injectivity of V and $(I + \widetilde{K})^{-1}$ we conclude that

$$\nu \cdot \left(h_q \circ h_r^{-1}\right) \frac{\partial}{\partial \nu} \{\Phi(\cdot\,,\zeta) + u\} = 0 \quad \text{on } \partial D.$$

This in turn implies $\nu \circ h_r \cdot h_q = 0$ since the normal derivative $\partial \{\Phi(\cdot\,,\zeta) + u\} / \partial \nu$ cannot vanish on open subsets of ∂D as a consequence of Holmgren's Theorem 6.7 and the boundary condition $\Phi(\cdot\,,\zeta) + u = 0$ on ∂D. Elementary calculations yield

$$\nu \circ h_r \cdot h_q = \frac{rq}{\sqrt{r^2 + [r']^2}} \tag{18.35}$$

and therefore $(\nu \circ h_r) \cdot h_q = 0$ implies that $q = 0$.

The above arguments also imply that the set

$$\left\{ (\nu \circ h_r) \cdot h_q \, \frac{\partial}{\partial \nu} \{\Phi(\cdot\,,\zeta) + u\} \circ h_r : q \in L^2[0, 2\pi] \right\}$$

is dense in $L^2[0, 2\pi]$. Then using the dense range of V and the surjectivity of $(I+\widetilde{K})^{-1}$ from (18.34) we can conclude that the range of $F'(r;\cdot)$ is dense. □

For the extension of the above analysis on Fréchet differentiability to the Helmholtz equation, we refer to [32, 194].

18.4 Iterative Methods

The Theorems 18.9 and 18.10 serve as theoretical foundation for the application of regularized Newton methods and related iteration schemes for the approximate solution of (18.5). In this method, given $u|_{\partial B}$, the nonlinear equation

$$F(r) = u|_{\partial B}$$

is replaced by its linearization

$$F(r) + F'(r;q) = u|_{\partial B}, \tag{18.36}$$

which has to solved for q in order to improve an approximate boundary given by r into a new approximation given by $\widetilde{r} = r + q$. As usual, Newton's method consists of iterating this procedure. From the form (18.34) of the Fréchet derivative it can be seen that the linear operator $q \mapsto F'(r;q)$ is compact from $L^2[0, 2\pi]$ into $L^2(\partial B)$. Therefore, for the solution of the linear equation (18.36) a regularization has to be incorporated, such as Tikhonov regularization. Injectivity and dense range of $F'(r;\cdot)$ as prerequisites for regularization schemes are guaranteed by Theorem 18.10.

For practical computations q is taken from a finite-dimensional subspace U_n of $L^2[0, 2\pi]$ with dimension n and equation (18.36) is approximately solved by collocating it at m points $x_1, \dots, x_m \in \partial B$. Then writing

$$q = \sum_{k=1}^{n} a_k q_k$$

where $q_1, \dots, q_n$ denotes a basis of U_n, one has to solve the linear system

$$\sum_{k=1}^{n} a_k \,(F'(r; q_k))(x_j) = u(x_j) - (F(r))(x_j), \quad j = 1, \dots, m, \tag{18.37}$$

for the real coefficients $a_1, \dots, a_n$. In general, i.e., when $m > n$, the system (18.37) is overdetermined and has to be solved approximately by a least squares method. In addition, since we have to stabilize the ill-posed linearized equation (18.36), we replace (18.37) by the least squares problem of minimizing the penalized defect

$$\sum_{j=1}^{m} \left| \sum_{k=1}^{n} a_k \,(F'(r; q_k))(x_j) - u(x_j) + (F(r))(x_j) \right|^2 + \alpha \sum_{k=1}^{n} a_k^2 \tag{18.38}$$

with some regularization parameter $\alpha > 0$, that is, we employ a Tikhonov regularization in the spirit of the Levenberg–Marquardt algorithm (see [173]). Assuming that the basis functions $q_1, \dots, q_n$ are orthonormal in $L^2[0, 2\pi]$, for example trigonometric monomials, the penalty term in (18.38) corresponds to L^2 penalization. However, numerical evidence strongly suggests to replace the L^2 penalization by a Sobolev penalization, i.e., by considering $F'(r; \cdot))$ as an operator from $H^p[0, 2\pi]$ into $L^2(\partial B)$ for $p = 1$ or $p = 2$ (see also Section 16.5).

In order to compute the right-hand sides of the linear system (18.37), in each iteration step the direct Dirichlet problem for the boundary ∂D given by the radial function r has to be solved for the evaluation of $(F(r))(x_j)$. For this, we can numerically solve the integral equation of the second kind from Theorem 6.25 via the Nyström method of Section 12.2 using the composite trapezoidal rule. Then the values at the collocation points can be obtained via numerical quadrature with the composite trapezoidal rule for the modified double-layer potential of Theorem 6.25. For the evaluation of the normal derivative entering the boundary condition (18.33) for the Fréchet derivatives one can use the approximation for the normal derivative of the double-layer potential as described in Section 13.5. To compute the matrix entries $(F'(r; q_k))(x_j)$ we need to solve n additional direct Dirichlet problems for the same boundary ∂D and different boundary values given by (18.33) for the basis functions $q = q_j$, $j = 1, \dots, n$, that is, we have to solve the linear system set up already for the evaluation of $(F(r))(x_j)$ again for n different right-hand sides.

For numerical examples for related inverse Dirichlet problems for the Helmholtz equation we refer to [96, 124, 141, 144] in the two-dimensional case and to [51, 88] in the three-dimensional case.

In closing our analysis on Newton iterations for the boundary to data operator F we note as main advantages, which also hold in more general situations, that this approach is conceptually simple and, as numerical examples indicate, leads to highly accurate reconstructions with reasonable stability against errors in the data. On the other hand, it should be noted that for the numerical implementation an efficient forward solver is needed for the solution of the corresponding direct boundary value problem for each iteration step. Furthermore, good a priori information is required in order to be able to choose an initial guess that ensures numerical convergence. In addition, on the theoretical side, although some progress has been made through the work of Hohage [97] and Potthast [195] the convergence of regularized Newton iterations for the operator F has not been completely settled. At the time this is being written it remains an open question whether the general convergence results on Newton type methods such as the Levenberg–Marquardt algorithm or the iteratively regularized Gauss–Newton iterations as available in the literature [22, 115] are applicable to inverse boundary value problems.

For the following discussion on modified Newton type iterations with reduced computational costs, for variety and in order to avoid technical difficulties due to the boundedness condition for the exterior Dirichlet problem in two dimensions, we consider the Neumann boundary condition (18.3). We also replace the potential approach by the direct approach based on Theorem 7.37, i.e, on Green's integral theorem and Green's representation formula. Adding Green's representation formula (6.12) applied to the solution u of the boundary value problem and Green's second integral theorem applied to $\Phi(\cdot\,,\zeta)$ and observing the boundary condition (18.3) we obtain that

$$v(x) = \Phi(x,\zeta) + \int_{\partial D} v(y)\,\frac{\partial\Phi(x,y)}{\partial\nu(y)}\,ds(y), \quad x \in \mathbb{R}^2 \setminus \bar{D},\ x \neq \zeta, \tag{18.39}$$

for the total field $v := u + \Phi(\cdot\,,\zeta)$. From this, letting x tend to the boundary, by the jump relation for the double-layer potential we arrive at the integral equation of the second kind

$$\varphi(x) - 2\int_{\partial D} \varphi(y)\,\frac{\partial\Phi(x,y)}{\partial\nu(y)}\,ds(y) = 2\Phi(x,\zeta), \quad x \in \partial D, \tag{18.40}$$

for the unknown boundary values $\varphi := v|_{\partial D}$. Given ∂D, by Theorem 6.21 and the Riesz theory the integral equation (18.40) is uniquely solvable. Evaluating (18.39) on ∂B we obtain that

$$\int_{\partial D} \varphi(y)\,\frac{\partial\Phi(x,y)}{\partial\nu(y)}\,ds(y) = u(x), \quad x \in \partial B. \tag{18.41}$$

In terms of our inverse problem, we now can interpret (18.40) and (18.41) as a system of two integral equations for the unknown boundary ∂D and the unknown bounday values φ of the total field v on ∂D. For the sequel it is convenient to call (18.41) the *data equation* since it contains the given trace of u on ∂B for the inverse problem and (18.40) the *field equation* since it represents the boundary condition.

Both equations are linear with respect to φ and nonlinear with respect to ∂D. Equation (18.41) is severely ill-posed (see Sections 16.6 and 18.2) whereas (18.40) is well-posed.

Obviously there are three options for an iterative solution of (18.40) and (18.41). In a first method, given an approximation for the boundary ∂D one can solve the well-posed integral equation of the second kind (18.40) for φ. Then, keeping φ fixed, equation (18.41) is linearized with respect to ∂D to update the boundary approximation. In a second approach, one also can solve the system (18.40) and (18.41) simultaneously for ∂D and φ by Newton iterations, i.e., by linearizing both equations with respect to both unknowns. Whereas in the first method the burden of the ill-posedness and nonlinearity is put on one equation, in a third method a more even distribution of the difficulties is obtained by reversing the roles of (18.40) and (18.41), i.e., by solving the severely ill-posed equation (18.41) for φ via regularization and then linearizing (18.40) to obtain the boundary update.

For a more detailed description of these ideas, using the parameterization (18.11) for starlike ∂D and recalling the mapping h_r from (18.29), we introduce the parameterized operators $A_1, A_2 : C^2_{2\pi} \times L^2[0, 2\pi] \to L^2[0, 2\pi]$ by

$$A_1(r, \psi)(t) := \int_0^{2\pi} 2\,[h_r'(\tau)]^\perp \cdot \operatorname{grad}_y \Phi(h_r(t), h_r(\tau))\psi(\tau)\, d\tau$$

and

$$A_2(r, \psi)(t) := \int_0^{2\pi} [h_r'(\tau)]^\perp \cdot \operatorname{grad}_y \Phi(\eta(t), h_r(\tau))\psi(\tau)\, d\tau$$

for $t \in [0, 2\pi]$, where we parametrized

$$\partial B = \{\eta(t) : t \in [0, 2\pi]\}.$$

Of course, the operators A_1 and A_2 are a renaming of the operators L_r (without the added constant) and V_r from Theorems 18.7 and 18.8. Then the equations (18.40) and (18.41) can be written in the operator form

$$\psi - A_1(r, \psi) = 2u^i \circ h_r \tag{18.42}$$

and

$$A_2(r, \psi) = u \circ \eta \tag{18.43}$$

for $\psi = \varphi \circ h_r$, where we have set $u^i := \Phi(\cdot\,, \zeta)$ indicating the incident field. The linearization of these equations requires the Fréchet derivatives of the operators A_1 and A_2 with respect to r which are given by Theorems 18.7 and 18.8, i.e.,

$$\begin{aligned}(A_1'(r, \psi; q))(t) = 2 \int_0^{2\pi} & [h_r'(\tau)]^\perp \cdot \Phi_y''(h_r(t), h_r(\tau))\,[h_q(\tau) - h_q(t)]\,\psi(\tau)\, d(\tau) \\ +2 \int_0^{2\pi} & [h_q'(\tau)]^\perp \cdot \operatorname{grad}_y \Phi(h_r(t), h_r(\tau))\,\psi(\tau)\, d(\tau)\end{aligned} \tag{18.44}$$

and

$$
\begin{aligned}
(A_2'(r,\psi;q))(t) = &\int_0^{2\pi} [h_r'(\tau)]^\perp \cdot \Phi_y''(\eta(t),h_r(\tau))\, h_q(\tau)\,\psi(\tau)\, d(\tau) \\
&+ \int_0^{2\pi} [h_q'(\tau)]^\perp \cdot \operatorname{grad}_y \Phi(\eta(t),h_r(\tau))\,\psi(\tau)\, d(\tau)
\end{aligned}
\tag{18.45}
$$

for $t \in [0,2\pi]$. Note that as opposed to the Fréchet derivative of the boundary to data operator F as given in Theorem 18.9 the derivatives of A_1 and A_2 are given in an explicit form as integral operators which offers computational advantages.

Now, given an approximation to the boundary parameterization r, the field equation (18.42) can be solved for the density ψ. Then, keeping ψ fixed, linearizing the data equation (18.43) with respect to r leads to the linear equation

$$
A_2'(r,\psi;q) = u \circ \eta - A_2(r,\psi) \tag{18.46}
$$

for q to update the radial function r via $r + q$ using the data $u \circ \eta$. This procedure can be iterated. For fixed r and ψ the operator $A_2'(r,\psi;\cdot)$ has a smooth kernel and therefore is severely ill-posed. This requires stabilization, for example via Tikhonov regularization. Injectivity of the linear operator $A_2'(r,\psi;\cdot)$ as the main prerequisite for regularization schemes is guaranteed by the following theorem.

Theorem 18.11. *Assume that r and ψ satisfy the field equation (18.42). Then the linear operator $A_2'(r,\psi;\cdot) : L^2[0,2\pi] \to L^2[0,2\pi]$ is injective.*

Proof. Analogous to the proof of Theorem 18.9 it can be seen that $\psi \in C^{1,\alpha}_{2\pi}$. Assume that $A_2'(r,\psi;q) = 0$. In view of (18.35) we can rewrite $h_q = \tilde{q}\nu \circ h_r$ with some scalar function $\tilde{q}$ such that $q = 0$ if and only if $\tilde{q} = 0$ (see also the concept of parallel curves on p. 85). Then we have

$$
\operatorname{div}_x \left\{\Phi(x,h_r(\tau))\,[h_q(\tau)]^\perp\right\} = \tilde{q}(\tau)\,\frac{\partial}{\partial\tau}\,\Phi(x,h_r(\tau))
$$

and from (18.32) we obtain that

$$
(A_2'(r,\psi;q))(t) = \int_0^{2\pi} \tilde{q}(\tau)\psi'(\tau)\,\frac{\partial}{\partial\tau}\,\Phi(\eta(\tau),h_r(\tau))\,d\tau, \quad t \in [0,2\pi].
$$

Now for the bounded harmonic function w in $\mathbb{R}^2 \setminus \bar{D}$ defined by

$$
w(x) := \int_0^{2\pi} \tilde{q}(\tau)\psi'(\tau)\,\frac{\partial}{\partial\tau}\,\Phi(x,h_r(\tau))\,d\tau, \quad x \in \mathbb{R}^2 \setminus \bar{D},
$$

by the uniqueness for the exterior Dirichlet problem and the analyticity of harmonic functions from $A_2'(r,\psi;q) = 0$, that is, from $w = 0$ on ∂B we can conclude that $w = 0$ in $\mathbb{R}^2 \setminus \bar{D}$. Since the set of functions $\{\Phi(x,\cdot) : x \in \mathbb{R}^2 \setminus \bar{D}\}$ is complete in $L^2(\partial D)$, from $w = 0$ in $\mathbb{R}^2 \setminus \bar{D}$ we deduce that the product $\tilde{q}\psi'$ has weak derivative identically equal to zero. Therefore $\tilde{q}\psi' = c$ in $[0, 2\pi]$ for some constant c and since the derivative of the 2π-periodic function ψ must have at least one zero, it follows that $c = 0$, that is, $\tilde{q}\psi' = 0$ in $[0, 2\pi]$.

Now assume that ψ is constant on some open subinterval of $[0, 2\pi]$, that is, v is constant on an open subset Γ of ∂D. Since $\partial v/\partial \nu = 0$ on ∂D for the total field v, from Holmgren's Theorem 6.7 we would have that v is constant in all of D which contradicts its singularity at the source point ζ. Hence we have $\tilde{q} = 0$ and consequently $q = 0$ which ends the proof. □

This approach for solving an inverse boundary value problem has been proposed by Johansson and Sleeman [111] first for an inverse obstacle scattering problem, i.e., an inverse boundary value problem for the Helmholtz equation. It can be related to the Newton iterations for the boundary to data operator F. From the derivation of (18.42) and (18.43) we have that

$$F(r) = A_2\Big(r, [I - A_1(r, \cdot)]^{-1}(2u^i \circ h_r)\Big). \tag{18.47}$$

By the product and chain rule this implies Fréchet differentiability of the boundary to data map F also in the case of the Neumann boundary condition (18.3) with the derivative given by

$$\begin{aligned} F'(r; q) = {} & A_2'\Big(r, [I - A_1(r, \cdot)]^{-1}(2u^i \circ h_r); q\Big) \\ & + A_2\Big(r, [I - A_1(r, \cdot)]^{-1} A_1'\Big(r, [I - A_1(r, \cdot)]^{-1}(2u^i \circ h_r); q\Big)\Big) \\ & + A_2\Big(r, [I - A_1(r, \cdot)]^{-1}(2(\operatorname{grad} u^i \circ h_r) \cdot h_q)\Big). \end{aligned} \tag{18.48}$$

Hence, the iteration scheme (18.46) can be interpreted as Newton iterations for F with the derivative of F approximated by the first term in the representation (18.48). As to be expected from this close relation, the quality of the reconstructions via (18.46) can compete with those of Newton iterations for the boundary to data operator F with the benefit of reduced computational costs.

Without proof we note that it is also possible to characterize the Fréchet derivative in the Neumann case analogous to Theorem 18.9 via $F'(r; q) = w \circ \eta$, where $w \in C^2(\mathbb{R}^2 \setminus \bar{D}) \cap C(\mathbb{R}^2 \setminus D)$ denotes the uniquely determined harmonic function vanishing at infinity and satisfying the Neumann boundary condition

$$\frac{\partial w}{\partial \nu} = \frac{d}{ds}\left(\nu \cdot \left(h_q \circ h_r^{-1}\right) \frac{d}{ds}\{\Phi(\cdot\,, \zeta) + u\}\right) \quad \text{on } \partial D, \tag{18.49}$$

where d/ds denotes differentiation with respect to arc length (see [93]). In particular, analogous to (18.34) this implies the representation

$$(F'(r; q)) \circ \eta^{-1} = 2V(I - K)^{-1}\left[\frac{d}{ds}\left(\nu \cdot \left(h_q \circ h_r^{-1}\right) \frac{d}{ds}\{\Phi(\cdot\,, \zeta) + u\}\right)\right] \tag{18.50}$$

and proceeding as in the proof of Theorem 18.10 this can be used to show that the linear operator $F'(r;\cdot) : L^2[0,2\pi] \to L^2[0,2\pi]$ is injective and has dense range also in the Neumann case.

Following ideas first developed by Kress and Rundell [149] for the inverse Dirichlet problem that we introduced at the end of Section 18.2 on p. 376, a second approach for iteratively solving the system (18.42) and (18.43) consists in simultaneously linearizing both equations with respect to both unknowns. In this case, given approximations r and ψ both for the boundary parameterization and the Dirichlet boundary values, the system of linear equations

$$-A_1'(r,\psi;q) - 2(\operatorname{grad} u^i \circ h_r)\cdot h_q + \chi - A_1(r,\chi) = -\psi + A_1(r,\psi) + 2u^i \circ h_r \quad (18.51)$$

and

$$A_2'(r,\psi;q) + A_2(r,\chi) = -A_2(r,\psi) + u \circ \eta \quad (18.52)$$

has to be solved for q and χ in order to obtain updates $r+q$ for the boundary parameterization and $\psi+\chi$ for the boundary values. This procedure again is iterated and coincides with traditional Newton iterations for the system (18.42) and (18.43). It has been analyzed including numerical examples by Ivanyshyn and Kress [108] for the inverse Neumann problem and also for inverse obstacle scattering problems (see [107, 109]). Due to the smoothness of the kernels in the second equation, the system (18.51) and (18.52) is severely ill-posed and requires regularization with respect to both unknowns.

The simultaneous iterations (18.51) and (18.52) again exhibit connections to the Newton iteration for (18.5) as expressed through the following theorem.

Theorem 18.12. *For fixed r let ψ be the unique solution of the field equation*

$$\psi - A_1(r,\psi) = 2u^i \circ h_r. \quad (18.53)$$

If q satisfies the linearized boundary to data equation

$$F'(r;q) = u \circ \eta - F(r) \quad (18.54)$$

then q and

$$\chi := [I - A_1(r,\cdot)]^{-1}\left(A_1'(r,\psi;q) + 2(\operatorname{grad} u^i \circ h_r)\cdot h_q\right) \quad (18.55)$$

satisfy the linearized data and field equations (18.51) and (18.52). Conversely, if q and χ satisfy (18.51) and (18.52) then q satisfies (18.54).

Proof. If q satisfies (18.54), from (18.47) and the equation (18.53) for ψ we have

$$F'(r;q) = u \circ \eta - A_2(r,\psi).$$

In view of (18.55), the representation (18.48) of the derivative of F yields

$$F'(r,q) = A_2'(r,\psi;q) + A_2(r,\chi)$$

and combining this with the previous equation establishes that (18.52) holds. From the definition (18.55) of χ we observe

$$\chi - A_1(r,\chi) - A_1'(r,\psi;q) - 2(\operatorname{grad} u^i \circ h_r)\cdot h_q = 0$$

Therefore, in view of (18.53) we also have that (18.51) is satisfied.

Conversely, the first equation (18.51) implies that χ can be expressed by (18.55) and inserting this into (18.52) leads to

$$A_2'(r,\psi;q) + A_2(r,[I - A_1(r,\cdot)]^{-1}\left(A_1'(r,\psi;q)+2(\operatorname{grad} u^i \circ h_r)\cdot h_q\right) = -A_2(r,\psi) + u\circ\eta$$

and via (18.48) this implies (18.54). □

Theorem 18.12 illustrates the difference between the iteration method based on (18.51) and (18.52) and the Newton iterations (18.54) for the boundary to data operator F. In general when performing (18.51) and (18.52) in the sequence of updates the relation (18.53) between the approximations r and ψ for the parameterization and the boundary values will not be satisfied. This observation also indicates a possibility to use (18.51) and (18.52) for implementing the Newton scheme (18.54) numerically. It is only necessary to replace the update $\psi+\chi$ for the boundary values by $2[I - A_1(r+q,\cdot)]^{-1}(u^i \circ (h_r + h_q))$, i.e., at the expense of throwing away χ and solving the field equation for the updated boundary with representation $r+q$ for new boundary values.

The third possibility, namely given an approximation r for the boundary parameterization first solving a regularized version of the severely ill-posed equation (18.41) for φ and then linearizing (18.40) to obtain the boundary update has not yet been explored in the literature. However, the hybrid method of Section 18.2 can be viewed as a modification of this approach (see [110]).

18.5 Sampling Methods

For the remainder of this chapter we explicitly indicate the dependence of the solution to (18.1)–(18.2) on the source point by considering the bounded solution to the Laplace equation

$$\Delta u(\cdot\,,y) = 0 \quad \text{in } \mathbb{R}^2 \setminus \bar{D} \tag{18.56}$$

satisfying the Dirichlet boundary condition

$$u(\cdot\,,y) = -\Phi(\cdot\,,y) \quad \text{on } \partial D \tag{18.57}$$

for $y \in \partial B$ and investigate the inverse boundary problem of determining ∂D from a knowledge of $u(x,y)$ for all $x, y \in \partial B$. Since by Theorem 18.1 the knowledge of $u(x,y)$ for all $x \in \partial B$ and only one $y \in \partial B$ already uniquely determines ∂D, the inverse problem with data for all $x, y \in \partial B$ is overdetermined.

Our subsequent analysis is based on the integral operator $U : L^2(\partial B) \to L^2(\partial B)$ with kernel given by the data of the inverse problems, i.e., U is defined by

$$(Ug)(x) := -\int_{\partial B} u(x, y)g(y)\, ds(y), \quad x \in \partial B. \tag{18.58}$$

From the well-posedness of the exterior Dirichlet problem (Theorem 6.30) we deduce that the solution u of (18.56)–(18.57) is continuous on $\partial B \times \partial B$. Hence, U has a continuous kernel and therefore is compact. In addition, we introduce an operator $H : L^2(\partial B) \to L^2(\partial D)$ by the single-layer potential

$$(Hg)(x) := \int_{\partial B} \Phi(x, y)g(y)\, ds(y), \quad x \in \partial D,$$

which also is compact. The adjoint $H^* : L^2(\partial D) \to L^2(\partial B)$ of H is given by

$$(H^*\varphi)(x) = \int_{\partial D} \Phi(x, y)\varphi(y)\, ds(y), \quad x \in \partial B.$$

By $R : v|_{\partial D} \mapsto v|_{\partial B}$we denote the operator that maps the boundary values on ∂D of bounded harmonic functions $v \in C(\mathbb{R}^2 \setminus D) \cap C^2(\mathbb{R}^2 \setminus \bar{D})$ onto their restriction on ∂B. From the discussion of the representation of the Fréchet derivative (18.34) we know already that we can write $R = 2V(I + \widetilde{K})^{-1}$, where V represents the modified double-layer potential (6.41) on ∂D and $\widetilde{K}$ is the modified double-layer integral operator as used in Theorem 6.25, and that R can be extended as a bounded injective operator $R : L^2(\partial D) \to L^2(\partial B)$ with dense range. For the solution of (18.56)–(18.57) we then can write

$$-u(\cdot\,, y)|_{\partial B} = R\Phi(\cdot\,, y)|_{\partial D}$$

for all $y \in \partial B$. From this, multiplying by $g \in L^2(\partial B)$ and integrating over ∂B, in view of the well-posedness of the exterior Dirichlet problem and the continuity of R we deduce that $Ug = RHg$. Hence we have the factorization

$$U = RH. \tag{18.59}$$

Now, for a constant unit vector e we define

$$\Psi(x, z) := e \cdot \operatorname{grad}_x \Phi(x, z), \quad z \in B, x \in \mathbb{R}^2 \setminus \{z\}, \tag{18.60}$$

and consider the integral equation of the first kind

$$Ug_z = \Psi(\cdot\,, z)|_{\partial B} \tag{18.61}$$

for a parameter point $z \in D$. Assume that $g_z \in L^2(\partial B)$ is a solution of (18.61). Then the single-layer potential

$$v_z(x) := \int_{\partial B} \Phi(x.y)g_z(y)\, ds(y), \quad x \in B,$$

with density g_z is harmonic in B and as consequence of (18.59) satisfies

$$Rv_z|_{\partial D} = RHg_z = \Psi(\cdot\,, z)|_{\partial B}.$$

From this, by the uniqueness for the exterior Dirichlet problem and analyticity we obtain that

$$v_z = \Psi(\cdot, z) \quad \text{on } \partial D. \tag{18.62}$$

From the latter we conclude that $\|g_z\|_{L^2(\partial B)} \to \infty$ if the point z approaches the boundary ∂D. Therefore, in principle, the boundary ∂D may be found by solving the integral equation (18.61) for z taken from a sufficiently fine grid in B and determining ∂D as the location of those points z where $\|g_z\|_{L^2(\partial B)}$ becomes large. However, in general, the unique solution of the interior Dirichlet problem in D with boundary condition (18.62) will have an extension as a harmonic function across the boundary ∂D only in special cases. Hence, the integral equation of the first kind (18.61), in general, will not have a solution. Nevertheless, a mathematical foundation of the above method can be provided. We will denote it as the *linear sampling method* since it is based on ideas of the linear sampling method in inverse obstacle scattering as first proposed by Colton and Kirsch [30].

However, before we discuss the linear sampling method any further, we first consider a factorization method motivated by Kirsch's [125] factorization method from inverse obstacle scattering. To this end we need to introduce the square root of a positive definite compact operator.

Theorem 18.13. *Let $A : X \to X$ be a compact and positive definite operator in a Hilbert space X. Then there exists a uniquely defined compact and positive definite operator $A^{1/2} : X \to X$ such that $A^{1/2}A^{1/2} = A$.*

Proof. In terms of the spectral decomposition from Theorem 15.12 in the form (15.13) we define

$$A^{1/2}\varphi := \sum_{n=1}^{\infty} \sqrt{\lambda_n}\,(\varphi, \varphi_n)\varphi_n \tag{18.63}$$

for $\varphi \in X$ with the positive square roots of the positive eigenvalues λ_n of A with orthonormal eigenelements φ_n. Then $A^{1/2} : X \to X$ is compact by Theorem 2.22 since for the partial sums $A_m^{1/2}\varphi := \sum_{n=1}^{m} \sqrt{\lambda_n}\,(\varphi, \varphi_n)\varphi_n$ we have norm convergence $\|A_m^{1/2} - A^{1/2}\| = \sqrt{\lambda_{m+1}} \to 0$ as $m \to \infty$. From

$$(A^{1/2}\varphi, \psi) = \sum_{n=1}^{\infty} \sqrt{\lambda_n}\,(\varphi, \varphi_n)(\varphi_n, \psi)$$

for all $\varphi, \psi \in X$ we observe that $A^{1/2}$ is positive definite. Clearly $A^{1/2}\varphi_n = \sqrt{\lambda_n}\,\varphi_n$ for $n \in \mathbb{N}$ and therefore, in view of the self-adjointness of $A^{1/2}$, we obtain

$$A^{1/2}A^{1/2}\varphi = \sum_{n=1}^{\infty} \sqrt{\lambda_n}\,(A^{1/2}\varphi, \varphi_n)\varphi_n = \sum_{n=1}^{\infty} \lambda_n\,(\varphi, \varphi_n)\varphi_n = A\varphi$$

for all $\varphi \in X$. From the fact that any operator B with the required properties of a square root satisfies $B^2 = A$ it follows that the eigenvalues of B are given by $B\varphi_n = \sqrt{\lambda_n}\,\varphi_n$. Therefore by the spectral theorem B must be of the form (18.63). □

Theorem 18.14. *Let $A : X \to X$ be a compact and positive definite operator and for $\alpha > 0$ let φ_α denote the Tikhovov regularized solution for the equation $A\varphi = f$ for $f \in X$. Then $\lim_{\alpha\to 0}(\varphi_\alpha, f)$ exists if and only if $f \in A^{1/2}(X)$. If $f = A^{1/2}g$ for $g \in X$ then*

$$\lim_{\alpha\to 0}(\varphi_\alpha, f) = \|g\|^2.$$

Proof. From the representation for φ_α given in the proof of Theorem 15.23 we obtain that

$$(\varphi_\alpha, f) = \sum_{n=1}^{\infty} \frac{\lambda_n}{\alpha + \lambda_n^2}\, |(f, \varphi_n)|^2 \tag{18.64}$$

in terms of a singular system $(\lambda_n, \varphi_n, \varphi_n)$ of A. Now assume that $f = A^{1/2}g$ for some $g \in X$. Then $(f, \varphi_n) = (g, A^{1/2}\varphi_n) = \sqrt{\lambda_n}\,(g, \varphi_n)$ and consequently

$$(\varphi_\alpha, f) = \sum_{n=1}^{\infty} \frac{\lambda_n^2}{\alpha + \lambda_n^2}\, |(g, \varphi_n)|^2 \to \sum_{n=1}^{\infty} |(g, \varphi_n)|^2 = \|g\|^2$$

as $\alpha \to 0$.

Conversely, assume that $\lim_{\alpha\to 0}(\varphi_\alpha, f)$ exists. Then from (18.64) we observe that there exists a constant $M > 0$ such that

$$\sum_{n=1}^{\infty} \frac{\lambda_n}{\alpha + \lambda_n^2}\, |(f, \varphi_n)|^2 \leq M$$

for all $\alpha > 0$. Passing to the limit $\alpha \to 0$, this implies that

$$\sum_{n=1}^{\infty} \frac{1}{\lambda_n}\, |(f, \varphi_n)|^2 \leq M.$$

Hence,

$$g := \sum_{n=1}^{\infty} \frac{1}{\sqrt{\lambda_n}}\, (f, \varphi_n)\varphi_n \in X$$

is well defined and $A^{1/2}g = f$. □

Theorem 18.15. *Let X and Z be Hilbert spaces and let $A : X \to X$ and $C : Z \to X$ be compact linear operators such that $A = CC^*$ with the adjoint $C^* : X \to Z$ of C. Assume further that A is positive definite and C is injective. Then*

$$A^{1/2}(X) = C(Z).$$

Proof. Since the operator A is compact and positive definite, by the spectral Theorem 15.12 there exists a complete orthonormal system $\varphi_n \in X$ of eigenelements of A with positive eigenvalues λ_n, i.e.,

$$A\varphi_n = \lambda_n \varphi_n, \quad n = 1, 2, \ldots.$$

Then the elements

$$g_n := \frac{1}{\sqrt{\lambda_n}} C^* \varphi_n, \quad n = 1, 2, \ldots, \tag{18.65}$$

form a complete orthonormal system for Z. The orthonormality of the g_n is an immediate consequence of the orthonormality of the φ_n and the property $A = CC^*$. To establish completeness, let $g \in Z$ satisfy $(g, g_n) = 0$, $n = 1, 2, \ldots$. Then

$$(Cg, \varphi_n) = (g, C^* \varphi_n) = \sqrt{\lambda_n}\,(g, g_n) = 0, \quad n = 1, 2, \ldots.$$

Hence, by Theorem 1.28, the completeness of the φ_n implies that $C\varphi = 0$. From this we obtain $\varphi = 0$, since C is injective. Therefore, again by Theorem 1.28, the g_n are complete. This together with

$$Cg_n = \frac{1}{\sqrt{\lambda_n}} A\varphi_n = \sqrt{\lambda_n}\,\varphi_n, \quad n = 1, 2, \ldots,$$

implies that $(\sqrt{\lambda_n}, g_n, \varphi_n)$ is a singular system for the compact operator C. Together with the singular system $(\sqrt{\lambda_n}, \varphi_n, \varphi_n)$ of the operator $A^{1/2}$, Picard's Theorem 15.18 now implies that $A^{1/2}(X) = C(Z)$. □

We want to apply Theorems 18.13 and 18.15 to the operator U and therefore we need to establish that it is positive definite together with the existence of an appropriate factorization. To this end we recall the single-layer operator $S : L^2(\partial D) \to L^2(\partial D)$ given by

$$(S\varphi)(x) := \int_{\partial D} \Phi(x, y)\varphi(y)\, ds(y), \quad x \in \partial D,$$

where, for convenience, we have dropped a factor of 2.

Theorem 18.16. *The operators U, R, and S are related through*

$$U = RSR^*, \tag{18.66}$$

where $R^* : L^2(\partial B) \to L^2(\partial D)$ is the adjoint of R.

Proof. By definition of the operators we have $H^* = RS$, whence $H = SR^*$ follows. Inserting this into (18.59) completes the proof of (18.66). □

Theorem 18.17. *Assume that the diameter of D is less than one. Then the operators S and U are positive definite.*

Proof. Since its kernel is real-valued and symmetric, the operator S is self-adjoint. Consider the single-layer potential v with density $\varphi \in L^2(\partial D)$ and assume that φ has mean value zero. Then

$$v(x) = O\left(\frac{1}{|x|}\right), \quad \operatorname{grad} v(x) = O\left(\frac{1}{|x|}\right), \quad |x| \to \infty,$$

and, using the jump relations for single-layer potentials with L^2 densities, an application of Green's second integral theorem in D and $\mathbb{R}^2 \setminus D$ yields

$$(S\varphi, \varphi) = \int_{\partial D} \bar{\varphi}\, S\varphi\, ds = \int_{\partial D} v \left\{ \frac{\partial \bar{v}_-}{\partial \nu} - \frac{\partial \bar{v}_+}{\partial \nu} \right\} ds = \int_{\mathbb{R}^2} |\operatorname{grad} v|^2 dx$$

whence $(S\varphi, \varphi) \geq 0$ follows. For arbitrary $\varphi \in L^2(\partial D)$ we write

$$\varphi = \psi_0 \int_{\partial D} \varphi\, ds + \widetilde{\varphi}$$

where ψ_0 is the natural charge from Theorem 6.21 with $\int_{\partial D} \psi_0\, ds = 1$. Then $\int_{\partial D} \widetilde{\varphi}\, ds = 0$ and analogous to φ we decompose v into the sum

$$v = v_0 \int_{\partial D} \varphi\, ds + \widetilde{v}$$

with the single-layer potentials v_0 and $\widetilde{v}$ with densities ψ_0 and $\widetilde{\varphi}$, respectively. By Theorem 6.21 for the potential v_0 we have that $v_0 = c$ on ∂D with some constant c and this, in particular, implies that $(S\psi_0, \widetilde{\varphi}) = 0 = (S\widetilde{\varphi}, \psi_0)$. From $\int_{\partial D} \psi_0\, ds = 1$ we observe that $v_0(x) \to -\infty$ as $|x| \to \infty$ and therefore, by the maximum-minimum principle for harmonic functions it follows that $v_0(x) \leq c$ for all $x \in \mathbb{R}^2 \setminus D$. This implies that $\psi_0 = -\partial v_+/\partial \nu > 0$ and therefore $c = S\psi_0 > 0$, since by assumption the diameter of D is less than one and therefore $\ln|x - y| < 0$ for all $x, y \in \partial D$ with $x \neq y$. Consequently

$$(S\varphi, \varphi) = (S\psi_0, \psi_0) \left| \int_{\partial D} \varphi\, ds \right|^2 + (S\widetilde{\varphi}, \widetilde{\varphi}) = c \left| \int_{\partial D} \varphi\, ds \right|^2 + (S\widetilde{\varphi}, \widetilde{\varphi}) \geq 0.$$

Equality implies that $\operatorname{grad} \widetilde{v} = 0$ in $\mathbb{R}^2$ und $\int_{\partial D} \varphi\, ds = 0$ whence $\varphi = 0$ follows by the jump relations. Therefore S is positive definite.

From (18.66) we obtain that U is self-adjoint and that $(Ug, g) = (SR^*g, R^*g) \geq 0$ for all $g \in L^2(\partial B)$. By the positive definiteness of S equality holds in this inequality only if $R^*g = 0$. As already noted the operator R has dense range and therefore R^* is injective. Hence $g = 0$ and therefore U is positive definite. □

As first proposed by Hähner [84] and analyzed for the three-dimensional case, in the factorization method for solving our inverse problem the integral equation (18.61) is now replaced by

$$U^{1/2} g_z = \Psi(\cdot\,, z)|_{\partial B}. \tag{18.67}$$

Our aim is to show that this equation is solvable for g_z in $L^2(\partial B)$ if and only if $z \in D$.

Since under the assumption of Theorem 18.17 the single-layer operator S is compact and positive definite, by Theorem 18.13 there exists a uniquely defined compact and positive definite operator $S^{1/2} : L^2(\partial D) \to L^2(\partial D)$ satisfying $S^{1/2}S^{1/2} = S$. Using the boundedness of S from $H^{-1/2}(\partial D)$ into $H^{1/2}(\partial D)$ from Theorem 8.25 and the duality pairing between $H^{-1/2}(\partial D)$ and $H^{1/2}(\partial D)$ we can estimate

$$\left\|S^{1/2}\varphi\right\|_{L^2}^2 = |(S\varphi, \varphi)| \le \|S\varphi\|_{H^{1/2}}\|\varphi\|_{H^{-1/2}} \le c\|\varphi\|_{H^{-1/2}}^2$$

for all $\varphi \in L^2(\partial D)$ and some positive constant c. Hence, by the denseness of $L^2(\partial D)$ in $H^{-1/2}(\partial D)$ it follows that $S^{1/2}$ is a bounded operator from $H^{-1/2}(\partial D)$ into $L^2(\partial D)$. Furthermore, by Corollary 2.12 we can estimate

$$\left\|S^{1/2}\varphi\right\|_{H^{1/2}} = \sup_{\|\psi\|_{H^{-1/2}}=1} \left|(S^{1/2}\varphi, \psi)\right| = \sup_{\|\psi\|_{H^{-1/2}}=1} \left|(\varphi, S^{1/2}\psi)\right| \le c\|\varphi\|_{L^2}$$

for all $\varphi \in L^2(\partial D)$. Therefore, finally, $S^{1/2}$ is also a bounded operator from $L^2(\partial D)$ into $H^{1/2}(\partial D)$. Under the assumption that the diameter of ∂D is less than one $S : H^{-1/2}(\partial D) \to H^{1/2}(\partial D)$ is an isomorphism (see p. 169). Hence, for each $f \in H^{1/2}(\partial D)$ there exists $\psi \in H^{-1/2}(\partial D)$ such that $S\psi = f$. Then for $\varphi := S^{1/2}\psi \in L^2(\partial D)$ we have $f = S^{1/2}\varphi$, that is, $S^{1/2} : L^2(\partial D) \to H^{1/2}(\partial D)$ also is an isomorphism.

Theorem 18.18. *Assume that the diameter of D is less than one. Then*

$$U^{1/2}\left(L^2(\partial B)\right) = R\left(H^{1/2}(\partial D)\right).$$

Proof. From Theorem 18.15 applied to the operators $U : L^2(\partial B) \to L^2(\partial B)$ and $RS^{1/2} : L^2(\partial D) \to L^2(\partial B)$ we obtain that that

$$U^{1/2}\left(L^2(\partial B)\right) = RS^{1/2}\left(L^2(\partial D)\right)$$

and the statement of the theorem follows from $S^{1/2}\left(L^2(\partial D)\right) = H^{1/2}(\partial D)$. □

Lemma 18.19. $\Psi(\cdot, z)|_{\partial B} \in R\left(H^{1/2}(\partial D)\right)$ *if and only if* $z \in D$.

Proof. If $z \in D$ then clearly $\Psi(\cdot, z)$ is a bounded harmonic function in $\mathbb{R}^2 \setminus D$ that is continously differentiable on ∂D. Therefore $\Psi(\cdot, z)|_{\partial D} \in H^{1/2}(\partial D)$ and $\Psi(\cdot, z)|_{\partial B} = R\Psi(\cdot, z)|_{\partial D}$.

Conversely, let $z \notin D$ and assume there exists $f \in H^{1/2}(\partial D)$ such that $Rf = \Psi(\cdot, z)|_{\partial B}$. Then, by the uniqueness for the exterior Dirichlet problem and analyticity, the solution v to the exterior Dirichlet problem with boundary trace $v|_{\partial D} = f$ must coincide with $\Psi(\cdot, z)$ in $(\mathbb{R}^2 \setminus \bar{D}) \setminus \{z\}$. If $z \in \mathbb{R}^2 \setminus \bar{D}$ this contradicts the analyticity of v. If $z \in \partial D$ from the boundary condition it follows that $\Psi(\cdot, z)|_{\partial D} \in H^{1/2}(\partial D)$ which is a contradiction to $\Psi(\cdot, z) \notin H_1(D)$ for $z \in \partial D$. □

Putting Theorem 18.18 and Lemma 18.19 together we arrive at the following final characterization of the domain D.

Corollary 18.20. *Assume that the diameter of D is less than one. Then $z \in D$ if and only if*

$$U^{1/2} g_z = \Psi(\cdot, z)|_{\partial B}$$

is solvable in $L^2(\partial B)$.

Corollary 18.20 can be used for a reconstruction with the aid of a singular system (λ_n, g_n, g_n) of the operator U. Then, by Picard's Theorem 15.18 as used in the proof of Theorem 18.15 we have that $z \in D$ if and only if

$$\sum_{n=1}^{\infty} \frac{|(g_n, \Psi(\cdot, z)|_{\partial B})|^2}{|\lambda_n|} < \infty. \tag{18.68}$$

At first glance Corollary 18.20 seems to imply that the nonlinear inverse problem has been completely replaced by a linear problem. However, determining a singular system of U is nonlinear and there is still a nonlinear problem involved for finding those points z where (18.68) is satisfied. Of course an obvious way to approximately solve this nonlinear problem is by truncating the series (18.68) through a finite sum for z on a grid in B and approximately determining ∂D as the location of those points z where this sum is large.

We also note that the norm $\|g_z\|_{L^2(\partial B)}$ of the solution to (18.67) tends to infinity as z approaches ∂D. Assume to the contrary that $\|g_{z_n}\|_{L^2(\partial B)}$ remains bounded for a sequence (z_n) in D with $z_n \to z \in \partial D$ for $n \to \infty$. Then, by Theorem 16.3, without loss of generality we may assume weak convergence $g_{z_n} \rightharpoonup g_z \in L^2(\partial B)$ as $n \to \infty$. The compactness of $U^{1/2}$ implies that (see Problem 16.5)

$$U^{1/2} g_z = \lim_{n\to\infty} U^{1/2} g_{z_n} = \lim_{n\to\infty} \Psi(\cdot, z_n)|_{\partial B} = \Psi(\cdot, z)|_{\partial B}$$

i.e., we have a contradiction.

We emphasize that in the derivation of Corollary 18.20 the unknown domain is not required to be connected, i.e., it may consist of a finite number of components that does not need to be known in advance. Furthermore, for the application of the factorization method it is not necessary to know the boundary condition. Using the above tools, in particular Theorem 18.15, it can be proven that Corollary 18.20 is also valid for the Neumann boundary condition (18.3) (see [147]).

From a historical point of view, it is notable that the factorization method was first developed by Kirsch [125] in inverse obstacle scattering although this case requires much deeper tools from functional analysis due to the fact that the far field operator as counterpart of our data operator U no longer is self-adjoint. For the factorization method in this more general setting we refer to [127] and for the application of the factorization method to impedance tomography we refer to [86, 87].

In concluding this section, we briefly return to the linear sampling method.

Corollary 18.21. *Assume that the diameter of D is less than one. For $z \in D$ denote by g_z the solution of the equation (18.67) of the factorization method and for $\alpha > 0$ and $z \in B$ let g_z^α denote the Tikhonov regularized solution of the equation (18.61) of* *the linear sampling method, i.e., the solution of*

$$\alpha g_z^\alpha + U^*Ug_z^\alpha = U^*\Psi(\cdot, z)|_{\partial B},$$

and define the single-layer potential

$$v_{g_z^\alpha}(x) := \int_{\partial B} \Phi(x, y) g_z^\alpha(y)\, ds, \quad x \in B,$$

with density g_z^α. *If* $z \in D$ *then*

$$\lim_{\alpha\to 0} e \cdot \operatorname{grad} v_{g_z^\alpha}(z) = \|g_z\|^2_{L^2(\partial B)},$$

and if $z \notin D$ *then* $\lim_{\alpha\to 0} e \cdot \operatorname{grad} v_{g_z^\alpha}(z) = \infty$.

Proof. In view of (18.60) we have $e \cdot \operatorname{grad} v_{g_z^\alpha}(z) = (g_z^\alpha, \Psi(\cdot, z))_{L^2(\partial B)}$ and the statement follows from Theorem 18.14 and Corollary 18.20. □

As pointed out above, the norm $\|g_z\|_{L^2(\partial B)}$ of the solution to (18.67) tends to infinity as $z \to \partial D$. Therefore, by Corollary 18.21 also the limit $\lim_{\alpha\to 0} |e \cdot \operatorname{grad} v_{g_z^\alpha}(z)|$ tends to infinity when z approaches the boundary, i.e., the main feature from the motivation for the linear sampling method on p. 392 is verified.

Problems

18.1. Show that the Dirichlet problem for the Laplace equation in the doubly connected domain $B \setminus \bar{D}$ (in the geometric setting of this chapter) is uniquely solvable (see Problem 6.3).
Hint: Find the solution in the form

$$u(x) = \int_{\partial B} \varphi_B(y) \frac{\partial \Phi(x, y)}{\partial \nu(y)}\, ds(y) + \int_{\partial D} \varphi_D(y) \left\{ \frac{\partial \Phi(x, y)}{\partial \nu(y)} + \Phi(x, y) \right\} ds(y), \quad x \in B \setminus \bar{D}.$$

18.2. Show that one pair of Cauchy data on ∂B of a harmonic function with zero trace on ∂D uniquely determines D (in the geometric setting of this chapter).

18.3. Analogous to Example 16.30, determine a singular value decomposition of the operator A defined by (18.8).

18.4. For the kernel of the Fréchet derivative of the double-layer operator in Theorem 18.7 show that

$$k'(t, \tau; r, q) = -\frac{2}{\pi} \frac{[h_r'(\tau)]^\perp \cdot \{h_r(t) - h_r(\tau)\}}{|h_r(t) - h_r(\tau)|^4} \{h_r(t) - h_r(\tau)\} \cdot \{h_q(t) - h_q(\tau)\}$$
$$+ \frac{1}{\pi} \frac{[h_q'(\tau)]^\perp \cdot \{h_r(t) - h_r(\tau)\} + [h_r'(\tau)]^\perp \cdot \{h_q(t) - h_q(\tau)\}}{|h_r(t) - h_r(\tau)|^2}$$

for $t \neq \tau$. Use this representation to show that $k'(\cdot, \cdot; r, q)$ is continuous and to verify the inequality (18.30).

18.5. Determine a singular system for the operator U defined by (18.58) for the case where D and B are concentric disks.

References

1. Abel, N.H.: Auflösung einer mechanischen Aufgabe. J. f. reine u. angew. Math. **1**, 153–157 (1826).
2. Adams, R.A., and Fournier, J.J.F.: *Sobolev Spaces,* 2nd ed. Elsevier/Academic Press, Amsterdam 2003.
3. Akduman, I., and Kress, R.: Electrostatic imaging via conformal mapping. Inverse Problems **18**, 1659–1672 (2002).
4. Anderssen, R.S., de Hoog, F.R., and Lukas, M.A.: *The Application and Numerical Solution of Integral Equations.* Sijthoff and Noordhoff, Alphen aan den Rijn 1980.
5. Anselone, P.M.: *Collectively Compact Operator Approximation Theory and Applications to Integral Equations.* Prentice-Hall, Englewood Cliffs 1971.
6. Anselone, P.M., and Moore, R.H.: Approximate solutions of integral and operator equations. J. Math. Anal. Appl. **9**, 268–277 (1964).
7. Arnold, D.N., and Noon, D.N.: Coercitivity of the single layer heat potential. J. Comput. Math. **7**, 100–104 (1989).
8. Aronszajn, N.: Theory of reproducing kernels. Trans. Amer. Math. Soc. **68**, 337–404 (1950).
9. Atkinson, F.V.: Normal solvability of linear equations in normalized spaces. Matem. Sbornik **28**, 3–14 (1951) (In Russian).
10. Atkinson, K.E.: Iterative variants of the Nyström method for the numerical solution of integral equations. Numer. Math. **22**, 17–31 (1973).
11. Atkinson, K.E.: *The Numerical Solution of Integral Equations of the Second Kind.* Cambridge Univ. Press, Cambridge 1997.
12. Aubin, J.P.: *Applied Functional Analysis,* 2nd ed. John Wiley & Sons, New York 2000.
13. Baderko, E.A.: Parabolic problems and boundary integral equations. Math. Meth. in the Appl. Sci. **20**, 449–459 (1997).
14. Baker, C.T.H.: *The Numerical Treatment of Integral Equations.* Clarendon Press, Oxford 1977.
15. Baumeister, J.: *Stable Solution of Inverse Problems.* Vieweg, Braunschweig 1986.
16. Beer, A.: *Einleitung in die Elektrostatik, die Lehre vom Magnetismus und die Elektrodynamik.* Vieweg, Braunschweig 1865.
17. Berger, M.S.: *Nonlinearity and Functional Analysis.* Academic Press, New York 1977.
18. Beylkin, G., Coifman, R.R., and Rokhlin, V.: Fast wavelet transforms and numerical algorithms. Comm. Pure Appl. Math. **44**, 141–183 (1991).
19. Brakhage, H.: Über die numerische Behandlung von Integralgleichungen nach der Quadraturformelmethode. Numer. Math. **2**, 183–196 (1960).
20. Brebbia, C.A., Telles, J.C.F., and Wrobel, L.C.: *Boundary Element Techniques.* Springer, Berlin 1984.
21. Brezis, H.: *Functional Analysis, Sobolev Spaces and Partial Differential Equations.* Springer, New York 2011.

R. Kress, *Linear Integral Equations*, Applied Mathematical Sciences 82,
DOI 10.1007/978-1-4614-9593-2,

22. Burger, M., Kaltenbacher, B., and Neubauer, A.: Iterative solution methods. In: *Handbook of Mathematical Methods in Imaging* (Scherzer, ed.). Springer, Berlin, 345–384 (2011).
23. Calderón, A.P.: The multipole expansion of radiation fields. J. Rat. Mech. Anal. **3**, 523–537 (1954).
24. Cakoni, F., and Kress, R.: Integral equation methods for the inverse obstacle problem with generalized impedance boundary condition. Inverse Problems **29**, 015005 (2013).
25. Cannon, J.R.: *The One-Dimensional Heat Equation.* Addison-Wesley, Menlo Park 1984.
26. Céa, J.: Approximation variationelle des problèmes aux limites. Ann. Inst. Fourier **14**, 345–444 (1964).
27. Chapko, R., and Kress, R.: A hybrid method for inverse boundary value problems in potential theory. Jour. on Inverse and Ill-Posed Problems **13**, 27–40 (2005).
28. Chen, G., and Zhou, J.: *Boundary Element Methods.* Academic Press, London 1992.
29. Christiansen, S.: Condition number of matrices derived from two classes of integral equations. Math. Meth. in the Appl. Sci. **3**, 364–392 (1981).
30. Colton, D., and Kirsch, A.: A simple method for solving inverse scattering problems in the resonance region. Inverse Problems **12**, 383–393 (1996).
31. Colton, D., and Kress, R.: *Integral Equation Methods in Scattering Theory.* SIAM, Philadelphia 2013.
32. Colton, D., and Kress, R.: *Inverse Acoustic and Electromagnetic Scattering Theory,* 3rd ed. Springer, Berlin 2013.
33. Constanda, C.: *Direct and Indirect Boundary Integral Equation Methods.* Chapman & Hall/CRC, Boca Raton 2000.
34. Costabel, M.: Boundary integral operators for the heat equation. Integral Eq. and Operator Theory **13**, 498–552 (1990).
35. Costabel, M., and Saranen, J.: Parabolic boundary integral operators symbolic representation and basic properties. Integral Equations Operator Theory **40**, 185–211 (2001).
36. Courant, R., and Hilbert, D.: *Methods of Mathematical Physics II.* Wiley-Interscience Publication, New York 1966.
37. Cryer, C.W.: *Numerical Functional Analysis.* Clarendon Press, Oxford 1982.
38. Dahmen, W.: Wavelet and multiscale methods for operator equations. Acta Numerica **6**, 55–228 (1997).
39. Davis, P.J.: On the numerical integration of periodic analytic functions. In: *Symposium on Numerical Approximation* (Langer, ed.). Univ. of Wisconsin Press, Madison, 45–59 (1959).
40. Davis, P.J.: *Interpolation and Approximation.* Blaisdell Publishing Company, Waltham 1963.
41. Davis, P.J., and Rabinowitz, P.: *Methods of Numerical Integration,* 2nd ed. Academic Press, San Diego 1984.
42. Delves, L.M., and Mohamed, J.L.: *Computational Methods for Integral Equations.* Cambridge Univ. Press, Cambridge 1985.
43. Dieudonné, J.: Sur les homomorphismes d'espaces normés. Bull. Sci. Math. **67**, 72–84 (1943).
44. Dieudonné, J.: The index of operators in Banach spaces. Integral Eq. and Operator Theory **8**, 580–589 (1985).
45. du Bois-Reymond, P.: Bemerkungen über $\Delta z = 0$. J. f. reine u. angew. Math. **103**, 204–229 (1888).
46. Elliott, D F., and Rao, K.R.: *Fast Transforms. Algorithms, Analyses, Applications.* Academic Press, New York 1982.
47. Engels, H.: *Numerical Quadrature and Cubature.* Academic Press, New York 1980.
48. Engl, H.W.: On the convergence of regularization methods for ill-posed linear operator equations. In: *Improperly Posed Problems and Their Numerical Treatment* (Hämmerlin and Hoffmann, eds.). International Series of Numerical Mathematics. Birkhäuser-Verlag, Basel **63**, 81–96 (1983).
49. Engl, H.W., Hanke, M., and Neubauer, A.: *Regularization of Inverse Problems.* Kluwer Academic Publisher, Dordrecht 1996.
50. Fairweather, G., and Karageorghis, A.: The method of fundamental solutions for elliptic boundary value problems. Adv. Comput. Math. **9**, 69–95 (1998).

51. Farhat, C., Tezaur, R., and Djellouli, R.: On the solution of three-dimensional inverse obstacle acoustic scattering problems by a regularized Newton method. Inverse Problems **18**, 1229–1246 (2002).
52. Fenyö, I., and Stolle, H.: *Theorie und Praxis der linearen Integralgleichungen, I–IV.* Birkhäuser-Verlag, Basel 1981–1984.
53. Fichera, G.: Linear elliptic equations of higher order in two independent variables and singular integral equations, with applications to anisotropic inhomogeneous elasticity. In: *Partial Differential Equations and Continuum Mechanics* (Langer, ed.). Univ. of Wisconsin Press, Madison, 55–80 (1961).
54. Folland, G.: *Introduction to Partial Differential Equations,* 2nd ed. Princeton Univ. Press, Princeton 1995.
55. Fredholm, I.: Sur une classe d'équations fonctionelles. Acta Math. **27**, 365–390 (1903).
56. Freud, R., Golub, F., and Nachtigal, N.: Iterative solutions of linear systems. Acta Numerica **1**, 57–100 (1992).
57. Fridman, V.: Method of successive approximation for a Fredholm integral equation of the first kind. Uspeki Mat. Nauk. **11**, 233–234 (1956) (In Russian).
58. Friedman, A.: *Partial Differential Equations of Parabolic Type.* Prentice-Hall, Englewood Cliffs 1964.
59. Gakhov, F.D.: *Boundary Value Problems.* Pergamon Press, Oxford 1966.
60. Galerkin, B.G.: Expansions in stability problems for elastic rods and plates. Vestnik inzkenorov **19**, 897–908 (1915) (In Russian).
61. Ganesh, M., and Graham, I. G.: A high-order algorithm for obstacle scattering in three dimensions. J. Comput. Phys. **198**, 211–242 (2004).
62. Gevrey, M.: Sur les équations aux dérivées partielles du type parabolique. J. de Math. **9**, 305–471 (1913).
63. Gilbarg, D., and Trudinger, N.S.: *Elliptic Partial Differential Equations of Second Order.* Springer, Berlin 2001.
64. Gilbert, R.P.: *Function Theoretic Methods in Partial Differential Equations.* Academic Press, New York 1969.
65. Golberg, M.A., and Chen, C.S.: *Discrete Projection Methods for Integral Equations.* Computational Mechanics Publications, Southampton 1997.
66. Golberg, M.A., and Chen, C.S.: The method of fundamental solutions for potential, Helmholtz and diffusion problems. In: *Boundary Integral Methods – Numerical and Mathematical Aspects* (Golberg, ed.). Computational Mechanics Publications, Southampton, 103–176 (1998).
67. Golub, G., and van Loan, C.: *Matrix Computations,* 4th ed. John Hopkins University Press, Baltimore 2013.
68. Graham, I. G., and Sloan, I. H.: Fully discrete spectral boundary integral methods for Helmholtz problems on smooth closed surfaces in $\mathbb{R}^3$. Numer. Math. **92**, 289–323 (2002).
69. Greenbaum, A.: *Iterative Methods for Solving Linear Systems.* SIAM, Philadelphia 1997.
70. Greengard, L., and Moura, M.: On the numerical evaluation of electrostatic fields in composite materials. Acta Numerica **3**, 379–410 (1994).
71. Greengard, L., and Rokhlin, V.: A fast algorithm for particle simulations. J. Comput. Phys. **73**, 187–207 (1987).
72. Greengard, L., and Rokhlin, V.: A new version of the fast multipole method for the Laplace equation in three dimensions. Acta Numerica **6**, 229–269 (1997).
73. Grisvard, P.: *Elliptic Problems in Nonsmooth Domains.* Pitman, Boston 1985.
74. Groetsch, C.W.: *The Theory of Tikhonov Regularization for Fredholm Equations of the First Kind.* Pitman, Boston 1984.
75. Hackbusch, W.: *Multi-Grid Methods and Applications.* Springer, Berlin 1985.
76. Hackbusch, W.: *Integral Equations: Theory and Numerical Treatment.* Birkhäuser-Verlag, Basel 1994.
77. Hackbusch, W., and Nowak, Z.P.: On the fast matrix multiplication in the boundary element method by panel clustering. Numer. Math. **56**, 229–245 (1989).

78. Hackbusch, W., and Sauter, S.: On the efficient use of the Galerkin method to solve Fredholm integral equations. Appl. Math. **38**, 301–322 (1993).
79. Hadamard, J.: *Lectures on Cauchy's Problem in Linear Partial Differential Equations.* Yale University Press, New Haven 1923.
80. Haddar, H., and Kress, R.: Conformal mappings and inverse boundary value problems. Inverse Problems **21**, 935–953 (2005).
81. Haddar, H., and Kress, R.: Conformal mapping and impedance tomography. Inverse Problems **26** , 074002 (2010).
82. Haddar, H., and Kress, R.: A conformal mapping method in inverse obstacle scattering. Complex Var. Elliptic Equ. (2013) (to appear).
83. Hähner, P.: Eindeutigkeits- und Regularitätssätze für Randwertprobleme bei der skalaren und vektoriellen Helmholtzgleichung. Dissertation, Göttingen 1990.
84. Hähner, P.: An inverse problem in electrostatic. Inverse Problems **15**, 961–975 (1999).
85. Hämarik, U., Avi, E., and Ganina, A.: On the solution of ill-posed problems by projection methods with a priori choice of the discretization level. Math. Model. Anal. **7**, 241–252 (2002).
86. Hanke, M., and Brühl, M.: Recent progress in impedance tomography. Inverse Problems **19**, 565–590 (2003).
87. Hanke, M., and Kirsch, A.: Sampling methods. In: *Handbook of Mathematical Methods in Imaging* (Scherzer, ed.). Springer, Berlin, 501–550 (2011).
88. Harbrecht, H., and Hohage. T.: Fast methods for three-dimensional inverse obstacle scattering problems. Jour. Integral Equations and Appl. **19**, 237–260 (2007).
89. Helms, L.: *Introduction to Potential Theory.* Wiley-Interscience, New York 1969.
90. Hemker, P.W., and Schippers, H.: Multiple grid methods for the solution of Fredholm integral equations of the second kind. Math. Comp. **36**, 215–232 (1981).
91. Henrici, P.: *Applied and Computational Complex Analysis.* Vol 3. Wiley-Interscience Publication, New York 1986.
92. Hestenes, M.R., and Stiefel, E.: Methods of conjugate gradients for solving linear systems. J. Res. Nat. Bur. Standards **49**, 409–436 (1952).
93. Hettlich, F.: Fréchet derivatives in inverse obstacle scattering. Inverse Problems **11**, 371–382 (1995). Erratum: Inverse Problems **14**, 209–210 (1998).
94. Heuser, H.: *Funktionalanalysis,* 4. Aufl. Teubner-Verlag, Stuttgart 2006.
95. Hilbert, D.: Über eine Anwendung der Integralgleichungen auf ein Problem der Funktionentheorie. In: *Verhandlungen des dritten internationalen Mathematiker Kongresses in Heidelberg 1904* (Krazer, ed.). Teubner, Leipzig, 233–240 (1905).
96. Hohage, T.: Logarithmic convergence rates of the iteratively regularized Gauss–Newton method for an inverse potential and an inverse scattering problem. Inverse Problems **13**, 1279–1299 (1997).
97. Hohage, T.: *Iterative Methods in Inverse Obstacle Scattering: Regularization Theory of Linear and Nonlinear Exponentially Ill-Posed Problems.* Dissertation, Linz 1999.
98. Hohage, T.: On the numerical solution of a three-dimensional inverse medium scattering problem. Inverse Problems **17**, 1743–1763 (2001).
99. Hohage, T.: Fast numerical solution of the electromagnetic medium scattering problem and applications to the inverse problem. J. Comp. Phys. **214**, 224–238 (2006).
100. Holmgren, E.: Sur l'équation de la propagation de la chaleur. Arkiv Mat. Fysik. 4, **14**, 1–11 (1908).
101. Holmgren, E.: Sur l'équation de la propagation de la chaleur II. Arkiv Mat. Fysik. 4, **18**, 1–28 (1908).
102. Hsiao, G.C., and MacCamy, R.C.: Solution of boundary value problems by integral equations of the first kind. SIAM Review **15**, 687–705 (1973).
103. Hsiao, G.C., and Saranen, J.: Boundary integral solution of the two-dimensional heat equation. Math. Meth. in the Appl. Sciences **16**, 87–114 (1993).
104. Hsiao, G.C., and Wendland, W. L.: *Boundary Integral Equations.* Springer, Berlin 2008.
105. Isakov, V.: *Inverse Problems for Partial Differential Equations,* 2nd ed. Springer, New York 2006.

106. Ivanov, V.K.: Integral equations of the first kind and an approximate solution for the inverse problem of potential. Soviet Math. Doklady **3**, 210–212 (1962) (English translation).
107. Ivanyshyn, O.: *Nonlinear Boundary Integral Equations in Inverse Scattering.* Dissertation, Göttingen, 2007.
108. Ivanyshyn, O., and Kress, R.: Nonlinear integral equations for solving inverse boundary value problems for inclusions and cracks. Jour. Integral Equations and Appl. **18**, 13—38 (2006). Corrigendum: Jour. Integral Equations and Appl. **22**, 647–649 (2010).
109. Ivanyshyn, O., and Kress, R.: Nonlinear integral equations in inverse obstacle scattering. In: *Mathematical Methods in Scattering Theory and Biomedical Engineering* (Fotiatis and Massalas, eds.). World Scientific, Singapore, 39–50 (2006).
110. Ivanyshyn, O., Kress, R., and Serranho, P.: Huygens' principle and iterative methods in inverse obstacle scattering. Adv. Comput. Math. **33**, 413–429 (2010).
111. Johansson, T., and Sleeman, B.: Reconstruction of an acoustically sound-soft obstacle from one incident field and the far field pattern. IMA J. Appl. Math. **72**, 96–112 (2007).
112. Jörgens, K.: *Lineare Integraloperatoren.* Teubner-Verlag, Stuttgart 1970.
113. Kabanikhin, S.I.: *Inverse and Ill-posed Problems: Theory and Applications.* de Gruyter, Berlin-Boston 2011
114. Kaltenbacher, B.: Regularization by projection with a posteriori discretization level choice for linear and nonlinear ill-posed problems. Inverse Problems **16**, 1523–1539 (2000).
115. Kaltenbacher, B., Neubauer, A., and Scherzer, O.: *Iterative Regularization Methods for Nonlinear Ill-Posed Problems.* de Gruyter, Berlin, 2008.
116. Kantorovic, L.V., and Akilov, G.P.: *Functional Analysis in Normed Spaces.* Pergamon Press, Oxford 1964.
117. Kantorovic, L.V., and Krylov, V.I.: *Approximate Methods of Higher Analysis.* Noordhoff, Groningen 1964.
118. Kanwal, R.P.: *Linear Integral Equations.* Reprint of 2nd (1997) ed. Birkhäuser, Basel 2013.
119. Kellogg, O.: *Foundations of Potential Theory.* Springer, New York 1967.
120. Kersten, H.: Grenz- und Sprungrelationen für Potentiale mit quadrat-summierbarer Dichte. Resultate d. Math. **3**, 17–24 (1980).
121. Kirsch, A.: *Generalized boundary value- and control problems for the Helmholtz equation.* Habilitationthesis, Göttingen 1984.
122. Kirsch, A.: Private communication, 1986.
123. Kirsch, A.: Surface gradients and continuity properties for some integral operators in classical scattering theory. Math. Meth. in the Appl. Sci. **11**, 789–804 (1989).
124. Kirsch, A.: The domain derivative and two applications in inverse scattering. Inverse Problems **9**, 81–96 (1993).
125. Kirsch, A.: Characterization of the shape of the scattering obstacle by the spectral data of the far field operator. Inverse Problems **14**, 1489–1512 (1998).
126. Kirsch, A.: *An Introduction to the Mathematical Theory of Inverse Problems,* 2nd. ed. Springer, New York 2010.
127. Kirsch, A., and Grinberg, N.: *The Factorization Method for Inverse Problems.* Oxford University Press, Oxford 2008.
128. Kirsch, A., and Kress, R.: On an integral equation of the first kind in inverse acoustic scattering. In: *Inverse Problems* (Cannon and Hornung, eds.). International Series of Numerical Mathematics. Birkhäuser-Verlag Basel **77**, 93–102 (1986).
129. Kirsch, A., Kress, R., Monk, P., and Zinn, A.: Two methods for solving the inverse acoustic scattering problem. Inverse Problems **4**, 749–770 (1988).
130. Kirsch, A., and Ritter, S.: The Nyström method for solving a class of singular integral equations and applications in 3D-plate elasticity. Math. Meth. in the Appl. Sci. **22**, 177–197 (1999) .
131. Král, J.: The Fredholm method in potential theory. Trans. Amer. Math. Soc. **125**, 511–547 (1966).
132. Krasnoselski, M.A., Vainikko, G.M., Zabreko, P.P., and Rutitskii, Y.: *Approximate Solution of Operator Equations.* Wolters-Noordhoff, Groningen (1972).

133. Kress, R.: Ein ableitungsfreies Restglied für die trigonometrische Interpolation periodischer analytischer Funktionen. Numer. Math. **16**, 389–396 (1971).
134. Kress, R.: Zur numerischen Integration periodischer Funktionen nach der Rechteckregel. Numer. Math **20**, 87–92 (1972).
135. Kress, R.: A boundary integral equation method for a Neumann boundary value problem for force-free fields. J. of Eng. Math. **15**, 29–48 (1981).
136. Kress, R.: On the Fredholm alternative. Integral Equations and Operator Theory **6**, 453–457 (1983).
137. Kress, R.: On constant-alpha force-free fields in a torus. J. of Eng. Math. **20**, 323344 (1986).
138. Kress, R.: A Nyström method for boundary integral equations in domains with corners. Numer. Math. **58**, 145–161 (1990).
139. Kress, R.: Fast 100 Jahre Fredholmsche Alternative. In: *Überblicke Mathematik. Jahrbuch 1994* (Chatteri et al., eds.). Vieweg, Braunschweig, 14–27 (1994).
140. Kress, R.: On the numerical solution of a hypersingular integral equation in scattering theory. J. Comp. Appl. Math. **61** , 345–360 (1995). Corrigendum: J. Comp. Appl. Math. **235**, 3109 (2011).
141. Kress, R.: Integral equation methods in inverse acoustic and electromagnetic scattering. In: *Boundary Integral Formulations for Inverse Analysis* (Ingham and Wrobel, eds.). Computational Mechanics Publications, Southampton, 67–92 (1997).
142. Kress, R.: *Numerical Analysis.* Springer, New York 1998.
143. Kress, R.: On the low wave number behavior of two-dimensional scattering problems for an open arc. J. for Anal. and Its Appl. **18**, 297–305 (1999).
144. Kress, R.: Newton's method for inverse obstacle scattering meets the method of least squares. Inverse Problems **19**, 91–104 (2003).
145. Kress, R.: Inverse problems and conformal mapping. Complex Var. Elliptic Equ. **57**, 301–316 (2012).
146. Kress, R.: A collocation method for a hypersingular boundary integral equation via trigonometric differentiation. Jour. Integral Equations and Appl. (to appear).
147. Kress, R., and Kühn, L.: Linear sampling methods for inverse boundary value problems in potential theory. Appl. Numer. Math. **43**, 161–173 (2002).
148. Kress, R., and Mohsen, A.: On the simulation source technique for exterior problems in acoustics. Math. Meth. in the Appl. Sci. **8**, 585–597 (1986).
149. Kress, R., and Rundell, W.: Nonlinear integral equations and the iterative solution for an inverse boundary value problem. Inverse Problems **21**, 1207–1223 (2005).
150. Kress, R., and Serranho, P.: A hybrid method for two-dimensional crack reconstruction. Inverse Problems **21**, 773–784 (2005)
151. Kress, R., and Sloan, I.H.: On the numerical solution of a logarithmic integral equation of the first kind for the Helmholtz equation. Numer. Math **66**, 193–214 (1993).
152. Kress, R., Sloan, I.H., and Stenger, F.: A sinc quadrature method for the double-layer integral equation in planar domains with corners. J. for Integral Eq. and Appl. **10**, 291–317 (1998).
153. Kupradze, V.D.: Dynamical problems in elasticity. In: *Progress in Solid Mechanics, Vol. III* (Sneddon and Hill, eds.). North Holland, Amsterdam, 1–259 (1963).
154. Kussmaul, R.: Ein numerisches Verfahren zur Lösung des Neumannschen Aussenraumproblems für die Helmholtzsche Schwingungsgleichung. Computing **4**, 246–273 (1969).
155. Landweber, L.: An iteration formula for Fredholm integral equations of the first kind. Amer. J. Math. **73**, 615–624 (1951).
156. Lax, P.D.: Symmetrizable linear transformations. Comm. Pure Appl. Math. **7**, 633–647 (1954).
157. Lax, P.D., and Phillips, R.S.: *Scattering Theory,* 2nd ed. Academic Press, New York 1989.
158. Little, S., and Reade, J.B.: Eigenvalues of analytic kernels. SIAM J. Math. Anal. **15**, 133–136 (1984).
159. Liu, Y.: *Fast Multipole Boundary Element Method. Theory and Applications in Engineering.* Cambridge University Press, Cambridge 2009.
160. Louis, A.K.: *Inverse und schlecht gestellte Probleme.* Teubner, Stuttgart 1989.
161. Martensen, E.: Über eine Methode zum räumlichen Neumannschen Problem mit einer Anwendung für torusartige Berandungen. Acta Math. **109**, 75–135 (1963).

162. Martensen, E.: *Potentialtheorie.* Teubner-Verlag, Stuttgart 1968.
163. Martensen, E.: Riesz theory without axiom of choice. Proc. Amer. Math. Soc. **99**, 496–500 (1987).
164. Maue, A.W.: Über die Formulierung eines allgemeinen Beugungsproblems durch eine Integralgleichung. Zeit. Physik **126**, 601–618 (1949).
165. McLean, W.: *Strongly Elliptic Systems and Boundary Integral Equations.* Cambridge University Press, Cambridge 2000.
166. Meinardus, G: *Approximation of Functions: Theory and Numerical Methods.* Springer, New York 1967.
167. Meister, E.: *Randwertaufgaben der Funktionentheorie.* Teubner-Verlag, Stuttgart 1983.
168. Meschkowski, H.: *Hilbertsche Räume mit Kernfunktion.* Springer, Berlin 1962.
169. Mikhlin, S.G.: *Mathematical Physics, an Advanced Course.* North-Holland, Amsterdam 1970.
170. Mikhlin, S.G., and Prössdorf, S.: *Singular Integral Operators.* Springer-Verlag, Berlin 1986.
171. Miranda, C.: *Partial Differential Equations of Elliptic Type,* 2nd ed. Springer, Berlin 1970.
172. Mönch, L.: On the numerical solution of the direct scattering problem for an open sound-hard arc. J. Comp. Appl. Math. **71**, 343–356 (1996).
173. Moré, J.J.: The Levenberg–Marquardt algorithm, implementatiion and theory. In: *Numerical analysis* (Watson, ed). Springer Lecture Notes in Mathematics **630**, Berlin, 105–116 (1977).
174. Morozov, V.A.: On the solution of functional equations by the method of regularization. Soviet Math. Doklady **7**, 414–417 (1966) (English translation).
175. Morozov, V.A.: Choice of parameter for the solution of functional equations by the regularization method. Soviet Math. Doklady **8**, 1000–1003 (1967) (English translation).
176. Morozov, V.A.: *Methods for Solving Incorrectly Posed Problems.* Springer, Berlin 1984.
177. Müller, C.: *Foundations of the Mathematical Theory of Electromagnetic Waves.* Springer, Berlin 1970.
178. Multhopp, H.: Die Berechnung der Auftriebsverteilung von Tragflügeln. Luftfahrt-Forschung **4**, 153–169 (1938).
179. Müntz, C.H.: Zum dynamischen Wärmeleitungsproblem. Math. Zeit. **38**, 323–337 (1934).
180. Muskhelishvili, N.I.: *Singular Integral Equations.* Noordhoff, Groningen 1953.
181. Nashed, M.Z., and Wahba, G.: Convergence rates of approximate least squares solutions of linear integral and operator equations of the first kind. Math. Comp. **28**, 69–80 (1974).
182. Natanson, I.P.: *Konstruktive Funktionentheorie.* Akademie-Verlag, Berlin 1955.
183. Natterer, F.: Regularisierung schlecht gestellter Probleme durch Projektionsverfahren. Numer. Math. **28**, 329–341 (1977).
184. Neumann, C.: *Untersuchungen über das logarithmische und Newtonsche Potential.* Teubner-Verlag, Leipzig 1877.
185. Noether, F.: Über eine Klasse singulärer Integralgleichungen. Math. Ann. **82**, 42–63 (1921).
186. Nussbaumer, H.J.: *Fast Fourier Transform and Convolution Algorithms.* Springer, Berlin 1982.
187. Nyström, E.J.: Über die praktische Auflösung von Integralgleichungen mit Anwendungen auf Randwertaufgaben. Acta Math. **54**, 185–204 (1930).
188. Petrov, G.I.: Application of Galerkin's method to a problem of the stability of the flow of a viscous fluid. Priklad. Matem. i Mekh. **4**, 3–12 (1940) (In Russian).
189. Phillips, D.L.: A technique for the numerical solution of certain integral equations of the first kind. J. Ass. Comp. Math. **9**, 84–97 (1962).
190. Plato, R., and Vainikko, G.: On the regularization of projection methods for solving ill-posed problems. Numer. Math. **28**, 63–79 (1990).
191. Plemelj, J.: Ein Ergänzungssatz zur Cauchyschen Integraldarstellung analytischer Funktionen, Randwerte betreffend. Monatshefte f. Math. u. Phys. **19**, 205–210 (1908).
192. Plemelj, J.: *Potentialtheoretische Untersuchungen.* Preisschriften der Fürstlich Jablonowskischen Gesellschaft. Teubner, Leipzig 1911.
193. Pogorzelski, W.: *Integral equations and their applications.* Pergamon Press, Oxford 1966.
194. Potthast, R.: Fréchet differentiability of boundary integral operators in inverse acoustic scattering. Inverse Problems **10**, 431–447 (1994).

195. Potthast, R.: On the convergence of a new Newton-type method in inverse scattering. Inverse Problems **17**, 1419–1434 (2001).
196. Potthast, R.: *Point-Sources and Multipoles in Inverse Scattering Theory.* Chapman & Hall, London 2001.
197. Prössdorf, S.: *Einige Klassen singulärer Gleichungen.* Birkhäuser-Verlag, Basel 1974.
198. Prössdorf, S., and Saranen, J.: A fully discrete approximation method for the exterior Neumann boundary value problem for the Helmholtz equation. Z. Anal. Anwendungen **13**, 683–695 (1994).
199. Prössdorf, S., and Silbermann, B.: *Projektionsverfahren und die näherungsweise Lösung singulärer Gleichungen.* Teubner, Leipzig 1977.
200. Prössdorf, S., and Silbermann, B.: *Numerical Analysis for Integral and Related Operator Equations.* Akademie-Verlag, Berlin 1991, and Birkhäuser-Verlag, Basel 1991.
201. Radon, J.: Über die Randwertaufgaben beim logarithmischen Potential. Sitzungsberichte der Akadademie der Wissenschaften Wien. **IIa 128**, 1123–1167 (1919).
202. Rao, K.R., Kim, D.N., and Hwang, J. J.: *Fast Fourier Transform – Algorithms and Applications.* Springer, Dordrecht 2010.
203. Lord Rayleigh: *The Theory of Sound.* London 1896.
204. Rieder, A.: *Keine Probleme mit Inversen Problemen.* Vieweg, Wiesbaden 2003.
205. Riesz, F.: Über lineare Funktionalgleichungen. Acta Math. **41**, 71–98 (1918).
206. Ritz, W.: Über eine neue Methode zur Lösung gewisser Variationsprobleme der mathematischen Physik. J. f. reine u. angew. Math. **135**, 1–61 (1908).
207. Rjasanow, S., and Steinbach, O.: *The Fast Solution of Boundary Integral Equations.* Springer, Berlin 2007.
208. Rokhlin, V.: Rapid solution of integral equations of classical potential theory. J. Comput. Phys. **60**, 187–207 (1985).
209. Rudin, W.: *Functional Analysis,* 2nd ed. McGraw-Hill, New York 1991.
210. Saad, Y.: *Iterative Methods for Sparse Linear Systems,* 2nd ed. PWS Publishing Company, Boston 2003.
211. Saad, Y., and Schultz, M.H.: GMRES: A generalized minimum residual algorithm for solving nonsymmetric linear systems. SIAM J. Sci. Stat. Comput. **7**, 856–869 (1986).
212. Saranen, J., and Vainikko, G.: Trigonometric collocation methods with product integration for boundary integral equations on closed curves. SIAM J. Numer. Anal. **33**, 1577–1596 (1996).
213. Saranen, J., and Vainikko, G.: *Periodic Integral and Pseudodifferential Equations with Numerical Approximation.* Springer, Berlin 2002.
214. Sauter, S.A., and Schwab, C.: *Boundary element methods.* Springer, Berlin 2011.
215. Schauder, J.: Über lineare, vollstetige Funktionaloperationen. Studia Math. **2**, 183–196 (1930).
216. Schippers, H.: *Multiple Grid Methods for Equations of the Second Kind with Applications in Fluid Mechanics.* Mathematical Centre Tracts 163, Amsterdam 1983.
217. Schönhage, A.: *Approximationstheorie.* de Gruyter, Berlin 1971.
218. Seidman, E.: Nonconvergence results for the application of least-squares estimation to ill-posed problems. J. Optimization Theory and Appl. **30**, 535–547 (1980).
219. Serranho, P.: A hybrid method for inverse scattering for shape and impedance. Inverse Problems **22**, 663–680 (2006).
220. Serranho, P.: A hybrid method for sound-soft obstacles in 3D. Inverse Problems and Imaging **1**, 691–712 (2007).
221. Sloan, I.H.: Error analysis for a class of degenerate-kernel methods. Numer. Math. **25**, 231–238 (1976).
222. Sloan, I.H.: Analysis of general quadrature methods for integral equations of the second kind. Numer. Math. **38**, 263–278 (1981).
223. Sloan, I.H.: Error analysis of boundary integral methods. Acta Numerica **1**, 287–339 (1992).
224. Sokhotski, Y.V.: On definite integrals and functions utilized for expansions into series. Doctor thesis, St. Petersburg (1873) (In Russian).

225. Steinbach, O.: *Numerical Approximation Methods for Elliptic Boundary Value Problems. Finite and Boundary Elements.* Springer, New York 2008.
226. Steinbach, O., and Wendland, W.L.: On C. Neumann's method for second-order elliptic systems in domains with non-smooth boundaries. J. Math. Anal. Appl.**262**, 733–748 (2001).
227. Stetter, H.J.: The defect correction principle and discretization methods. Numer. Math. **29**, 425–443 (1978).
228. Symm, G.T: Integral equation methods in potential theory. II Proc. Roy. Soc. A **275**, 33–46 (1963).
229. Taylor, A.E.: *Introduction to Functional Analysis.* John Wiley & Sons, New York 1967.
230. Taylor, M.E.: *Pseudodifferential Operators.* Princeton Univ. Press, Princeton 1981.
231. Taylor, M.E.: *Partial Differential Equations.* Springer, New York 1996.
232. Tikhonov, A.N.: On the solution of incorrectly formulated problems and the regularization method. Soviet Math. Doklady **4**, 1035–1038 (1963) (English translation).
233. Tikhonov, A.N.: Regularization of incorrectly posed problems. Soviet Math. Doklady **4**, 1624–1627 (1963) (English translation).
234. Tikhonov, A.N., and Arsenin, V.Y.: *Solutions of Ill-Posed Problems.* Winston and Sons, Washington 1977.
235. Trefethen, L.N., and Bau, D.: *Numerical Linear Algebra.* SIAM, Philadelphia 1997.
236. Treves, F.: *Basic Linear Partial Differential Equations.* Academic Press, New York 1975.
237. Treves, F.: *Introduction to Pseudodifferential Operators and Fourier Integral Operators.* Plenum, New York 1980.
238. Vainikko, G.: Fast solvers of the Lippmann–Schwinger equation In: *Direct and Inverse Problems of Mathematical Physics* (Gilbert, Kajiwara and Xu, eds). Kluwer, Dordrecht, 423–440 (2000).
239. Vainikko, G., and Hämarik, U.: Projection methods and self-regularization in ill-posed problems. Izv. Vyssh. Ucebn, Zaved, Mat, **10**, 3–17 (1985) (in Russian).
240. Vekua, N.P.: *Systems of Singular Integral Equations.* Noordhoff, Groningen 1967.
241. Verchota, G.: Layer potentials and regularity for the Dirichlet problem for Laplace's equation in Lipschitz domains. J. Func. Anal. **39**, 572–611 (1984).
242. Volterra, V.: Sulla inversione degli integrale definiti. Rom. Acc. Linc. Rend. **5**, 177–185 (1896).
243. Wang, Y., Yagola, A.G., and Yang, C.: *Optimization and Regularization for Computational Inverse Problems and Applications.* Springer, Berlin 2011.
244. Wegmann, R.: Methods for numerical conformal mapping. In: *Handbock of Complex Analysis: Geometric Function Theory, Vol 2* (Kühnau ed.). North Holland, Amsterdam, 351–477 (2006).
245. Wendland, W.L.: Die Fredholmsche Alternative für Operatoren, die bezüglich eines bilinearen Funktionals adjungiert sind. Math. Zeit. **101**, 61–64 (1967).
246. Wendland, W.L.: Die Behandlung von Randwertaufgaben im $\mathbb{R}^3$ mit Hilfe von Einfach- und Doppelschichtpotentialen. Numer. Math. **11**, 380–404 (1968).
247. Wendland, W.L.: Bemerkungen über die Fredholmschen Sätze. Meth. Verf. Math. Phys. **3**, 141–176 (1970).
248. Wendland, W.L.: *Elliptic Systems in the Plane.* Pitman, London 1979.
249. Wienert, L.: Die numerische Approximation von Randintegraloperatoren für die Helmholtzgleichung im $\mathbb{R}^3$. Dissertation, Göttingen 1990.

Index

R. Kress, *Linear Integral Equations*, Applied Mathematical Sciences 82,
DOI 10.1007/978-1-4614-9593-2,

Printed in the United States
By Bookmasters

Printed in the United States
By Bookmasters